Igneous Petrology

IGNEOUS PETROLOGY

SECOND EDITION

Anthony Hall

Longman Group Limited
Longman House, Burnt Mill, Harlow
Essex CM20 2JE, England
and Associated Companies throughout the world

First published 1987
Second edition 1996

British Library Cataloguing in Publication Data
A catalogue entry for this title is available from the
British Library.

ISBN 0-582-23080-2

Library of Congress Cataloging-in-Publication data
A catalog entry for this title is available from the Library
of Congress.

Typeset by 3 in 10/12pt Times

Produced through Longman Malaysia, GPS

CONTENTS

PREFACE

The purpose of this book is to review what is known of the origin of igneous rocks and magmas. I have assumed that the student for whom the book is written will already have some knowledge of mineralogy and of the petrography and classification of igneous rocks.

The nature of an igneous rock is determined not only by how its magma is formed, but by what happens to the magma during its ascent to the surface and its eruption or intrusion. Therefore the philosophy of this book is to pay equal attention to the processes of magma formation and emplacement. It is important that the geochemistry of igneous rocks should not be considered in isolation from their tectonic setting and field relationships.

The arrangement of the book is as follows. Chapters 1–7 discuss general aspects of the occurrence, composition and evolution of igneous rocks and magmas. Chapters 8–12 deal with magmas similar to those which can be seen erupting from present-day volcanoes. These chapters are grouped by magma type rather than tectonic setting. In discussing the origin of the various magma types I have tried to avoid over-emphasizing fractionation as a petrogenetic process simply because it is more easily quantifiable than other processes, although it is obviously an important mechanism in basic magmas. Chapters 13 and 14 deal with plutonic rocks that are never erupted as lavas but are nevertheless igneous.

ACKNOWLEDGEMENTS

We are grateful to the following for permission to reproduce copyright material:

Carnegie Institution of Washington for fig. 24 from fig. 29 (Williams 1942); Chapman & Hall Ltd for fig. 305 from fig. 15.3 (Kjarsgaard and Hamilton 1989); Elsevier Science Ltd. for figs 239 and 240 from figs 12 and 13 reprinted from *Geochimica et Cosmochimica Acta*, Vol 58, McCulloch and Bennett, Progressive growth of the Earth's continental crust and depleted mantle: geochemical constraints, pp. 4717–38, copyright © 1994, with kind permission from Elsevier Science Ltd., The Boulevard, Langford Lane, Kidlington OX5 1GB, UK; Elsevier Science Publishers BV for fig. 255 from fig. 107 (Mehnert 1968), fig. 249 from fig. 1 (Shervais 1982), fig. 242 from fig. 1 (Meschede 1986); the Geochemical Society for figs 189 and 248 from Mysen, ed., Geochemical Society Special Publication No. 1, 1987; The Geologist's Association for fig. 212 from fig. 16 (Wells and Wooldridge 1931); Gebrüder Borntraeger Verlagsbuchhandlung for fig. 302 from fig. 1 (Knudsen 1991); Hawaii Natural History Association Ltd. for fig. from fig. 4 (Macdonald and Hubbard 1975); the editor, Mineralogical Magazine for fig. 100 from fig. 1 (Groome and Hall 1974); Mineralogical Society of America for fig. 187 from fig. 2 (Wager 1963) copyright © Mineralogical Society of America; the author, Dr J S Myers for fig. 49 from fig. 7 (Myers 1975); New York State Museum and Science Service for fig. 355 from fig. 2 (De Waard and Romey 1969); The Open University for fig. 94 from fig. 20 (Gass and Thorpe 1976); copyright © 1976 The Open University Press; Oxford University Press for fig. 107 from fig. 1 (Macdonald and Katsura 1964); Prentice-Hall Inc. for fig. 38 (Macdonald 1972) copyright © 1980 Princeton University Press; Springer-Verlag for fig. 320 from fig. 3 (Dawson 1980); the Royal Society of Edinburgh for fig. 272 from fig. 6 reproduced by permission of the Royal Society of Edinburgh and R J Pankhurst, M J Hole and M Brook from *Transactions of the Royal Society of Edinburgh: Earth Sciences*, Vol 79, parts 2 and 3 (1988), pp. 123–33; John Wiley & Sons Inc. for fig. 178 from fig. 17 (Kanasewich 1968) copyright © by John Wiley &

Sons Inc.; John Wiley & Sons Inc. for fig. 8 from fig. 8 (Sugimura 1968) copyright © 1968 John Wiley & Sons Inc., fig. 64 from fig. 8.8 (Pitcher and Berger 1972), fig. 181 from fig. 2.1 (Doe and Zartman 1979).

CHAPTER 1

Igneous activity at the present day

There are two reasons for examining recent volcanic activity before turning to the igneous rocks that have formed in the geological past. Firstly, it is possible to place recent volcanism in its tectonic setting, whereas paleogeographic and paleotectonic reconstructions of ancient igneous provinces are more difficult. Secondly, recent volcanic rocks are obviously magmatic, whereas ancient plutonic rocks include compositions which do not correspond to those of any magma which can be directly observed, and include cumulates and melting residua as well as consolidated magmas.

Figure 1. The distribution of the Earth's active volcanoes. A full list is given by Simkin *et al.* (1981).

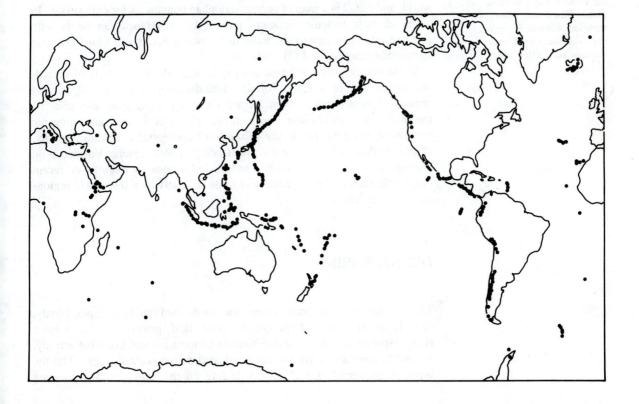

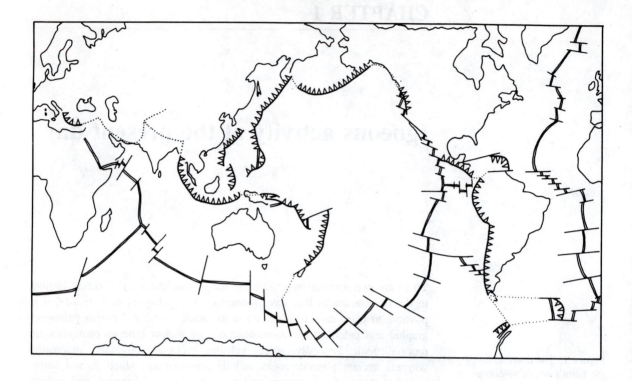

Figure 2. The boundaries of the Earth's major tectonic plates. The double lines represent oceanic spreading axes and the barbed lines indicate subduction zones.

Figure 1 and Table 1 show the distribution of the active volcanoes of the world, and Fig. 2 is a map of present-day plate boundaries for comparison. In terms of their tectonic situation, present-day volcanoes can be broadly grouped into three categories: those of (1) ocean basins; (2) island arcs and continental margins; and (3) continents.

In the ocean basins, volcanic activity is seen along the crests of spreading ridges and in isolated volcanic islands away from the ridges, and intrusive igneous rocks which formed along spreading axes also underlie much of the inaccessible ocean floor. The island arcs and orogenic continental margins are the site of the most concentrated volcanic activity, about two-thirds of all active volcanoes being in such regions. Volcanism on continents is much more feeble, and includes some in regions of recent orogenesis such as the Caucasus and some in relatively stable shield regions such as East Africa.

OCEAN BASINS

The magmatism of the ocean basins may be divided into three types. Firstly, there is the substance of the oceanic crust itself, produced at the oceanic ridges (spreading axes) by a combination of intrusive and extrusive activity. Secondly, there are active volcanoes situated on the oceanic ridges. Thirdly, there are isolated volcanic cones standing on the flanks of the spreading

Table 1. Location of volcanic eruptions in the five-year period 1976–80 inclusive (from Simkin *et al.* 1981)

At divergent plate boundaries	
Iceland	5
Submarine eruptions	not seen
At convergent plate boundaries	
Lesser Antilles	4
Andes	17
Central America	42
Cascades	7
Alaska	20
Kamchatka–Kuriles	15
Japan–Ryukyu	29
Izu–Bonin–Marianas	16
Philippines–Sulawesi	15
South-West Pacific	52
New Zealand	13
Sumatra–Java	24
Southern Italy	12
Oceanic intraplate volcanism	
Hawaii	2
Galapagos	5
Réunion	4
Comoro Islands	1
Marion Island	1
Continental intraplate volcanism	
East Africa	11
Antarctica	5

ridges or resting on the deep ocean floor; the majority of oceanic islands and seamounts are of this type.

THE OCEAN FLOOR

The oceanic crust has been divided by seismologists into three layers:

Layer 1 – average thickness 0.4 km, seismic velocity V_p = 1.5–2.0 km/s
Layer 2 – average thickness 2.1 km, seismic velocity V_p = 2.5–6.6 km/s
Layer 3 – average thickness 4.9 km, seismic velocity V_p = 6.6–7.6 km/s

Layer 1 is the surface layer of oceanic sediment and varies a great deal in thickness, being very thin on the ocean ridges and thickest near the continental margins. Layers 2 and 3 are essentially igneous in composition, and have been investigated by means of dredging (mainly on the oceanic ridges) and by the boreholes of the Deep Sea Drilling Project (DSDP, 1968–85) and Ocean Drilling Project (ODP, from 1985 onwards). The rocks that have been recovered so far fall into the following categories: basalt, gabbro, metabasalt, metagabbro, peridotite, serpentinite and breccias. Layer 2 appears to be mainly basaltic, while layer 3 is thought to be gabbroic.

The basalts of the oceanic crust are rather uniform in character. They are predominantly low-K tholeiites. This type of basalt is known as abyssal tholeiite or mid-ocean ridge basalt (MORB) because it is generated at the mid-ocean ridges (oceanic spreading axes).

Some of the basalts, and nearly all of the gabbros, from the ocean floor have undergone hydrothermal alteration. Enormous hydrothermal convection systems occur in the young crust of the oceanic spreading axes where the crust is still hot, and as a result there is extensive metasomatism of the volcanic rocks by seawater. The majority of the metamorphosed basic rocks are in the greenschist and amphibolite facies of metamorphism; some are in the zeolite facies. These metabasic rocks are not usually schistose, although some samples from fracture zones show cataclasis.

In places the sedimentary and volcanic layers of the oceanic crust are very thin, and the underlying mantle peridotite is exposed (Lagabrielle and Cannat 1990). Peridotite and serpentinite are also exposed along the scarps of oceanic fracture zones (transform faults). Peridotite–mylonites occur above sea-level at St Paul's Rocks, which are on the mid-Atlantic ridge just north of the Equator (Roden *et al.* 1984). Breccias are common along fracture

Figure 3. Geological map of the northern part of Macquarie Island (after Griffin and Varne 1980).

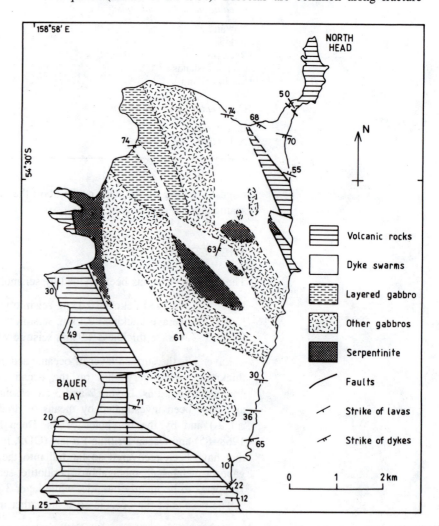

zones and may contain each of the rock types described above, i.e. basalts, metagabbros and serpentinites.

One of the few places where oceanic crust may be seen above sea-level is Macquarie Island, which is 1100 kilometres south-west of New Zealand. In the northern part of the island (Fig. 3) are pillow lavas with interstitial *Globigerina* ooze. These rocks are faulted and tilted. Palaeontological studies suggest that the *Globigerina* ooze is of Pliocene age and was deposited in 2000–4000 metres of water. The lavas are followed downwards in the succession by dolerite dykes from 1 to 3 metres thick, intruded in swarms so dense that they locally occupy 100% of the ground and represent an overall dilation of more than 1 in 1. Below the dyke swarms are massive gabbro, layered gabbro and harzburgite. These exposures on Macquarie Island enable us to reconstruct a section through the upper few kilometres of the oceanic crust, shown in Fig. 4.

Figure 4. Schematic section through the Macquarie Island ophiolite complex (after Griffin and Varne 1980).

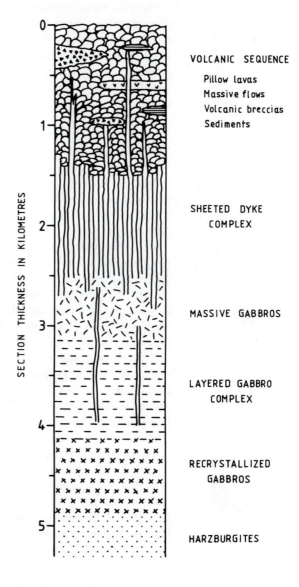

Macquarie Island is situated on a rise linking the continental mass of New Zealand with the actively spreading Indian–Antarctic ridge. Although it is the best-exposed section of oceanic crust that is available for examination, it is somewhat anomalous in its present elevation compared with the rest of the ocean floor. Other possible exposures of oceanic crust are on the islands of Yap in the western Pacific (Shiraki 1971) and Zargabad in the Red Sea (Bonatti 1988).

Several borehole sections through the upper part of the oceanic crust were obtained during the Deep Sea Drilling Project. For example, a 1350-metre section through the sea floor in the eastern Pacific (hole 504B) showed about 300 metres of sediment, followed by 800 metres of pillow lavas (often brecciated), followed by 300 metres of dykes (Anderson *et al.* 1982). The dykes varied from steeply inclined (50–60°) to vertical and normally only showed one chilled margin, i.e. they were injected into one another.

There are many similarities between the rocks dredged or drilled from the ocean floor and those found in ophiolite complexes on land (Chapter 13), and this suggests that the ophiolites are in fact fragments of oceanic crust which have been tectonically emplaced into the continental crust during orogenic activity. Using the evidence of the ophiolites we can reconstruct a complete section through typical oceanic crust, as shown in Fig. 340 in Chapter 13.

OCEANIC ISLANDS

Iceland is the only volcanic island to be situated right on an actively spreading oceanic ridge, although some other volcanoes are situated close to ridges on very young oceanic crust (for example Ascension Island and the Galapagos Islands).

Iceland is entirely oceanic in character; there is no reason to suppose that it is underlain by any continental crust, and its rocks are entirely volcanic. The main present-day volcanism is concentrated in a NE–SW belt through the middle of Iceland, lying parallel to the Reykjanes ridge, which is the axis of spreading in the North Atlantic Ocean. There are older Quaternary and Tertiary lavas on either side of this belt (Fig. 5). Basalts predominate among the exposed igneous rocks and there are smaller amounts of intermediate and acid rocks. According to Jakobsson (1979), the proportions of different magma types in all the postglacial volcanic rocks of Iceland are: basalt 92%, basaltic andesite and trachybasalt 4%, andesite and benmoreite 1%, dacite and rhyolite 3%. The basalts of the main active volcanic zones are all tholeiitic, but alkali basalts occur in the flank zones away from the main volcanic axis.

Those volcanic islands which are situated away from the oceanic ridges represent magmatism not associated with oceanic spreading, although their activity may in some cases have been initiated when the islands were situated on the ridge crests. The exposed lavas of these islands are petrographically variable, but they are normally more alkaline in character than those of the oceanic crust or of Iceland. A selection of rock assemblages is given below:

Figure 5. The geology of Iceland (after Saemundsson 1979).

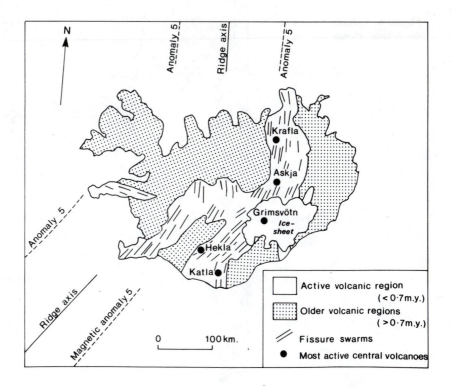

Ascension:	olivine tholeiite (dominant) + hawaiite + mugearite + trachyte (minor) + pantellerite (minor)
Azores:	olivine basalt + hawaiite + trachyte
Fernando de Noronha:	alkali basalt + nephelinite + trachyte + phonolite
St Helena:	alkali basalt + mugearite + hawaiite + trachyte + phonolite
Trinidade:	nephelinite + phonolite
Tristan da Cunha:	ankaramitic basanite + tephrite + phonolite (minor)
Réunion:	olivine tholeiite (dominant) + mugearite + intrusive alkali syenite
Mauritius:	alkali olivine basalt (dominant) + mugearite + phonolitic trachyte
Hawaii:	tholeiite (dominant) + alkali basalt + mugearite + hawaiite + trachyte
Tahiti:	alkali basalt + tahitiite (predominant) + phonolite + trachyte
Galapagos:	tholeiitic basalt + alkali basalt + icelandite (minor) + quartz trachyte (minor)

The abundance of alkali basalt in the oceanic islands is in marked contrast to the tholeiitic character of ocean floor basalts, and there are also many trace element and isotopic differences between them. The origin of these

Figure 6. Volcanic islands of the Atlantic Ocean.

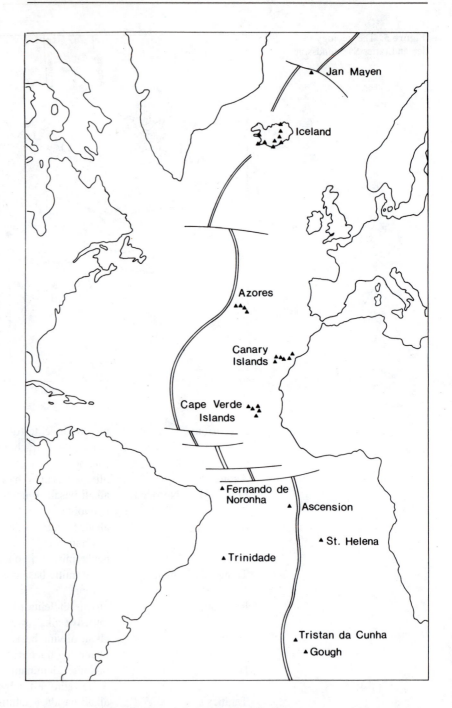

differences will be discussed in Chapter 8. It has been suggested that there is a tendency for oceanic island lavas to show the most strongly alkaline character the further away they are from the axis of the mid-ocean ridge. In the Atlantic Ocean this idea is plausible. Oversaturated lavas occur on the ridge (Iceland, Ascension) while very undersaturated lavas occur well away from the ridge (Trinidade). In the Pacific Ocean there is no such relationship.

The Hawaiian Islands are predominantly tholeiitic even though they are a long way from a ridge axis. On the whole it seems that tholeiite is abundant where the scale of magma production is greatest (Iceland, Hawaii), but alkali basalts are more important in the smaller and less productive volcanic islands.

In addition to the volcanic islands which rise above sea-level, many more oceanic volcanoes are represented by seamounts whose summits are below present sea-level. Batiza (1982) has reviewed the distribution of seamounts and estimates that there are between 22,000 and 55,000 in the Pacific Ocean, of which 1500–2000 are presently active. Their activity is thus second only to new crustal formation at oceanic spreading axes as a site of present-day magmatism. Batiza estimated that the volcanic seamounts constitute from 5% to 25% of the volume of volcanic rock in the oceanic crust.

OCEAN BASIN FLOOD BASALTS

In addition to the isolated occurrences of intraplate volcanism represented by the oceanic islands, there are also some areas of extensive submarine volcanism which post-date the oceanic crust on which they rest. These might be regarded as analogous to the flood basalts which occur on land, for example the Columbia River basalts, but at present they are not very well known. One of the largest is the Ontong Java plateau in the western Pacific, which covers an area of about 1.5 million km^2 (Floyd 1989).

ISLAND ARCS AND CONTINENTAL MARGINS

Volcanic activity is more intense in island arcs and along unstable continental margins than in other exposed regions of the Earth's surface. More than two-thirds of the world's active volcanoes are in this environment. The unifying feature of these regions is that they are situated over inclined seismic zones (Benioff zones) in the underlying mantle (Fig. 7), and there is a very strong presumption that the volcanism originates within or adjacent to these seismic zones.

In terms of plate tectonics, volcanic regions of this type may be divided into: (1) island arcs bounded on either side by oceanic crust; (2) continental margins bounded by oceanic crust on one side and continental crust on the other; and (3) intra-continental fold belts, bounded by continental crust on both sides. The term 'orogenic' may conveniently be used to denote all these regions of convergent plate boundaries.

ISLAND ARCS

Examples of island arc volcanism are the New Hebrides, Tonga, the Marianas and the Kermadec Islands in the Pacific, the Lesser Antilles in the

Figure 7. Volcanism and inclined seismic zones of the Western Pacific. Volcanoes are shown as black dots and Benioff zones are marked by 100 kilometre depth contours.

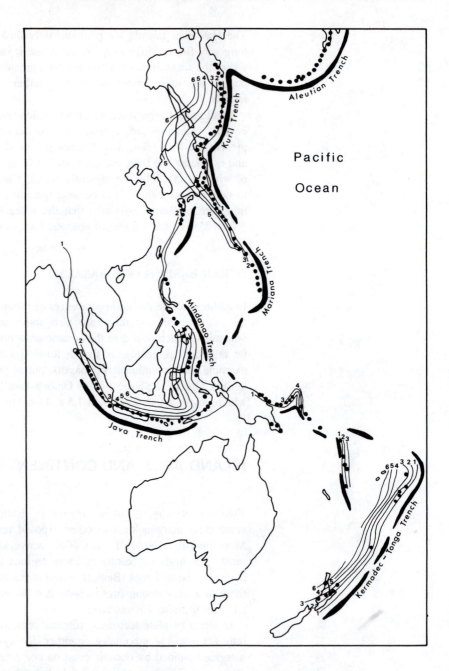

Atlantic, and the South Sandwich Islands in the Antarctic Ocean. Some examples of the volcanic rock association in such areas are given below.

In the New Hebrides the volcanic rocks are olivine basalts (some nepheline-normative and some quartz-normative), picrites, ankaramites and andesites, but no acid lavas are present. In the Tonga and Kermadec Islands the lavas are basalts (either tholeiitic or high-alumina type), andesites and dacites, with a large compositional gap between 57 and 65% SiO_2, i.e. between the andesites and dacites. In the Solomon Islands the lavas are olivine-basalts, picrites, ankaramites and andesites, but again there are no acid lavas. In the Lesser Antilles the lavas are basalts, andesites, dacites and rhyolites, with andesites being preponderant and rhyolites being present only in small amount; a continuous range of compositions is present. In the South Sandwich Islands there are basalts (tholeiites), andesites, dacites and rhyolites; basalt is preponderant and rhyolite is again present in only small amounts.

OROGENIC CONTINENTAL MARGINS

Volcanism occurs along those continental margins which are underlain by inclined seismic zones; in some cases they are the continuation of island arcs. Examples are New Zealand, the Philippines, Japan, southern Alaska, part of central America (southern Mexico–Costa Rica), and much of the west coast of South America, Sumatra and Java in the Indian Ocean, and the southern Aegean Sea in the eastern Mediterranean.

As a general rule, the volcanic rock types of continental margins are the same as those of island arcs (basalt–andesite–dacite–rhyolite), but they occur in different proportions. The acid rocks are more abundant in orogenic continental margins than in island arcs. The relative abundances for several provinces are given in Table 48 (Chapter 10).

Among the Quaternary volcanic rocks of Japan there are basalts, andesites, dacites and rhyolites, of which the andesites are the most abundant and the dacites and rhyolites are mostly represented by pyroclastic rocks (Fig. 8). During the Quaternary and Neogene the volume of dacite and rhyolite erupted has been approximately equal to the volume of andesite and basalt. Small amounts of trachyandesite and trachyte also occur. There is a general tendency for the basaltic lavas to become progressively more alkaline in a north-westerly direction, i.e. away from the ocean margin and towards the deeper part of the underlying Benioff zone. A similar tendency is observed in other continental margin environments, for example in Indonesia.

The chain of islands in SW Indonesia stretching from Sumatra to Sumbawa lies on the continental margin between south-east Asia and the Indian Ocean. It is one of the most active volcanic regions in the world and includes the celebrated island volcano of Krakatoa. The lavas in this volcanic chain are basalt, andesite, trachyandesite, dacite and rhyodacite, of which andesite and trachyandesite are the most abundant. More alkaline lava types occur further east in Indonesia in the Molucca and Banda Seas, but in a different tectonic situation.

In New Zealand, the most active volcanic area is the Taupo volcanic zone in North Island, which is on a direct continuation of the Kermadec volcanic island arc. Unlike the Kermadec volcanics, which are mainly basaltic and andesitic, the magmas of the Taupo province are mainly rhyolitic, with andesite and basalt in subordinate amounts. A high proportion of the acid material consists of ash and welded tuff. In the Auckland area, to the north-west of the Taupo province, i.e. overlying a deeper level of the presumed extension of the Kermadec Benioff zone, are lavas whose compositions include alkali basalts (basanites).

Figure 8. Relative volumes of eruptive products of the Quaternary land volcanoes of Japan (after Sugimura 1968). B = basaltic, A = andesitic, D = dacitic, R = rhyolitic rocks.

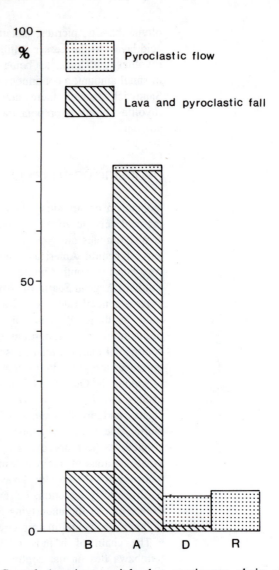

On the Central American mainland a continuous chain of volcanoes stretches from Costa Rica to Mexico, erupting predominantly basalt and andesite lavas, dacitic pumice, and rhyolitic tuffs and ignimbrites. This line of subduction-related volcanoes ends in western Mexico, where the oceanic spreading axis of the East Pacific Rise runs into the west coast of the American continent. It resumes briefly in the Cascade range, and there is then another break between the Cascades and Alaska.

Compared with central America, volcanic eruptions in the Cascade range of Washington to northern California are relatively infrequent (Fig. 9). The small amount of acid magma and relatively subordinate role of andesite is not typical of an orogenic continental margin. Earlier in the Cenozoic, the volcanism of the western United States was very different, and the eruptions of basalt, andesite and rhyolite were of typical continental margin type, as elsewhere round the Pacific Ocean. McBirney and White (1982) give the volume percentages of different rock types in the Cascades as follows:

		Basalt	Andesite	Dacite–Rhyolite
Oligocene–Miocene	(central Oregon)	10	45	45
M. and U. Miocene	(central Oregon)	39	41	20
Pliocene	(central Oregon)	90	9	1
Quaternary	(central Oregon)	85	13	2
Quaternary	(northern California)	69	29	2

Figure 9. Volcanism of the Cascades and neighbouring regions, north-west United States (after McBirney and White 1982).

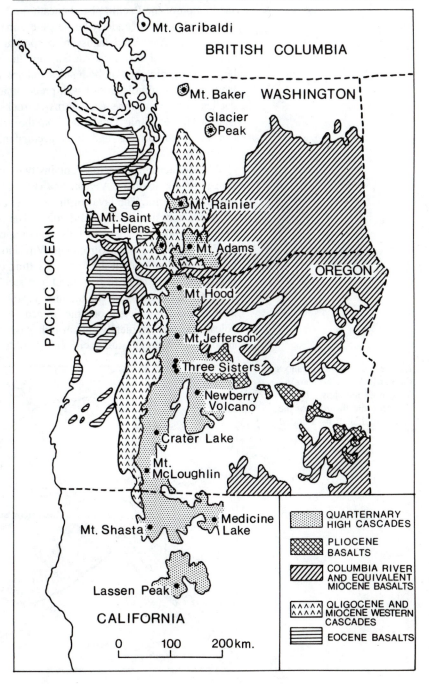

Before the Pliocene, andesite was the most abundant magma erupted in the Cascades, although great volumes of basalt were erupted during the Miocene in the Columbia River province to the east. Andesitic volcanism did not cease in the Quaternary, and andesite is a major constituent of the Quaternary volcanic cones of the High Cascades. Greater volumes of basalt were erupted at the same time but are topographically less conspicuous.

The volcanism of the continental margin of southern Alaska is related to subduction of the Pacific plate beneath the North American continent, and is associated with an underlying seismic zone and with the Aleutian trench. The eastern Aleutian Islands are a continuation of the continental margin volcanic belt of southern Alaska, but the western Aleutians are situated between oceanic crust on both sides and are a true island arc (Fig. 10). In the western Aleutian volcanoes basalt predominates, but there are also andesites, dacites and even some acid intrusive rocks in the more deeply dissected islands (for example Unalaska). On the Alaskan peninsula and mainland andesite predominates and it is accompanied by substantial amounts of rhyolite (e.g. Katmai).

The Mediterranean region contains two continental margin volcanic arcs. One is in the southern Aegean Sea (Fig. 11) and includes the well-known volcano Santorini. The lavas of the Aegean arc include the usual assemblage of basalt, andesite, dacite and rhyolite. The other active volcanic area in the Mediterranean is in southern Italy. This region includes the volcanoes of the Aeolian Islands (Lipari, Stromboli, Vulcano), Vesuvius and Etna. There is much more doubt about the tectonic setting of these volcanoes. The Aeolian volcanoes are thought to be related to a subduction zone dipping westwards under Calabria, although the inclined seismic zone is very discontinuous and there is no associated oceanic trench. Furthermore the lavas of these volcanoes are atypically alkali-rich and include leucite-bearing rocks, although basalt, andesite, dacite and rhyolite are all present (Keller 1982).

Figure 10. Active volcanoes of the northern Pacific margin. Volcanoes are shown by triangles and the edge of the continental shelf by a dotted line.

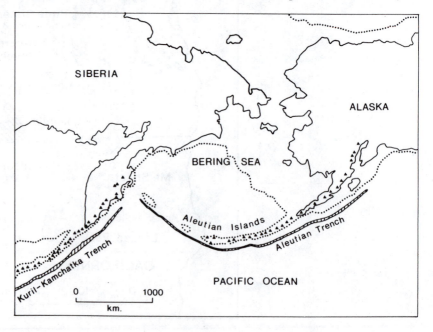

Figure 11. Recent volcanoes of the Middle East. Dotted lines show the edges of the Russian and Arabian continental platforms.

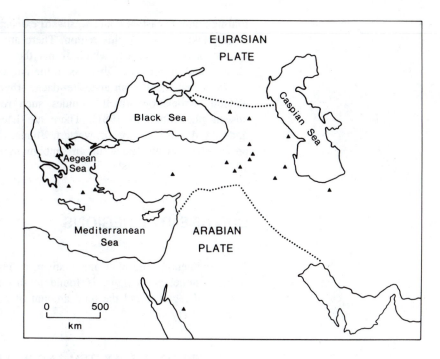

INTRACONTINENTAL OROGENIC BELTS

The type of volcanism observed at zones of continental collision is of particular interest because this is the presumed tectonic setting of many ancient orogenic belts, with their attendant magmatism. The only volcanoes which are currently active in such a tectonic situation are those of Turkey and Iran, where the Arabian and Eurasian plates come into contact.

The crustal structure of the Arabian–Eurasian collision zone is very complicated (Fig. 11). On the Eurasian side there is oceanic crust in the Black Sea and Caspian Sea, and on the Arabian side there is oceanic crust in the Eastern Mediterranean and the Arabian Sea. In the west, the volcanic activity of peninsular Turkey might be regarded as an extension of that in the Aegean arc, i.e. a continental margin between a northern continent (Greece and Turkey) and a southern ocean (the eastern Mediterranean). In the east, the volcanic activity of south-east Iran might be similarly regarded. The least ambiguous area in this region is that along the border between Turkey and Armenia, where there is undoubted continental crust to both north and south.

The best known volcano in this province is Mount Ararat, which has not been active in historic times but has erupted during the Pleistocene. Its lavas are andesites, dacites and rhyodacites. Other late Cainozoic volcanic rocks in this area are dominated by andesitic lava and in Armenia by a large extent of ignimbrite. Quaternary alkaline lavas ranging from basanite to trachyte also occur in Eastern Turkey and Armenia (Innocenti *et al.* 1976, 1982).

There was late Tertiary magmatism at many locations in the Himalayan fold belt and its eastward and westward extensions, but very little volcanism is now taking place in this region. There are Quaternary volcanoes in the Kunlun mountain range, which forms the northern edge of the Tibetan plateau, and in northern Tibet. As in the Armenian region this volcanism is of two types: there is an andesite–dacite–rhyolite association, and also an alkaline association which includes such rocks as leucite–tephrite and nosean–phonolite (Jin 1981). There are late Quaternary–Recent basalt–andesite–dacite volcanoes in western Yunnan, but they lie to the east of the present-day seismic zone in an ambiguous tectonic setting (Zhu *et al*. 1983).

CONTINENTAL REGIONS

True continental intra-plate magmatism, as opposed to that of seismically active continental margins, is found at the present day in only a small number of regions, and the total amount of magma being erupted in these regions is very small.

EAST AFRICAN RIFT SYSTEM AND RED SEA

The East African rift system extends for 3500 kilometres in a N–S direction through East Africa (Fig. 12). It appears to be a continuation of the Red Sea and of the Jordan–Dead Sea rift beyond, but although it shows tensional features it should not necessarily be regarded as an incipient oceanic rift. The igneous activity of the region has been continuous since the Miocene or even earlier, and if the continental crust has not split to form a new ocean basin during this period there is no certainty that it will do so in the future.

The volcanoes in East Africa are of great petrographic interest, especially in the region around Lake Victoria, where the rift system splits into two branches, the Western Rift and the Eastern Rift. The Western Rift Valley contain Nyiragongo, one of the few volcanoes to have held a long-lasting lava lake for much of this century, and the Eastern Rift Valley contains Oldoinyo Lengai, the only volcano known to have erupted carbonatite lava in historic time.

In the Western Rift Valley, the lavas are predominantly mafic and extremely alkaline in character. They vary from volcano to volcano, often quite markedly between volcanoes which are close together, but include nephelinites, leucitites and melilitites as well as alkaline olivine-basalts and trachybasalts. There are a few trachytes and latites. A small amount of tholeiitic basalt occurs in the South Kivu area and there is a little rhyolite in the same area.

In the Eastern Rift Valley, the lavas are predominantly alkaline olivine-basalts, accompanied by trachyte and phonolite. There are some basalts with nepheline and melilite, but few of the potassium-rich basic rock types typical of the Western Rift. Tholeiitic basalts plus trachyte and rhyolite occur in the Afar Rift at the northern end of the East African Rift Valley system.

Figure 12. The volcanoes and rift zones of Africa.

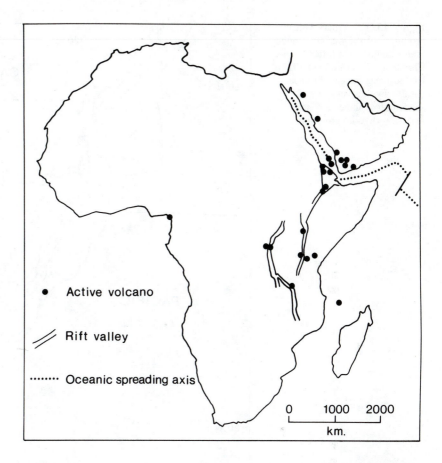

The Red Sea and Gulf of Aden lie on the extension of the western Indian Ocean ridge, and are newly developed ocean basins resulting from spreading at their axes. Recent volcanism is most intense adjacent to the triple junction at the southern end of the Red Sea, where the East African rift system reaches the Red Sea and Gulf of Aden spreading axes. It extends up into the western part of the Arabian peninsula. The Recent basalts and associated felsic rocks of this region range from tholeiitic to strongly alkaline (Mohr 1983). Quaternary volcanism in the Afar rift in Ethiopia is bimodal in character, with basaltic lavas and silicic ignimbrites; during the mid-Tertiary the surrounding area was the site of extensive flood basalt eruptions.

WESTERN UNITED STATES

The western side of North America is a region of unusual tectonic complexity (Fig. 13). In Alaska and in Central America there is active subduction of the Pacific and Cocos plates beneath the continent, and this is accompanied by typical continental margin volcanism. In the intervening region there is a small area of probable subduction and associated magmatism in the Cascade province of the NW United States. During the twentieth century the Cascade volcanoes have been the only ones to erupt in

Figure 13. Quaternary volcanic provinces of the western United States and Mexico (after Christiansen and Lipman 1972). Dotted areas indicate bimodal basalt–rhyolite volcanism; shaded areas indicate basalt–andesite–rhyolite provinces.

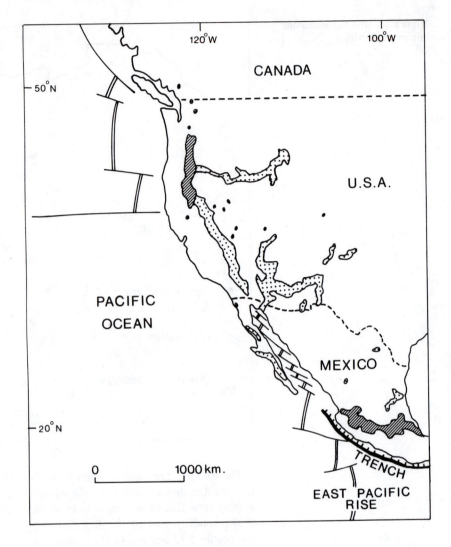

the mainland USA and Canada, but there are in fact many other Recent volcanoes in the western United States which are not related to continental-margin subduction. They extend from Yellowstone Park in the north to Baja California in the south.

It is probable that up to mid-Tertiary times the Pacific plate was being subducted beneath North America over most of the length of the American Pacific coast, and this is reflected in the presence of orogenic basalt–andesite–rhyolite volcanism on the American mainland. During the late Tertiary this subduction appears to have come to an end in the south-west United States as the East Pacific Rise approached the continental margin. The compressional tectonic regime associated with subduction then gave way to an extensional regime in the region lying along the subcontinental extension of the East Pacific Rise (Christiansen and Lipman 1972; Snyder *et al.* 1976; Leeman 1982). This region is characterized by high heat-flow and

crustal thinning, which may reflect upwelling and spreading in the underlying mantle (Smith 1978; Lachenbruch and Sass 1978).

The Recent volcanism of the south-western United States and Baja
California is varied in character. A bimodal basalt–rhyolite association
predominates, but there are also local associations of tholeiite with andesite
and dacite (e.g. northern Arizona, Gulf of California) or alkali basalt with
hawaiite (e.g. south-west Utah). During the Quaternary, the Yellowstone
Plateau volcanic field was the site of very large rhyolitic eruptions,
accompanied by a very small amount of basalt but no intermediate lavas.

OTHER CONTINENTAL PROVINCES

Apart from the two major regions described above, very few truly continental volcanoes are presently active.

An isolated volcano in an interesting location is Balagan-Tas in northeastern Siberia. It is situated on line with the trend of the Nansen–Gakkel
ridge from the Arctic Ocean on to the Asian continent, and together with
neighbouring subvolcanic intrusions shows a basalt–rhyolite association. An
analogy might be made with some of the Quaternary volcanoes of the south-
west United States, which are aligned with the trend of the East Pacific Rise
on to the North American continent.

In West Africa a chain of volcanic peaks extends across the continental
margin from the Cameroun mountains into the Atlantic. The north-east part
of this chain (the 'Cameroun line') is on continental crust and the south-
western end is on the ocean floor. In the Cameroun mountains the volcanic
rocks are very alkaline and include leucitite, nephelinite, tephrite, trachybasalt, trachyte and phonolite. The off-shore islands in this chain (São
Tomé and Principe) show alkali basalts, basanites, tephrites, phonolites and
trachytes.

There are two similar volcanic provinces in eastern Antarctica. A line of
volcanoes, including the active Mount Erebus, extends along the west side of
the Ross Sea on continental crust to the Balleny Islands in the Southern
Ocean. The lavas in this province are alkali basalt, trachyte and phonolite.
Volcanism also extends across the continental margin from Heard Island in
the Southern Ocean (alkali basalt and trachybasalt) to Gaussberg on the
Antarctic continent (leucitite).

Isolated occurrences of continental volcanism are widespread in a large
area of north-east Asia which stretches from North Korea through Manchuria
and Mongolia and includes the area around Lake Baykal in Russia. The
number of historically active volcanoes in this region is small. The lava
suites are dominated by alkali basalts, but nephelinites, leucite-basalts,
trachytes and pantellerites also occur. The volcanism in this part of the world
is relatively poorly known, but has been summarized by Whitford-Stark
(1983).

Although the presently active areas of continental intraplate volcanism are
few in number, there are additional areas which have been active during the
Quaternary. The Tertiary–Recent volcanic province of eastern Australia is
very extensive, and contains many different lava types, predominantly

basaltic but ranging from quartz-tholeiite to basanite during the Quaternary. The Quaternary–Recent volcanism of the Massif Central in France and the Eifel in Germany is highly alkaline in character, giving mainly alkali basalt with trachyte and phonolite, but with extremely alkaline mafic lavas in some of the Eifel volcanoes.

The relatively subordinate role of intraplate igneous activity, both continental and oceanic, compared with that at plate margins, is indicated by the estimates of present-day magma production given in Table 2.

Table 2. Magma production rates (in km³/year) in different tectonic environments (after Fisher and Schmincke 1984)

	Extrusive	*Intrusive*
Divergent plate boundaries	3	18
Convergent plate boundaries	0.6	8
Oceanic intra-plate	0.4	2
Continental intra-plate	0.1	1.5

LOCATION OF INTRAPLATE VOLCANISM

HOT-SPOTS AND MANTLE PLUMES

Apart from at subduction zones, where volcanism is most intense, active volcanism is rather sporadically distributed, and there is great interest in why it occurs exactly where it does. There must be magmatism all along the divergent plate margins to create new oceanic crust, but it is particularly intense in certain places, such as Iceland, which forms a large mass of mainly basaltic rock rising above the general level of the mid-Atlantic ridge. Away from the spreading axes, oceanic volcanism is concentrated in clusters of activity, such as in the Hawaiian Islands. In one of the first applications of plate tectonics to igneous geology, Wilson (1963) proposed that linear chains of volcanoes in oceanic regions owe their origin to stationary or slow-moving heat sources underlying the comparatively fast-moving lithospheric plates at the Earth's surface.

Figure 14 shows the Hawaiian volcanic chain. The presently active volcanoes of Hawaii lie at the south-eastern end of the chain which extends for over 6000 kilometres across the north-western Pacific. The earliest-formed volcanoes in this chain are represented by the Emperor seamounts, none of which now rise above the ocean surface. The seamounts and islands of the chain are progressively younger towards the south-east, and the correlation of position with age is consistent with the Pacific plate having migrated over a lower mantle 'hot-spot' which is now beneath the island of Hawaii. If this is so, the alignment of the chain must mark the relative movement directions of the Pacific plate and the underlying hot-spot. The locus of active volcanism has migrated 5000 kilometres during the last 60 million years.

The abrupt change in alignment west of Midway Island is presumed to reflect a change in this relative movement at some time during the Tertiary.

Figure 14. The Hawaiian Islands and Emperor seamounts (after Heezen and Fornari 1978). The 2000 metre submarine contours are outlined, and ages of volcanic activity shown.

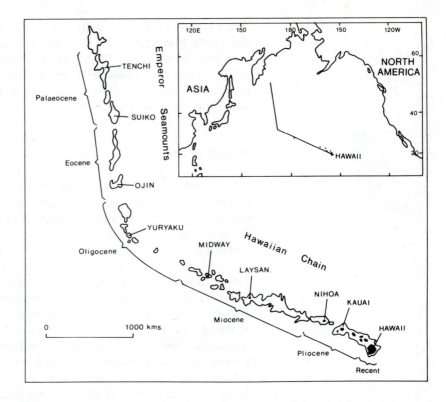

There are two other volcanic lines in the southern Pacific which may be possible hot-spot markers, the Tuamotu and Austral chains, both of which are similar to the Hawaiian–Emperor chain in length and have a similar kink. The Austral chain also resembles the Hawaiian one in having its only presently active volcanism at the south-eastern end (Macdonald seamount).

Sleep (1990) listed other possible sites of hot-spots, including such important volcanic areas as Iceland, the Azores, Yellowstone Park and the Galapagos Islands. Some of the postulated hot-spots are isolated occurrences of volcanic activity; others form part of an alignment similar to the Hawaiian chain. Some are in oceanic areas and others are continental. Many are of long standing; for example Sutherland (1981) showed that during the past 53 million years the loci of igneous activity in eastern Australia have been migrating southwards, at about the same rate as the Australian plate has been moving northwards (as shown by magnetic anomalies in the Southern Ocean). Apparent hot-spots within one tectonic plate may even remain in the same positions relative to one another for periods as long as 100 million years (Duncan 1981).

As an explanation for hot-spot volcanism, Morgan (1971) suggested that the relatively intense magmatism at these sites is due to a 'plume' of hot material rising from the lower mantle, and this explanation has been widely accepted. Moberly and Campbell (1984) drew attention to a correlation between Hawaiian volcanic activity and episodes of normal geomagnetic polarity, from which they inferred that hot-spot plumes arise from the boundary between the mantle and the Earth's core.

In addition to volcanic activity, the oceanic hot-spots are characterized by abnormally high heat flow and by elevation of the sea floor compared to other parts of the ocean (White and McKenzie 1989; Davies and Richards 1992). Many petrologists have incorporated mantle plumes into their petrogenetic models, for instance by assuming the mixing of materials from different levels in the mantle penetrated by a plume.

FLOOD BASALTS

Ancient volcanic provinces represent the products of several different tectonic environments. Most of them correspond to types of volcanic activity seen at the present day, with the exception of the continental flood basalts. These are the most voluminous outpourings of lava that have taken place in continental areas, and yet there is no modern example. The great plateau lava provinces are few in number (see Table 7 in Chapter 2). They include the Columbia River, Karroo and Deccan plateaux, and the smaller but well-studied lavas of the Tertiary volcanic province in the British Isles. Basalts predominate in all these provinces, but are accompanied by a great diversity of minor rock types. Some of these examples may be related to major continental rifting. For example the British Tertiary province and comparable igneous rocks in Greenland are situated close to the margins of the North Atlantic Ocean, and date from close to the time of its opening in the early Tertiary. On the other hand, the Columbia River basalts do not appear to be related to a plate boundary.

The large volumes of basaltic magma involved in plateau volcanism imply

Figure 15. Map of the South Atlantic showing the relationship of the Parana and Etendeka flood basalts (130–120 million years) to the Rio Grande Rise and Walvis Ridge, interpreted as lying along the track of a mantle plume (after White and McKenzie 1989).

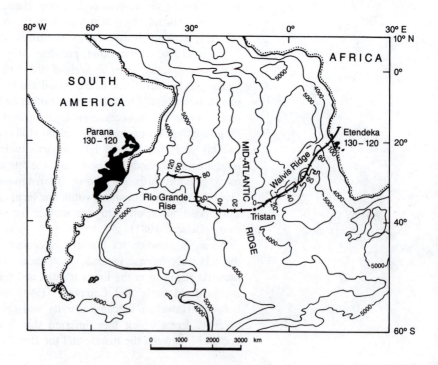

the periodic occurrence of unusually extensive melting in the underlying mantle. According to White and McKenzie (1989), all the continental flood basalt provinces can be related to rifting above an upwelling mantle plume. Early Tertiary volcanism in the Brito-Arctic province of Greenland, Scotland and Ireland is associated with the opening of the North Atlantic Ocean above the Icelandic plume. The Parana basalts are associated with the opening of the South Atlantic above the plume responsible for the formation of the Walvis ridge and which is now centred on Tristan da Cunha (Fig. 15).

Hawkesworth and Gallagher (1993) agreed that large-scale melting events denote high temperatures in the upper mantle, but questioned whether such melting is proof of the arrival of a deep-seated mantle plume and suggested alternatively that it could be triggered by regional extension within the lithosphere.

Recently it has been shown that there are some very large outpourings of basalt in oceanic areas which are comparable in scale to the continental flood basalts. The largest of these are the Ontong Java plateau in the western Pacific and the Kerguelen plateau in the Southern Ocean, both of which were formed during the Cretaceous and are largely submerged below sea level. Coffin and Eldholm (1993) estimated that during its formation the rate of magma emplacement in the Ontong Java plateau may have exceeded that of the whole mid-ocean ridge system at the time.

CHAPTER 2

Volcanism

The products of volcanic eruptions are lavas and pyroclastic rocks. Magmas with low viscosity and a low content of dissolved gas are erupted quietly as lava flows, but magmas with a high viscosity or a high content of dissolved gas are erupted explosively and give rise to a high proportion of pyroclastic products and relatively little lava.

The best-known volcanoes of the world, such as those in Hawaii, are ones in which lava is abundantly erupted, and the appearance of the flows is familiar from films and photographs. Pyroclastic deposits are not so easily observed in eruption, and once erupted are more readily removed by erosion. Thus lava tends to be thought of as the principal product of volcanic eruptions. This is highly debatable, since by far the commonest subaerial eruptions are those of the convergent plate boundaries (Table 1), where pyroclastic products are often more voluminous than lavas. Much of this pyroclastic material is rapidly converted into sediment, but it is not too difficult to recognize redeposited sedimentary rocks of volcanic parentage from their mineralogy and chemical composition, after making due allowance for weathering and diagenesis. The thick greywacke sequences of orogenic regions are major repositories of redeposited pyroclastic material.

The two most important factors governing whether extrusion is quiet, giving rise mainly to lava, or violent, giving rise mainly to pyroclastic rocks, are the viscosity of the magma and its content of dissolved gases. Basaltic magma is usually less viscous than rhyolitic magma and contains less gas in solution and therefore it is normally erupted as lava, whereas rhyolitic magma is more likely to give rise to rhyolitic ash than to an actual rhyolite. Some basic magmas do have a high gas content, and some rhyolites are erupted quietly after the magmas have lost their dissolved volatiles, but these are exceptions to the rule.

FACTORS CONTROLLING ERUPTION

THE RISE OF MAGMA

As a first approximation, the ability of magma to rise to the surface is determined by its density and the pressure at its source. The maximum height of the magma column in the conduit of a volcano is controlled by the hydrostatic pressure at the base of the magma column being equal to the lithostatic pressure at the same depth. For example in Fig. 16 the pressure exerted by the overburden of rock on the magmatic source region just equals the pressure exerted by a column of magma reaching to the top of the volcano (17.4 kilobars). Vogt (1974) showed that there is a positive correlation between the height of oceanic island volcanoes and the thickness of the lithosphere on which they stand, implying that the magmas are driven from the base of the lithospheric plate.

Some magmas do not rise directly from their source to the site of eruption, but collect in a magma chamber at an intermediate depth. In Hawaii, magma can sometimes be seen to erupt from the summit of Mauna Loa while a lava lake stands quietly in the nearby but lower crater of Kilauea, suggesting that the Kilauea crater is not directly connected to the magmatic source region. Seismic studies have confirmed that there is a shallow magma reservoir

Figure 16. Hypothetical cross-section through a region of oceanic crust, showing hydrostatic equilibrium between the magma column in a volcano and the overburden pressure exerted by the surrounding crust and mantle.

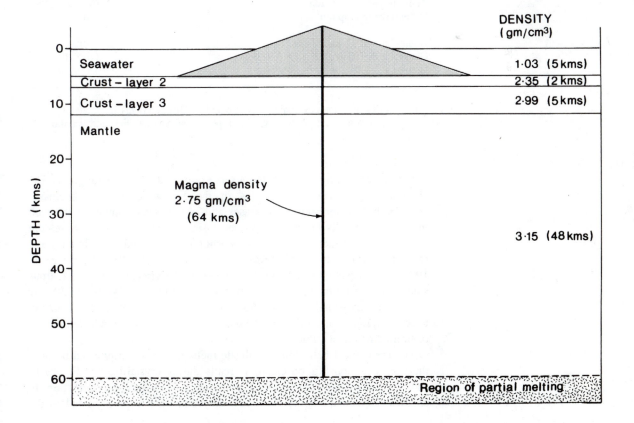

beneath Kilauea. On the other hand the presence of mantle-derived xenoliths in some basalts shows that they have ascended from their source too rapidly for the xenoliths to have been deposited. The major flood basalt eruptions are likewise so voluminous as to require a continuous flow of magma from the mantle source.

VISCOSITY OF MAGMAS

Some information on viscosities is given in Table 3. As far as lavas are concerned, field observations show that acid magmas are much more viscous than basic magmas, and this is confirmed by laboratory measurements on melted rocks. Experiments by Kani and Hosokawa on molten basalt with various additives have shown that Si, Al and K contribute to a high viscosity, whereas Na, Ca, Fe and Mg contribute to a low viscosity (Murase 1962).

Table 3. Viscosities of magmas

	Viscosity (Pa.s)
Basalt melt at 1200 °C	$10–10^2$
Andesite melt at 1200 °C	$10^3–10^4$
Rhyolite melt at 1200 °C	$10^5–10^6$
Basalt lava in eruption	$10^2–10^4$
Andesite lava in eruption	$10^4–10^6$
Dacite lava in eruption	$\sim 10^{10}$
Granite at 800 °C, anhydrous	$\sim 10^{11}$
Granite at 800 °C with $P_{H_2O} = 0.5$ kb	2.6×10^5
Granite at 1200 °C with $P_{H_2O} = 2$ kb	0.5×10^2

Data from various sources, mainly Murase and McBirney (1973) for melts at 1200 °C, Macdonald (1972) for lavas in eruption, and Shaw (1963) and Persikov (1977) for granites.

In practice, the figures in Table 3 serve only as a very general guide. The viscosity of silicate melts varies in the following way: it decreases considerably as temperature increases and as the water content of the melt increases; it usually but not always decreases by a small amount as pressure increases; and it increases as the proportion of suspended crystals increases (McBirney and Murase 1984; Richet 1984; Dingwell 1987; Scarfe *et al.* 1987). It is possible to calculate approximately the viscosity of an anhydrous silicate melt from its chemical composition (Bottinga and Weill 1972; Shaw 1972; McBirney and Murase 1984), but since most magmas contain an uncertain proportion of dissolved volatile constituents these calculations are of limited practical value.

The rheological behaviour of silicate melts is rather complicated because their physical constitution changes greatly during crystallization. Figure 17 shows some of the ways in which materials behave during deformation. In an ideal Newtonian liquid (Fig. 17), the strain rate is proportional to the

Figure 17. The relationship between strain rate and shear stress in various types of liquid.

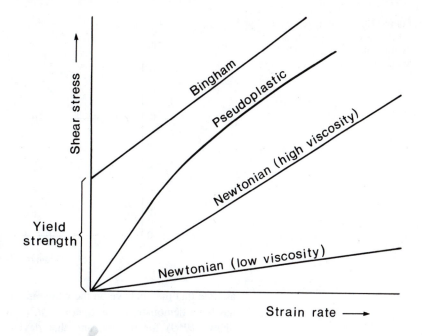

applied stress, but in a Bingham material deformation does not take place until the applied stress reaches a critical value, i.e. there is a finite yield strength which must be exceeded before it will move at all. Laboratory studies of completely molten silicates have mostly shown that they behave as Newtonian liquids, and the same is true of partially crystalline lavas (Murase 1962), but field measurements on crystallizing magmas sometimes indicate non-Newtonian (Bingham-like) behaviour, and various authors have attempted to calculate their yield strengths (Shaw *et al.* 1968; Hulme 1974; Pinkerton and Sparks 1978; McBirney and Murase 1984; Murase *et al.* 1985).

To understand fully the behaviour of silicate melts it is necessary to consider their atomic structure. The atoms which compose silicate liquids are not arranged randomly, but into clusters which prefigure the more orderly arrangement of crystalline silicates. In particular, each silicon atom is surrounded by four oxygens in the same tetrahedral arrangement that characterizes the structures of silicate minerals. As a silicate liquid cools towards the liquidus temperature (at which crystallization starts) the SiO_4 tetrahedra are increasingly linked by the other cations into a continuous framework, making it more and more difficult for the liquid to flow, but some flow continues to be possible until the solidus temperature is reached, by which time crystallization is complete. The effective viscosity increases so rapidly in the temperature interval between the liquidus and solidus that viscosities measured above the liquidus are of hardly any relevance to the flow behaviour of a magma whose crystallization is nearly complete. The morphology of a lava flow which ranges from a partly liquid interior to a largely crystalline crust cannot be modelled by any single value of the viscosity or yield strength. There are still insufficient laboratory data even on wholly liquid silicate melts, but until there are it would be reasonable to

Figure 18. The pseudoplastic behaviour of an experimental rhyolite melt at 1125 °C (from Spera *et al.* 1982).

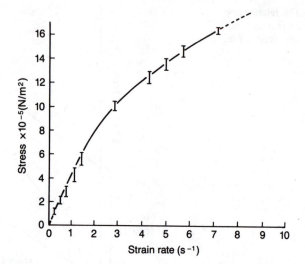

assume that they behave on the whole in a pseudoplastic manner (Fig. 17), as has been demonstrated by Spera *et al.* (1982) for rhyolites (Fig. 18) and by Shaw (1969) for basalts, and that deviations from Newtonian behaviour become more pronounced as cooling and crystallization proceed.

MAGMATIC GASES

All magmas contain dissolved gases. At high pressure, the amount of water and other volatile constituents that can be dissolved in a magma can be several percent (see Fig. 144 in Chapter 5), but as the magma rises to the surface the pressure decreases and dissolved gases are exsolved. Nearly all erupting magmas contain bubbles of gas in the process of exsolution, and several daring volcanologists have approached erupting lava to collect samples of the magmatic gases as they are emitted. In most cases the gas samples are contaminated by air and meteoric water, and by the gaseous products of oxidation and reduction of the lava during eruption. Thus reliable analyses of volcanic gases are few. An alternative method of finding the composition of volcanic gases is to analyse the contents of fluid inclusions in solidified lava, and of bubbles in pumice.

Table 4 gives some of the more reliable data on the composition of gas collected from erupting lava. These are all from lavas of broadly basaltic composition being erupted relatively quietly. There is very little information on the more explosively erupted andesitic and rhyolitic magmas, but studies of gaseous inclusions in acid lavas suggest that CO_2 and SO_2 are less important constituents of acid than of basic magmas and that H_2O predominates. In addition to the gases listed in Table 4, other gases which are frequently reported as being emitted from lavas are S_2, H_2S, Cl_2, HCl, N_2, inert gases, and F_2. Greenland *et al.* (1985) estimated that 90% of the CO_2 in Kilauea basalt magmas is lost by degassing in a shallow magma chamber before eruption, whereas most of the H_2O is evolved only when the lava actually erupts, so measurements of evolved gases must obviously be interpreted with caution.

Table 4. Compositions of magmatic gases collected from basaltic lavas during eruption (from Anderson 1975 and Rose *et al.* 1982)

Gas	Minimum (vol %)	Maximum (vol %)	Mean (vol %)	Mean (wt %)
H_2O	21	94	59.2	34.9
H_2	0.1	4.7	1.5	0.1
CO_2	0.5	40.9	17.0	24.5
CO	0.04	2.4	0.9	0.8
SO_2	1.3	59	18.8	39.5

These figures are a summary of analyses of eight 'least-contaminated' samples from Kilauea (4), Surtsey, Erta Ale, Nyiragongo and Etna.

TEMPERATURE OF ERUPTION

The temperature of eruption of lavas has been measured by many investigators, but mostly on lavas of basaltic and andesitic composition. Generally the results are in the range 1000–1200 °C (Macdonald 1972). These values are near or below the silicate liquidus temperatures for melts of similar compositions (tholeiite 1210 °C, alkali olivine basalt 1200 °C, andesite 1240 °C) determined by Murase and McBirney (1973). Variations in temperature of as much as 200 °C have been found within Hawaiian lava lakes, no doubt reflecting variation in the degree of crystallization as well as the heating effect of reaction between magmatic gases and the atmosphere.

There are no comparable measurements on acid lavas, mainly because no acid lavas have ever been closely observed in eruption. There have been no observations at all of rhyolite flows, and what measurements exist are on obsidian domes measured by optical pyrometer from a distance. These are of little value because the outer surface of an obsidian dome would be much cooler than the same magma when it was extruded through the vent, and in any case obsidian undergoes supercooling far below its solidus temperature without crystallizing. It is known from experimental studies that granite magma can exist at temperatures as low as 700 °C at depths where the pressure is sufficient to keep a high H_2O content in solution in the magma (Chapter 9), but rhyolite magma would probably have to be at a considerably higher temperature to be able to reach the Earth's surface.

The estimated extrusion temperatures given in Table 5 are based partly on field observations and partly on calculations from the mineralogy.

Table 5. Estimated extrusion temperatures of various lavas (after Carmichael *et al.* 1974)

Locality	Rock type	Temperature (°C)
Kilauea, Hawaii	Tholeiitic basalt	1150–1225
Paricutin, Mexico	Basaltic andesite	1020–1110
Nyiragongo, Zaire	Nephelinite	980
Nyamuragira, Zaire	Leucite basalt	1095
Taupo, New Zealand	Rhyolite lava and pumice	735–890
Mono Craters, California	Rhyolite lava	790–820
Iceland	Rhyodacite obsidian	900–925
New Britain	Andesite pumice	940–990
New Britain	Dacite lava and pumice	925
New Britain	Rhyodacite pumice	880

CRATERS AND CALDERAS

At the centre of many volcanoes there is a depression. Small depressions, less than a kilometre across, are usually called craters. Large depressions, more than a kilometre across, are called calderas (Table 6). Most craters, particularly those at the summit of volcanic cones, are the result of the explosive activity by which the vent is opened or kept open. However, some craters and nearly all calderas appear to be the result of subsidence, resulting from the presence of underlying magma.

Table 6. Diameters of some calderas (from Macdonald 1972)

Name	Location	Diameter (km)
Aso	Japan	20
Crater Lake	Oregon	10
Katmai	Alaska	5
Kauai	Hawaii	18
Kilauea	Hawaii	4
Krakatoa	Indonesia	8
La Garita (San Juan Mts)	Colorado	45
Santorini	Greece	14
Somma (Vesuvius)	Italy	3
Valles	New Mexico	21

Smith and Bailey (1968) classified calderas into two types:

Group 1. Calderas associated with mafic shield volcanoes, and whose origin is independent of pyroclastic eruptions.
Group 2. Calderas associated with felsic or mixed volcanoes, and whose formation is preceded or accompanied by voluminous eruptions of pumice and ash. These can be further subdivided into those in which collapse was piecemeal or chaotic, and those in which a coherent crustal block subsided along ring fractures.

Group 1 calderas have been subdivided by Williams and McBirney (1979) according to the nature of the collapse and the mode of eruption of the magma. The *Masaya* type (named after Masaya volcano in Nicaragua) are formed by piecemeal collapse over an area much broader than the existing volcano, without eruptions from rifts outside the caldera. In those of the *Hawaiian* type (Fig. 19), the summit block of a large shield volcano subsides along steeply inclined ring fractures following tumescence of the shield and drainage of magma into rift zones, with or without flank eruptions. The *Galapagos* type is also formed by collapse during the later stages of growth of a large shield volcano: engulfment results chiefly from injection of sills and from eruption of lavas from circumferential fractures near the summit.

Group 2 calderas are generally larger than those of Group 1. Several attempts have been made to distinguish between different types of Group 2 caldera, but it is obviously difficult to relate the topographic features of modern examples ('Krakatoa type', 'Katmai type') to the intrusive relation-

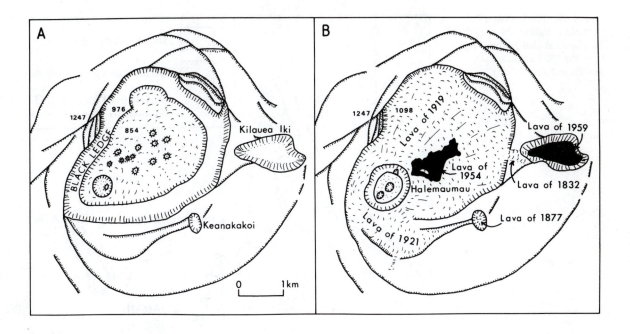

Figure 19. Kilauea caldera in 1825 (A) and 1965 (B), after Macdonald and Hubbard (1975).

ships of eroded ancient examples ('Glencoe type', 'Valles type'). Furthermore, the modern eruptions which are often taken as typical examples of caldera formation (Krakatoa in 1883, Katmai in 1912) were not actually observed at close range.

The *Krakatoa* eruption of 1883 has been described by many authors, whose work was brought together in a beautifully illustrated centenary volume by Simkin and Fiske (1983). The volcano had previously had a long history of both effusive and explosive activity, producing lavas and tephra of andesitic and basaltic compositions, but had been inactive since 1681. Prior to 1883 the Krakatoa island group consisted of one large island and two smaller ones, lying around and within a large prehistoric caldera, as shown in Fig. 20. The events of the 1883 eruption involved the removal of two of the three volcanic cones constituting Krakatoa Island, thus restoring the approximate outline of the prehistoric caldera. After the eruption the area between the three remaining islands was filled by the sea, with a depth of 250 metres at the site of the vanished Danan cone.

The 1883 eruption started with small explosions on 20 May, but nothing dramatic happened until 26 August. The paroxysmal phase of the eruption occurred on 26 and 27 August, and minor activity continued until February 1884. The volcano was then quiescent until 1927. The explosive activity of 26 and 27 August was of extreme violence. The larger explosions could be heard like thunder at a distance of 1000 kilometres. Tsunamis up to 40 metres high killed 36,000 people in Java and Sumatra. Clouds of ash were so dense as to result in total darkness in Djakarta, 160 kilometres away, at midday on 27 August. Ash rose to a height of 80 kilometres, and fell over an area of 800,000 square kilometres. Large pieces of pumice fell over an area of 100 square kilometres.

The features of this eruption most relevant to the problems of caldera

Figure 20. The Krakatoa (Krakatau) caldera showing the geographical changes resulting from the 1883 eruption (after Simkin and Fiske 1983).

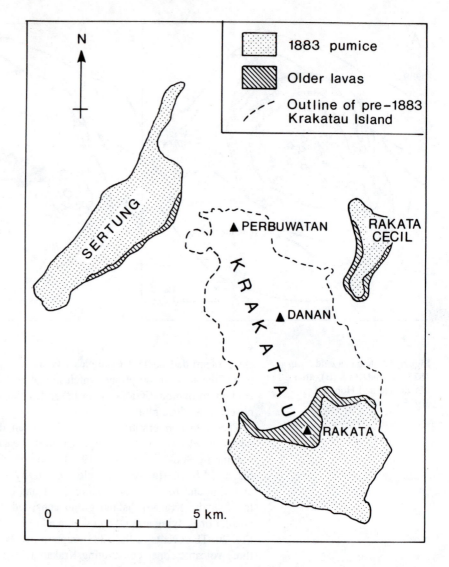

formation are the material budget and the explosion chronology. The volume of ejecta has been calculated at 18 cubic kilometres, of which 95% was dacitic pumice (newly erupted magma) and only 5% was composed of old rock fragments (basalt and andesite) from the former cones which had been destroyed in the eruption. It is inferred from this that the former cones were not blown away by the explosions but sank downwards. The vesiculation (conversion to pumice) of the underlying magma body, accompanied by eruption of some of the magma, must have caused the roof of the magma chamber to founder, as shown below in Fig. 24.

The chronology of this eruption is unusually well documented. There are many first-hand accounts, and the shock waves from the larger explosions were recorded on the pressure gauge of the Djakarta gas works. Their timing and relative intensities are shown in Fig. 21. The episodic nature of the

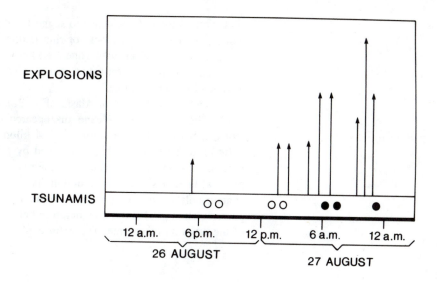

Figure 21. Chronology of the 1883 eruption of Krakatoa (after Self and Rampino 1981). The relative intensities of the explosions are indicated by the heights of the arrows. Large and small tsunamis are indicated by filled and open circles respectively.

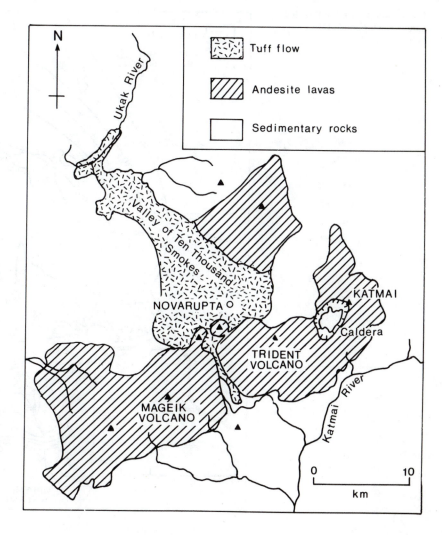

Figure 22. Geology of the Katmai area, Alaska (after Curtis 1968).

explosions is the only evidence to suggest that the subsidence was piecemeal rather than the result of block- or ring-faulting. Because the present caldera occupies the site of an older cone, and because it is covered by the sea, its exact shape is not determinable, and it is not possible to say whether it is bounded by ring fractures.

The caldera of *Katmai* in Alaska (Fig. 22) was formed in the eruption of 1912. The top of the volcano disappeared during the eruption and was replaced by a caldera measuring 3 × 4 kilometres, bounded by precipitous cliffs up to 1000 metres high and filled by a lake. The material which was missing amounted to 6 cubic kilometres and was mainly andesite. The pyroclastic deposits of the eruption have a greater volume than this but contain only a small proportion of lithic andesite fragments and are mainly pumice. The top of Katmai is therefore believed to have subsided as a result of the eruption of magma. The main explosive phases of the eruption were

Figure 23. Distribution of the Crater Lake pumice (after Williams 1942).

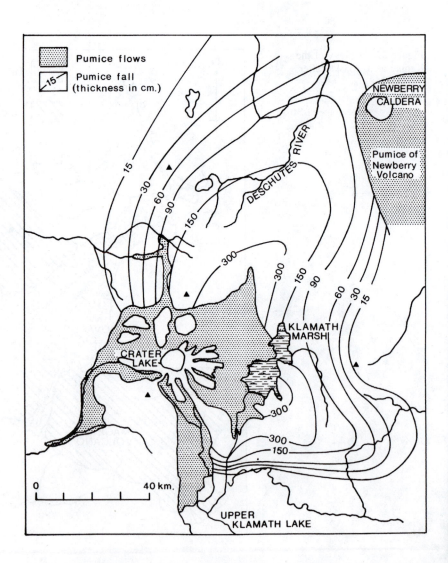

spread out over a period of 3 days and stoping of the summit is thought to have been gradual or piecemeal rather than a single event. The rate of subsidence might be estimated from the rate of expulsion of magma, which has been calculated as averaging 1 cubic kilometre per hour during the daylight hours of 6 June, although there is not necessarily a direct equation between the volume of subsidence and the rate of emission.

The erupted material of the 1912 eruption consists of an ash flow (11 cubic kilometres) and nine principal layers of air-fall ash (17 cubic kilometres), together equivalent on compaction to a total of 12 cubic kilometres of new material. The ash flow consists of andesitic pumice. The air-fall ash is mainly rhyolitic, but contains some banded pumice in which rhyolite and andesitic glass are both present. The ash flow, and most of the air-fall ash, were erupted

Figure 24. The suggested sequence of events in the formation of the Crater Lake caldera, Oregon (after Williams 1942): (a) an early stage in the eruption; (b) Vulcanian activity with ash clouds and ash flows from the summit; (c) the climactic phase of the eruption, with ash flows issuing from ring fractures while the summit starts to collapse; (d) the new caldera after the collapse; and (e) subsequent filling of the caldera by a lake, with some late eruptive material on the caldera floor.

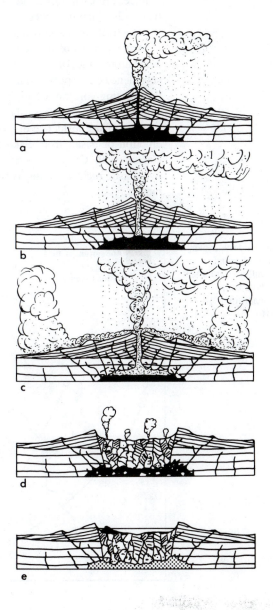

not from Katmai itself but from the new vent of Novarupta, 10 kilometres to the west. Curtis (1968) suggested that there was a rhyolitic magma chamber underlying Novarupta connected to an andesitic magma chamber underlying Katmai and that mixing of magma was taking place during the eruption. Mixed lava can be seen in the dome of Novarupta, which contains streaks of dark brown andesitic glass in the predominant lighter coloured rhyolite.

Crater Lake, Oregon, is a caldera formed about 5000 BC by the collapse of the pre-existing, mainly andesitic, volcanic cone known as Mount Mazama. The pyroclastic deposits associated with this collapse are spread over a wide area of Oregon and the neighbouring states. The initial deposits of airborne pumice and lithic ash were followed by rhyolitic pumice and basaltic scoria flows, to a total volume of about 75 km^3, of which at least 30 km^3 were glassy pumice, representing liquid magma. The caldera is 10 km across and more than 1200 metres deep. The distribution of pyroclastic deposits and hypothetical cross-sections are given in Figs 23 and 24.

Measurements of the subsidence involved in caldera formation have been made around the *Aira* caldera in Japan (Fig. 25). It can be seen that the area of subsidence corresponds to the outline of the caldera and not to the position of the erupting volcano, which is situated near its rim. This map also shows that calderas are not necessarily circular or elliptical, but can have irregular margins. The subsidence of the Aira caldera in 1914 was associated with the eruption of basalt from Sakurajima volcano, but the caldera as a whole is classed with Group 2 because it is surrounded by a plateau of mainly pyroclastic rocks covering 1650 square kilometres to an average depth of 100 metres.

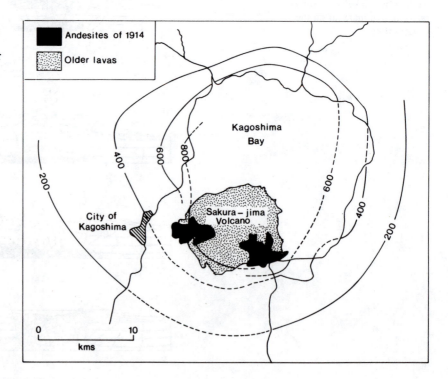

Figure 25. Aira caldera, Japan, showing the lines of equal subsidence (in millimetres) after the eruption of lava from Sakura-jima volcano in 1914 (after Koto 1916).

The emission of magma in a caldera-forming eruption can be from within the site of the caldera, as was the case at Krakatoa, or from a vent outside the caldera, as in the case of Katmai (Novarupta). Subsequent activity mostly takes place within the caldera. The formation of the *Long Valley* caldera in California (Fig. 26) was associated with the eruption of the Bishop Tuff 0.7 million years ago; most of the later rhyolites and basalts were erupted within the caldera, but Recent rhyolite and rhyodacite domes extend northwards from the caldera towards Mono Lake. A substantial body of magma underlies Long

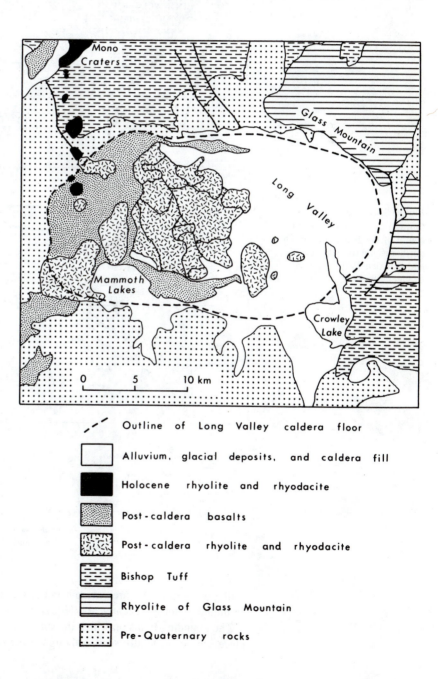

Figure 26. The Long Valley caldera, northern California (after Bailey *et al.* 1976).

Outline of Long Valley caldera floor

Alluvium, glacial deposits, and caldera fill

Holocene rhyolite and rhyodacite

Post-caldera basalts

Post-caldera rhyolite and rhyodacite

Bishop Tuff

Rhyolite of Glass Mountain

Pre-Quaternary rocks

Figure 27. The Glencoe cauldron subsidence, Scotland (after Clough *et al.* 1909). Dyke swarms are omitted (see Fig. 100). The thick black lines are the ring-faults.

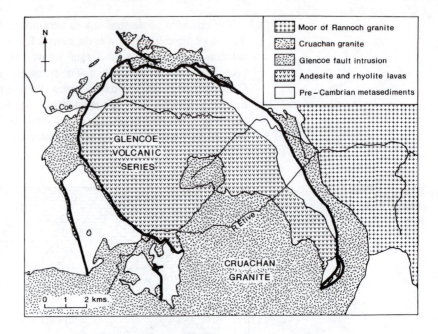

Figure 28. Schematic section through the youngest volcanic centre of the Burra complex in northern Nigeria (after Turner and Bowden 1979). Vertical exaggeration × 2.

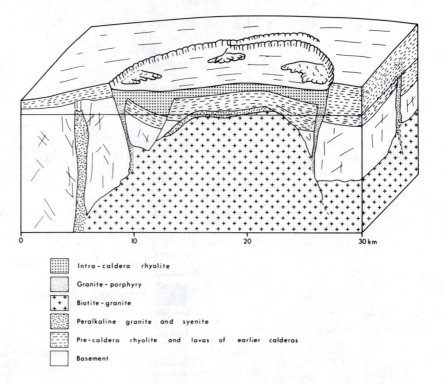

Valley caldera at the present day and has caused doming and many minor earthquakes during recent years (Hill *et al.* 1985).

The bounding faults on which the subsidence of modern calderas has taken place are not generally seen, although some calderas are bounded by steep cliffs which may be fault scarps. Some calderas may be the result of piecemeal subsidence and not be bounded by faults at all. The role of ring faulting is revealed by ancient volcanic complexes which have been denuded sufficiently to show their subsurface structure without having lost all their overlying volcanic rocks. The classic example is the cauldron subsidence of Glencoe (Fig. 27). There are many good examples in the Tertiary volcanic province of western Scotland, the Newer granite province of northern Nigeria (Fig. 28), and the Georgetown inlier in northern Queensland (Branch 1966).

A feature of several of the ancient sub-volcanic ring complexes is that there does not appear to have been a pre-existing volcanic cone on the site of the caldera (unlike Krakatoa, Katmai, Santorini or Crater Lake). In northern Queensland, the ignimbrites simply poured straight into the caldera basins.

LAVAS

LAVA FLOWS

The flow of lava from a vent is mainly determined by its viscosity. The lavas which are normally fluid enough to flow away from the vent under gravity are basalts and some andesites. More viscous lavas, such as rhyolites, trachytes, phonolites and some other andesites, tend to form either lava domes if low in volatiles, or ash and pumice flows if rich in volatiles.

Three types of flow are recognized in lavas of basic composition. *Pahoehoe* flows are those with a smooth, rounded, undulating surface, sometimes with a 'ropy' or 'draped' appearance. *Aa* flows have a very rough fragmented ('clinkery') top, and their surface is extremely jagged and abrasive. *Block* lava flows also have a fragmental top and front, but the individual fragments are smooth-surfaced polyhedral blocks, and not rough as in aa lava. The differences between the three types of flow are mainly related to the viscosities of the magmas, and hence to their compositions and temperatures. The basalts of oceanic shield volcanoes and of continental lava plateaux are mainly erupted as aa flows, whereas the orogenic basalts and andesites are more often block lavas. Pahoehoe lava is the least viscous, aa is intermediate, and block lava is the most viscous. Many flows change from pahoehoe to aa or block type as they progress away from the vent, whereas others are already of aa or block type when first erupted. In general, andesite magma is more viscous than basalt and is more likely to show the block form. Many flows are actually made up of separate portions of lava that have poured over one another during a single eruption as the lava advanced; these can be described as flow units.

Lava flows of felsic composition, such as rhyolite or trachyte, are rare, many of the apparent examples preserved in the geological record being compacted ash flows or lava domes. Where felsic lava flows do occur they are short and relatively thick. Among the very few historically recorded flows of acid lava are the 1771–75 rhyolite flow of Vulcano, Italy, and the 1953 dacite flow of Trident, Alaska, both of which owe their movement to extrusion on a

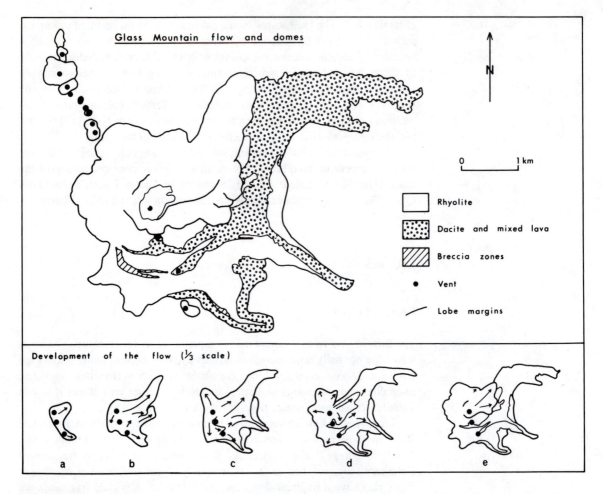

Figure 29. The rhyolite flows and domes of Glass Mountain, California (after Eichelberger 1975).

very steep slope. Other well-known examples of true rhyolitic lava flows are those of Glass Mountain, California (Fig. 29) and of Yellowstone Park.

The Southern Coulee of Mono Craters, California (Fig. 30) is a rhyolitic flow 3.6 kilometres long and 75 metres thick. Its relatively extensive flowage is attributable to its high degree of vesicularity. The actual specific gravity ranges from 1.75 to 0.65, compared with a theoretical value of 2.3 for massive obsidian. In other words at its most vesicular it is a pumice with a cavity:glass ratio of 3:1. Any greater degree of vesiculation would no doubt have caused the lava to dissociate into a spray, and indeed there was some particulate material erupted at an early stage in the eruption.

In contrast to the high viscosity of acid lavas, the highly alkaline mafic lavas are exceptionally mobile. In an eruption of Niragongo volcano in Zaire in January 1977, a 20 million cubic metre lava lake of melilite–nephelinite was discharged over an area of 20 square kilometres in less than an hour; the lava was of such low viscosity that in places the flow was less than a millimetre thick (Tazieff 1977). Carbonatite lavas may also be very mobile. Dawson *et al.* (1990) measured viscosities in the order of 1–100 Pa.s in erupting natrocarbonatite at Oldoinyo Lengai, depending on the flow rate and degree of vesicularity.

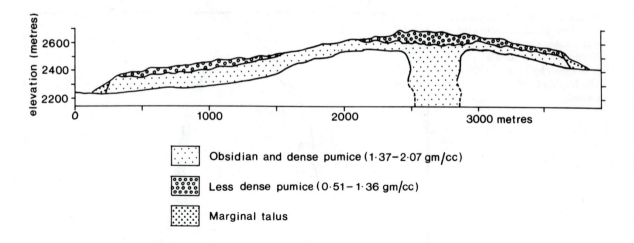

Obsidian and dense pumice (1·37−2·07 gm/cc)

Less dense pumice (0·51−1·36 gm/cc)

Marginal talus

Figure 30. The Southern Coulee rhyolite flow, Mono Craters, California (after Loney 1968).

LAVA DOMES

The most viscous types of lava do not easily flow unless they are highly vesicular. Instead they grow into domes. Most domes are of intermediate or felsic composition, and indeed this is the only way in which anyone has ever seen massive, non-vesicular rhyolite being erupted. The most viscous lavas do not flow under the influence of gravity at all, and can only move if they are subjected to upward pressure from the conduit. Lavas of this kind are forced upwards through the vent to form a protruding plug, as in the eruptions of Mont Pelée in 1902 and 1929–32 (Fig. 31).

There is a limit to the height that such a plug can attain before it starts to collapse. A growing spine continually loses material from its sides and top. Fragments break off and come crashing down around the base of the dome. Chilling by rain helps to spall off the outer layers. Sometimes the spines tilt or bend under their own weight. Because of their slow eruption they are cool on the outside, but occasionally a crack opens to reveal the hot material within, and sometimes a small amount of relatively less viscous hot lava exudes from the inside. The spine is eventually reduced to a pile of talus.

Somewhat less viscous domes are able to expand outwards as well as upwards by inflation with hot magma inside, and with increasing lateral spread there is a gradation from domes into coulees. There are excellent sections through the rhyolite domes of the Tarawera district in New Zealand, where they have been blown open by later explosions. The Tarawera complex consists of 11 domes and short flows (Fig. 32), forming a part of the Okataina volcanic centre in which there are many more rhyolite domes. Figure 33 shows the sequence of eruption in a cross-section along the line of the 1886 craters.

The term 'dome' is used to describe not only bodies of lava that have grown by addition of magma from below ('endogeneous domes'), such as the Pelée and Tarawera examples, but others that have piled up after issuing from a vent on the top ('exogeneous domes'). So-called domes of basaltic lava are normally of the exogeneous kind, for example one on Vesuvius which was extruded in 1895–99 and built up to a height of 150 metres.

Figure 31. The spine of almost solid lava which rose from the summit dome of Mont Pelée in November 1902 (from a photograph by Lacroix 1904). The height of the spine was 350 metres.

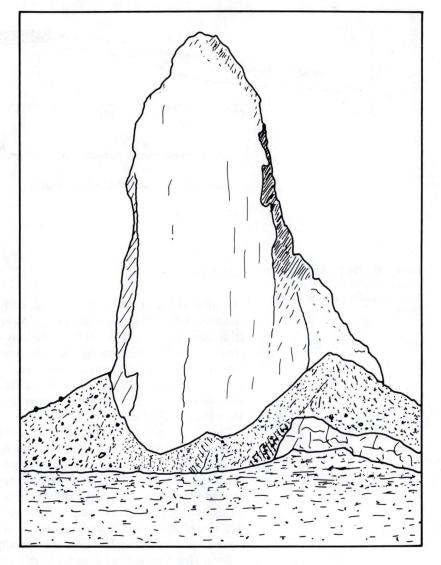

In addition to extrusion at the surface or in a crater, viscous magma may also grow into dome-like bodies and laccoliths beneath the surface of a volcano, as well as forming plugs in the volcanic conduits. After erosion of the surrounding material, which may be soft ash, intrusive plugs and domes are very difficult to distinguish from extrusive ones.

Lava domes may have any of a wide range of felsic compositions. Rhyolite lava is normally extruded in this way, as in the examples from California and New Zealand shown in Figs 29, 32 and 33. Intermediate and alkaline lavas may also form domes. Three small domes of dacite were formed in the 1866 eruption of Santorini in Greece, and another one in the 1925 eruption. There are several dacite domes around the summit of Lassen Peak, California, in addition to dacitic flows and domes of andesite. The

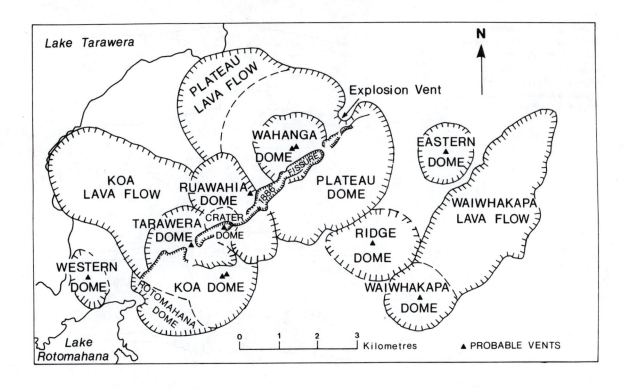

Figure 32. Rhyolite domes and flows of the Tarawera complex, New Zealand (after Cole 1970).

peak itself consists of a very large dacite dome, 800 metres high and 2500 metres across, bounded on its south side by cliffs which have been striated and polished by their upward movement against the walls of the vent. Among domes of andesitic composition may be mentioned those produced in the 1902 eruption of Mont Pelée and the 1980 eruption of Mount St Helens. Domes of alkaline composition are well represented in the volcanoes of the Atlantic islands and of the European mainland. Examples of trachyte and phonolite domes are those of Ascension Island and St Helena respectively. Intrusive domes of trachyte and phonolite are present in the Quaternary volcanoes of the Massif Central in France, e.g. the trachyte of Puy-de-Dôme.

The growth of lava domes is very slow compared with the rate of movement of basaltic lava flows. The great pillar of lava which rose above Pelée after the 1902 eruption grew at up to 10 metres per day, reaching a height of 310 metres after six months. The dome of Santa Maria volcano in Guatemala grew to a height of 500 metres and a diameter of 1200 metres within two years of its appearance in August 1922, and continued to grow thereafter. In general, the formation of a lava dome takes place at a late stage in a period of eruptive activity, following a much more vigorous phase in which the products are pyroclastic rocks, for instance at Pelée in 1902 or Novarupta in 1912, although there is often some overlap of the explosive and extrusive activity.

Figure 33. Diagrammatic section through the Tarawera domes along the line of the 1886 fissure, showing the sequence of eruption of rhyolite domes, ashes and pumice flows (after Cole 1970).

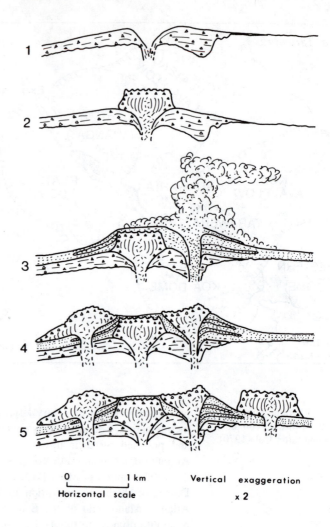

0   1 km   Vertical   exaggeration

Horizontal scale   x 2

SUBMARINE LAVAS

Many volcanic eruptions take place on the sea floor, and the amount of submarine lava probably exceeds the amount of subaerial lava. Submarine lavas differ from subaerial ones in both flow form and vesicularity.

The characteristic flow form of subaqueous eruptions is pillow lava. Pillow lava is made up of elliptically shaped masses, usually less than a metre across, although occasionally as much as 10 metres, whose interstices may be filled with sediment (in contrast to tongues of pahoehoe lava which may also have elliptical cross-sections). The outer layers of fresh pillows are normally glassy, showing that they chilled from the margin inwards. Individual pillows are moulded on to the surfaces of the underlying ones showing that the flows grew by superposition. Pillow lavas frequently pass laterally or downwards into unpillowed lava. It is generally thought that the pillows grow by a process of bulbous budding.

Most pillow lavas are basaltic or spilitic, but some are known of andesitic or trachytic composition. Pillow lavas have never been observed to form on land, and the majority are presumed to have been erupted on to the sea floor, but there is also the possibility that some could have been surrounded by soft sea-floor sediment rather than actual sea water. Photographs of Recent pillow lavas have been taken on the sea floor of the mid-Atlantic ridge, and off the coast of Hawaii at a depth of 2000 metres (Ballard and Moore 1977; Fornari *et al.* 1978), and some have even been filmed in the course of eruption. A few pillow lavas which occur among the Tertiary plateau basalts of Mull have been interpreted as having been erupted into a crater lake, and others in Iceland were evidently erupted subglacially or into glacial melt water.

The vesicularity of submarine lavas is governed by the pressure of overlying water, which inhibits the release of dissolved gases from the lava. This has been shown by dredging basalts from the sea floor off Hawaii (Moore 1965). In shallow water the lavas are as vesicular as on land, but with increasing depth the vesicles become smaller and less abundant. At a depth of 4000 metres vesicles are rare and are very small (average diameter <0.1 mm).

An important change in the properties of seawater takes place at a depth of about 2200 metres. The pressure at this depth equals the critical pressure of seawater. Above this pressure there is no such phenomenon as boiling. A lava flow would not convert seawater to steam, and in the absence of a steam jacket the lava would be much more effectively chilled. At supercritical pressures seawater must expand to a very fluid and rapidly convecting liquid which would cool the new lava very rapidly. Any volatiles (water or CO_2) which were exsolved by the magma would similarly be dense fluids quite unlike the gases released by subaerial or shallow-water flows, and would be powerful metasomatizing agents.

PYROCLASTIC ROCKS

Basaltic magmas have relatively low viscosity and gas content, and in predominantly basaltic volcanoes such as Kilauea the effusion of lava is quiet and the amount of pyroclastic material produced is small. There are two main exceptions to this rule. One is that new eruptions may be initiated by an explosive phase as the magma and its exsolved gases force a way through to the surface. The other is that explosions occur if surface water, groundwater, seawater or ice come into contact with hot volcanic rock (phreatic activity) or with the magma (phreatomagmatic activity). A good example of the latter was the 1963 eruption of the new volcano Surtsey off the south coast of Iceland.

Magmas with a higher viscosity and higher dissolved gas content than basalt have a less placid and less uniform style of eruption. The release of dissolved gas from the magma may be spasmodic, and a characteristic feature is the eruption of lava bombs and the formation of spatter cones. This type of activity was well displayed during the building of another new volcano, Paricutin, Mexico, between 1943 and 1952. The Paricutin lavas are andesite and basaltic andesite.

Felsic magmas such as rhyolite and phonolite have the highest viscosities and dissolved gas contents. The lowering of pressure as the magma approaches the surface causes rapid exsolution of the dissolved gas with an expansion in volume of the vesiculating magma. Consequently most magmas of this composition are explosively erupted. The vesiculating magma is driven rapidly to the surface, producing very high eruption plumes and the extrusion of pyroclastic flows as well as clouds of magma droplets and gas. In these eruptions a collapse caldera may be formed.

Some authors have attempted to classify eruptions according to their explosive character as Hawaiian, Strombolian, Pelean etc. This is a somewhat misleading system of nomenclature, because many volcanoes change character as they develop, even within a single eruption. Thus Surtsey started with phreatomagmatic explosions and continued with quiet effusion of lava. The historic eruptions of Hekla, also in Iceland, have each started with the explosive eruption of acid or intermediate ash and pumice followed by the quieter eruption of basalt. Large composite volcanoes like Santorini or Krakatoa have had rare, short paroxysmal eruptions, interspersed with longer periods of slow eruption of viscous magma as domes, and even longer periods of inactivity.

The pyroclastic material erupted from a volcano may be emitted either in fragmental form, resulting in scoria cones and ash fall deposits, or more coherently, resulting in ash flow deposits. Individual particles or fragments may be classified on the basis of grain size, the limiting diameters being:

>64 mm   – bombs (ejected as liquid) or blocks (ejected as solid)
64–2 mm – lapilli
<2 mm   – ash

In reality there is often a mixture of different grain sizes, more so than in detrital sedimentary rocks, and in volcanic stratigraphy many formations described as ash contain a high proportion of particles coarser than 2 mm.

SCORIA CONES

Bombs cannot be ejected far from the vent because of their weight. An eruption in which a large proportion of the erupted material is thrown out as bombs and lapilli gives rise to an accumulation known as a scoria cone or cinder cone. The term 'scoria' is poorly defined, being used by English-speaking geologists for lava bombs and lapilli but by some other geologists for the crusted material on top of a lava flow, or even for any vesicular basic lava.

Scoria cones are most commonly made up of basaltic material. Liquid lumps of acid magmas cannot be thrown up in the same way because they are too viscous, and lava lakes are not present in the vents of erupting acid volcanoes. Many variations are observed in the nature and shape of lava bombs, blocks and lapilli, and in the way in which they accumulate. For instance they may weld themselves together around the vent to form a spatter cone. Further away from the vent they are often mixed with finer ash and are usually in a loose condition.

A good example of a large scoria cone is the initial phase of Paricutin, which rose to a height of 400 metres in its first year of activity.

ASH FALLS

The finer-grained products of an explosive eruption which are thrown into the air fall to the ground to give a layer of ash. The extent of the ash fall depends mainly on how high the ash is projected into the air. Most ash clouds are thrown up to a height of less than a kilometre and fall back on to the volcano so as to build up a cone. Very powerful upward blasts which carry the ash column to a height of 10 kilometres or more are rare, occurring perhaps a few tens of times per century. The ash from these eruptions falls over a wide area as a sheet deposit. Walker (1973) suggested that the nature of the ash deposit should be used to distinguish between Strombolian (cone-building) and Plinian (sheet-forming) types of eruption. Figure 34 contrasts the distribution of the ash deposited in an eruption of each type. In the 1973 eruption of Eldjfell, ash fell very thickly around the crater (>400 cm) but only spread over a radius of 2 or 3 kilometres; thus the immediate effect of the ash fall was to raise and steepen the volcanic cone. In the AD 79 eruption of Vesuvius the eruption plume rose to a great height above the volcano, and the ash was distributed over a much larger area and showed a greater influence of wind direction. The thickness of the Vesuvius ash around the crater was less than at Eldjfell, but the area covered by ash to a depth of over a metre was about 450 km² around Vesuvius compared with about 2.5 km² at Eldjfell.

Many ash falls land in water, and others are rapidly eroded from land and redeposited in water. All volcanic ashes are very susceptible to weathering and diagenetic alteration, especially those of basic composition. They are

Figure 34. Distribution of ash deposits from: (A) the January 1973 eruption of Eldjfell, Iceland; and (B) the AD 79 eruption of Vesuvius (after Booth 1979 and Lirer *et al.* 1973 respectively). Thicknesses are given in centimetres.

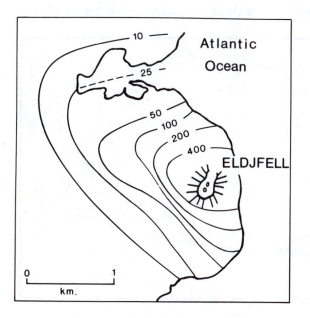
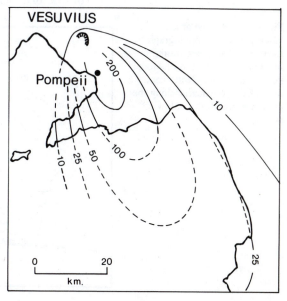

rapidly converted into a mixture of low temperature alteration products, such as zeolites, chlorite, and smectites.

Some of the very biggest ash falls have been associated with eruptions of rather unusual types. A very great ash fall accompanied the Laki fissure eruption in Iceland in 1783, covering the whole island with ash and causing great destruction of the vegetation and consequent famine. The dust cloud from this eruption spread over Europe and North America and persisted for several months. The explosions of Krakatoa in 1883, believed to have been caused by caldera collapse, threw ash so high into the atmosphere that the finer dust eventually spread over most of the inhabited world, taking 14 days to make its first complete circuit of the globe. A similar eruption at Tambora in 1815 is estimated to have ejected 100 cubic kilometres of pyroclastic debris.

ASH FLOWS

This is the group of volcanic deposits which has proved to be the most difficult to understand and which has given rise to the greatest controversy. It has gradually come to be realized that most, and perhaps all, of the world's flat-lying accumulations of rhyolitic or dacitic rock are not lava flows but are of pyroclastic origin. True rhyolite or dacite lava has been seen to erupt only as domes, confined to the area of the volcanic vent; more extensive flows of acid lava are known from only a very few localities.

The eruption of Mont Pelée in 1902 greatly influenced geologists in the following decades by drawing attention to the phenomenon of a nuée ardente (fiery cloud). It was a nuée ardente of this eruption that destroyed the city of St Pierre with great loss of life. Observation of nuées ardentes has shown that they are invariably associated with a flow of pyroclastic material along the surface of the ground (Fig. 35). Such flows have three distinct origins: (1) collapse of a dense eruption cloud; (2) avalanching of hot ash already

Figure 35. A nuée ardente, based partly on a photograph by Lacroix (1904) taken during the January 1903 eruption of Mont Pelée, Martinique. Note that a flow of dense material along the ground precedes the billowing hot cloud.

deposited on a steep surface; and (3) flow of magma undergoing extreme vesiculation or frothing.

The study of ancient volcanic provinces has revealed many examples of welded tuffs, in which the glassy particles were deposited at such a high temperature that they were still sticky. These are obviously not ash fall deposits, because air-fall ash is never hot enough for welding to occur, so the suspicion arose that welded tuffs were the products of nuées ardentes. In the 1930s, Marshall renamed the rhyolitic rocks of the Taupo–Rotorua district, New Zealand, as ignimbrites ('fiery cloud rocks'), and this name has come to be widely used. Some authors use the name 'ignimbrite' as a synonym for 'welded tuff', in which case it is superfluous, but others emphasize its genetic connotations, which is unfortunate since no modern nuée ardente has actually produced a welded deposit. Indeed no observed eruption of any kind has left a welded deposit. The cloud that destroyed St Pierre in 1902 left only a layer of ash and lapilli.

In a Pelean eruption, a distinction must be made between the incandescent avalanche which travels along the ground and the great cloud of dust and gas which billows up from it. The two may produce overlapping deposits or they may become separated; the avalanche necessarily travels downhill whereas the cloud can spread over topographic ridges or be blown further by the wind (Fig. 36). It is the ground-level avalanche which could potentially form a welded tuff, and not the cloud of dust and gas.

It is now generally accepted that a large amount of pyroclastic material has been deposited after flowing along close to ground level, but there is still some uncertainty as to what proportion may have been transported as a 'particulate-flow' and what proportion as a 'froth-flow'. Probably there is every gradation between an intact pumice, a partially disrupted pumice and a completely disrupted pumice, in addition to material which was in a particulate form at the time of its emission from the vent. Hausen (1954) described the Dorena welded tuff in Oregon as having been formed from a vesiculating lava flow. The top and bottom of the flow were fragmented, but not the interior, which shows a flow structure attributed to 'flowage along planes of minute vesicles within the partially expanded lava'. Boyd (1961) suggested that the Canyon rhyolite flow in Yellowstone Park was emplaced as a froth which partially collapsed but was not completely disrupted and continued to move as a lava.

The viscosity of acid magma is such that it can readily flow only when vesiculated by escaping gas. Such escaping gas will expand the lava and convert it either to pumice or to a shower of liquid droplets. Since an accumulation of such material is prone to compression under its own weight, the pumice may be collapsed and the glassy particles become welded together. Thus sheet-like bodies of acid volcanic rock commonly show some of the following characteristics: gradation of ash into welded tuff; gradation of tuff into lava or lava-like rock; presence either of pumice or of rock with the texture of collapsed pumice. Welded tuffs are very often compressed under their own weight and assume the appearance of streaky lava. They can easily be recognized by the shape of some of the contained glass shards (collapsed bubbles), but in the past many of them have been misidentified as coherent lava flows.

Figure 36. Map of Mont Pelée, Martinique, showing the area devastated by the eruptions of 8 and 20 May 1902. The hot avalanches followed the valley of the Rivière Blanche to the sea. The movement of the accompanying hot cloud has been interpreted in two ways: (1) Lacroix (1904) believed that the ash cloud left the valley at a sharp bend and continued southwards directly to St Pierre; (2) Fisher *et al.* (1980) believed the cloud expanded outwards from the main course of the ground-level flow.

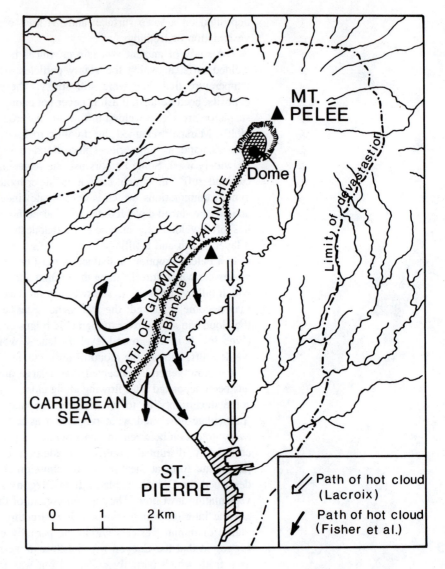

The final products of a pyroclastic flow do not necessarily reveal its method of transportation. A froth flow will usually be flattened, its bubbles squeezed out and the glassy fraction welded together. A particulate flow may not necessarily be welded together when it settles, and may be difficult to distinguish from an ash fall. Many tuffs have been found which are partially welded and partially unwelded. An ash flow needs to be emplaced rapidly if it is to stay hot enough to become welded. Wilson's (1985) calculations on the great ignimbrite of the AD 186 Taupo eruption indicate the speed and scale of such eruptions. This ignimbrite was formed as a single flow unit, 30 cubic kilometres being erupted in less than 10 minutes, and it flowed for 80 kilometres in all directions, crossing mountains in its path. Its initial flow speed was estimated as 250–300 metres per second (560–670 miles per hour).

Not all ash flows are of the froth-flow type, originating as vesicular lava flows. Ash flows may form simply by the avalanching of loose material on a steep slope, and these flows may also be accompanied by a nuée ardente (hot dust cloud). The ash flows ('nuée ardente deposits') on Mayon volcano shown in Fig. 37 are of this type.

PYROCLASTIC SURGE DEPOSITS

This category of pyroclastic deposits is associated with Plinian eruptions, in which a very large volume of pumice is propelled high into the atmosphere, forming a tall plume. Such a plume, with a large burden of suspended solid, inevitably collapses, and the falling material is swept outwards as a low-density suspension of solid particles in hot gas and air.

Pyroclastic surges have some of the character of ash falls and some of the character of ash flows. The solid material is in the form of discrete particles as in an ash fall, but the material is transported across the ground to its site of deposition rather than falling vertically from the air. Unlike ash falls, surge deposits do not blanket the ground with a uniform thickness but are concentrated in valleys and depressions. They often show directional sedimentary features such as cross-bedding and dune structures.

Figure 37. Map of Mayon Volcano, Philippines, showing the deposits of the 1968 eruption (after Moore and Melson 1970).

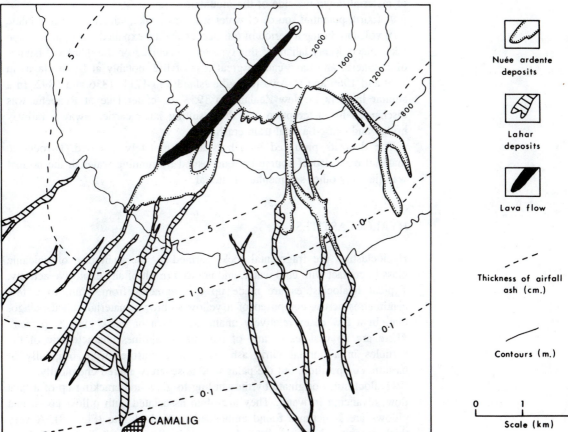
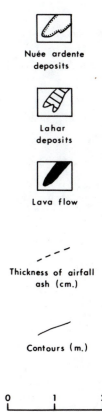

Nuée ardente deposits

Lahar deposits

Lava flow

Thickness of airfall ash (cm.)

Contours (m.)

Scale (km)

LAHARS

Lahars are mudflows composed of pyroclastic material. They arise in various ways, and can occur anywhere where the ground is sloping and water is available. The water has several possible sources. In tropical areas of high rainfall, such as Indonesia or the Philippines, thick ash deposits may become saturated with water during wet weather. Loose water-saturated material cannot rest on such a steep slope as loose dry material, and a mudflow may result. This will be particularly coherent and fast-moving if the water accumulates at the base of the ash deposit to form a lubricating layer. Figure 37 shows the distribution of ash-flow, air-fall and lahar deposits from the 1968 eruption of Mayon volcano in the Philippines.

In temperate regions torrential rainstorms are often precipitated by violent eruptions, particularly those of the Plinian type in which ash is thrown high into the atmosphere where the particles can serve as nuclei for water droplets or ice.

In the high volcanoes of the Andes, such as Cotopaxi (Equador) or Villarica (Chile), lahars are caused by melting of snow and ice around the summit. During the eruption of Mount St Helens in May 1980, two lahars which started to flow during the first few minutes of the eruption and were fed by water from melting snow and ice discharged 14 million cubic metres of material down the side of the mountain.

A fourth potential source of water is a crater lake, which may accumulate on a volcano during a period of quiescence and be expelled when an eruption takes place. Many lahars of this type have occurred on the inhabited islands of Indonesia and have caused great loss of life, notably at Kelut (Java) in 1919 and 1966 and at Awu (Sangihe Islands) in 1711, 1856 and 1892. In a similar incident in New Zealand in 1953 the crater lake of Ruapehu was discharged by an eruption and the resulting lahar carried away a railway bridge and caused a fatal train crash.

The deposits produced by lahars are completely unsorted. Pyroclastic material may also of course be redeposited by running water in the normal way like any other sedimentary material.

HYALOCLASTITES

Hyaloclastites are fragmental rocks formed by the shattering of volcanic glass in contact with cold water, and not as a result of any explosive activity. Typical hyaloclastites are made up of fragments from 1 mm to a few centimetres across composed of a yellowish-brown material called palagonite. In a few cases relatively unaltered brown or black glass is present. There are no vesicles or any of the curved outlines characteristic of the particles in a normal vitric ash. Most hyaloclastites were originally of basaltic composition, but the palagonite is severely altered chemically.

Hyaloclastites originate by the chilling to glass and cracking up of a lava flow advancing in water. They are often associated with pillow lavas, and pillows are sometimes found embedded in hyaloclastite (Fig. 38). A very thick deposit of hyaloclastite and associated pillow lavas in Iceland has been

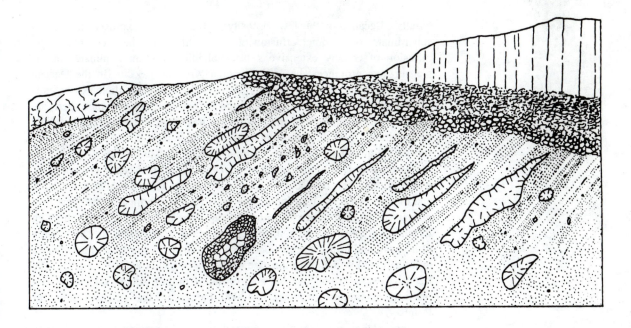

Figure 38. Diagrammatic cross-section of a hyaloclastite flow formed in a lake in the Columbia River region, Oregon, showing bedding of the hyaloclastite and elongated pillows (after Macdonald 1972).

called the 'Palagonite Formation' and appears to have formed by the extrusion of lava beneath a glacial cover, but most hyaloclastites are of submarine origin.

MAJOR VOLCANIC MASSIFS

The four main types of volcanic landform are: (1) lava plateaux; (2) shield volcanoes; (3) strato-volcanoes; and (4) pyroclastic sheet deposits. Each type of massif contains lava, pyroclastic rocks and intrusions, but differing in their relative proportions and compositions. Lava plateaux are extensive flat-lying accumulations of lava sheets, mainly basaltic, and exemplified by the Recent and Tertiary lavas of Iceland or the Columbia plateau. Shield volcanoes are conical structures consisting mainly of lava, exemplified by the volcanoes of the Hawaiian Islands. Strato-volcanoes are composite volcanoes, containing both lava and pyroclastic materials, often changing their shape as a consequence of their more explosive character, and exemplified by Vesuvius or Mount St Helens. Pyroclastic sheet deposits are extensive and relatively flat-lying accumulations of pyroclastic rocks, such as those of the North Island of New Zealand or of Yellowstone Park.

LAVA PLATEAUX

Plateau lavas are the result of the eruption of very large volumes of extremely fluid basaltic magma which has flowed readily enough to develop sheet-like flows covering large areas. They are often described as 'flood basalts'. Because of the low viscosity of the magma, explosive activity is subordinate to the quiet effusion of lava, which may be accompanied by intrusion of equally extensive sills. Tholeiitic basalt predominates in the plateau lava provinces, but other magma types also occur. In the Deccan plateau the range of minor rock types associated with the tholeiites extends from nephelinite to rhyolite, and in the British Tertiary province there are tholeiitic and alkali basalts and minor rhyolites but few rocks of intermediate composition. The Ethiopian flood basalt province is of mixed character, with substantial amounts of both tholeiitic and alkali basalts.

There are only a limited number of really large flood-basalt provinces, of which perhaps the best known are the Columbia River basalts in the USA and the Deccan plateau in India. The most recently formed of the great lava plateaux is Iceland, where volcanicity continues at the present day, although Iceland is atypical in being situated in an oceanic environment. The volume of lava emitted in the largest flood basalt provinces is immense, more than 1 million cubic kilometres in the case of the Karroo and the Siberian traps. The Deccan basalts extend over an area of half a million square kilometres and reach a maximum thickness of 2000 metres in the neighbourhood of Bombay.

Table 7 lists some of the largest flood-basalt provinces. According to Yoder (1988) the time taken to accumulate the great lava plateaux was sufficiently long that they do not represent exceptional rates of magma supply. For example, the Columbia River basalts took a little over 10 million years to build up to a volume of about $200,000 \, km^3$, a rate averaging $0.02 \, km^3/yr$, which compares with an annual rate of emission of $2 \, km^3/yr$ for all present-day subaerial volcanoes. The very large size of individual flows implies the eruption of large volumes of magma at long intervals, in contrast with the relatively frequent but small eruptions of the present-day Hawaiian volcanoes. The immense size of the lava plateaux listed in Table 7 may be emphasized by comparing them with Etna, the largest and most active volcano in Europe, which has a volume of about $500 \, km^3$.

It is often said that plateau lavas are fed from fissure eruptions, but this is an oversimplification. In Iceland, eruption of lavas from elongated fissures has been observed, and in many places the lavas are cut by great parallel dyke swarms, which could well have been feeders for lavas higher in the succession. There are huge dyke swarms associated with the Columbia River basalts, one of which has been estimated to contain over 21,000 dykes. On

Table 7. Major flood basalt provinces

Region	Area before erosion (km^2)	Original volume (km^3)	Age
Siberian plateau	2,500,000	2,000,000	Permian–Triassic
Karroo lavas, South Africa	2,000,000	1,400,000	Jurassic
Parana plateau, South America	1,200,000	800,000	Cretaceous
Ethiopian plateau, Ethiopia	750,000		Oligocene–Miocene
Deccan, India	500,000	500,000	Cretaceous–Eocene
Columbia River, USA	160,000	170,000	Miocene

All the figures are very approximate, especially the volume estimates.

the other hand, the plateau lavas of the British Tertiary province are underlain by a number of large intrusive centres which have the appearance of being the roots of large shield volcanoes, although they are also the foci of extensive dyke swarms (see Fig. 99 in Chapter 3). In Iceland too, the flat-lying Tertiary lavas are overlain by large central volcanoes such as Hekla.

Iceland lies on the Mid-Atlantic Ridge, and consists entirely of volcanic and related intrusive rocks, apart from superficial Quaternary deposits, mainly glacial. The present-day volcanism is concentrated in rift zones elongated parallel to the trend of the oceanic ridge (Fig. 5) and older volcanic rocks are symmetrically arranged on either side. The oldest rocks in north-west and eastern Iceland are between 13 and 16 million years old. The Tertiary volcanic rocks of Iceland are predominantly subaerial flood basalts (80–85%), among which are several discrete local accumulations of acid and intermediate rocks (10%), and a small proportion of pyroclastic rocks (5–10%). The Quaternary lavas are also mainly basalts, but there are abundant basaltic hyaloclastites, some or most being the result of subglacial eruptions. In eastern Iceland, where there are sections through the older basalts they are seen to be cut by thousands of gently inclined basaltic sheets and some larger gabbro intrusions.

Iceland is made up of layer upon layer of plateau lavas with concentrations of activity at a number of volcanic centres. Hekla is an example of a large central volcano, while the Laki eruption of 1783 is often considered as the type example of a fissure eruption. The central volcanoes have erupted acid and intermediate lavas as well as basalt, whereas the fissure eruptions are solely basaltic, and unlike the central volcanoes each fissure erupts only once. An appreciable part of the lava plateau in Iceland (and elsewhere) is made up of intrusions within the lava pile, particularly of sills.

The fissure swarms are up to 100 km in length. On the surface they are marked by open fissures, graben structures, and crater rows. At depth, they are seen as dykes and as normal faults. About half of the magma eruption of Iceland during historic times has come from two particularly large fissure eruptions, those of Eldgja in about 930 and Laki in 1783. The Laki fissure (Fig. 39) was about 25 km long, and lava flowed for 80 kilometres down the Skafta valley, filling it in places to a depth of more than 200 metres. A substantial amount of ash was erupted from the fissure, covering most of Iceland and doing great damage to crops and livestock.

The average rate of eruption of Icelandic volcanoes in historic time has been approximately $4 \, km^3$ per century.

The *Columbia Plateau* in the north-western United States (Fig. 40) was built up during the period from 17 to 6 million years ago (mostly between 17 and 13 million years ago) by a series of fissure eruptions. The succession of basalt lavas is well exposed by river erosion, and it is possible to trace the flows back to their source and relate individual flows to their feeder dykes (Hooper 1982). The average thickness of the basalt pile is about 1 kilometre, but drilling has revealed a thickness as great as 3 kilometres in the centre of the plateau. The number of flows is estimated to be between 120 and 150, ranging in thickness up to 120 metres with an average of 15 to 30 metres.

The stratigraphy of the Columbia River basalts is well known from mapping, with the help of magmatic polarity measurements. The first group

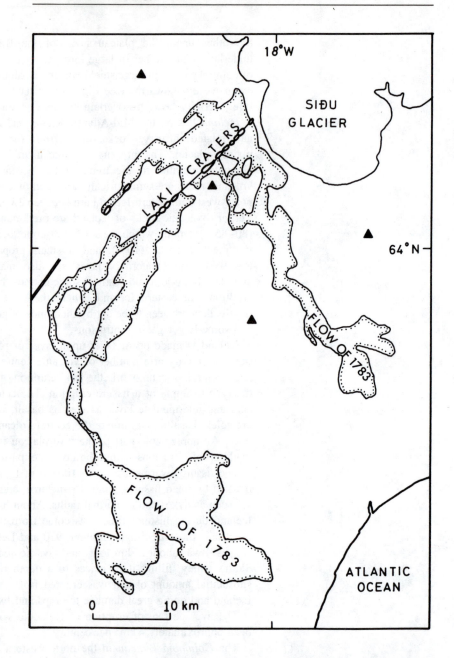

Figure 39. The Laki fissure eruption of 1783 (after Thorarinsson 1970).

of lavas to be erupted were the Imnaha basalts. These were formed in 20 to 25 eruptions at intervals averaging about 20,000 years. They were followed by the Grande Ronde basalts which are the most voluminous group of lavas. The Wanapum basalts, erupted at about 14 million years ago, were the last of the major lavas to be erupted, but smaller flows continued for another 8 million years. Some of the later eruptions had the effect of damming valleys and producing lakes, and eruption of lava into these lakes produced complexes of pillow lava and hyaloclastites (Fig. 38).

The Roza flow of the Wanapum formation is one of the largest lava flows

Figure 40. The Columbia River basalts, north-west United States. The location, orientation and concentration of feeder dykes are represented approximately (after Hooper 1982).

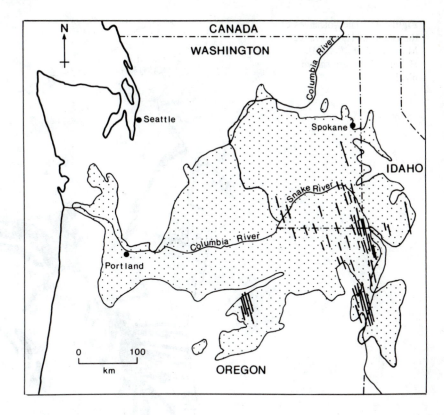

known. It can be traced over 40,000 km². It was erupted from a belt of fissures 5 km wide and 200 km long in the eastern part of the plateau, along which some spatter cones are preserved. Hooper (1982) has pictured the lava front as 30 metres high and 100 kilometres wide, advancing across the ground at an average speed of 5 km/hour. The maximum rate of effusion was estimated at 1 km³ of lava per day per kilometre of fissure, lasting for about 7 days.

SHIELD VOLCANOES

A shield volcano is one which is built around an eruptive centre, from which the accumulation of volcanic rocks thins outwards. Shield volcanoes are predominantly basaltic, with a low proportion of pyroclastic rocks. The *Hawaiian Islands* contain typical examples of shield volcanoes. In a 'Hawaiian-type' eruption, lava is erupted quietly from a crater or fissure and flows outwards for some distance. The building of a shield volcano is a long process, and each individual eruption makes only a small contribution to the total volume of the volcano (Fig. 41). The characteristic shape of a shield volcano is a broad cone with a gentle outward slope.

Most of the shield volcanoes are in oceanic areas. Small examples occur

Figure 41. The volcanoes of the island of Hawaii (after Macdonald *et al.* 1983). Historic flows are shown in black.

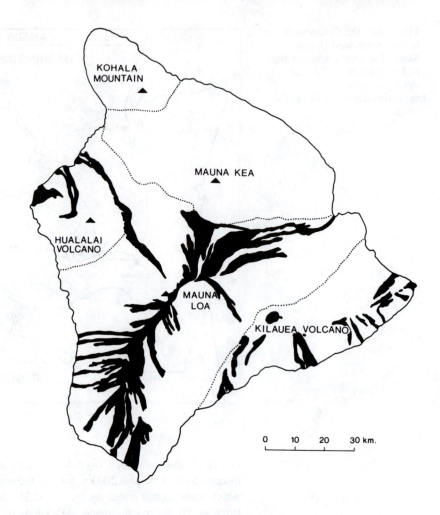

KOHALA
MOUNTAIN

MAUNA KEA

HUALALAI
VOLCANO

MAUNA
LOA

KILAUEA VOLCANO

0   10   20   30 km.

on Iceland, superimposed on the more extensive lava plateau. There are a number of shield volcanoes in East Africa.

With a slightly more explosive style of activity and a more viscous type of magma, the flat cone typical of shield volcanoes passes to the steep cone typical of composite volcanoes. A transitional stage is represented by *Etna*, the largest of European volcanoes. It rises from sea level to a height of 3300 metres and is 40 km across at the base. It has accumulated from the products of very frequent but small eruptions since the early Pleistocene. During historic times, there have been about 15 eruptions per century, each adding only a small contribution to the bulk of the volcano (Fig. 42). Many of the lavas are erupted from the 260 or more small cones on the flanks of the volcano, some of which are aligned and most of which are thought to be fed from dykes and sills. They are most commonly trachybasalts. Some of the eruptions are mildly explosive in character, but pyroclastic rocks are estimated to make up only about 12% of the volcanic products.

Figure 42. Historic lava flows of Mt Etna, Sicily (after Pichler 1970).

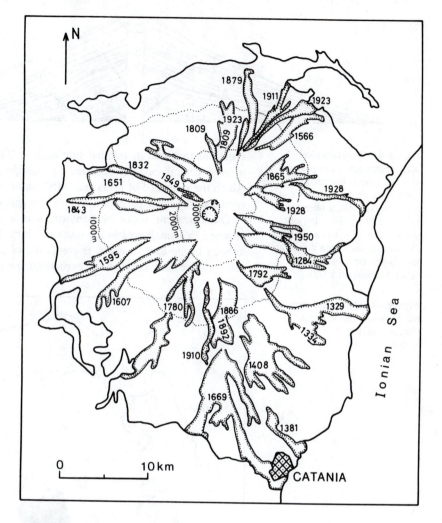

STRATO-VOLCANOES

Strato-volcanoes (or composite volcanoes) are formed by the alternation of lava flows and pyroclastic material. A steep cone with a summit crater is the simplest type of volcanic structure, but the alternation of quiet effusion with explosive activity causes frequent changes in the shape of the crater, and the opening of new fissures and vents causes the formation of new craters, both at the summit and on the flanks of the original cone. The internal structure of these volcanoes is very complex. There are typically many dykes and sills within the volcanic pile as well as lavas and ashes. Figure 43 shows an idealized cross-section through a composite volcano. There are many strato-volcanoes in the circum-Pacific region, for example Fujiyama or Mount St Helens. Andesite is a typical product of these volcanoes, but more acid and basic magmas are also erupted.

The *Santorini* group of islands in the Aegean Sea (Fig. 44) are the remains of a large composite volcano which has been opened up by paroxysmal

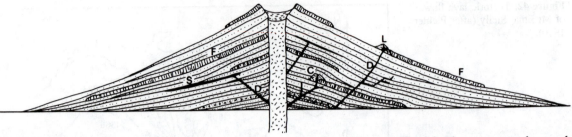

Figure 43. Diagrammatic section through a composite volcano (strato-volcano) showing a central conduit and crater, dykes (D), a lateral cone (L), lava flows (F), a buried cinder cone (C), and a sill (S). Pyroclastic layers are dotted and lava flows are cross-hatched (after Macdonald 1972).

Figure 44. Geological map of the Santorini group of islands, Greece (after Pichler *et al*. 1980).

eruptions to reveal its internal stratification. The islands are arranged around a large central caldera, bounded by precipitous cliffs 350 metres high in which the sequence of lavas and tuffs can be examined. The existing islands represent the outer parts of a former volcanic cone.

The basement rocks of Santorini are schists and marbles, which are exposed in a small area in the south of Thira. The bulk of the volcanic cone, as seen in the caldera walls, is composed of alternating agglomerates, tuffs, ash, pumice and lavas, with pyroclastic rocks predominant in the earlier

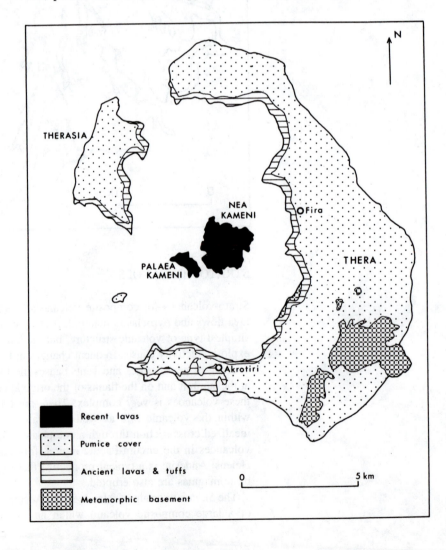

Figure 45. Geological map and diagrammatic cross-section of Tristan da Cunha (after Baker *et al.* 1964).

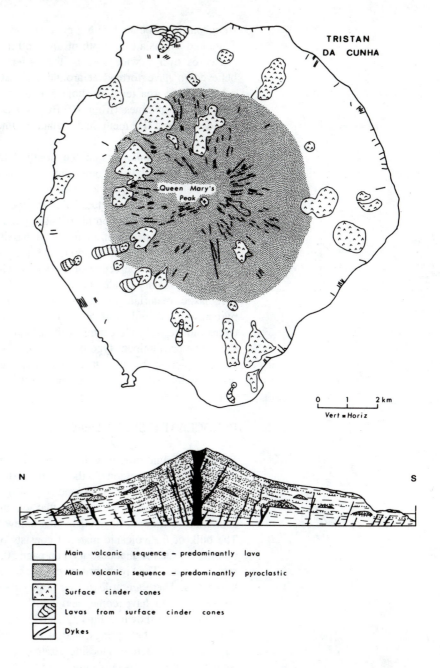

volcanic succession but lavas becoming more abundant later. In the north-east wall of the present caldera is the infilled cross-section of a previous caldera. Isotopic age determinations indicate an age of about 1.6 million years for the earliest volcanic rocks, with subsequent eruptive activity continuing up to the present century. Basalt, andesite, dacite and rhyodacite are all present in the volcanic succession. Many dykes are present, some of which are seen to have fed lava flows.

The great caldera-forming event which removed the centre of the stratocone is associated with a great spread of pumice which covers most of Thira and Thirasia to a depth of up to 60 metres and can be traced in deep-sea cores over a wide area of the eastern Mediterranean. The caldera is believed to have formed at around 1470 BC. The maximum depth of the caldera below sea level is about 500 metres. Eruptions within the caldera during historic times (from 197 BC to AD 1950) have built up the new islands of Nea Kaimeni and Palaia Kaimeni, consisting of overlapping dacitic domes.

Tristan da Cunha lies in the South Atlantic 400 km east of the Mid-Atlantic Ridge. It is a composite cone rising from the ocean floor 3000 metres below sea level to a height of 2100 metres above sea level (Fig. 45). It was studied in some detail following the eruption of October 1961 which necessitated the evacuation of its entire population (Baker *et al*. 1964). Most of the island consists of interbedded basaltic lavas and pyroclastic rocks derived from the central conduit. The proportion of pyroclastic material to lava is greatest in the centre of the island, close to the vent. There are more than 30 secondary cinder cones on the flanks of the volcano, mainly of pyroclastic material, but some with thin lava flows, and many small pyroclastic accumulations are buried within the stratified sequence. Many dykes radiate from the centre of the island, and some volcanic necks can be seen. The predominant type of lava is trachybasalt, but there are also alkali basalts, picritic basalts, trachyandesites and trachytes. Trachyte mainly occurs as intrusive bodies, and only rarely as lava flows.

PYROCLASTIC SHEET DEPOSITS

In some volcanic provinces large areas of land are blanketed by pyroclastic deposits. Such a region is the central part of the North Island of New Zealand, the *Taupo* volcanic zone (Fig. 46). This is a downfaulted basin, in which the Mesozoic sediments that outcrop to the east and west are depressed to a depth of 4000 metres below the Quaternary volcanic rocks. The bulk of the volcanic material consists of very large sheets of rhyolitic ignimbrite, such as that described on page 50, together with air-fall ashes and minor lavas. The ignimbrites cover an area of about 20,000 square kilometres. There are at least four large calderas from which the ignimbrites are believed to have come, and which also contain later rhyolite domes including those shown in Figs 32 and 33. The area is still volcanically active. There are many hot springs, and at the southern end of the zone are three large and very active andesitic strato-volcanoes (Ruapehu, Tongariro and Ngauruhoe). Basalt occurs only rarely in this province, as isolated flows and scoria cones.

The Taupo pyroclastic deposits are so widely spread that their rate of accumulation has not compensated for the amount of subsidence at the central site of eruption, leading to a topography which is depressed in the centre and elevated around the rim. Wilson and Walker (1985) described this as an 'inverse volcano', challenging the familiar mental picture of volcanoes as being mountains.

Volcanism in the *Jemez Mountains*, New Mexico, began in the late

Figure 46. The Taupo volcanic zone, North Island, New Zealand.

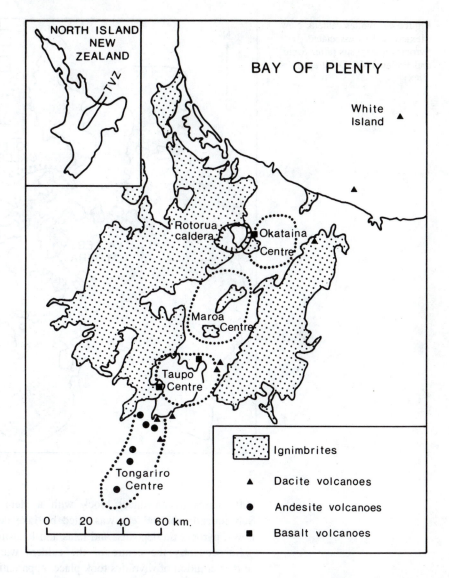

Miocene or early Pliocene with the eruption of first a basalt–rhyolite sequence, then two basalt–andesite–rhyolite sequences that took about 10 million years to accumulate (Smith and Bailey 1968). In the middle Pleistocene there were two very large outbursts of pyroclastic volcanism, which produced the Bandelier tuff (Fig. 47). In the first of these episodes a series of rhyolite ash flows were deposited, burying the older topography under a gently sloping cap of mainly welded tuffs about 200 cubic kilometres in volume. A collapse at the summit of this pile formed the Toledo caldera. After an interval of 300,000 years a second series of ash flows were erupted, again with a volume of about 200 cubic kilometres. This series is about 250 metres thick near its crest and spreads out over an area of 60 km × 100 km. The caldera associated with the second pyroclastic outburst is called the Valles caldera. The area within the caldera subsided by 600 to

Figure 47. Valles caldera, New Mexico, and its associated pyroclastic deposits (after Smith and Bailey 1966; Doell *et al.* 1968).

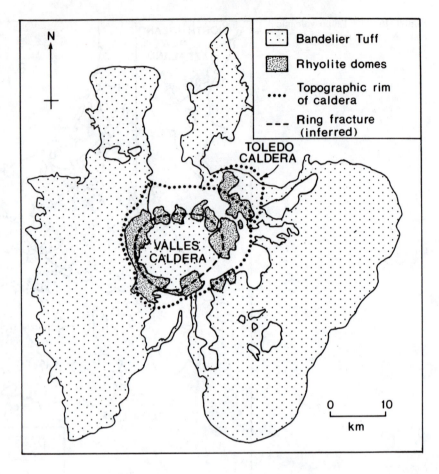

900 metres as an intact block with a stepped zone around the edge. Subsequently the caldera was filled by lake deposits, rhyolite domes and flows, pyroclastic deposits, and talus and landslides from the caldera walls.

Subsequently the centre of the caldera was uplifted and domed, and further eruption of rhyolites took place, apparently along a new ring fracture. Calderas of this kind which have been domed by the upward pressure of new post-caldera magma have been called resurgent cauldrons by Smith and Bailey (1968), and the Valles caldera is the type example.

Yellowstone Park contains one of the world's largest rhyolite plateaux. The volume of rhyolite magma which has been erupted during the last 2 million years has been estimated at over $6000 \, km^3$. This would have formed a substantial mountain had it accumulated into a cone; for comparison Vesuvius has a volume of about $12 \, km^3$ and Mauna Loa has a volume of about $42,000 \, km^3$.

The Yellowstone plateau is made up mainly of rhyolite flows and welded tuffs. There are pumices and pumice-breccias representing all degrees of vesiculation from coherent lava to pyroclastic flows. Small flows of basalt are intercalated in the rhyolitic lava and tuff succession, and there are also rhyolite domes, basaltic cinder cones and small bodies of rhyolite–basalt mixed lava. The basalt flows are numerous but only amount to about 1% of the total volume. To the south-west they merge with the basalts of the Snake River Plain. The mixed lava has been found at four localities, the largest mass being a flow 30–60 metres thick. Some of the mixed lava appears to have formed from the mixing of liquid basalt and rhyolite, while elsewhere it consisted of solid basaltic inclusions in rhyolitic magma. There is no reason to suppose that volcanic activity in Yellowstone Park is at an end; hot-spring activity is continuing and there may be some magma present at depth.

CHAPTER 3

Intrusion

The most important role in the emplacement of plutonic igneous rocks is played by gravity. Their mode of intrusion is largely determined by the difference in density between the magma and its country rocks, and other factors such as the viscosity of the magma or the stress regime in the country rocks have only a modifying influence.

DENSITY OF ROCKS AND MAGMAS

The densities of common rock types are given in Table 8, but the densities of magmas are not easy to measure and must be estimated indirectly. There are very few density determinations on molten silicate rocks, and none at all on magmas with all their volatile constituents still present.

Three ways have been used to estimate magma densities. The first is to extrapolate low temperature density and thermal expansion data on volcanic glasses up to their probable liquidus temperatures. The second is to measure the actual densities of melted rocks. The third is to calculate the densities of the melts from the partial molar volumes of the magma constituents.

The first method is unsatisfactory because the thermal expansion of silicate glasses is not linear up to high temperatures. Low temperature measurements of the coefficient of thermal expansion cannot be extrapolated beyond the glass transition temperature. This is the temperature at which a glass starts to soften rapidly when heated, and for igneous rock compositions it is at around 700–800 °C (Carmichael *et al.* 1974). At this temperature many of the physical properties of a glass change rapidly, including the coefficient of thermal expansion, which greatly increases.

The second method is experimentally very difficult. Murase and McBirney (1973) succeeded in measuring the densities of melts of the most important igneous compositions at a range of temperatures, and their results at 1250 °C are given in Table 9. Although experimental measurement is more direct than the other methods of estimating density it suffers from a serious

Table 8. Densities of various rock types (in g/cm³)

	Mean	Range
Granite	2.67	2.52–2.81
Granodiorite	2.72	2.67–2.78
Diorite	2.84	2.72–2.96
Gabbro, dolerite	2.97	2.80–3.12
Peridotite	3.23	3.15–3.28
Dunite	3.28	3.20–3.31
Syenite	2.76	2.63–2.90
Anorthosite	2.73	2.64–2.92
Rhyolitic obsidian	2.37	2.33–2.41
Andesitic glass	2.47	2.40–2.57
Basaltic glass	2.77	2.70–2.85
Trachytic obsidian	2.45	2.43–2.47
Leucite–tephrite glass	2.55	2.52–2.58
Shale	2.38	2.06–2.66
Sandstone	2.40	2.17–2.70
Limestone	2.57	2.26–2.80
Gneiss	2.70	2.61–2.84
Schist	2.79	2.76–2.82
Granulite	2.83	2.63–3.10
Amphibolite	2.99	2.79–3.14
Eclogite	3.39	3.34–3.45

The data are from Daly *et al.* 1966 (the means given here are unweighted averages of a large number of values).

Table 9. Densities of molten rocks: experimental measurements by Murase and McBirney (1973)

	Density at 1250°C (g/cm³)	Density at room temp. (g/cm³)	Silicate liquidus temp. (°C)
Rhyolite	2.17	2.28	—
Andesite	2.41	2.59	1240
Tholeiitic basalt	2.60	2.76	1210
Alkali olivine basalt	2.68	2.83	1200

The densities at room temperature are for rapidly chilled melts.

drawback, namely that artificially melted rocks differ from magmas in not containing their original dissolved volatile constituents.

The prediction of magma densities from partial molar volumes has been described by Bottinga and Weill (1970), with further refinements by Bottinga *et al.* (1984). The density of the magma (ρ) is calculated from the equation:

$$\rho = \frac{\sum X_i M_i}{\sum X_i V_i}$$

where X_i is the mole fraction of component i, M_i is its molecular mass (gram formula weight), and V_i is its partial molar volume. Partial molar volumes of the common oxide constituents, mainly obtained from synthetic silicate glasses, are given in Table 10. The main problem with this approach is that there is not much information on the partial molar volume of H_2O in silicate glasses, and the partial molar volume of this constituent is strongly dependent on pressure. Bottinga and Weill suggested using the data on the partial molar volume of H_2O in the system albite–water obtained by Burnham and Davis (1971). Additional estimates of the partial molar volume of water are given by Lange and Carmichael (1990).

Table 10. Partial molar volumes of the principal oxide constituents of silicate melts

	Partial molar volume at 1300 °C (cm³/mole)	Mol. mass	Mol. mass/Mol. vol.
SiO_2	26.92	60.09	2.23
TiO_2	22.43	79.90	3.56
Al_2O_3	36.80	101.94	2.77
Fe_2O_3	41.44	159.70	3.85
FeO	13.35	71.85	5.38
MgO	11.24	40.32	3.59
CaO	16.27	56.08	3.45
Na_2O	28.02	61.98	2.21
K_2O	44.61	94.20	2.11
H_2O	34.81	18.02	0.52

The data for all constituents other than H_2O are from Lange and Carmichael (1987). The figure for H_2O is extrapolated from the data of Burnham and Davis (1971) and is for zero confining pressure. Partial molar volumes of all the oxide constituents are pressure-dependent.

The calculation of densities from partial molar volumes is particularly difficult to apply to granitic melts because there are insufficient data on partial molar volumes at the low temperature of granite magmas. For basaltic magmas the calculated densities agree well with the few direct measurements, and they enable various predictions to be made of density effects in petrogenesis. If the molecular mass of each oxide is divided by its partial molar volume, as in Table 10, it is possible to predict which constituents would have most effect on the density of a magma. Constituents with a low ratio will cause the magma to be low in density, and constituents with a high ratio will lead to a high density. For example the high molar fraction of water in a hydrous magma could lead to a very substantial reduction in density compared with the equivalent dry magma.

To compare the densities of rocks and magmas in relation to the intrusive behaviour of magmas, we next need to know the density of common rock types at magmatic temperatures. These are relatively easy to calculate by combining the low temperature densities with data on the coefficients of thermal expansion. Thermal expansions from low to magmatic temperatures are not linear, especially because of phase changes, such as the polymorphic transitions of SiO_2, the dissociation of carbonates, and the metamorphic reactions in sedimentary rocks. Thermal expansion data for minerals, such as those in Table 11, can be used to predict rock densities at high temperatures if one assumes that the expansion of a rock approximately equals the weighted average volume expansion of its constituent minerals.

Table 11. Volume thermal expansion (i.e. density decrease) per cent for various minerals from 20 °C to 800 °C and 1000 °C (after Skinner 1966)

	20 °C to 800 °C	*20 °C to 1000 °C*
Albite	2.07	2.75
Andalusite	2.68	3.61
Anorthite	1.10	1.45
Augite	1.98	2.67
Cordierite	0.43	0.64
Enstatite	2.28	—
Forsterite	2.92	3.86
Hornblende	2.22	2.84
Orthoclase	1.91	2.61
Quartz	4.42	4.29
Sillimanite	1.41	1.98

The densities quoted in Tables 8 and 9 are at low pressure. The densities of magmas, like those of rocks, increase with pressure, although identical magma and rock do not necessarily have the same compressibility. Figure 48 shows the density variation for rock and melt of olivine–tholeiite composition. At 10 kilobars (equivalent to a depth of about 35 kilometres) the melt would be 4–5% denser than at the surface. It is interesting to note that the basaltic melt undergoes a particularly rapid change in density between about 12 and 15 kilobars. This matches the rapid change that would be undergone

Figure 48. The variation of density with pressure of a tholeiitic melt (Kilauea 1921 olivine tholeiite) and crystalline tholeiite at 1200 °C (after Kushiro 1980a).

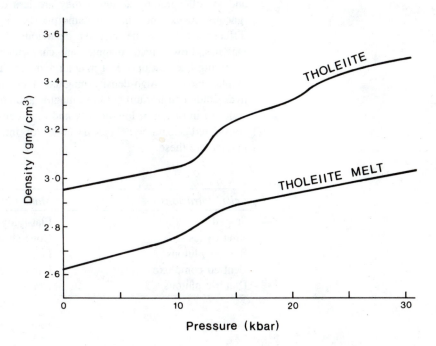

Table 12. Approximate densities of common rock and magma types (g/cm³)

	Magmas	*Rocks*
Sandstone	—	2.4–2.6
Shale	—	2.4–2.6
Limestone	—	2.6–2.7
Granite	2.2–2.3	2.6–2.7
Mica-schist	—	~2.7
Basalt, gabbro	2.6–2.7	~2.9
Amphibolite	—	2.8–3.0
Peridotite	—	3.1–3.2

The magmas are assumed to be at their liquidus temperatures. The rocks are assumed to be well-compacted if sediments, and at contact metamorphic temperatures (500–1000 °C).

by solid basalt at about the same pressure in changing to garnet–granulite and then to eclogite.

The figures in Table 12 show approximately the differences in density between the commoner types of magma and their country rocks. It is assumed that the magmas concerned would be entirely liquid, although in practice some plutonic rocks may be intruded as crystal–liquid mushes, and many more would contain a proportion of suspended phenocrysts. As a rough approximation, one can assume that a magma of a given composition would be about 10–15% less dense than the equivalent solid rock.

The common magma types can be divided into two classes. Felsic magmas (granite, syenite, nepheline syenite) are less dense than the most frequently occurring country rocks. In contrast, mafic magmas (basalt, andesite) are similar in density to many sedimentary and metamorphic rocks and to solid granite, although they are less dense than basic or ultrabasic igneous rocks and their metamorphosed equivalents. There is a great difference between the intrusive behaviour of low-density and high-density magmas. Low-density magmas are characterized by intrusion mechanisms involving buoyant uprise of magma and subsidence of country rocks. In the emplacement of high-density magmas, there is a much closer approach to hydrostatic equilibrium between magma and country rock. The characteristic styles of intrusion of low-density and high-density magmas will therefore be considered separately. Typical modes of intrusion of the two classes of magma are these:

Felsic intrusions	*Mafic intrusions*
Stoped stocks	Flat-lying sheets
Ring dykes	Cone sheets
Bell-jar plutons	Funnels
Centred complexes	Funnel-dykes
Diapiric plutons	Ring dykes and ring complexes

FELSIC INTRUSIONS

Before discussing the various modes of intrusion, there is an important matter of terminology to consider. A *stock* (or *pluton*) is a body of intrusive igneous rock emplaced by a specific mechanism at a specific time and place. A *batholith* is a collection of plutons, not necessarily of the same age, magma type or emplacement mechanism. I will discuss batholiths after I have distinguished between the individual types of pluton.

STOPED STOCKS

Stoping is a miners' term meaning the prising of masses of ore from the roof and walls of an underground excavation. Daly applied the term to describe a magma rising by breaking off joint blocks from the overlying country rock. The term suggests that the magma forces its way into cracks in its roof and helps to detach lumps of overlying material, but in view of the density contrast and the size of many intrusions it is likely that gravity alone is sufficient to cause collapse of the cover.

Figure 49 illustrates the stoping of roof rocks into the xenolith-rich *Corralillo* tonalite in Peru. The country rocks were first fractured, then penetrated by tonalite veins, and then detached into the magma, where the fragments became progressively rounded and reduced in size.

Many of the intrusions which contain abundant xenoliths do not show any indication of the presence of the roof. A particularly interesting example is the *Thorr* pluton in Donegal (Fig. 50), in which the southern half of the outcrop is xenolith-rich. Many of the xenoliths are very large (up to several hundred metres), and their parentage (for example limestone, quartzite, schist, dolerite) is still recognizable despite marginal reaction with the magma. Pitcher (1952) mapped the composition of the xenoliths over an area of about 30 square kilometres in the southern part of the intrusion, and was able to correlate their stratigraphy and structure with that of the surrounding country rocks. More northerly outcrops contain fewer and fewer xenoliths as the rock passes from granodiorite or quartz–diorite into a relatively leucocratic granite. In order for the structure and stratigraphy ('ghost stratigraphy') of the country rocks to be preserved in the xenoliths, they must have moved very little in the surrounding magma.

The *Beinn-an-Dubhaich* granite in Skye (Fig. 51) is an example of a stock which has almost certainly been emplaced by stoping. The undisturbed nature of the country rocks, revealed by the continuity and lack of deflection of dykes, shows that the magma has not pushed its country rocks aside, while the irregular shape contrasts with the sheet, ring or cylinder-form that might result from other modes of intrusion. The large masses of country rock that are seen within this particular intrusion are interpreted as roof-pendants rather than xenoliths because they have not been tilted or rotated in the magma. An irregular undulating roof may be considered characteristic of a stoped intrusion.

Although it is easy to find examples of granite apophyses penetrating

Figure 49. Stoping, break-up, and rounding of andesite xenoliths (shaded) in the Corralillo tonalite, Peru: (a) at the contact; (b) 100 metres below the contact; (c) 200 metres below the contact (after Myers 1975).

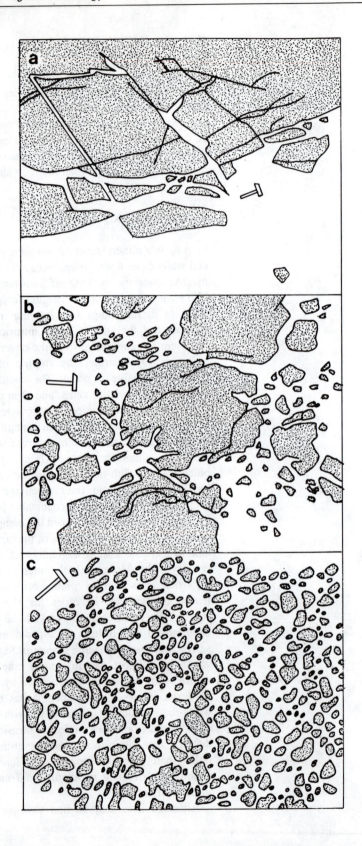

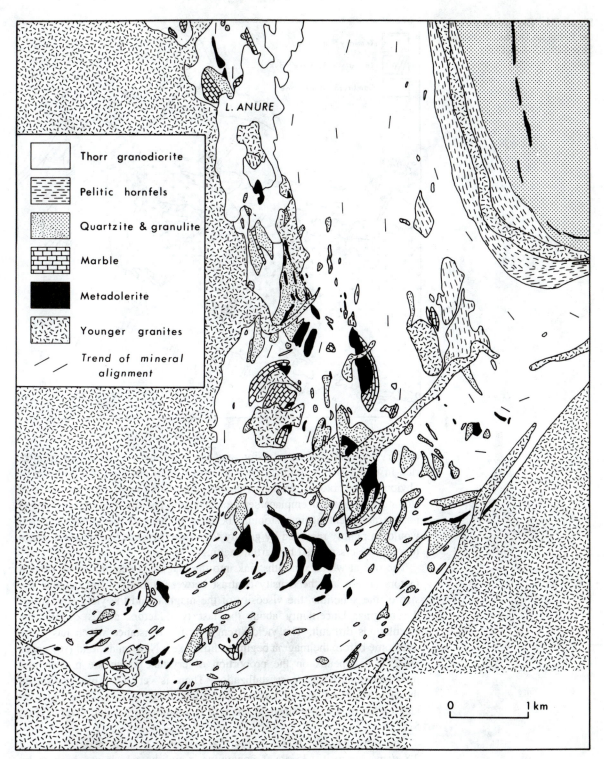

Figure 50. The xenolith-rich
Thorr granodiorite, Donegal,
Ireland (after Pitcher 1952).

Figure 51. The eastern part of the Beinn an Dubhaich granite, Skye (after Tilley 1951).

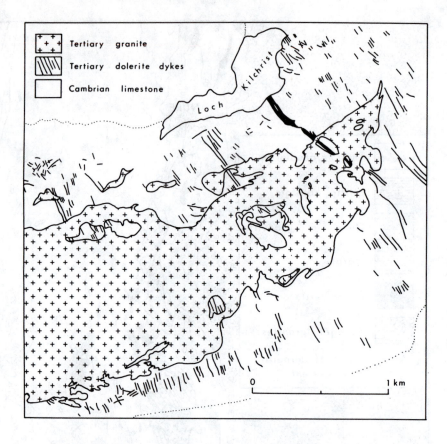

country rocks around an intrusion, and there are many examples of granites rich in undigested xenoliths, most granites do not contain an abundance of obvious stoped blocks, so there remains some doubt as to the importance of stoping as a general emplacement mechanism. The absence of xenoliths is not an argument against stoping, because in general the stoped material would be rapidly removed by gravity to a lower level of the intrusion where it would be dissociated by partial melting.

The rate at which xenoliths sink into the magma is determined by several factors. These are the density contrast between xenoliths and magma, the size of the xenoliths, the viscosity of the magma, and the yield strength of the magma. Uncertainty about the last two factors makes quantitative calculations difficult. The yield strength is the resistance which has to be overcome before the magma begins to flow like a Newtonian fluid (see pages 26–27); it depends on the proportion of crystals which are present and increases rapidly during crystallization. There is very little information on the yield strengths of acid magmas, but calculations by Spera (1980) give an idea of the importance of yield strength in relation to basaltic magma (Fig. 52).

The settling rates shown in Fig. 52 assume a basaltic magma (density $2.8 \, g/cm^3$, viscosity 35 Pa.s) containing xenoliths which are heavier by $0.65 \, g/cm^3$. Actual yield strengths measured for basalt magma in a Hawaiian lava lake are in the order of $1000 \, dyne/cm^2$ ($100 \, N/m^2$) with about 25%

Figure 52. The influence of yield strength on the settling rates of xenoliths. The rates are calculated for basaltic magma (density 2.8 g/cm^3, viscosity 350 poises) containing xenoliths which are heavier by 0.65 g/cm^3 (after Spera 1980). Note: 10 dynes/cm^2 = 1 N/m^2.

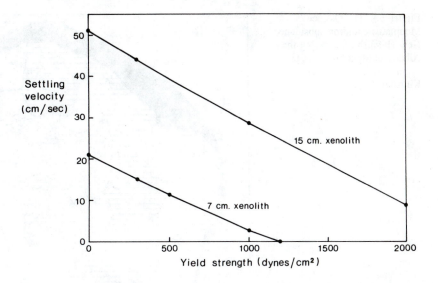

crystals and 2–5% gas bubbles present, but estimates for flows undergoing rapid crystallization are up to 100 times higher (Yoder 1976). Therefore in relation to the time taken for intrusion, xenolith settling rates in a gabbro could easily change from very fast to zero as crystallization proceeds.

RING DYKES AND BELL-JAR PLUTONS

Ring dykes have the form of a hollow cylinder surrounding a subsided block of country rock. Many ring dykes have been mapped, of which the Albany ring dyke in New Hampshire (Fig. 53) and the Loch Ba felsite in Mull (Fig. 54) are classic examples. Not all ring dykes are so uniform in outcrop width, and there are many arcuate intrusions which resemble discontinuous or incomplete rings.

Some, but not all, ring dykes have contacts which dip outwardly at a steep angle. This observation implies that subsidence of the central block within a ring fracture is the mechanism which opens the ring dyke and allows it to be filled with magma. Comparison with volcanic calderas suggests that in many cases a proportion of the magma has actually passed through the ring dyke en route to the surface, where it has been erupted. Volcanic examples provide evidence of subsidence of the central block, and in some plutonic examples there is additional evidence from stratigraphic comparison of the lavas and sediments outside and inside the ring.

There is a possibility that in some instances the subsidence may have taken place sufficiently far below the surface that the ring structure had an intact roof, and that the magma never made its way up to the surface. Section 1 in Fig. 55A might well give a ring outcrop like that of Fig. 54, but section 2 would show a circular pluton whose origin by ring faulting would not be so obvious; the central plug would not be visible, and there would be no indication of subsidence. Figure 56 shows just such an interpretation of the

Figure 53. The Ossipee Mountains cauldron subsidence, New Hampshire, showing the Albany nordmarkite (quartz–syenite) ring dyke (after Kingsley 1931).

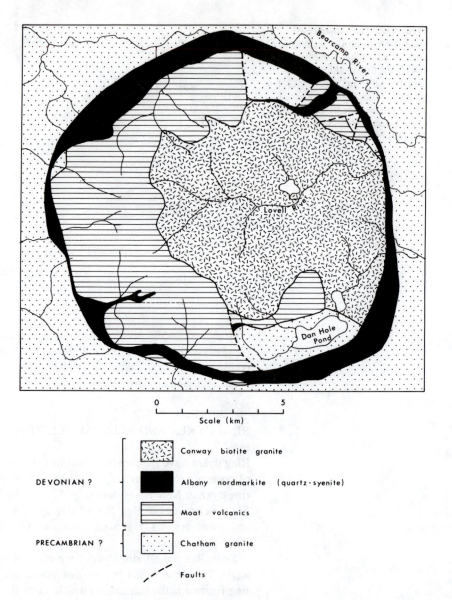

Scale (km)

	Conway biotite granite
DEVONIAN ?	Albany nordmarkite (quartz-syenite)
	Moat volcanics
PRECAMBRIAN ?	Chatham granite
	Faults

Brandberg pluton in Namibia. Such plutons might be quite common if subsidence were to take place to the depth indicated in Fig. 55B. In favourable circumstances, i.e. exposures with great topographic relief in mountainous country, it may be possible to see both the top and sides of such plutons, of which an example is the Anta-Julquillas pluton in the Coastal Batholith of Peru (Knox 1974).

An alternative interpretation of ring dykes is that their initiating fractures might be due to the presence of the underlying magma body alone, and not necessarily related to the overlying presence of the ground surface. In this case, the ring dykes might be initiated by a paraboloidal fracture system above the main magma chamber, as shown in Fig. 55C. Again, a large vertical subsidence of the central block would result in the formation of a

Figure 54. Ring dykes in the Tertiary igneous complex of Mull, Scotland (after Bailey *et al.* 1924). Patches of 'Central-type' basalt which are preserved within the Loch Ba felsite ring dyke are at a lower level than this lava type occurs outside the ring. The basaltic lava pile (unornamented) contains many minor intrusions, especially cone sheets and vent agglomerates, which are not shown.

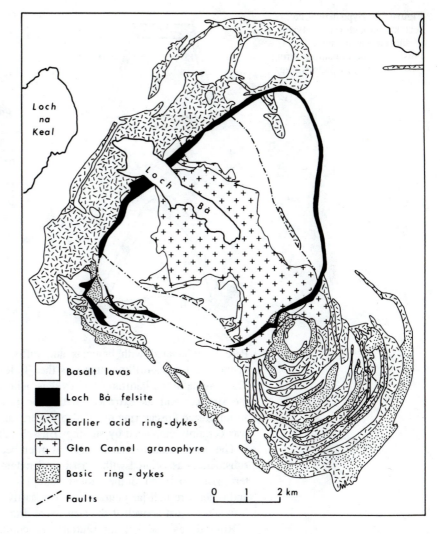

Basalt lavas

Loch Bà felsite

Earlier acid ring-dykes

Glen Cannel granophyre

Basic ring-dykes

Faults

circular pluton of '*bell-jar*' form (Fig. 55D) in which the role of ring-faulting might be difficult to recognize. Many plutons of roughly circular plan with undeformed country rocks may be of this type. Bussell (1985) has described examples from the Peruvian Andes of bell-jar plutons in which both the top and bottom surfaces of the 'bell' are exposed.

A serious objection to subsidence models of ring-dyke formation is posed by the observation that many ring dykes are essentially vertical or even inward-dipping, and they can not therefore be occupying spaces opened up by the subsidence of the central block (Taubeneck 1967). At the classic locality of Glencoe (Fig. 27), the central block appears to have subsided because it contains lavas which are not preserved outside the ring fault, but the ring fault itself has a generally inward dip of about 85°. In the case of the Albany quartz–syenite shown in Fig. 53, Billings (1945) suggested that ring faulting provided a zone of weakness, but that stoping provided the means by which the magma actually entered the ring. The Patirumy ring dyke in

Figure 55. Hypothetical cross-sections through various types of subsidence structure, illustrating the formation of ring dykes and bell-jar plutons (see text).

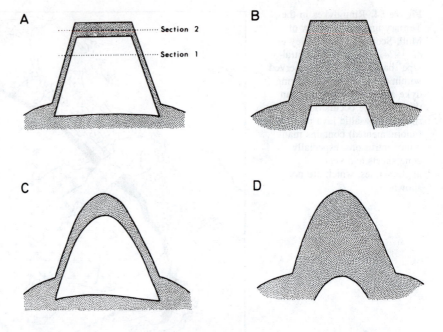

Peru is associated with breccias and tuffisites which suggest that it was opened up by gas-drilling; this was then followed by stoping (Knox 1974). The magma of the Patirumy ring dyke is believed to have eventually reached the surface, and its extruded pyroclastic products can be identified. The xenolith-rich Puscao ring dyke, also in Peru, actually terminates upwards and was certainly emplaced by stoping (Bussell 1985).

The name *cauldron-subsidence* has often been used to describe large-scale subsidences associated with intrusion of magma, but it is not a well-defined term (Branch 1966). Some quoted examples are related to caldera formation and others are bell-jar plutons not necessarily associated with volcanism, so it would be just as well if the term were dropped.

Ring dykes and bell-jar plutons are commonly associated in clusters. Where they are arranged concentrically, the whole assemblage of intrusions can be described as a *ring complex*. Sometimes they overlap asymmetrically or are spread out along a line as the centre of activity migrated between successive episodes of intrusion. The *Sara-Fier* complex in Nigeria (Fig. 57) beautifully displays a succession of overlapping ring intrusions and bell-jar plutons aligned over a distance of 45 kilometres from north to south.

CENTRED COMPLEXES

It is very common for the individual components of composite intrusions to be arranged in concentric rings. Many such examples have been described as 'ring complexes'. This term carries the possible implication that the 'rings' may be ring dykes, which in many cases is not correct. There are undoubtedly concentrically arranged complexes in which some of the individual components genuinely are ring dykes, for example the Bagstowe

Figure 56. Geological map and hypothetical cross-section through the Brandberg pluton, Namibia (after Korn and Martin 1954; Hodgson 1973).

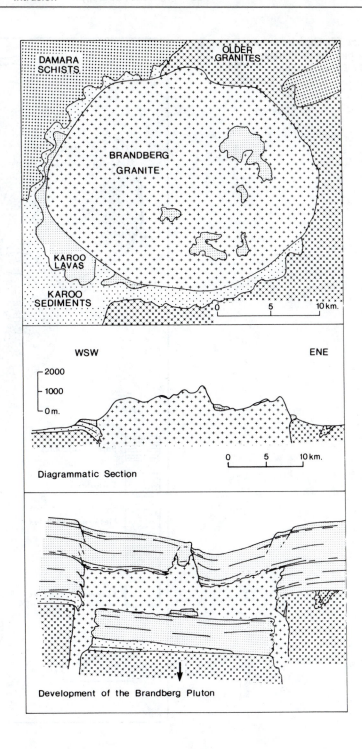

DAMARA SCHISTS

OLDER GRANITES

BRANDBERG GRANITE

KAROO LAVAS

KAROO SEDIMENTS

0   5   10 km.

WSW

ENE

2000
1000
0 m.

0   5   10 km.

Diagrammatic Section

Development of the Brandberg Pluton

Figure 57. Geology of the
Sara-Fier complex, Nigeria
(after Turner 1963).

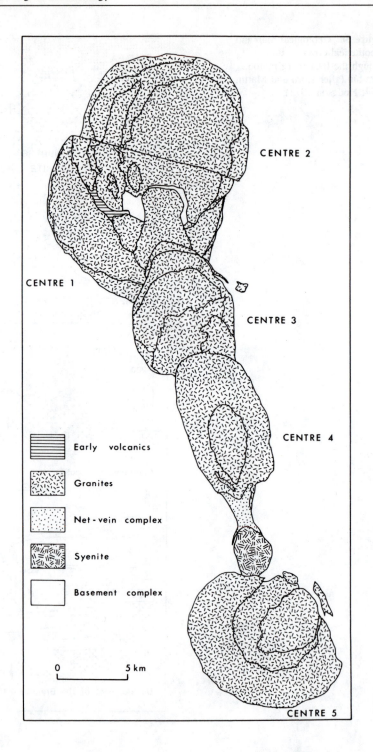

CENTRE 2

CENTRE 1

CENTRE 3

CENTRE 4

Early   volcanics

Granites

Net - vein   complex

Syenite

Basement   complex

0   5 km

CENTRE 5

ring dyke complex in Queensland (Branch 1966) or the ring dykes of the Beinn Chaisgidle centre in Mull (Fig. 54), both of which show screens of country rock between the ring dykes, but many others are likely to be simply cored stocks in which the present ring-shaped outcrops result from sequential intrusion at a focus, and where there was never an internal ring dyke contact. In order to avoid the implication that all rings are ring dykes, Pitcher and Berger (1972) referred non-committally to concentrically arranged plutons as 'centred complexes'.

An example of a centred complex in which the individual rings have no internal contacts against earlier rocks is the *Rosses* complex in Donegal (Fig. 58). This is roughly circular, and about 10 kilometres across. It consists of four successive units (G1–G4), each with a steeply dipping external contact, and intruded in sequence with the oldest around the edge and the youngest in the middle. The country rock is an earlier granodiorite of very different appearance, i.e. foliated and with abundant xenoliths.

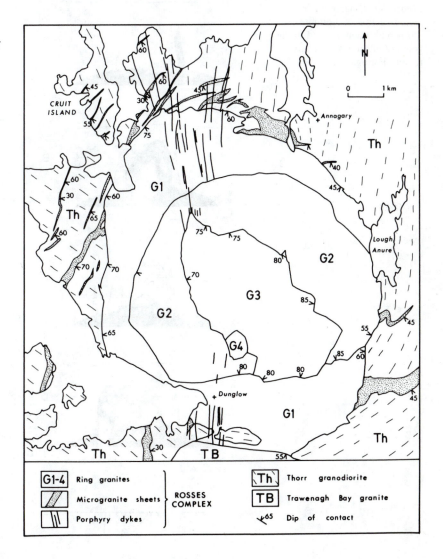

Figure 58. The Rosses centred complex, Donegal, Ireland (after Pitcher 1953).

The granites of the Rosses complex are each very homogeneous in composition, apart from some local post-magmatic alteration, and each of the units is petrographically very similar. The slightly coarser grain size of G2, compared with G1 and G3, enables the contacts to be followed easily, but where G1 and G3 are in contact it is necessary to peer closely at the rock surface to see where they adjoin. In any area where the exposure was less than excellent it would not be possible to detect internal contacts such as are seen in the Rosses complex, and many of the intrusions referred to in the literature as being 'zoned' may in fact have a concentric composite structure. The contact between the various units of the Rosses complex, while mapping out as rings or ovals on a large-scale map, are straight on the scale of the individual exposure, being deflected from place to place in such a way as to indicate that they were controlled by the joint system of the earlier rocks.

Figure 59. Diagrammatic reconstruction of a granite intrusion sequence based on the geology of the Castro Daire region, Portugal (after Oen 1960).

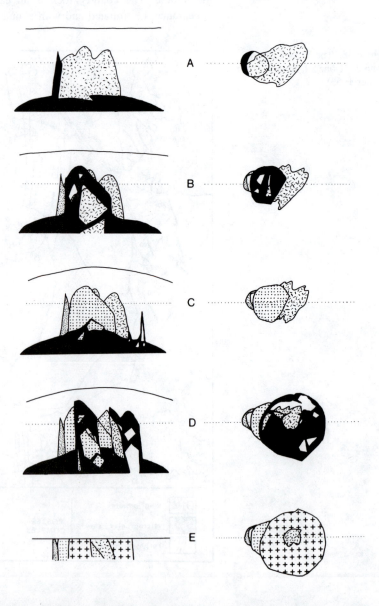

The width of the units in the Rosses complex, combined with the steepness of the internal contacts, precludes their having been formed as ring dykes.

A possible reason for the concentric arrangement of centred complexes is that a magma which rises initially by stoping or large-scale subsidence of the overlying country rocks may complete its intrusion by stoping or subsidence of its own initial products into the shrinking core of the magma reservoir. A more complex pattern of repeated intrusion at a single centre is illustrated by Oen's interpretation of the granite sequence in the Castro Daire region of northern Portugal (Fig. 59).

SHEETED INTRUSIONS

Many mafic intrusions have the form of flat-lying sheets. Felsic intrusions do not generally have this shape overall, but many stock-like bodies have a sheeted component. The *Tregonning-Godolphin* granite in Cornwall (Fig. 60) is about 3 kilometres across and has a gently undulating upper surface, but where its contacts fall away at the margins there are several flat-lying sheets extending from the main body for about a kilometre into the surrounding country rocks. The relatively flat top of the intrusion, and lack of xenoliths, suggest that the pluton was not emplaced by piecemeal stoping. The marginal sheets may be envisaged as wedges of magma which have been frozen in the process of prising large flat slabs of country rock off the roof and walls of the magma body.

There is an alternative interpretation of flat-lying sheets as a late-stage intrusion feature. When the bulk of the magma making up the stock crystallized, the increase in density (granitic rock 15% heavier than granitic magma) would cause a drag on the overlying country rocks, and might open up flat-lying cracks which could be filled with a residual melt fraction. In the Tregonning-Godolphin example the marginal sheets are in fact mainly composed of pegmatite and aplite, in common with the top couple of metres of the main intrusive body.

Another example of a sheeted intrusion is the *Barnesmore* granite in Donegal (Figs 61 and 62). In the sequence of intrusions making up this complex, the third and latest component (G3) forms a large mass extending into a series of sheets cutting earlier components (G1 and G2). It is assumed

Figure 60. Coastal section through the Tregonning–Godolphin granite, Cornwall (after Stone 1975).

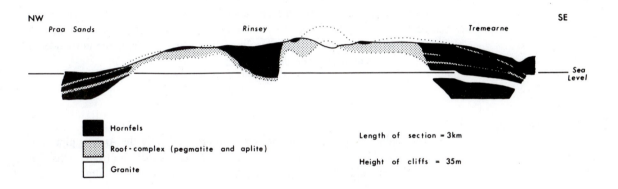

NW   Praa Sands   Rinsey   Tremearne   SE   Sea Level

■ Hornfels

▨ Roof-complex (pegmatite and aplite)

□ Granite

Length of section = 3km

Height of cliffs = 35m

Figure 61. Geological map of the Barnesmore granite pluton, Donegal, Ireland (after Walker and Leedal 1954).

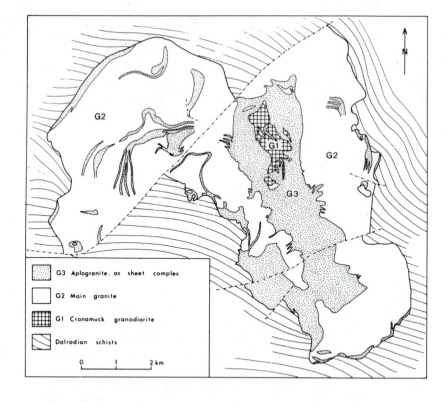

Figure 62. Diagrammatic cross-section through the Barnesmore granite pluton (see Fig. 61).

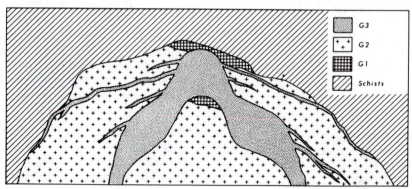

that after a large amount of magma had consolidated to give G1 and G2, the central part of these newly consolidated granites collapsed into the still unconsolidated magma below, peeling off in slabs, of which the unfoundered remnants remain between the sheets of G3.

DIAPIRIC PLUTONS

The main features of a diapiric intrusion are shown by the *Ardara* pluton in Donegal (Figs 63 and 64), which was described by Akaad (1956). This is a roughly circular intrusion about 10 kilometres across, with an almost vertical

Figure 63. The Ardara pluton, Donegal, Ireland (after Akaad 1956).

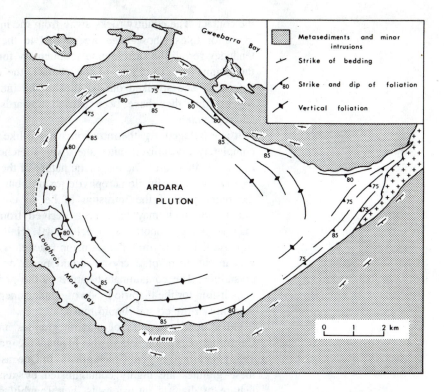

Figure 64. Generalized section through the north-western part of the Ardara pluton, showing the structures associated with diapiric intrusion. Within the pluton, foliation intensifies and xenoliths are progressively flattened as the outer contact is approached. Outside the pluton, the country rocks show intense deformation parallel to the contact (after Pitcher and Berger 1972).

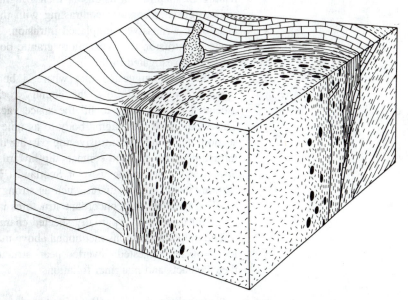

sharp contact against its metasedimentary country rocks. The structures in both granite and country rocks indicate that the granite has forced its way into place. The granite is unfoliated in its central part but becomes strongly foliated towards its margin, and the numerous xenoliths in the outer part of the intrusion become progressively more flattened towards the outer contact. Both the foliation and the plane of flattening of the xenoliths are parallel to the contact. The country rocks away from the intrusion show a development of minor folds whose axes are parallel to the contact, so that there is a tendency for all the surrounding sedimentary formations to strike parallel to the granite margin. Close to the contact, the country rocks are intensely deformed, and streaked out parallel to the intrusion in such a way as to suggest that they have been carried upwards by the movement of the adjacent granite.

It is envisaged that the intrusion swelled like a balloon, as its partially or completely crystallized outer shell was stretched by the inflow of magma into its molten core. The magmatic nature of the unfoliated central granite is clearly shown by its glomerophyric texture, but geochemical study indicates the outer part of the intrusion to be so contaminated that the present grandioritic rock may have been derived from dissociated xenoliths and magma in the proportions of 3:1 or 4:1 (Hall 1966a). It is reasonable to suppose that at the time the intrusion reached its present level the outer part was in the form of a crystal–liquid mush or perhaps almost completely crystalline. The contaminated nature of the granite may indeed have been responsible for the diapiric mode of emplacement, by virtue of the increased viscosity of the crystal–liquid mush.

Other good examples of diapiric granites are the Bald Rock pluton in California (Compton 1955), the Flamanville granite in Normandy (Martin 1953), and the Cannibal Creek granite in Queensland (Bateman 1985). All of these granites are highly contaminated by xenolithic material. A general feature of diapiric intrusions is their roughly circular shape in plan, with smooth curving contacts contrasting with the angular joint-controlled contacts of a permissively emplaced intrusion. This feature suggests a diapiric mode of intrusion for many more granite bodies whose structures have not yet been investigated in detail.

It has been suggested that where a body of buoyant magma moves upwards by internal flow, the country rocks might converge below the ascended magma body to fill the space vacated by its upward movement. This would lead the magma body to assume an inverted pear-shaped cross-section, the dome-like intrusive top passing down into a narrowing stalk. An intrusion whose exposed contacts dip inwards in this expected manner is the *Strontian* granite in the west of Scotland (Fig. 65). This has many similar features to the Ardara diapir, such as strong marginal foliation in the outer tonalitic member and concordant structures in the country rocks. It is cut by a late intrusion with a partially sheeted character.

The individual plutons mentioned above may represent erosion to different levels of the suggested 'inverted pear' structure, as indicated by the dips of their contacts and marginal foliations:

Cannibal Creek	40–60° outwards
Bald Rock	60–80° outwards
Flamanville	70–90° outwards
Ardara	nearly vertical
Strontian	50–70° inwards

A different type of diapiric intrusion is represented by the *North Arran*

Figure 65. The Strontian granite, Scotland (after Sabine 1963). Foliation in the tonalite and the surrounding schists is indicated by alignment of the ornament.

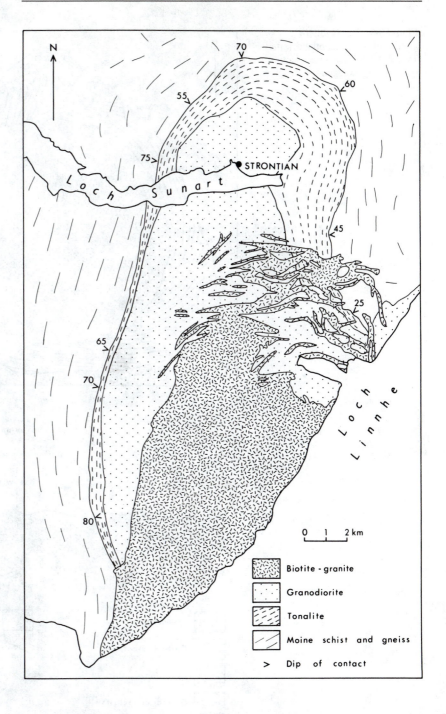

granite, shown in Fig. 66. The structures in the country rocks show that they have been pushed upwards by the intrusion (England 1992), but the granite itself shows no sign of marginal foliation or contamination. Rather than swelling like a balloon while partially crystallizing, this granite appears to have domed the country rocks by simple vertical pressure. A similar lifting of the roof is shown by the *Little Chief* stock in southern California (Fig. 67).

Figure 66. The Northern granite of Arran, Scotland (after Tyrrell 1928). The cross-section is vertically exaggerated × 2.

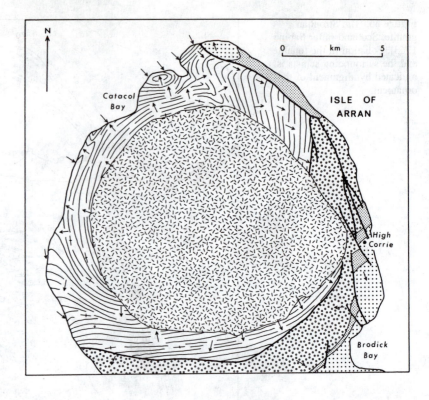

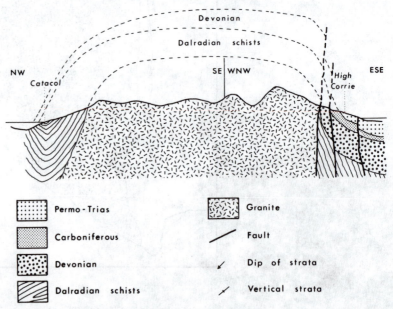

Permo-Trias		Granite	
Carboniferous		Fault	
Devonian		Dip of strata	
Dalradian schists		Vertical strata	

This granite porphyry appears to have uplifted its country rocks asymmetrically within a bounding ring-fracture.

There is some uncertainty about the driving mechanism for diapiric rise of magma. One school of thought is that the magma remains connected to the melt remaining in its source region, and is fed mainly by 'overburden

Figure 67. The Little Chief stock, California, showing displacements on faults. The numbers give the fault throw in thousands of feet; the number is on the upthrown side. X indicates a throw of less than 250 feet (after McDowell 1974).

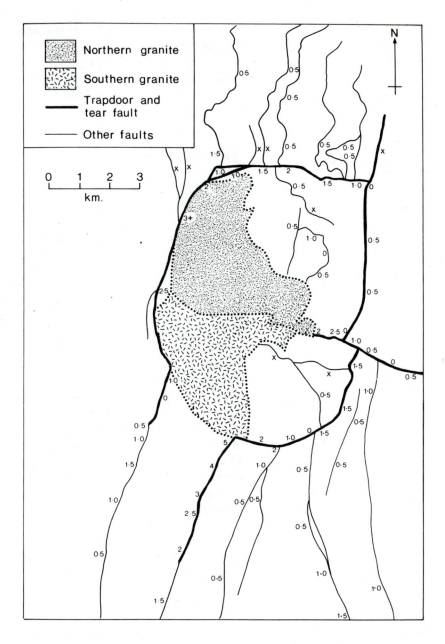

squeeze' (Yoder 1976). The other view is that even if the rising body of magma is detached from its source it can still force its way up by reason of the density contrast between itself and its surroundings, i.e. by buoyancy.

The mechanics of diapiric intrusion have been analysed mathematically by Marsh (1982a). He showed that the rate of ascent of a diapir is crucially dependent on the temperature and viscous behaviour of the country rocks into which it is intruded. The country rock overlying the diapir needs to be heated to its solidus temperature to permit a reasonable rate of ascent. It follows from this that a single body of magma cannot ascend very far because of the limited amount of heat that it can supply to the country rock, but that if magma ascends in successive batches the second and later batches will ascend much more rapidly than the first. This is well illustrated by the Ardara pluton, in which a second, inner intrusion has penetrated through the frozen outer carapace of the first intrusion (Fig. 64).

BATHOLITHS

Granites are more abundant than any other type of intrusion, and they occupy larger areas. In many parts of the world granites extend for hundreds of kilometres, and these large masses of intrusive rock are described as batholiths. Some of the older maps and publications describing such areas give the misleading impression that these batholiths were intruded as single large bodies of magma. This is not the case. Every batholith which is well exposed, and which has been mapped in detail, has been found to be composite.

Figure 68 shows a small example, the Donegal batholith in north-west Ireland, which has been studied in very great detail (Pitcher and Berger 1972). This consists of five separate intrusions each with a different mode of emplacement, plus several additional outlying masses. Four of the individual components (Thorr, Rosses, Barnesmore and Ardara) have already been quoted as examples of particular intrusion types.

The individual plutons making up a batholith vary in size, but the largest of them are rarely more than about 30 kilometres in lateral extent and many are much smaller. Thirty kilometres is also the approximate diameter of the largest modern calderas, and many well-known ones (e.g. Katmai, Krakatoa) are much smaller than this. It is perhaps significant that 30 kilometres is the approximate thickness of the Earth's crust in many continental areas. It would be difficult to envisage any single body of acid magma being more than about 30 kilometres across unless one were to suppose that large areas of the crust had once been almost completely molten from top to bottom. However, detailed field studies show that only a very few individual granite intrusions approach this size. Furthermore, the individual components in batholiths are often quite different from one another in their mode of intrusion and may have been separated by thousands or even millions of years in their times of emplacement. The Coastal batholith of Peru, shown in Fig. 69, was emplaced over a span of 60–70 million years (Pitcher *et al.* 1985), and although the intrusion mechanisms of its roughly 1000 individual plutons are not all known, there are some whose form is obviously distinctive, for example the partial ring dykes indicated in Fig. 69B.

The massive appearance of batholiths is a consequence of the repeated and voluminous production of granite magma in the same region during a period of orogenic activity. A batholith is thus not a type of intrusion, to be compared with a stock or a dyke, but a type of area, where individual plutons are so numerous as to overlap or intersect one another. Figure 70 shows another example, the Berridale batholith in Australia.

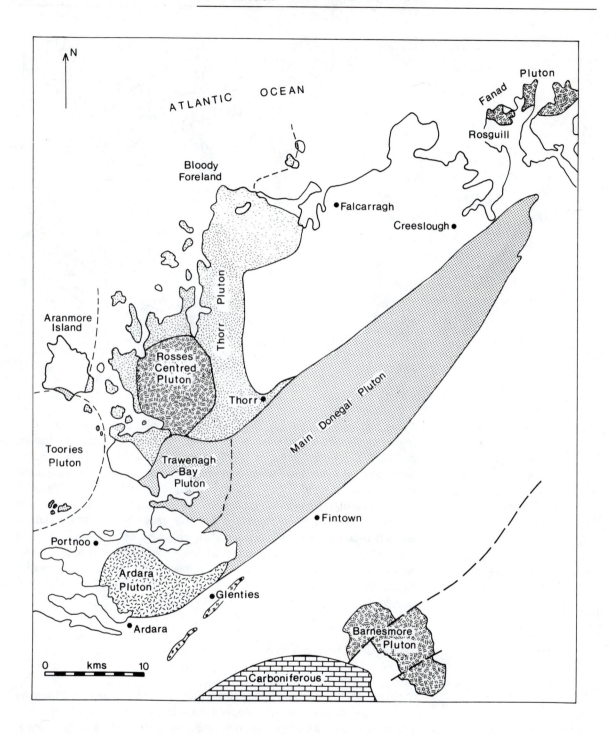

Figure 68. The Donegal batholith, Ireland, showing its constituent plutons (after Pitcher and Berger 1972).

A general observation which can be made of terrains in which acid intrusions are abundant is that the predominant intrusion mechanism changes as one progresses from the surface downwards. Near-surface, sub-volcanic intrusions, exposed in areas where the overlying volcanic rocks are still preserved, show a predominance of ring structures such as ring

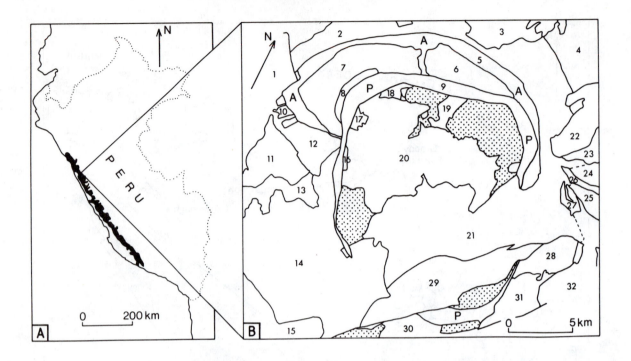

Figure 69. The Coastal batholith of Peru (A), with details of the internal intrusive contacts in one small area (B). The plutons in (B) are numbered from 1 to 32; shaded areas are patches of volcanic country rocks or roof rocks (after Knox 1974). The Patirumy and Anta-Julquillas ring dykes are specifically identified by the letters P and A respectively.

dykes and bell-jar plutons. There are many classic examples in the Oslo region of Norway, the west of Scotland, northern Nigeria, Queensland and New Hampshire, some of which have already been mentioned. Forcefully intruded plutons are more common, and ring structures less common, at deeper levels in the crust. This vertical gradation has been elaborated, with particular reference to granites, by Buddington (1959) and Hutchison (1970). It is possible to recognize a whole range of structural relationships between granite magmas and their enclosing country rocks as their exposed remains are traced from the deepest levels of the crust to the ground surface, where they may even be intruded into their own extrusive equivalents (Fig. 71).

MAFIC INTRUSIONS

Whereas acid magmas are lighter than virtually all well-consolidated country rocks, basic magmas are heavier than some country rocks and lighter than others. It is only when intruded into relatively dense country rocks, such as basalts or amphibolites, that basic magmas can adopt intrusion forms similar to those normally adopted by acid magmas. Characteristic shapes assumed by mafic intrusions are flat-lying sheets, cone sheets, funnels, funnel-dykes and ring dykes, of which the first is much the commonest form in the sedimentary rocks of the upper crust.

Figure 70. Simplified geological map of the Berridale batholith, Australia, showing the contacts between its numerous granitic components (after White *et al*. 1974).

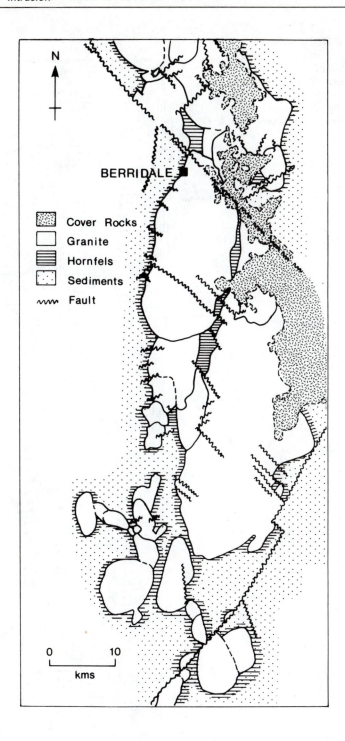

Figure 71. Diagrammatic section through the Coastal batholith of Peru, showing the emplacement of high-level granites into their own volcanic ejecta by repeated cauldron subsidence (after Myers 1975).

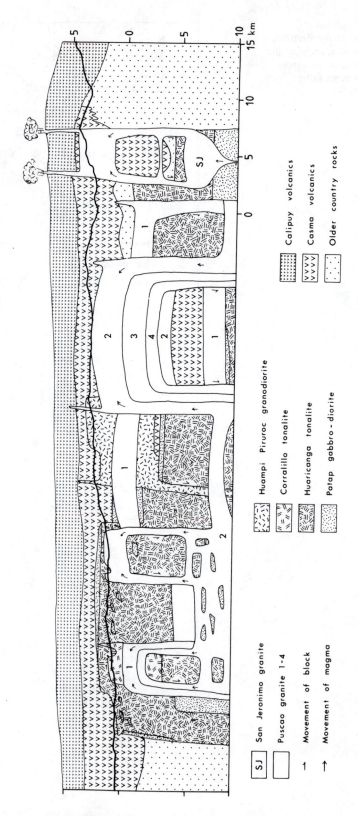

Mafic intrusions are surprisingly uncommon in the middle and lower crust, considering the abundance of basic magmas, and Glazner (1994) suggested that this could be explained by the density increase that takes place when magma consolidates. He calculated that a crystallized mafic pluton would sink at a rate in the order of several kilometres per million years in middle and lower crust of typical viscosity.

FLAT-LYING SHEETS

Flat-lying sheets can range in size from small sills less than a metre thick up to major intrusions underlying thousands of square kilometres and comparable in volume to the larger felsic intrusions. The Whin Sill in northern England is up to 100 metres thick and underlies an area of at least 5000 square kilometres. The Palisades sill in New Jersey, of which the scarp extends for 60 kilometres along the west side of the Hudson River, is up to 300 metres thick. The Gettysburg sill in Pennsylvania is 600–800 metres thick and extends for about 250 kilometres, with an outcrop averaging 2 kilometres in width. These are a few of the largest individual examples, but in total thickness they are matched by the large number of smaller sills which occur in some provinces, for example the Karroo dolerites of South Africa. All of these major dolerite sills occur in undeformed sedimentary successions.

In addition to their occurrence in sedimentary terrains, it is common for sills to be injected within volcanic piles, and cross-sections through some eroded volcanoes show that sills can constitute as much as half of the total thickness of igneous rock (Fig. 72). Actual injection of sills at the present day may well be responsible for the periodic inflation of volcanic cones that has been revealed by tilt-meters and other geophysical instruments on volcanoes such as Kilauea.

The term 'sill' is usually taken to denote a body which is concordant with the intruded country rocks, but in practice all large sills vary in thickness and transgress the stratigraphic succession when mapped over a large area. These transgressions take the form of abrupt steps rather than angular discordance, as shown by the Midland Valley Sill in the Stirling coalfield (Fig. 73). Here the sill steps through 200 metres of the stratigraphic succession in a distance of 2 kilometres.

The largest flat-lying sheets very often have a gently inward-dipping basinal form. This is illustrated by the Triassic dolerite sheet of the *Cornwall* district in Pennsylvania (Fig. 74). This is about 300 metres thick, and

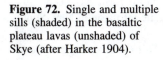

Figure 72. Single and multiple sills (shaded) in the basaltic plateau lavas (unshaded) of Skye (after Harker 1904).

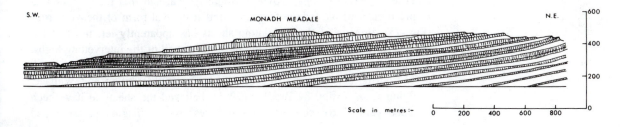

S.W.   MONADH MEADALE   N.E.

Scale in metres :-   0   200   400   600   800

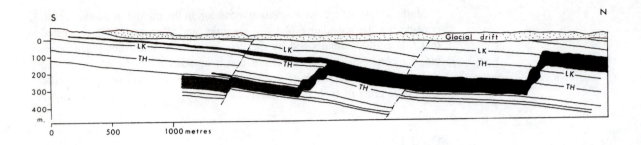

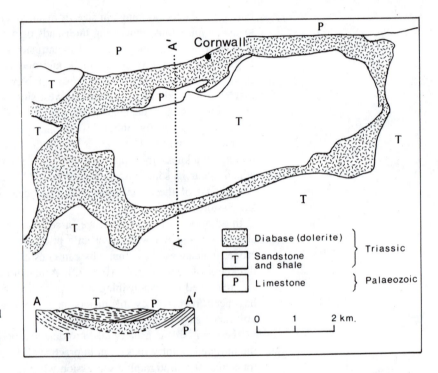

Figure 73. Section through the Midland Valley sill as revealed by mine workings and boreholes in the Stirlingshire coalfield, Scotland (after Dinham and Haldane 1932).

Figure 74. Geological map and cross-section of the Cornwall diabase (dolerite), Pennsylvania (after Hotz 1952).

outcrops in an oval pattern. Diamond drilling has shown that it extends continuously within the oval, flattening out at a level several hundred metres below that of the outcrop. A smaller example whose form has been fully revealed by quarrying is the Prospect dolerite–picrite intrusion in New South Wales (Wilshire 1967).

Meyboom and Wallace (1978), following Du Toit (1920), suggested that an undulating discordant sheet is the commonest form assumed by the dolerites of eastern *Cape Province*, South Africa, and that their undulating nature is responsible for the frequent circular or oval form of their outcrops (Fig. 75). Erosion of the undulating sheets has apparently left most of the saucer-shaped lows while removing most but not all of the intervening highs. The angle of dip of the sheets is usually low (up to 20°) but rises to as much as 65° on the edges of the steepest-sided saucers. Meyboom and Wallace, applying a suggestion by Bradley (1965), believed the sheets to have been intruded parallel to 'compensation surfaces' where magma pressure and

Figure 75. The outcrop pattern of dolerite sheets in the Queenstown area, South Africa (after Meyboom and Wallace 1978).

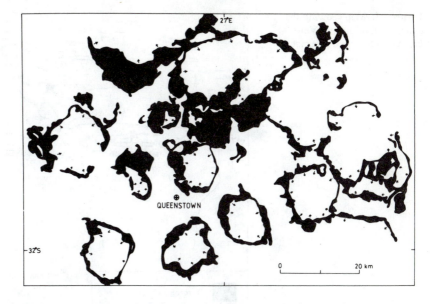

lithostatic pressure were equal. According to this hypothesis there is a relationship between the undulations of the dolerite sheets and the form of the original ground surface, such that low points of the undulating sheet would underlie high points of the overlying topography (Fig. 76). The form of each sheet would be a compromise between the need to propagate along a compensation surface and the ease of propagation along bedding planes, and stepping is the means by which the sill regains its preferred course when it is deflected too far from the compensation surface by a particular bed.

Figure 76. The relationship between magma pressure (isopiestic surfaces shown as dotted lines) and lithostatic pressure (isopiestic surfaces shown as continuous lines). The 'compensation surface' (heavy line) corresponds to where the magma and lithostatic pressures are equal.

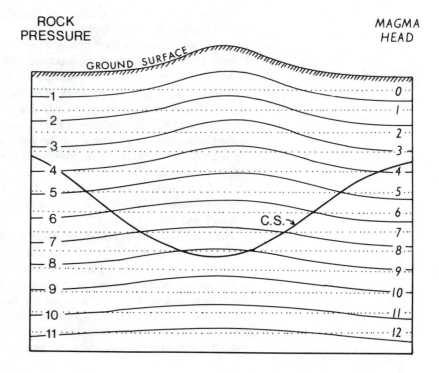

Figure 77. Map showing the extent of the Midland Valley and Whin Sills and associated tholeiitic dykes.

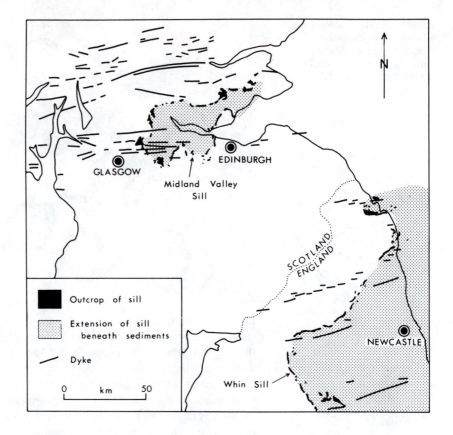

The Whin Sill in northern England (Fig. 77) has given its name to this class of intrusion (sills). Because it is intruded into an actively worked coalfield succession, its form can be fully studied in three dimensions, and Francis (1982) has mapped its structural and stratigraphic relationships in detail. It underlies an area of 5000 square kilometres, and with offshoots ('leaves') is up to 100 metres thick. The known volume of the intrusion is about 215 cubic kilometres. It was intruded in late Carboniferous times into the Carboniferous succession, and Francis mapped its stratigraphic position in the Carboniferous by reference to the Westphalian B/C boundary, which is the nearest approximation to the ground surface at the time of intrusion. His mapping revealed a basinal form such that the sill is thickest in the lowest parts of the basinal structure (Fig. 78). There is a similar relationship in the contemporaneous Midland Valley Sill which lies 100 kilometres to the north (Figs 77 and 79).

There are four major east–west dykes of similar material traversing the area occupied by the Whin Sill, but not in the part where it is thickest. Francis's model of the emplacement mechanism is shown in Fig. 80. He envisaged the intrusion of one or more of the dykes to within 1 kilometre of the ground surface to act as a feeder for the sill, and to provide a small hydrostatic head for intrusion. Two noteworthy features of the model are: (1) that the sediments were recently deposited and there was plenty of pore water to turn to steam and ease the passage of magma; and (2) the syn-

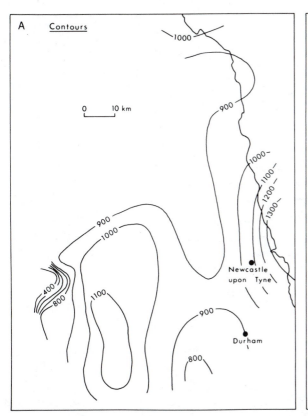

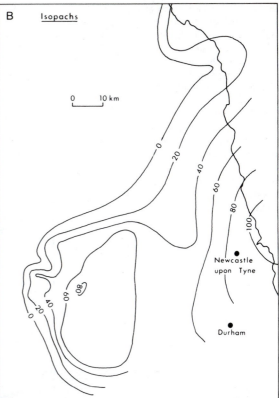

Figure 78. (A) The depth of the Whin Sill below the stratigraphic horizon Westphalian B/Westphalian C, contoured in metres; (B) the thickness of the Whin Sill (or total thickness where split into more than one leaf) in metres (after Francis 1982).

sedimentary basinal form of the country rocks is the cause of the basinal form of the intrusion. Francis's model of sill intrusion differs from Bradley's in that it is the structure of the sedimentary basin and not the shape of the ground surface which determines the form of the intrusion. Moreover, it is implicit in Francis's model that undulating sills such as those shown in Fig. 75 may never have been continuous, i.e. there were no former dome-like areas between the saucer-like areas which are preserved.

Laccoliths are concordant lens-shaped intrusions produced when magma is injected in a sill-like way between sedimentary layers, but is too viscous to flow readily and consequently swells into a bulge. This is most likely to happen to dioritic magmas, which are dense enough to share some of the intrusion characteristics of basalt or gabbro but which are considerably more viscous. Corry (1988) reviewed the emplacement mechanism of laccoliths, giving many examples.

Figure 79. Cross-section through the Midland Valley sill, Scotland, after restoration of the Westphalian B/C stratigraphic boundary to the horizontal (after Francis 1982). Vertical scale is 4 × horizontal scale. Stratigraphic divisions are D = Dinantian, N = Namurian, W = Westphalian.

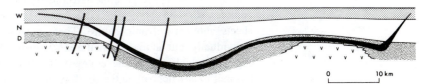

Figure 80. The mechanism of sill intrusion (after Francis 1982). (A) A dyke or fault intrusion reaches to within 0.5–1.0 km of the surface and above the optimum level for lateral flow; (B) lateral intrusion under a head leads to gravitational down-dip flow, and accumulation at the bottom of a syn-sedimentary basin; (C) magma ascends from the bottom of the basin to achieve hydrostatic equilibrium. More than one feeder dyke may be present.

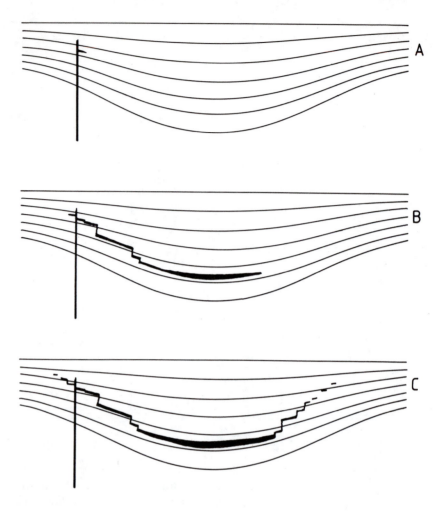

CONE SHEETS

Cone sheets are thin sheets having a circular outcrop, and inclined inwards, so that they have the form of a downward-pointing cone. Individual cone sheets are only a few metres thick, but they occur in large numbers so as to constitute a substantial part of the complexes in which they are present. They are concentrically arranged, and the outer ones of a set dip at lower angles than the inner ones so that they all converge towards a common focus at depth (Fig. 81).

Outstanding examples of cone-sheet development occur in the Mull and Ardnamurchan Tertiary igneous complexes in Scotland, each of which contains hundreds if not thousands of nested cone sheets. The map of the Ardnamurchan complex (Fig. 82) shows in a diagrammatic way the disposition of the cone sheets about a volcanic centre. There are too many cone sheets to show individually, but an idea of their abundance can be gained from the section in Fig. 83, which shows the cone sheets along a short section (less than 1 kilometre) at Kilchoan on the south coast of Ardnamurchan.

Figure 81. Idealized section through a complex of ring dykes and cone sheets (after Richey 1932).

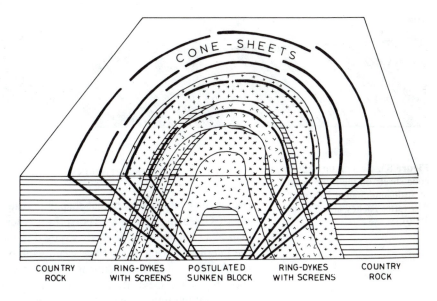

Figure 82. Highly simplified map of the Tertiary igneous complex of Ardnamurchan, Scotland, showing the disposition of cone sheets about the intrusive centres (after Richey and Thomas 1930).

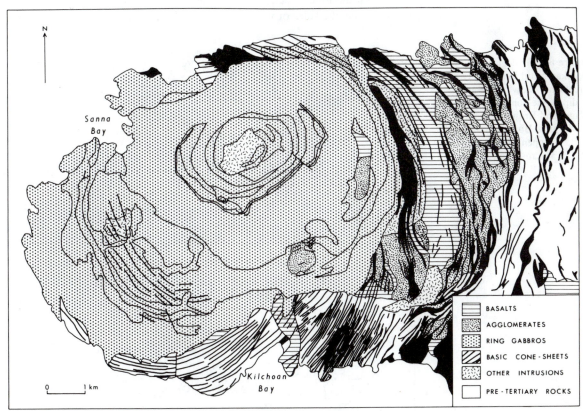

The cone sheets of centre 2 of the Ardnamurchan complex at Kilchoan total 1000 metres in thickness, and dip at angles of 35–45°, so the intrusion of the cone sheets will have necessitated an uplift of the conical block overlying the cone sheets by several hundred metres.

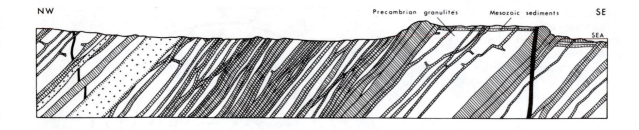

FUNNEL-SHAPED INTRUSIONS

The classic example of a funnel-shaped intrusion is the *Skaergaard* intrusion in east Greenland (Fig. 84). The surface outcrop of the intrusion measures 7 km × 11 km, and it is perfectly exposed in mountainous glaciated country. The internal layering of the Skaergaard gabbros, together with the dip of the neighbouring lavas and sediments, shows the intrusion to have been tilted since its emplacement. If allowance is made for this tilting, the form of the intrusion at the time of emplacement is as shown in Figs 85 and 86. It is an inverted cone, the sides of which converge downwards. The axis of the cone

Figure 83. Section through cone sheets on the south coast of Ardnamurchan at Kilchoan Bay (after Richey and Thomas 1930).

Figure 84. Geological map of the Skaergaard intrusion, east Greenland, showing the angle of dip of the contact (after Wager and Brown 1968). MBG = Marginal Border Group, BS = Basistoppen sheet (gabbro), MD = Macro-dyke (gabbro).

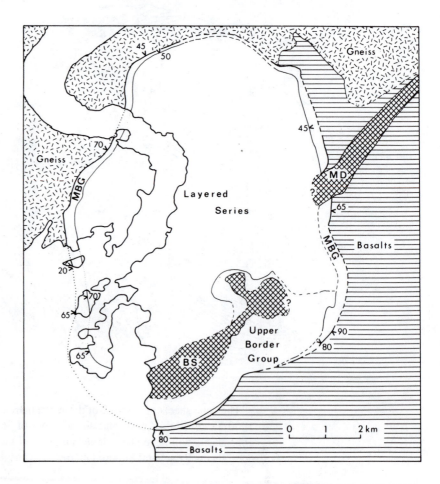

Figure 85. Outline of the Skaergaard intrusion showing the original dips of the contact after correction for post-intrusion tilting (after Wager and Brown 1968).

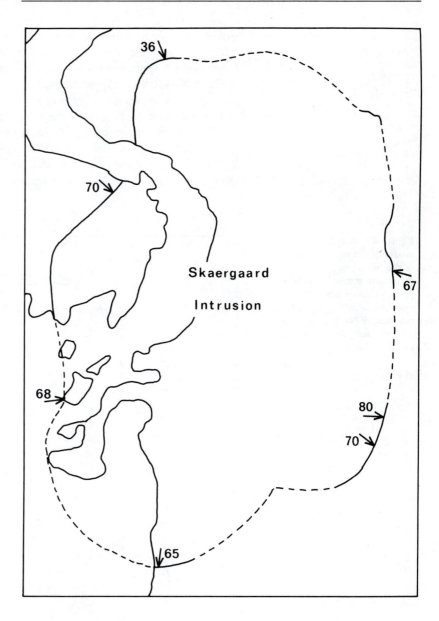

Figure 86. Cross-section through the Skaergaard intrusion, after correction for post-intrusion tilting (after Wager and Brown 1968).

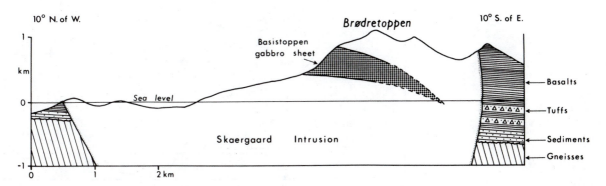

would originally have been inclined at about 20° to the vertical. The internal differentiation of the Skaergaard intrusion, which is described in Chapter 7, suggests that it was filled with magma in a single event.

Figure 87 shows a hypothetical downward projection of the Skaergaard intrusion from its presently exposed contacts. This reconstruction is consistent with gravity measurements over the intrusion, but it should be remembered that the form of subsurface bodies cannot be uniquely determined by gravity data alone, and other models may also fit the gravity data. The shape implied by Fig. 87 is a shallow cone or funnel 7–11 km across and 3 km deep. The term 'funnel-shaped' implies the presence of a feeder pipe at the base of the

Figure 87. A hypothetical model of the subsurface shape of the Skaergaard intrusion consistent with the unpublished gravity data of Blank and Gettings (after Taylor and Forester 1979). The contours are elevations of the contact in metres, relative to sea level.

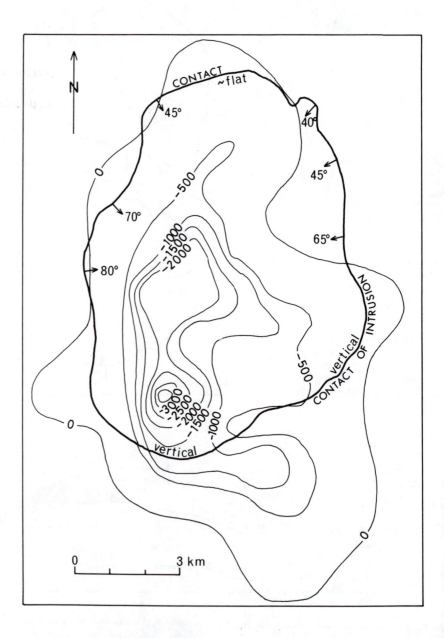

cone, but this cannot be seen. Obviously, formation requires an initial conical fracture very similar to that of cone sheets, but the country rock inside the cone was apparently removed completely, either blown out or floated up on the magma. There is no indication that the magma was particularly volatile-rich, so the latter process seems more probable. Norton *et al.* (1984) reviewed the evidence for the shape of the Skaergaard magma chamber, and suggested that the basalt–gneiss unconformity in the country rocks strongly influenced the course of intrusion, and Irvine (1992) postulated that extensional faulting associated with the opening of the north Atlantic may have played a part in opening up space for the intrusion.

Since the Skaergaard intrusion was first described, other intrusions have been described as funnel-shaped, although none is as well exposed. An intrusion which links cone sheets with funnels is the *Eureka* dolerite in Tasmania (Fig. 88). This has the form of a cone sheet but is much thicker than most. It is of particular interest in that uplift of the central cone of overlying country rocks can be demonstrated from the absence within the dolerite ring of the Permian deposit which dips gently towards the intrusion on its northern side.

Figure 88. Map and hypothetical section through the Eureka dolerite, Tasmania (after Spry 1958).

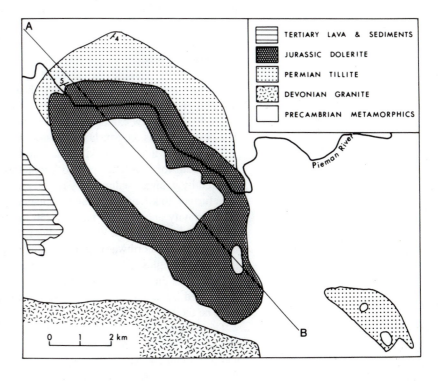

TERTIARY LAVA & SEDIMENTS

JURASSIC DOLERITE

PERMIAN TILLITE

DEVONIAN GRANITE

PRECAMBRIAN METAMORPHICS

Pieman River

0   1   2 km

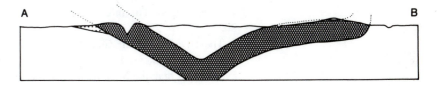

Apart from the Eureka dolerite, there is no satisfactory example of a funnel-shaped intrusion in which the remains of the uplifted cone of pre-existing country rocks can be found. In all probability, such a cone would be tilted, if not completely disrupted, and it is possible that funnel intrusions, like cone sheets, are formed sufficiently close to the ground surface that their uplifted cone would be raised into the volcanic superstructure and subsequently removed by erosion. Mineralogical studies (McBirney and Nakamura 1974) have shown that the total pressure during crystallization of the Skaergaard magma was about 500 bars at the Sandwich horizon (top of the Layered Series), corresponding to a cover of only 1800 metres, which is small in relation to the diameter of the intrusion (8000 metres) or to the exposed thickness of the Layered Series (2500 metres).

In certain respects, the *Birds River* complex (see Fig. 93 below) shows the sort of relationships that might be expected at the top of a funnel intrusion. There is a central plug of country rocks which have been elevated, and which have inward-dipping lower surfaces, and the outer contact of the gabbro itself is inwardly dipping (although only on its eastern side). In other ways, such as the steeply outward-dipping western contact, the Birds River gabbro does not conform to the ideal funnel shape, but then neither does the Skaergaard intrusion, which also has an outward dip on one side.

FUNNEL-DYKES

This name is given to a small number of intrusions with an elongated outcrop like that of a dyke but with a V-shaped cross-section that narrows downward. There are only a few examples of this class of intrusion, but all are very large. The best known is the Muskox intrusion in northern Canada.

The *Muskox* intrusion (Irvine and Smith 1967; Irvine 1975, 1980) extends north–south for a distance of 110 kilometres. At its northern end it is partly covered by younger sedimentary rocks, but the buried intrusion can be traced for a further 30 kilometres as an aeromagnetic anomaly. This is continued as a major gravity anomaly for another 250 kilometres. The southern part of the intrusion has the form of a vertical dyke 200–500 metres wide (Fig. 89). The outcrop widens northwards into an elongated funnel shape whose walls dip inwards at angles of 20–35°. The overlying sediments and the internal layering both dip northwards at about 5°, showing that the intrusion must have been tilted by about this amount, so the funnel-shaped part of the intrusion must be the highest part.

The southern dyke-like part of the intrusion is structurally the lowest part and so is considered to be the feeder intrusion. It consists mainly of gabbro and has a chilled margin of tholeiitic basalt composition. The funnel part of the intrusion consists predominantly of ultrabasic rocks with some gabbro, and there is a granophyric roof zone. Both rhythmic and cryptic layering are well developed. The inferred cross-section of the intrusion is shown in Fig. 90.

The *Great Dyke* of Zimbabwe (Fig. 91) and the Jimberlana intrusion in western Australia are two other examples of funnel dykes. The Great Dyke is even longer than the Muskox intrusion (>500 kilometres), and gravity surveys

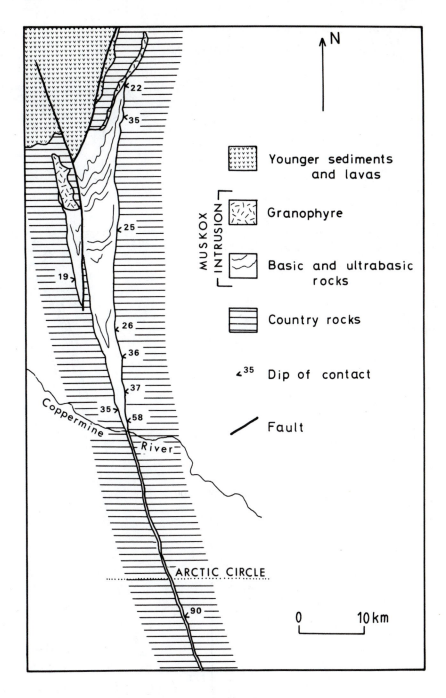

Figure 89. The Muskox intrusion, nothern Canada (after Smith and Kapp 1963; Irvine 1975).

suggest that it has a Y-shaped cross-section with a dyke-like feeder 1 km wide (Podmore and Wilson 1987). There are very few exposures of its contacts, but where they are seen they dip inwards at about 45°. It shows well-developed internal differentiation, which is described in Chapter 7.

Figure 90. Hypothetical cross-section through the Muskox intrusion (after Irvine and Smith 1967).

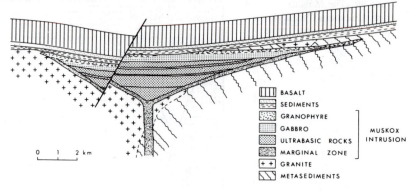

Figure 91. The Great Dyke of Zimbabwe (after Worst 1958).

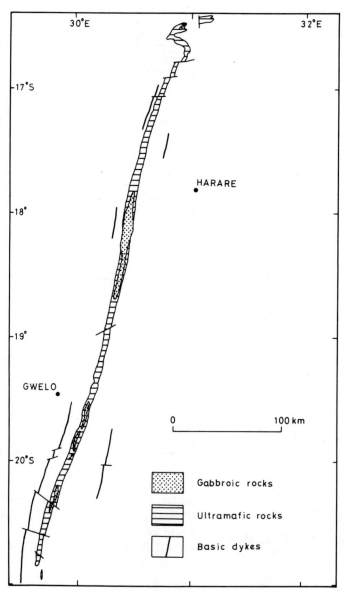

RING DYKES AND RING COMPLEXES

Basic ring dykes are much less common than acid ones. This is not surprising, since it must be relatively uncommon for basic magmas to have a lower density than the intruded country rocks which this form of emplacement implies. An excellent example of a basic ring dyke is the one described by Upton (1960) from the *Kungnat* complex in Greenland (Fig. 92). This is a gabbro, in contrast to the mafic syenites forming the bulk of the complex, and was intruded after the various syenites. This particular ring dyke has contacts which dip outwards at angles of between 20° and 90°, but along most of the southern and eastern part of the ring the outward dip is close to 50°. Although the outward dip means that space for the gabbro magma could have been made by subsidence alone, there are actually many large angular inclusions of gneiss in the western part of the dyke which suggest that it was at least partly emplaced by stoping.

The *Beinn Chaisgidle* centre in Mull (Fig. 54) contains several arcuate intrusions of gabbro which have the appearance of incomplete ring dykes. Not only are they associated with acid ring dykes, but several of them actually show an upward transition from gabbro into granophyre in steep mountain-side sections. This is perhaps an indication that gabbroic magma has followed acid magma in a mode of emplacement initiated by the latter.

Figure 92. The Kungnat quartz–syenite complex, Greenland (after Upton 1960).

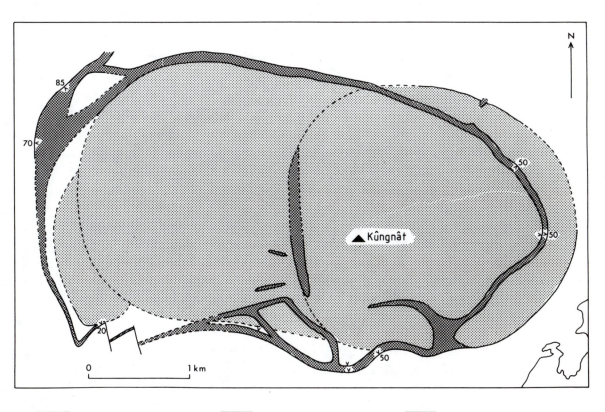

Gneiss   Syenites   Gabbro

It is difficult to know whether there are any gabbroic plutons analogous to the bell-jar intrusions which would result from wholesale subsidence on a ring fracture, such as has been described for many acid intrusions. The *Birds River* complex in South Africa (Fig. 93) was the original example cited by Du Toit when he introduced the term 'bell-jar'. This intrusion has certainly formed on a ring fracture, but the blocks of country rock preserved within the ring can be shown by their stratigraphy to have been raised by up to 730 metres from their previous positions. The outer contact of the main intrusion dips at a high angle outwards in the west, but at a low angle inwards in the east. The relationship between the gabbro and the internal masses of country rock is complex. They are well exposed in mountainous country and it can be seen that some of them are both underlain and overlain by gabbro. There are many other interesting features in this intrusion, such as the presence of ferrogabbros and granophyres underneath included masses of country rock, and evidence for localized partial melting of the latter, but as an example of a bell-jar pluton it is not entirely satisfactory. Upward displacement of country rocks within a ring fracture is also a feature of the basic–ultrabasic plutonic complex of Rhum, Scotland (Wager and Brown 1968), and suggests the possibility that a body of basic magma might be able to float its roof up within the confines of a cylindrical fault zone.

A major basic intrusive complex which has the appearance of being composed mainly of overlapping ring dykes is that of *Ardnamurchan* in Scotland (Fig. 82). The ring intrusions of Ardnamurchan, together with

Figure 93. The Birds River complex, South Africa (after Eales and Booth 1974).

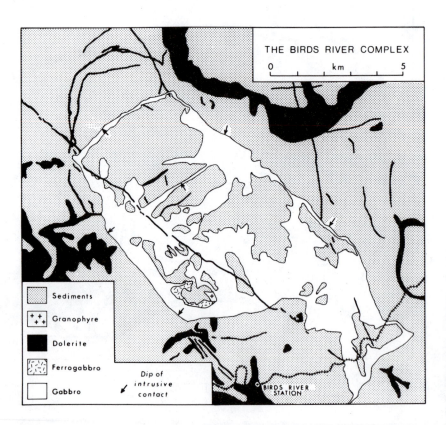

THE BIRDS RIVER COMPLEX

0   km   5

Sediments

Granophyre

Dolerite

Ferrogabbro

Gabbro

Dip of intrusive contact

BIRDS RIVER STATION

associated cone sheets, are arranged around two or three distinct centres, indicating a shift in the focus of intrusive activity. The difficulty of accounting for a ring subsidence might here be mitigated by the predominantly basic (and therefore high density) nature of the country rocks in which most of the Ardnamurchan ring dykes were emplaced. Some of the internal contacts of the Ardnamurchan complex are vertical or nearly vertical, and it is doubtful if the larger units have really been emplaced by the block-subsidence mechanism implied in Fig. 81. Several of the rings are xenolithic and could have been opened partly by stoping. The Great Eucrite of centre 3 is certainly too wide to be explained by a single ring subsidence, and Wager and Brown (1968) suggested a 'lopolithic' or funnel shape as being more likely.

The *Cuillin* complex of Skye (Fig. 94) contains concentric rings of gabbro and was intruded into a pile of basaltic lavas. The outermost ring overlies lava along an inward-dipping contact, so that a funnel shape is most likely for this intrusion.

Figure 94. Simplified geological map of the Cuillin complex, Skye (after Gass and Thorpe 1976).

LOPOLITHS

The very largest basic intrusions have often been described as 'lopoliths'.

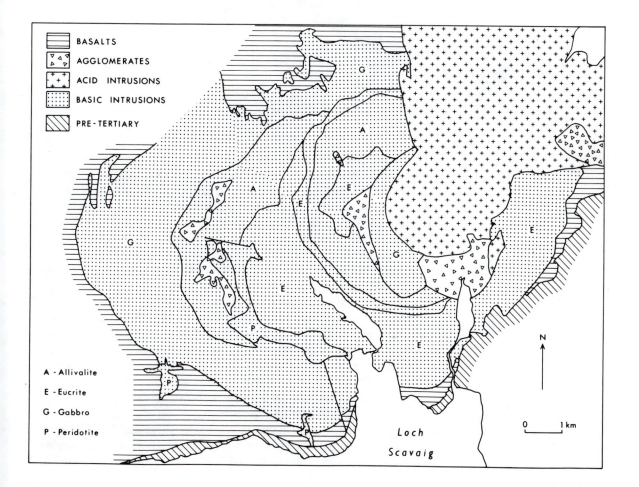

BASALTS

AGGLOMERATES

ACID INTRUSIONS

BASIC INTRUSIONS

PRE-TERTIARY

A - Allivalite

E - Eucrite

G - Gabbro

P - Peridotite

N

Loch

Scavaig

0   1 km

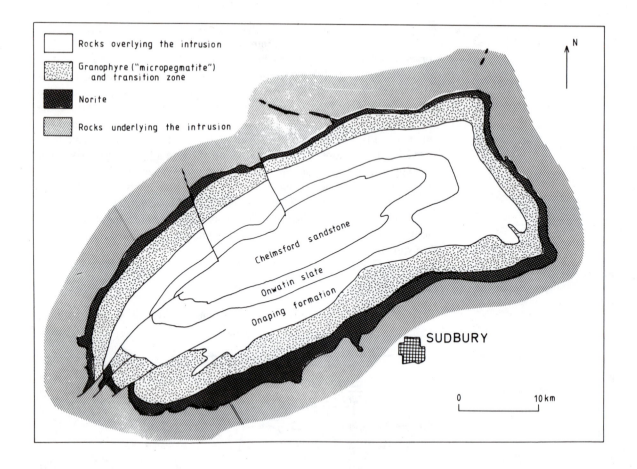

Figure 95. Geological sketch map of the Sudbury basin, Ontario.

Lopoliths are postulated as having the form of a saucer, being thick, very extensive, and roughly concordant with the country rocks. The classic examples are the Sudbury intrusion in Ontario (Fig. 95) and the Bushveld complex in South Africa (Fig. 96). The Bushveld complex is nearly 500 kilometres across. Apart from their size, these gabbros (or norites) are distinguished from smaller basic intrusions by being overlain in each case by a thick layer of more acid material – granite or granophyre.

The lopoliths are too large for their complete form to be determined by mapping or from boreholes, and the postulated shape may not be correct. The shape has mainly been suggested by the basinal form of the sedimentary country rocks, and by the broad disposition of internal variation in the intrusions. The postulated shape differs only in scale from the form of some well-established flat-lying sheet intrusions and especially the downward curving parts of some of the undulating sheet intrusions (e.g. those in Figs 74, 75 and 79), and Corry (1988) regarded them as no different from laccoliths except in having been intruded deeper in the crust.

Wilson (1956) doubted whether the bottom contacts of a supposed lopolith are in fact parallel to the internal layering, and suggested that they were funnel-shaped with external contacts dipping steeply inwards. This may be the case for the Sudbury intrusion (Naldrett *et al.* 1970), and according to

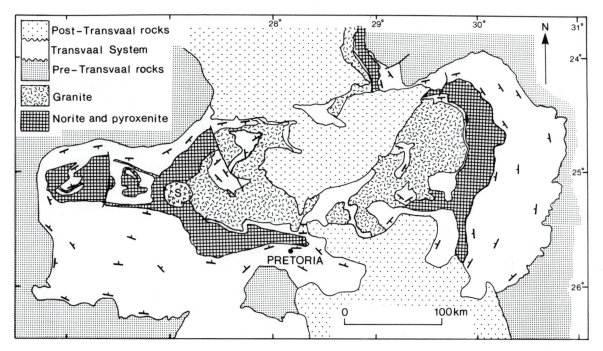

Figure 96. Geological sketch map of the Bushveld complex, South Africa (after Hall 1932). Conventional symbols indicate dip and strike in the Transvaal Beds. The letter 'S' denotes the Pilanesberg syenite.

Figure 97. The cross-cutting relationships of the floor of the Bushveld complex (after Button 1976). The dotted lines represent isotherms of contact metamorphism.

Wager and Brown (1968) the lower contacts of the Bushveld intrusion are also more steeply dipping than the internal layering, although the lower surface of the Bushveld complex is very irregular in detail (Fig. 97). However, a conical form presents even greater difficulties than a lopolithic one in that it could not be accommodated within the average thickness of the crust (*c.* 30 kilometres), and that uplift of the displaced cone would require a surface upheaval of spectacular proportions. These difficulties could perhaps be overcome if the intrusions were composite. There is some evidence that the Bushveld magma was injected in several pulses, but the regular

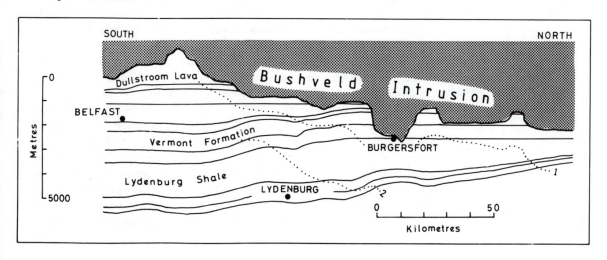

distribution of granophyre and norite in the Sudbury intrusion is more suggestive of a single magma body.

A novel suggestion for the formation of the Sudbury structure was proposed by Dietz (1964), who interpreted it as an astrobleme, i.e. the impact site of a large meteorite or asteroid. He suggested that the magma produced by impact heating filled an open pool at the Earth's surface. He interpreted the Onaping tuffs which overlie the norite and granophyre members of the intrusion as an additional product of this magma. A similar origin has been proposed for the Bushveld complex, and simultaneous melting of mantle and crustal rocks at the impact sites has been suggested as the reason for the unusually high degree of contamination shown by the basic rocks of these two intrusions.

DYKES AND DYKE SWARMS

Dykes and sills are often thought of as being similar types of intrusion, but there is a fundamental difference between them. The magma which enters a sill is in hydrostatic equilibrium with its country rocks and in most cases is coming to rest in place, but the magma flowing through a dyke is en route to a higher destination. Much lava may flow through a dyke fissure before the dyke consolidates. Sometimes a dyke fissure may be reopened, giving a multiple dyke, or a composite dyke if the new magma is different from the old, but these are much rarer than simple dykes. Dykes range in thickness from less than 10 cm to more than 100 metres, but thicknesses in the order of a metre or two are most common.

Some dykes are known which close upwards, but many undoubtedly reached the Earth's surface. Fissure eruptions, which must be the surface expressions of dykes, can be seen on Hawaii and in Iceland. In the Laki fissure eruption of 1783, lava poured from a fissure more than 20 kilometres in length (Fig. 39), and about 800 years earlier the Eldgja fissure, also in Iceland, erupted continuously along a length of 300 kilometres. In some volcanic areas dykes may be marked by rows of spatter cones, as on Etna. In Hawaii, the eruptive fissures are concentrated in zones, known as rift zones, which radiate from the volcanic centres. Erosion of some of the older Hawaiian volcanoes has revealed large numbers of thin dykes underlying the rift zones. The Klyuchevskaya volcano in Kamchatka has about 80 small cones on its flanks arranged along 12 lines radiating from the summit, in addition to arcuate lines which possibly represent cone sheets.

Dykes may be found in all regions in which igneous activity has taken place, but are not always abundant. They reach their highest concentration in dyke swarms, which may be radial or parallel. Radial dykes invariably converge on a volcanic centre or an igneous intrusion (Fig. 98). Each fissure was used by the magma only once, except in the rare multiple dykes, and reached no higher than the volcanic surface at the time of dyke intrusion. The fissures visible on the surface of an active volcano represent only the most recently opened dykes and are much less numerous than the dykes of plutonic radial swarms.

Figure 98. The syenodiorite porphyry intrusion of Dike Mountain, Colorado, and its associated dyke swarm (after Johnson 1961).

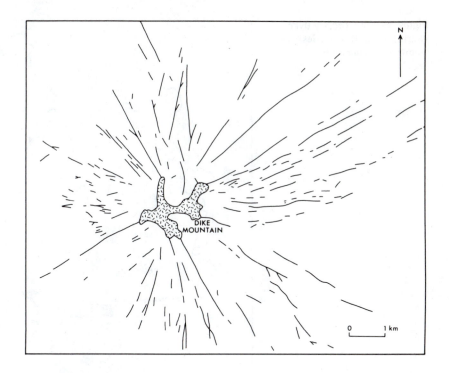

Linear dyke swarms extend for much greater distances, but are also concentrated around large intrusions or volcanic centres (Figs 99 and 100). The dyke swarms of the British Tertiary igneous province have a regional NW–SE alignment, but become most numerous in the vicinity of the former volcanoes, especially those of Skye, Mull and Arran. Some dykes stretch great distances from the volcanic centres. The Cleveland dyke, which is one of the Mull swarm, extends for 400 kilometres to the south-east.

Unlike the radial swarms, whose distribution results from cracking associated with the upward injection of the magma, the linear swarms are so extensive in relation to the thickness of the Earth's crust that they must be indicative of crustal tension over a large region. In Mull, Bailey *et al*. (1924) counted 375 dykes with an average thickness of 1.8 metres in a distance of nearly 20 kilometres, indicating crustal stretching of 1 in 26 at right angles to the elongation of the swarm. The dilation reaches nearly 25% in the vicinity of the Skye intrusive centre (Fig. 99), but averages only 3% in the west of Scotland as a whole (Speight *et al*. 1982).

The most intense linear dyke swarms are those of the ophiolite complexes, such as Troodos (see Chapter 13), which are thought to be fragments of former oceanic crust. The parallel dykes in the Troodos complex are so closely packed that their host rocks occupy less than 10% of the ground (Fig. 101). Very intense regional dyke swarms also occur in Iceland, and parallel the trend of the Mid-Atlantic Ridge. Observations such as these have led to the suggestion that oceanic crustal formation and sea-floor spreading take place by a mechanism of repeated dyke injection. A very interesting dyke swarm follows the coast of east Greenland parallel to the continental margin, and intensifies rapidly towards the coast (Fig. 102).

Figure 99. The Tertiary dyke swarms of the British Isles (shown diagramatically).

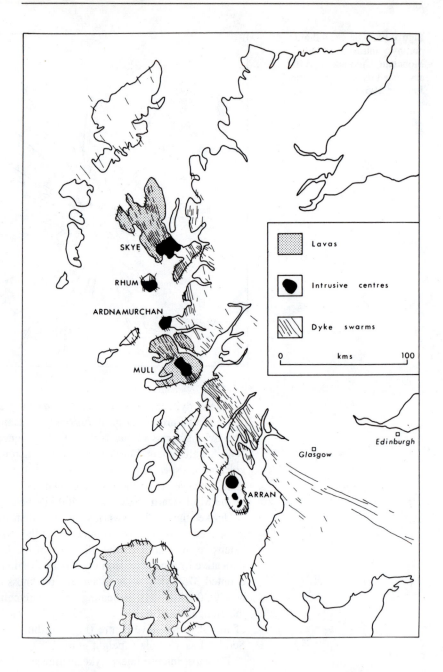

Some authors restrict the term 'dyke' to vertical sheet-like intrusions, but others apply it to include inclined sheets which may dip at quite low angles to the horizontal. Inclined sheets of granite or microgranite are associated with many large granite intrusions and take a variety of forms. A small proportion of dykes are not dilational in origin. These dykes do not have parallel walls (Fig. 103), and their form can be explained by either stoping or gas-drilling.

Figure 100. The Devonian dyke swarms of the west of Scotland (from Groome and Hall 1974).

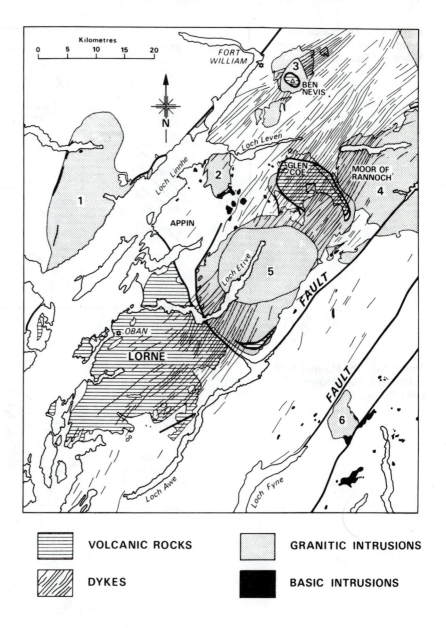

	VOLCANIC ROCKS		GRANITIC INTRUSIONS
	DYKES		BASIC INTRUSIONS

Figure 101. A traverse through the mafic dyke swarm of the Troodos complex, Cyprus. Later dykes are intruded into earlier dykes so that the chilled edges of the latter become separated and are difficult to match up (after Moores and Vine 1971).

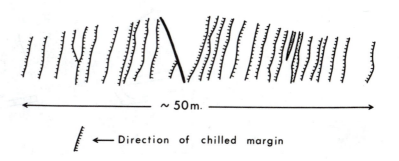

~ 50 m.

⫢⟵ Direction of chilled margin

Figure 102. The coastal dyke swarm of east Greenland (after Wager and Deer 1938).

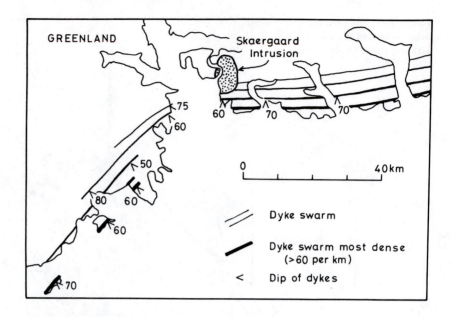

Figure 103. The Medford dyke at Pine Hill, Massachusetts (after Billings 1925).

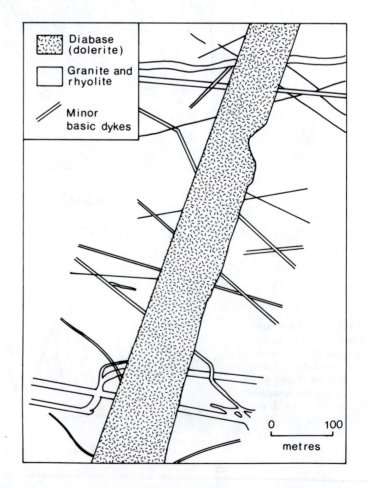

DIATREMES

Diatremes are vertical pipe-like bodies filled with breccia, or containing both igneous rocks and breccia. The breccia fragments may include wall rocks and igneous material. Diatremes appear to have been drilled by gas under pressure. In most gases the gas has been exsolved from the magma, but in some shallow examples the gas may be heated groundwater. Diatremes are a characteristic intrusive form of the hyperalkaline magmas such as kimberlite and melilitite, as well as carbonatites. These magmas are thought to be rich in CO_2 and H_2O, which are released by falling pressure as the magma approaches the surface. Several examples have been described of dyke-like intrusions passing upwards or laterally into diatremes, the change corresponding to the appearance of an exsolved gas phase (see Fig. 319 in Chapter 12). The maars (crater lakes) of the Eifel district in Germany are the probable surface expressions of diatremes. They fill small craters which are surrounded by narrow pyroclastic cones of alkali basalt ejecta and xenoliths, presumably expelled from the diatremes at high speed.

Other volatile-rich magmas are sometimes emplaced as diatremes, including diorites and granites. Breccias are particularly associated with granitic magmas that have exsolved water on their approach to the surface, and many examples are to be found in the copper–porphyry intrusions of the south-western United States.

CHAPTER 4

The chemical composition of igneous rocks

MAJOR ELEMENTS

The average compositions of the major types of igneous rock are given in Table 13. Averages are given for volcanic rather than plutonic rocks because volcanic rocks are less often the products of crystal accumulation or contamination, and are therefore closer to actual magma compositions. Even so, porphyritic lavas may still have undergone some crustal accumulation and only the wholly glassy rocks can be closely equated with liquid magmas.

Figure 104. Alkali–silica diagram showing the composition of the major igneous rock types.

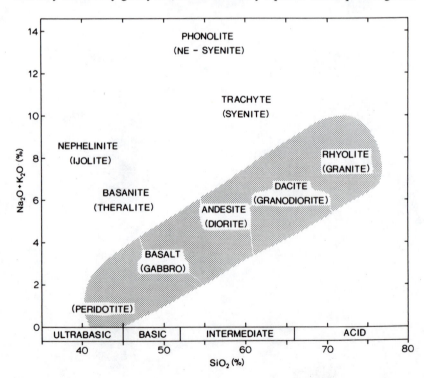

Table 13. Average compositions and CIPW norms of extrusive igneous rocks (after Le Maitre 1976)

	Rhyolite	Trachyte	Phonolite	Andesite	Basalt	Nephelinite
SiO_2	72.82	61.21	56.19	57.94	49.20	40.60
TiO_2	0.28	0.70	0.62	0.87	1.84	2.66
Al_2O_3	13.27	16.96	19.04	17.02	15.74	14.33
Fe_2O_3	1.48	2.99	2.79	3.27	3.79	5.48
FeO	1.11	2.29	2.03	4.04	7.13	6.17
MnO	0.06	0.15	0.17	0.14	0.20	0.26
MgO	0.39	0.93	1.07	3.33	6.73	6.39
CaO	1.14	2.34	2.72	6.79	9.47	11.89
Na_2O	3.55	5.47	7.79	3.48	2.91	4.79
K_2O	4.30	4.98	5.24	1.62	1.10	3.46
H_2O+	1.10	1.15	1.57	0.83	0.95	1.65
H_2O-	0.31	0.47	0.37	0.34	0.43	0.54
P_2O_5	0.07	0.21	0.18	0.21	0.35	1.07
CO_2	0.08	0.09	0.08	0.05	0.11	0.60
Q	32.87	5.00	—	12.37	—	—
C	1.02	—	—	—	—	—
Or	25.44	29.41	30.96	9.60	6.53	3.16
Ab	30.07	46.26	35.48	29.44	24.66	—
An	4.76	7.05	1.50	26.02	26.62	7.39
Lc	—	—	—	—	—	13.57
Ne	—	—	16.50	—	—	21.95
Di	—	2.14	6.89	4.84	14.02	32.36
Wo	—	—	0.73	—	—	—
Hy	1.34	2.06	—	9.49	15.20	—
Ol	—	—	—	—	1.50	2.32
Mt	2.14	4.33	4.05	4.74	5.49	7.95
Il	0.54	1.34	1.18	1.65	3.49	5.05
Ap	0.17	0.49	0.41	0.50	0.82	2.51
Cc	0.17	0.20	0.17	0.11	0.26	1.37

All igneous rocks differ from magmas in having lost some or all of their volatile constituents on cooling.

Figure 104 shows how the major types of igneous rock differ in their SiO_2 and alkali contents. The vast majority of igneous rocks lie in the 'main spectrum' of compositions represented by the shaded area in Fig. 104. The alkaline rocks, such as phonolites and nephelinites and their plutonic equivalents, are very diverse in composition and mineralogy and have been the subject of much petrographic attention, but they are of comparatively rare occurrence.

MAGMA VARIATION

Variation diagrams are used to represent the compositional variation in igneous rock suites. The most commonly used is the Harker diagram, in which the individual constituents are plotted against SiO_2. A typical example is shown in Fig. 105. From the diagram one can see the range of compositions, the trend of variation, and the degree of scatter of individual constituents about the trend. In this example, the rocks with about 50% SiO_2

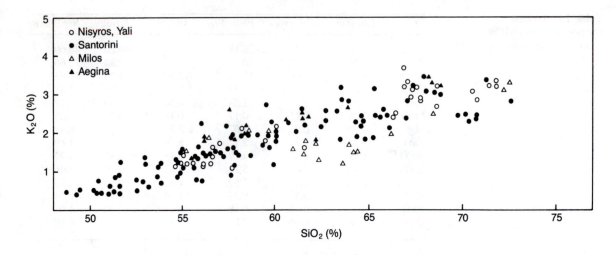

Figure 105. A plot of K₂O against SiO₂ for the lavas of the Aegean volcanic arc, Greece (after Keller 1982).

are basalts, those with about 70% SiO$_2$ are rhyolites, and the intermediate compositions are andesites and dacites.

One of the main aims of igneous petrogenesis is to decide what is the relationship between such diverse magma compositions as those represented in Fig. 105. Are they primary magmas, with different compositions generated individually at the source of the magma? Can one magma be regarded as parental, with the others being secondary or derivative, i.e. resulting from such processes as fractional crystallization or assimilation of country rocks? Could they be mixtures of two or more primary magmas? All these possibilities are discussed in subsequent chapters, particularly Chapter 7.

During the last 40 years, experimental petrologists have shown that a large range of magma compositions can be primary. Liquids with the composition of rhyolite, dacite, andesite, alkali and tholeiitic basalt, peridotite and carbonatite have all been produced by the partial melting of common rock types under appropriate conditions. Nevertheless, secondary processes are still thought to contribute to the diversity of igneous rock suites. Various petrographic and geochemical techniques have been applied to discovering the causes of magma diversity, and isotope geochemistry has proved to be particularly valuable.

NORMATIVE COMPOSITIONS

The interpretation of a chemical analyses, such as those in Table 13, is often assisted by converting them into normative compositions. In the normative system of calculation, the analysis is recalculated into an equivalent assemblage of hypothetical minerals. These are weight percentages of idealized end members, as listed in the Appendix. They do not necessarily correspond to the actual mineralogy of the rock, although there are obvious similarities.

One application of norms is to enable rock compositions to be compared easily with the phase assemblages found in phase equilibrium studies.

Another is that they quickly reveal the degree of silica- or alumina-saturation of a rock, enabling distinctions to be made between alkali and tholeiitic basalts or between peralkaline and peraluminous rhyolites.

SILICA SATURATION

The mineralogy of a rock depends on its chemical composition. Rocks with a very high silica content contain free silica, i.e. quartz. Those with a lower silica content may not contain enough SiO_2 to combine with the cations into silicate minerals and leave any surplus of free silica. If there is a large deficiency of SiO_2 there will be a further limitation on the choice of silicate minerals that can be present.

Silica can combine with MgO to form either Mg_2SiO_4 (olivine) or $Mg_2Si_2O_6$ (orthopyroxene), and if there is insufficient silica in the rock the MgO may occur as olivine because orthopyroxene requires a greater proportion of SiO_2 to MgO. One would not expect to find magnesian olivine in the same rock as quartz, because under equilibrium conditions the quartz would react with the olivine until either the latter was completely converted

Table 14. Chemical analyses and CIPW norms of basalts showing different degrees of silica saturation

	1	2	3
SiO_2	46.59	49.16	51.02
TiO_2	2.26	2.29	2.03
Al_2O_3	15.19	13.33	13.49
Fe_2O_3	2.96	1.31	3.22
FeO	9.89	9.71	8.12
MnO	0.18	0.16	0.17
MgO	8.74	10.41	8.42
CaO	10.02	10.93	10.30
Na_2O	3.01	2.15	2.10
K_2O	0.96	0.51	0.40
H_2O+	0.05	0.04	0.21
H_2O-	0.00	0.05	0.28
P_2O_5	0.29	0.16	0.26
Q	—	—	4.26
Or	5.56	2.78	2.22
Ab	20.96	17.82	17.82
An	25.30	25.30	26.13
Ne	2.27	—	—
Di	18.51	22.93	18.60
Hy	—	15.35	21.27
Ol	18.21	9.14	—
Mt	4.41	2.09	4.64
Il	4.26	4.41	3.80
Ap	0.67	0.34	0.67

1. Alkali basalt, 1801 eruption of Hualalai, Hawaii (Yoder and Tilley 1962) – undersaturated.
2. Olivine–tholeiite, 1921 eruption of Kilauea, Hawaii (Muir *et al.* 1957) – saturated.
3. Tholeiite, 1859 eruption of Mauna Loa, Hawaii (Macdonald and Katsura 1964) – oversaturated.

to orthopyroxene or the former was used up. Similarly, nepheline is incompatible with quartz, since they can react to form albite:

$$NaAlSiO_4 + 2SiO_2 \rightarrow NaAlSi_3O_8$$

Other minerals which cannot coexist with silica are leucite, sodalite, melilite and perovskite.

Rocks which contain an excess of SiO_2 are described as silica-over-saturated. They contain normative quartz, which corresponds either to actual quartz or to a silica-rich glass. Rocks which have so little SiO_2 that nepheline or leucite is present are described as silica-undersaturated. Rocks containing enough SiO_2 to eliminate nepheline, but not enough to eliminate magnesian olivine or generate quartz, are described as silica-saturated.

Table 14 gives examples of basaltic rocks showing different degrees of silica saturation. It is possible to estimate the degree of silica saturation of a rock from its norm even if it is too fine-grained, glassy or even met-

Figure 106. The classification of basalts in terms of their normative constituents (after Yoder and Tilley 1962).

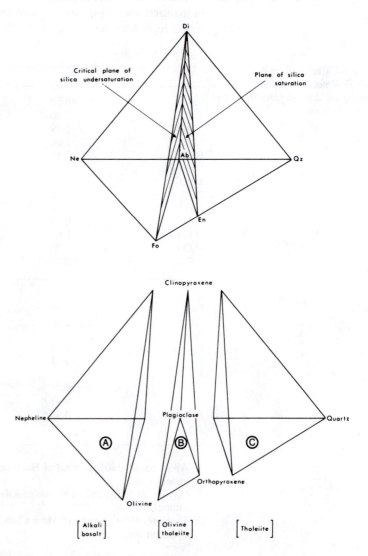

Figure 107. Alkali–silica diagram for Hawaiian basalts, showing the boundary between the tholeiitic and alkalic fields (after Macdonald and Katsura 1964). Dots = tholeiites, circles = alkali basalts.

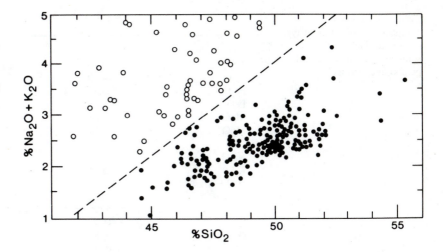

amorphosed to identify its original mineralogy. Figure 106 shows how the basalts can be classified according to their normative mineralogy, which in fresh, completely crystalline rocks would not be much different from their actual mineralogy.

In Yoder and Tilley's classification (Fig. 106), the basalts of regions A, B and C are distinguished as the alkali basalt group, the olivine–tholeiite group, and the tholeiite group respectively. This is the most logical classification, but there are many geologists who distinguish only between alkali basalts and tholeiitic basalts, separating them on the basis of alkali and silica contents alone, regardless of their petrography and mineralogy. Figure 107 shows the chemical distinction between the two categories, as defined by Macdonald and Katsura (1964), whose boundary has been arbitrarily adopted by many later workers.

ALUMINA SATURATION

The most important group of minerals to contain alumina and alkalis are the feldspars and feldspathoids, containing equal cation proportions of Na and Al, or K and Al:

Orthoclase	$= KAlSi_3O_8$
Albite	$= NaAlSi_3O_8$
Anorthite	$= CaAl_2Si_2O_8$
Nepheline	$= NaAlSiO_4$
Leucite	$= KAlSi_2O_6$

Any imbalance in the proportions of Al and Na or K in an igneous rock will lead to the formation of either additional aluminous minerals or additional alkali-bearing minerals.

Peralkaline rocks are those with the cation proportions Al < (Na + K). Such rocks will have alkali ferromagnesian minerals such as aegirine or riebeckite, and will contain acmite (Ac) in the norm. Peralkaline rocks are

Table 15. Average compositions of different types of granite, including those showing peraluminous and peralkaline character

	1	2	3
SiO_2	73.84	70.56	73.05
TiO_2	0.16	0.40	0.24
Al_2O_3	14.29	14.00	10.62
Fe_2O_3	0.34	0.91	3.04
FeO	0.75	2.41	2.98
MnO	0.05	0.06	0.21
MgO	0.21	0.48	0.10
CaO	0.69	1.63	0.60
Na_2O	3.61	3.56	4.23
K_2O	5.21	5.39	4.48
H_2O+	0.60	0.50	0.37
P_2O_5	0.25	0.10	0.08
Q	31.7	24.5	29.9
Or	30.6	31.7	26.7
Ab	30.4	29.9	29.3
An	1.7	6.4	—
C	2.1	—	—
$CaSiO_3$	—	0.3	1.0
$MgSiO_3$	0.5	1.2	0.3
$FeSiO_3$	0.9	3.0	4.6
Ac	—	—	5.5
Mt	0.5	1.4	1.6
Il	0.3	0.8	0.5
Ap	0.6	0.3	0.2

1. Muscovite granite (peraluminous).
2. Biotite–hornblende granite (metaluminous).
3. Riebeckite granite (peralkaline).
The averages are from Nockolds (1954).

not necessarily alkaline in terms of their total alkali and silica contents, as may be seen from the riebeckite granite in Table 15, which is not very different from the other granites except in its low aluminium content.

Peraluminous rocks are those with the cation proportions: Al > (Na + K + 2Ca). Such rocks may contain muscovite, corundum, or andalusite, and will contain corundum (C) in the norm. Muscovite–granite is a typical peraluminous rock, but many biotite–granites are also slightly peraluminous. The degree of alumina saturation of granites is often expressed as the aluminium saturation index, A.S.I., which equals the cation ratio Al/(Na + K + 2Ca) and the molecular ratio Al_2O_3/(Na_2O + K_2O + CaO).

TRACE ELEMENTS

The average abundance of the elements in igneous rocks is shown in Table 16. Trace elements occur in igneous rocks in the following ways:

1. Most substitute isomorphously for major elements in the crystal structures

Table 16. Abundance of the elements in basalts and granites (in parts per million)

Element	Basalt	Granite	Element	Basalt	Granite
1 H	1060	720	37 Rb	21	215
3 Li	14	52	38 Sr	285	256
4 Be	1	5	39 Y	25	48
5 B	11	12	40 Zr	109	246
6 C	300	140	41 Nb	12	27
7 N	3	20	42 Mo	5	3
8 O	447,000	491,000	50 Sn	2	6
9 F	273	1366	51 Sb	0.7	0.5
11 Na	21,600	27,300	55 Cs	1	7
12 Mg	40,600	4,300	56 Ba	268	698
13 Al	83,300	75,800	57 La	16	61
14 Si	230,000	333,000	58 Ce	32	120
15 P	1530	520	59 Pr	4	15
16 S	96	118	60 Nd	15	52
17 Cl	135	193	62 Sm	4	10
19 K	9,100	33,800	63 Eu	1	1
20 Ca	67,700	13,200	64 Gd	4	8
21 Sc	35	6	65 Tb	0.7	1
22 Ti	11,030	1860	66 Dy	4	9
23 V	284	34	67 Ho	1	2
24 Cr	208	24	68 Er	2	5
25 Mn	1550	390	69 Tm	0.4	1
26 Fe	81,900	21,200	70 Yb	2	5
27 Co	43	8	71 Lu	0.3	1
28 Ni	101	11	72 Hf	3	8
29 Cu	102	31	73 Ta	0.8	3
30 Zn	90	71	74 W	3	2
31 Ga	19	22	81 Tl	0.2	1
32 Ge	1	1	82 Pb	7	39
33 As	3	1	90 Th	4	37
35 Br	2	0.2	92 U	0.9	7

For trace elements, the data are the averages of international geostandard rocks of appropriate composition as compiled by Govindaraju (1989); the data for N are from Hall *et al.* (1991, 1994); all other elements not listed in the table have abundances below 1 ppm.

of rock-forming minerals. For example, most of the nickel in igneous rocks is present in ferromagnesian minerals such as olivine, where it occupies magnesium positions in the olivine structure. Table 17 lists some of the commonly observed substitutions of trace elements for major elements in the rock-forming minerals.

2. Some are concentrated in accessory minerals (e.g. zirconium mostly as zircon, boron as tourmaline, and copper as chalcopyrite or bornite).

3. Some of the rarer trace elements do not substitute for any of the major elements but are insufficiently abundant to form accessory minerals of their own. As magma crystallizes these elements may be taken into the rock-forming minerals in very small amounts without occupying regular lattice positions.

Table 17. Observed substitutions of trace elements for major elements in the rock-forming minerals

	Major elements	*Co-ordination*	*Trace elements*
Feldspars	Ca, Na, K	6–9	Ba, Eu, Pb, Rb, Sr
	Al, Si	4	Ge
Olivine	Mg, Fe	6	Co, Cr, Mn, Ni
	Si	4	Ge
Clinopyroxenes	Ca, Na	8	Ce, La, Mn
	Mg, Fe	6	Co, Cr, Ni, Sc, V
	Si	4	Ge
Micas	K	12	Ba, Cs, Rb
	Al, Mg, Fe	6	Co, Cr, Li, Mn, Sc, V, Zn
	Si, Al	4	Ge
Apatite	Ca	7–9	Ce, La, Mn, Sr, Th, U, Y
	P	4	As, S, V
Zircon	Zr	8	Ce, Hf, La, Lu, Th, Y, Yb
	Si	4	P

GEOCHEMICAL AFFINITY

Goldschmidt classified the elements into those which usually occur in silicate or oxide minerals, those which occur as sulphides, and those which occur as native metals. He described these three categories as lithophile, chalcophile, and siderophile respectively. The partitioning of the different elements between coexisting silicate, sulphide and metal phases can be studied in artificial smelting products (silicate slag, sulphide matte, molten metal), or in meteorites (nickel–iron metal, silicate minerals, sulphide accessory minerals), and Goldschmidt used analyses of these materials to categorize the various elements. Examples are:

Siderophile	–	Au, Ir, Pt
Chalcophile	–	Cu, Pb, Zn
Lithophile	–	Li, Na, Sr

The significance of this classification can be seen by reference to Fig. 108. For any element one can specify the levels of f_{O_2} and f_{S_2} needed to convert the native element into a sulphide or oxide respectively. The lithophile elements are those which occur as oxide (or silicate) phases even at low oxygen fugacities; the chalcophile elements are those which occur as sulphides even at low sulphur fugacities; the siderophile elements are those which remain in the form of native metal under all but the very highest oxygen or sulphur fugacities. The critical values of f_{O_2} and f_{S_2} for conversion of metal to oxide or sulphide depend on temperature, and can be calculated quantitatively for each element if the free energies of formation of the oxides and sulphides are known (Ernst 1976; Robie *et al.* 1978). It can be seen that under the oxygen and sulphur fugacities corresponding to field **A** in Fig. 108, copper would occur as a sulphide, tin as an oxide, and lead as a metal. Thus, under these

Figure 108. The stability fields of metal, oxide, and sulphide compounds of Cu, Pb and Sn in terms of oxygen and sulphur fugacities at 1000 °K (after Holland 1959). Dashed lines for Cu and Pb are projections of the oxide–sulphide boundaries across the stability fields of CuSO₄ and PbSO₄ (not shown).

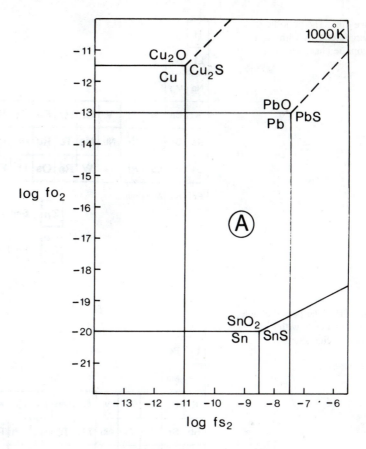

conditions, Cu, Sn and Pb are relatively chalcophile, lithophile and sidero-phile respectively.

At 1000 °C, the order of affinities of some of the more familiar trace elements are as follows (data from Arculus and Delano 1981):

Affinity for sulphur	Affinity for oxygen
Au (least)	Au (least)
Ge	Cu
Pb	Pb
Sn	Ni
Ni	Sn
Fe	Ge
Cu	Fe
Ga (most)	Ga (most)

The terms lithophile, chalcophile and siderophile are very useful for describing the general distribution of individual elements. To say that a particular element is chalcophile, for instance, is to give an immediate indication of which sort of rocks it is likely to be abundant in, and in which minerals it will be concentrated. The rarity of the siderophile elements in igneous rocks is generally attributed to their segregation into the Earth's

Figure 109. The periodic table, showing elements with chalcophile character.

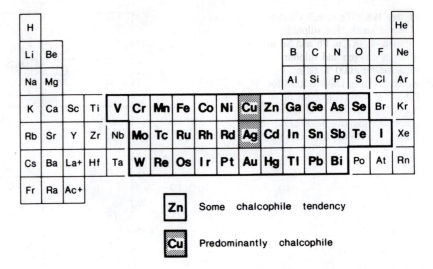

Figure 110. The periodic table, showing elements with siderophile character.

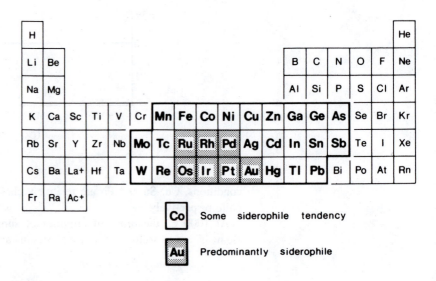

metallic core. As can be seen from Figs 109 and 110, the chalcophile and siderophile elements are those which lie in the middle part of the periodic table.

The chalcophile elements

The oxygen fugacity of most magmas is sufficiently low, and their sulphur content often sufficiently high, for sulphide minerals to crystallize. These sulphides will be the main carriers of the strongly chalcophile elements. Copper, in particular, does not enter silicate minerals to any significant extent, and most of the copper in igneous rocks is held by accessory sulphides.

Sulphide contents are higher in basic than in acid igneous rocks, and magmas of basic composition are relatively rich in several of the chalcophile elements, including nickel, copper and zinc. In the Skaergaard intrusion, which has been studied in detail, the accessory sulphides include pyrrhotite (FeS), marcasite (FeS_2), chalcopyrite ($CuFeS_2$), bornite (Cu_5FeS_4), and digenite (Cu_9S_5). Copper is obviously concentrated in the sulphides in this intrusion, as are silver and to some extent cobalt, but nickel and zinc are mainly accommodated by the silicates (Wager and Brown 1968).

The siderophile elements

All the siderophile elements have some chalcophile tendency as well, and in igneous rocks they are concentrated in accessory sulphide minerals and sulphide segregations. For instance, the platinum group of elements (Ru, Rh, Pd, Os, Ir, Pt) occurring as native metals, sulphides and arsenides in the Bushveld intrusion are concentrated in sulphide segregations, which probably separated as immiscible liquids from the gabbroic magma. The strongly siderophile elements show no ability to enter the crystal structures of silicates at all.

In the development of magmas, the separation of an immiscible sulphide phase, if it occurs, is the most important factor in determining the distribution of the strongly chalcophile and siderophile elements.

The lithophile elements

The majority of the lithophile trace elements in igneous rocks substitute isomorphously for major elements in the rock-forming silicate minerals. It is this group of elements which is of most value in the study of igneous petrogenesis. The substitutions of lithophile trace elements for major elements can provide evidence to show which crystals have participated in crystal fractionation, or which crystals have been left behind in melting residues.

Hydrophile elements

There is an important group of elements whose normal distribution in igneous rocks is very much modified by the presence of water. When a silicate magma coexists with an aqueous fluid, these 'hydrophile' elements are strongly partitioned into the aqueous phase. Among the elements which show pronounced hydrophilic behaviour are Au, Be, Cl, Li, Mo, Nb and Sn.

In some cases the hydrophile elements appear to have an affinity for water itself, for example Li (and other alkali elements) whose ions have a strong tendency to be coordinated by water molecules. In other cases the elements enter the aqueous phase in the form of complex ions (for example, $AuCl_4^-$ or $SnCl_6^{2-}$) because the aqueous phase contains elements such as Cl or S with which they can form complexes. This affinity for water is almost certainly responsible for the concentration of many rare elements and their minerals in pegmatites and hydrothermally altered rocks. Note that the elements described here as 'hydrophile' would be classified in the absence of water as lithophile (for example, Li), chalcophile (for example, Mo) or siderophile (for example, Au).

DISTRIBUTION OF TRACE ELEMENTS BETWEEN CRYSTALS AND LIQUIDS

In situations where crystals coexist with melt, for example in partial melting or crystal fractionation, we need to know which crystals a particular trace element will enter, and the relative concentrations of the trace element in the crystals and the liquid. In most cases, a trace element enters a crystal by substituting for a major element, i.e. by occupying a position in the atomic structure that would normally be occupied by the major element.

Substitution of trace elements for major elements

Isomorphous substitution of a trace element for a major element can occur if the following conditions are met:

(1) Their atoms must be approximately the same size. The substituting trace element atom has to fill a space in the crystal structure normally occupied by the element which it replaces. If it is too large it cannot do this, and if it is too small the structure will be distorted. Nickel easily enters the structure of ferromagnesian minerals because its ionic radius in silicates is very similar to that of magnesium and iron (Ni 0.77 Å; Mg 0.80 Å; Fe^{2+} 0.86 Å). A small discrepancy in size between the substituting atoms can be tolerated, especially at high temperatures. See from the alkali feldspar phase diagram with its asymmetrical solvus (Fig. 146 in Chapter 5) how potassium feldspar accepts the small Na^+ ion more readily than sodium feldspar accepts the large K^+ ion, and how the extent of possible substitution increases with temperature.

(2) The atoms must be of similar ionic charge. For example, Rb^+ replaces K^+, and Ni^{2+} replaces Mg^{2+}; but also Ba^{2+} replaces K^+ and Ce^{3+} replaces Ca^{2+}. The substitution of elements of different valency must be compensated for, either by a coupled substitution at another site in the crustal structure, for example

$$[Ba^{2+}Al^{3+}]\text{ substitutes for }[K^+Si^{4+}]\text{ in feldspars}$$

or by an atomic position being left vacant, for example

$$[2Al^{3+} + \text{vacancy}]\text{ substitutes for }[3Mg^{2+}]\text{ in micas}$$

An extreme case is the substitution of lithium for aluminium in lithian muscovite:

$$[2Li^+ + Al^{3+}]\,[Si_4]\text{ substitutes for }[2Al^{3+} + \text{vacancy}]\,[Si_3Al]$$

(3) The type of bonding between the trace element and neighbouring anions must be similar to that between the major element and the anions, i.e. it must have a similar degree of ionic or covalent character. Thus, cadmium (radius Cd^{2+} 1.03 Å) does not substitute for calcium (radius

Ca^{2+} 1.08 Å), despite their similar charge and size, because the Cd–O bond is much less ionic than the Ca–O bond. The likelihood that two elements will form similar types of bonds can be estimated from their electronegativity.

Electronegativity

The electronegativity of an element is the power of its atoms to attract electrons. Sodium has a low electronegativity, and fluorine a high electronegativity, i.e. Na has a very low tendency to form Na^- ions, but F has a very high tendency to form F^- ions. There are two quantitative measures of the ease of ionization: the ionization energy, and the electron affinity. The ionization energy is the energy needed for an electron to be removed from the atom (e.g. Na $\rightarrow$ Na^+). The electron affinity is the energy released when an electron is added to the atom (e.g. F $\rightarrow$ F^-). An empirical scale of electronegativities was devised by Pauling, based on measurements of bond strength in actual compounds, and electronegativities on Pauling's scale are given in Table 18.

Table 18. Electronegativities and atomic radii of some of the substituting elements in rock-forming minerals

		Electro-negativity	Covalent radius (Å)	'Ionic' radius (Å)	Co-ordination
Rb	1+	0.8	—	1.81	12
K	1+	0.8	1.96	1.63	9
K	1+	0.8	1.96	1.68	12
Ba	2+	0.9	—	1.55	9
Ba	2+	0.9	—	1.68	12
Li	1+	1.0	1.34	0.82	6
Ca	2+	1.0	—	1.08	6
Ca	2+	1.0	—	1.20	8
Sr	2+	1.0	—	1.33	8
Mg	2+	1.3	1.45	0.80	6
Al	3+	1.6	1.30	0.47	4
Al	3+	1.6	1.30	0.61	6
Cd	2+	1.7	—	1.03	6
Fe	2+	1.8	1.25	0.86	6
Fe	3+	1.8	1.25	0.73	6
Ni	2+	1.9	1.21	0.77	6
Si	4+	1.9	1.18	0.34	4
Pb	2+	2.3	—	1.41	9
S	2-	2.6	1.02	—	—
O	2-	3.4	0.73	1.30	—
F	1-	4.0	0.71	1.23	—

The electronegativity data are from Allred (1961). Covalent radii are from Huheey (1975) and 'ionic' radii are from Whittaker and Muntus (1970). The 'ionic' radii are the effective radii in compounds in which the main 'anion' is oxygen (this includes silicates). Inverted commas are used because the bonding is not completely ionic in character. In sulphides, the covalent character of the bonding is greater, and the ionic radii quoted here would not be applicable.

Figure 111. The relationship between the electronegativity difference of two elements bonded together and the degree of ionic character of the bond between them (after Pauling 1970).

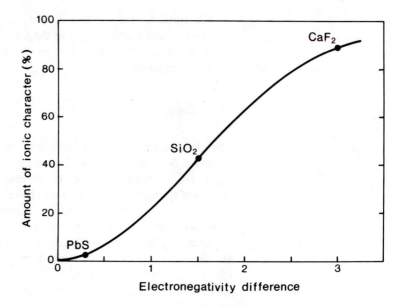

The greater the difference between the electronegativities of two elements, the greater will be their tendency to form ionic bonds. Thus CaF_2 will be mainly ionically bonded (Ca 1.0, F 4.0), whereas PbS will be predominantly covalent (Pb 2.3, S 2.6). The differences in physical properties between silicates and sulphides are partly a reflection of the more covalent bonding of the latter. Figure 111 indicates the degree of ionic or covalent character to be expected when two elements of different electronegativity are bonded together.

Ionic radii

The effective size of an atom in a crystal will depend on whether it is bonded ionically or covalently. Assuming that the atoms are closely packed together, the interatomic distance C–C in a completely covalent compound like diamond will be the sum of the covalent radii of two carbon atoms. The interatomic distance in an ionic crystal such as CaF_2 will be the sum of the two 'ionic radii' of Ca and F. Actually, there are no crystals in which the bonding is completely ionic in character. What are quoted as 'ionic radii', for example the data in Table 19, are in fact the actual radii as found in silicates, whose bonding is partly ionic and partly covalent in character. The silicates and oxides have oxygen as the principal 'anion'. In sulphides, which contain an 'anion' of lower electronegativity, the bonding is more covalent than in silicates, and the effective radii of the metal atoms are larger. The actual radius also depends on the coordination of each atom by oppositely charged atoms (see Table 20).

Table 19. Ionic radii (in Å) of the elements in silicates and oxides (after Whittaker and Muntus 1970)

	Ionic charge	Co-ordination number					
		[4]	*[6]*	*[7]*	*[8]*	*[9]*	*[12]*
Al	3+	0.47	0.61	—	—	—	—
B	3+	0.20	—	—	—	—	—
Ba	2+	—	1.44	1.47	1.50	1.55	1.68
Be	2+	0.35	—	—	—	—	—
Ca	2+	—	1.08	1.15	1.20	1.26	1.43
Cd	2+	0.88	1.03	1.08	1.15	—	1.39
Ce	3+	—	1.09	—	1.22	1.23	1.37
Co	2+	0.65	0.83	—	—	—	—
Cr	3+	—	0.70	—	—	—	—
Cs	1+	—	1.78	—	1.82	1.86	1.96
Cu	2+	0.70	0.81	—	—	—	—
Dy	3+	—	0.99	—	1.11	—	—
Er	3+	—	0.97	—	1.08	—	—
Eu	2+	—	1.25	—	1.33	—	—
Eu	3+	—	1.03	1.11	1.15	—	—
F	1−	1.23	1.25	—	—	—	—
Fe	2+	0.71	0.86	—	—	—	—
Fe	3+	0.57	0.73	—	—	—	—
Ga	3+	0.55	0.70	—	—	—	—
Gd	3+	—	1.02	1.12	1.14	—	—
Ge	4+	0.48	0.62	—	—	—	—
Hf	4+	—	0.79	—	0.91	—	—
Ho	3+	—	0.98	—	1.10	—	—
K	1+	—	1.46	1.54	1.59	1.63	1.68
La	3+	—	1.13	1.18	1.26	1.28	1.40
Li	1+	0.68	0.82	—	—	—	—
Lu	3+	—	0.94	—	1.05	—	—
Mg	2+	0.66	0.80	—	0.97	—	—
Mn	2+	—	0.91	—	1.01	—	—
Mo	6+	0.50	0.68	0.79	—	—	—
Na	1+	1.07	1.10	1.21	1.24	1.40	—
Nb	5+	0.40	0.72	0.74	—	—	—
Nd	3+	—	1.06	—	1.20	1.17	—
Ni	2+	—	0.77	—	—	—	—
O	2−	1.30	1.32	—	1.34	—	—
P	5+	0.25	—	—	—	—	—
Pb	2+	1.02	1.26	—	1.37	1.41	1.57
Pr	3+	—	1.08	—	1.22	—	—
Ra	2+	—	—	—	1.56	—	1.72
Rb	1+	—	1.57	1.64	1.68	—	1.81
Sb	5+	—	0.69	—	—	—	—
Sc	3+	—	0.83	—	0.95	—	—
Si	4+	0.34	0.48	—	—	—	—
Sm	3+	—	1.04	—	1.17	—	—
Sn	4+	—	0.77	—	—	—	—
Sr	2+	—	1.21	1.29	1.33	—	1.48
Ta	5+	—	0.72	—	0.77	—	—
Tb	3+	—	1.00	1.10	1.12	—	—
Th	4+	—	1.08	—	1.12	1.17	—
Ti	4+	—	0.69	—	—	—	—
Tl	1+	—	1.58	—	1.68	—	1.84
Tm	3+	—	0.96	—	1.07	—	—
U	4+	—	—	1.06	1.08	1.13	—

Table 19 (continued)

	Ionic charge	Co-ordination number					
		[4]	[6]	[7]	[8]	[9]	[12]
V	3+	—	0.72	—	—	—	—
W	6+	0.50	0.68	—	—	—	—
Y	3+	—	0.98	—	1.10	1.18	—
Yb	3+	—	0.95	—	1.06	—	—
Zn	2+	0.68	0.83	—	0.98	—	—
Zr	4+	—	0.80	0.86	0.92	—	—

Table 20. Co-ordination by oxygen of the major elements in rock-forming minerals

	[4]	[6]	[7]	[8]	[9]	[12]
Quartz	Si	—	—	—	—	—
Sanidine	Si, Al	—	—	—	K	—
Anorthite	Si, Al	Ca	Ca	—	—	—
Nepheline	Si, Al	—	—	Na	K	—
Analcite	Si, Al	Na	—	—	—	—
Olivine	Si	Mg, Fe^{2+}	—	—	—	—
Diopside	Si	Mg	—	Ca	—	—
Enstatite	Si	Mg	—	—	—	—
Aegirine	Si	Fe^{3+}	—	Na	—	—
Tremolite	Si	Mg	—	Ca	—	—
Phlogopite	Si, Al	Mg	—	—	—	K
Muscovite	Si, Al	Al	—	—	—	K
Garnet	Si	Al	—	Ca, Mg	—	—
Sphene	Si	Ti	Ca	—	—	—
Zircon	Si	—	—	Zr	—	—
Magnetite	Fe^{3+}	Fe^{2+}, Fe^{3+}	—	—	—	—
Ilmenite	—	Fe^{2+}, Ti	—	—	—	—
Perovskite	—	Ti	—	Ca	—	—
Apatite	P	—	Ca	—	Ca	—

Note: In many structures, the anions surrounding each cation are not equidistant from it. For example, of the eight oxygen atoms nearest to each sodium atom in aegirine, six are at a distance of 2.40–2.43 Å and two at a distance of 2.83 Å (Cameron *et al.* 1973). In spodumene ($LiAlSi_2O_6$), which has a similar structure, the lithium atoms are adjoined by six oxygens at a distance of 2.11–2.28 Å and two oxygens at a distance of 3.14 Å, and the coordination would be described as six-fold.

PARTITION COEFFICIENTS

Much quantitative information has been obtained on the actual partitioning of trace elements between crystals and melts, and between crystals and one another. It is assumed that at equilibrium, the ratio of the concentration of a trace element in a solid phase (C_S) to its concentration in the liquid (C_L) is a constant:

$$D = \frac{C_S}{C_L}$$

The constant D is called the *partition coefficient*. The partition coefficient depends on temperature, pressure, the composition of the liquid, and the composition of the solid, but not on the concentration of the trace element as long as it is small. Table 21 gives representative crystal/liquid partition coefficients for minerals in equilibrium with a basaltic liquid, and Rollinson (1993) gives a summary of the available partition coefficient data for a wide range of elements.

Most of the early information on partition coefficients was obtained by analysing coexisting phenocrysts and groundmass (especially glassy groundmass) in porphyritic rocks. There are problems in making use of such data because the conditions during equilibration are not usually known, and there is sometimes uncertainty as to whether equilibrium has been attained. Partition coefficients can be measured experimentally by equilibrating crystals and melt at a high temperature, quenching them and analysing the coexisting solid and glass phases. Four methods of analysis can be used:

1. Separation of the solid and liquid phases by hand-picking, heavy liquid separation, or differential solution. This is the most straightforward method, but it is not easy to check whether the crystalline and melt phase are homogeneous to confirm that equilibrium has been achieved. There is also a real possibility that the separated phases will still contain minute inclusions of trace element-rich accessory minerals.

2. Addition of a radioactive isotope of the trace element being studied to a synthetic starting material, followed by autoradiography of the equilibrated crystal and liquid phases to determine the distribution of the added isotope. This has the advantage of enabling the homogeneity of the

Table 21. Mineral/melt partition coefficients for basaltic liquids, mainly after Rollinson (1993)

	Ol	Opx	Cpx	Ho	Bi	Plag	Gar	Mt
Rb	0.01	0.02	0.03	0.29	3.06	0.07	0.04	
Sr	0.01	0.04	0.06	0.46	0.08	1.83	0.01	
Ba	0.01	0.01	0.03	0.42	1.09	0.23	0.02	
Y	0.01	0.18	0.90	1.00	0.03	0.03	9.00	0.20
Ti	0.02	0.10	0.40	1.50	0.90	0.04	0.30	7.50
Zr	0.01	0.18	0.10	0.50	0.60	0.05	0.65	0.10
Ce	0.01	0.02	0.12	0.20	0.03	0.12	0.03	2.2
Nd	0.01	0.03	0.27	0.33	0.03	0.08	0.07	2.0
Yb	0.01	0.34	0.58	0.49	0.04	0.07	11.50	1.4
Lu	0.02	0.42	0.53	0.43	0.05	0.06	11.90	
Ni	13.1	5.0	4.6	6.8				29.0
Co	6.6	3.0	1.0	2.0		0.05	1.0	7.4
V	0.1	0.6	1.4	3.4		0.03		26.0
Cr	0.7	10.0	34.0	12.5		0.08	1.3	153.0

Abbreviations: Bi biotite, Cpx clinopyroxene, Gar garnet, Ho hornblende, Mt magnetite, Ol olivine, Opx orthopyroxene, Plag plagioclase.

equilibrated phases to be monitored, but the disadvantage that not all the trace elements have radioactive isotopes suitable for the purpose.

3. Supplementation of the trace element concentration in synthetic starting materials to the levels which will permit the crystals and glass to be analysed by electron microprobe. This is the quickest and most widely used method, but because either the crystals or the liquid may be strongly depleted in the trace element, the concentration in the starting material needs to be in the order of percentages rather than parts per million. There is then a danger that the trace element will be too concentrated in some of the phases to conform to Henry's Law solution behaviour, i.e. its activity will no longer be proportional to its concentration and the distribution coefficient will not be a constant.

4. *In situ* analysis of phenocrysts and glassy matrix in porphyritic rocks using an ion microprobe, taking advantage of the greater sensitivity of this technique.

The effects of temperature, pressure, and liquid composition on partition coefficients put many difficulties in the way of their use in petrogenetic interpretation.

The influence of temperature on the partition coefficient has been investigated for a large number of trace elements in different minerals, and is substantial (Irving 1978). This dependence may be of value in geothermometry. For instance, Leeman and Lindstrom (1978) found a five-fold increase in the partition coefficient of nickel between olivine and basaltic melt between 1400 °C and 1100 °C and suggested that crystallization temperatures could be estimated to within 50 °C by this method. In a rock containing no glass or quenched groundmass, one can make use of the partition of a trace element between two coexisting minerals, if it is assumed that both were originally in equilibrium with the magma, and if the crystal-melt partition coefficients are known for each mineral. Thus Häkli and Wright (1967) measured the Ni distribution coefficients for olivine–glass, clinopyroxene–glass and olivine–clinopyroxene in the Makaopuhi lava lake over a range of crystallization temperatures; Häkli (1968) then applied the olivine–clinopyroxene distribution coefficients to calculate the crystallization temperatures in a Finnish gabbro intrusion.

The influence of pressure on partition coefficients has not been investigated in such detail, but the available evidence suggests that pressure has an appreciable effect. The partition coefficient for the distribution of Ni between olivine and a basic silicate liquid has been found to vary by a factor of more than two between 5 and 20 kilobars (Mysen and Kushiro 1979). Partition coefficients derived from phenocrysts and groundmass in volcanic rocks are therefore inapplicable to mantle conditions.

The liquid composition has a large influence on partition coefficients. Watson (1977) studied the partitioning of manganese between olivine and melt for a range of melt compositions at various temperatures, and found that at 1350 °C the partition coefficient varied by a factor of two between melts with 45 and 65% SiO_2. The partition coefficients of elements of variable oxidation state (notably Eu) are particularly affected by changes in the oxygen fugacity of the melt (Drake and Weill 1975).

Knowing the partition coefficients of a trace element between various crystals and melts enables us to calculate how the trace element would behave during the processes of partial melting and fractional crystallization. However, as a result of the many factors which affect them, the partition coefficients of trace elements may differ by as much as 100-fold between magmas of different compositions, as can be seen from the data in Table 22. Therefore these calculations should only be regarded as semi-quantitative.

Table 22. Clinopyroxene-matrix partition coefficients observed in igneous rocks of various compositions

	Ba	*Ce*	*Co*
Basalts and andesites	0.002–0.39	0.17–0.65	0.7–2.8
Dacites and rhyolites	0.02–0.06	0.6–1.2	6.0–11.0
High-SiO$_2$ rhyolite (Bishop Tuff)	0.5–2.3	10.3–11.3	56.0–88.0

Data are from Henderson (1982) and Mahood and Hildreth (1983).

Onuma diagrams

Onuma *et al.* (1968) had the idea of plotting crystal–liquid partition coefficients against ionic radii, as a means of comparing the efficiency with which different competing ions are incorporated into crystal structures. These diagrams have been widely adopted, and may be referred to as Onuma diagrams. Figure 112 shows the example of augite in an alkaline olivine basalt from Japan.

Jensen (1973) compiled diagrams of this kind from published data on a variety of minerals and rocks, and showed that:

1. Elements of the same valency fall on a smooth curve, with one or more peaks.
2. The curves for different valencies peak at approximately the same ionic radius.
3. A particular mineral always shows the same peaks, no matter what the host rock.

The augite plotted in Fig. 112 shows two peaks for the trivalent elements and possibly also for the divalent elements. These correspond to ionic radius values of 0.79 Å and 1.01 Å. The fixed positions of these peaks for all specimens of the same mineral show that the entry of the trace elements into the crystal structure is overwhelmingly controlled by the structure itself, and that ionic radius is the most important factor in trace element substitution. The two peaks for augite correspond to the six-fold and eight-fold coordinated sites in the clinopyroxene structure. The anomalous position of Cr in the diagram can be attributed to crystal field effects.

Jensen has suggested the following applications of Onuma diagrams:

1. To determine which sites in a mineral an element is occupying, and which it prefers.
2. To deduce partition coefficients for additional elements of known valency and ionic radius.

Figure 112. The relationship between crystal–liquid partition coefficients and ionic radii for augite and groundmass in an alkali olivine basalt from Takasima, Japan (after Matsui *et al.* 1977).

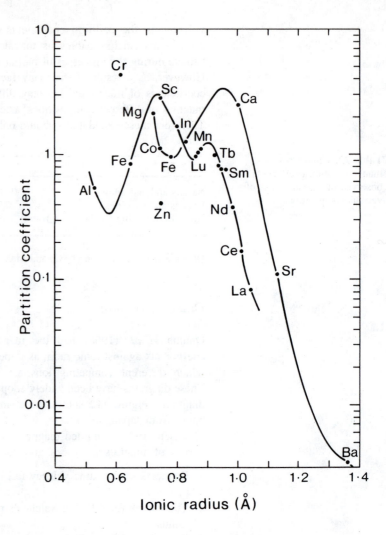

3. To deduce the valency of an element whose partition coefficient is known, but which can occur in different oxidation states, e.g. Ti or Mn; or to estimate how much of an element is present in each of two oxidation states when both are present, e.g. Eu^{2+} and Eu^{3+}.
4. In the case of transition elements, to determine whether an element is in a high spin or low spin state, where a difference in spin state makes an appreciable difference to the ionic radius (Cr, Mn, Fe, Co).

Rare earth elements

The rare earth elements (REE) form a coherent group of trace elements which because of their ionic radii substitute most readily for Ca in the structures of rock-forming minerals. They are usually concentrated in apatite, monazite and sphene, but significant amounts also occur in feldspars, pyroxenes and amphiboles. Because of the progressive variation in ionic radius in the rare-earth series, they have a tendency to become segregated

Figure 113. Generalized crystal–liquid partition coefficients for the rare-earth elements in some of the more common rock-forming minerals (after Zielinski 1975).

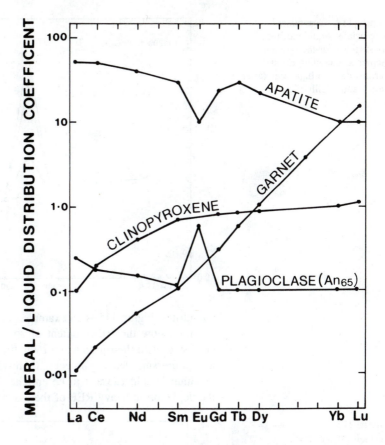

into different minerals (Fig. 113), with garnet and pyroxenes taking more of the heavy REE (which have low ionic radii), and feldspars and apatite taking the light REE (which have high ionic radii).

The distribution of europium is particularly interesting because unlike the other REE, which are always trivalent in igneous rocks, it can occur in two oxidation states, Eu^{2+} or Eu^{3+}. Under oxidizing conditions it is trivalent and behaves like the other rare earth elements, but under reducing conditions it occurs as Eu^{2+}, which has a greater ionic radius as well as a different ionic charge from Eu^{3+}. Divalent Eu is preferentially incorporated in feldspars, as can be seen from Fig. 114, which shows the partitioning of REE between plagioclase and liquid at various oxygen fugacities.

One can use the observed REE distributions of igneous rocks to deduce which minerals have been involved in partial melting or fractional crystallization processes. The presence or absence of a Eu anomaly particularly indicates the role of feldspars, and may give information on oxidation conditions in the magmatic source region. Fractionation of garnet from the magma, or the presence of garnet as a residual phase in the source region, may be indicated by depletion of the magma in heavy REE.

To show the differentiation of REE between different rock types, Coryell *et al.* (1963) suggested plotting their abundance in rocks as a ratio of their abundance in chondritic meteorites, on the assumption that the Earth would originally have contained the REE in similar relative abundance to the

Figure 114. Generalized plagioclase–liquid partition coefficients for the rare-earth elements, including Eu at various oxygen fugacities (after Drake and Weill 1975).

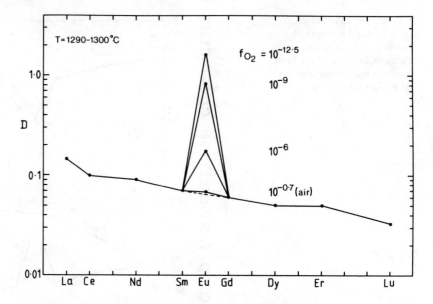

chondrites. Figure 115 is an example of a chondrite-normalized REE pattern. It shows how the REE content of some Archaean tonalites and dacites is consistent with their production by partial melting of tholeiitic metabasalts at a high pressure, leaving an eclogitic residuum. The presence of garnet in the residuum would cause it to be enriched in the heavy REE, thus explaining the depletion in heavy REE of the assumed complementary liquid.

Figure 115. Chrondrite–normalized rare-earth abundances of the Saganaga tonalite and dacite, Minnesota, compared with a partial melting model based on a tholeiitic source material (after Arth and Hanson 1975).

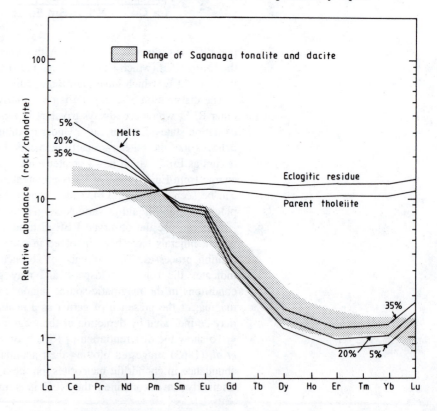

DISTRIBUTION OF TRACE ELEMENTS IN IGNEOUS ROCKS

Goldschmidt (1937) attempted to predict from crystal chemistry how readily individual trace elements would substitute for major elements in a crystallizing magma. He believed that ionic radius and ionic charge were the important factors, and that of two atoms with the same charge the smaller one would enter crystals preferentially, or of two atoms with the same radius the one with the higher ionic charge would enter crystals preferentially. These principles, often described as Goldschmidt's rules, unfortunately do not take account of electronegativity or crystal field effects, and correctly predict the behaviour of only a few elements. It can be seen from the Onuma diagram in Fig. 112, for instance, that Ca^{2+} enters the augite structure in preference to both Mn^{2+}, which is smaller and Sr^{2+}, which is larger; also Ca^{2+} enters the structure more readily than Sm^{3+}, which is about the same size. Goldschmidt's rules are therefore fundamentally unsound, and geochemists now prefer to interpret element distribution on the empirical basis of observed partition coefficients, rather than on theoretical considerations of crystal chemistry.

Partition coefficients can be used for geochemical modelling. By making an assumption about the mineralogy of the magmatic source region one can predict how the trace element content of the melt would vary as the degree of melting increased. Alternatively, starting from parent magma of an assumed composition one can predict the changes in trace element content of the liquid as fractional crystallization of particular minerals takes place. Needless to say, the uncertainties in starting composition, course of melting or crystallization, and appropriate values of the partition coefficients are too

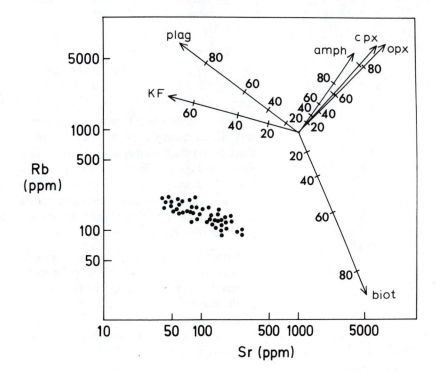

Figure 116. Variation of Rb and Sr in the Tertiary granitic rocks of Mull, Scotland (after Walsh *et al.* 1979). The arrows show the changes in magma composition that would result from fractional crystallization of plagioclase, alkali feldspar (KF), biotite, amphibole, clinopyroxene and orthopyroxene; the figures indicate the percentages of crystals removed.

great for such calculations to have any real quantitative significance. Geochemical modelling may, however, show which of several petrogenetic mechanisms are possible and which are not; for example, the rocks shown in Fig. 116 are not likely to be related by any process involving mainly the separation of pyroxene crystals.

Crystal fractionation

If one starts with a magma containing a certain concentration of a trace element, and if the crystallizing minerals and their distribution coefficients for that element are known, it is possible to calculate how the trace element content of the liquid will vary as crystallization proceeds. Take for instance the data for Ba and Ni set out in Table 23. From the individual mineral distribution coefficients we can calculate a bulk distribution coefficient for each element, which is obtained by adding together the mineral coefficients after multiplying each one by the weight fraction of that mineral. A bulk distribution coefficient >1.0 means that the element is concentrated in the crystallizing mineral assemblage as a whole and is depleted in the residual liquid. It can be seen from Table 23 that crystallization of the olivine–pyroxene–plagioclase assemblage from a basic liquid would cause depletion of the liquid in Ni and enrichment of the liquid in Ba.

Table 23. Calculation of bulk distribution coefficients

| *Crystallizing mineral assemblage* | *Mineral distribution coefficients* | |
wt %	*Ba*	*Ni*
Olivine 5%	0.01	13.1
Orthopyroxene 15%	0.01	5.0
Clinopyroxene 20%	0.03	4.6
Plagioclase 60%	0.23	0.01

Bulk distribution coefficients

$D_{Ba} = (0.05 \times 0.01) + (0.15 \times 0.01) + (0.20 \times 0.03) + (0.60 \times 0.23) = 0.15$
$D_{Ni} = (0.05 \times 13.1) + (0.15 \times 5.0) + (0.20 \times 4.6) + (0.60 \times 0.01) = 2.33$

Under conditions of perfect fractional crystallization, such that the crystallized minerals cannot react with the remaining liquid, the concentrations of the trace elements in the residual liquid would change in the way shown in Fig. 117. Elements with a very high bulk distribution coefficient would be rapidly removed from the liquid, while elements with a very low bulk distribution coefficient would be enriched up to very high levels in the last remaining fraction of liquid. In practice, the situation would not be as simple as this for the following reasons:

1. Crystallization rarely proceeds to completion in a closed system – actual magma chambers are likely to be tapped or replenished while crystal fractionation is going on, and in fact volcanic rocks can only be erupted if the magma reservoir is tapped.
2. Crystallization may not be of perfectly fractional type – there is likely to be some degree of equilibration between early separated crystals and residual liquid.

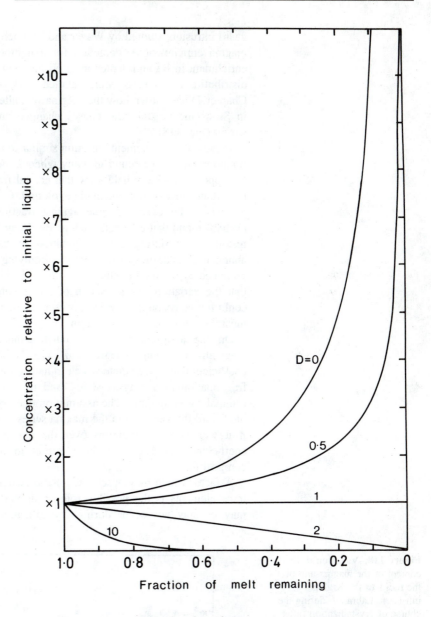

Figure 117. The changes in trace element content of a liquid undergoing Rayleigh fractionation (after Wood and Fraser 1976). Various values of the bulk distribution coefficient (D) are indicated.

3. The crystallizing mineral assemblage is unlikely to remain the same during the whole period of crystallization.
4. The distribution coefficients alter as the liquid composition changes and the crystallization temperature falls.
5. Small quantities of accessory minerals may have a large effect on the distribution of certain trace elements – for example the crystallization of apatite would have a drastic effect on the distribution of Ce, because Ce is very effectively removed from the liquid by apatite, i.e. the partition coefficient for Ce between apatite and liquid is very much greater than 1.

The trace element variation brought about by crystal fractionation has been documented for several differentiated igneous intrusions. The Skaergaard intrusion, studied by Wager and Mitchell (1951), shows clear trends of magma enrichment or depletion for a number of elements, including the enrichment in Ba and depletion in Ni that would be expected from the bulk distribution coefficients. More detailed work on the Kiglapait intrusion (see Chapter 7) has shown how the magma was alternately depleted and enriched in Sr as the crystallizing phases changed and their partition coefficients varied (Fig. 118).

Trends of trace element variation similar to those which would result from fractionation can be found in many igneous rock series, and have been used to support the widely held view that crystal fractionation is one of the most important processes in magmatic evolution. Conversely, the absence of such trends can be used to argue against fractionation. For example, Taylor (1968b) found that elements such as Ba, Li and Zr which are enriched in the residual melt during the fractionation of basaltic magma are no more abundant in andesites than in basalts, implying that andesite magmas are not produced by the fractionation of basalts. Likewise, Hall *et al.* (1993) showed that the variation in Cs content of the granites of the Sardinian batholith could not be accounted for by any combination of fractionating minerals and must have been primary in origin.

On the assumption that crystal fractionation is the main source of magmatic variation, Allègre *et al.* (1977) have proposed using partition coefficient data to calculate which minerals are responsible for trace element fractionation in a magma series, and to estimate the composition of the original parent magma. The assumption that crystal fractionation is responsible for all the variation in the magma series can hardly be justified, but even if it were, the uncertainties over the appropriate values of the partition coefficients could lead to large errors in the calculated parent magma compositions.

A factor which may upset fractionation calculations for a large number of trace elements is the possible role of sulphide liquid immiscibility, which may or may not occur during crystallization, depending on the sulphur

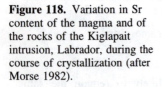

Figure 118. Variation in Sr content of the magma and of the rocks of the Kiglapait intrusion, Labrador, during the course of crystallization (after Morse 1982).

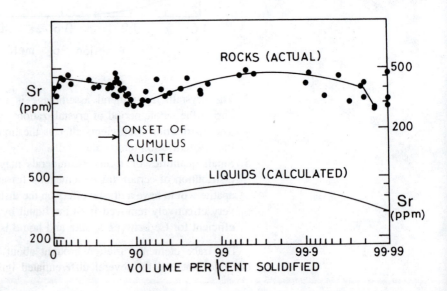

content of the magma and the relative fugacities of sulphur and oxygen. This would affect all those elements that have a partly chalcophile character (for example Co, Ni, Pb, Sn, Zn).

Partial melting

The distribution of trace elements in partial melting is more difficult to study quantitatively, because we are dealing with the melting of an unseen source material, the composition and mineralogy of which can only be approximately inferred. As in the case of crystallization, melting can be of fractional or equilibrium type. Perfect *fractional melting* involves the removal of infinitesimally small amounts of magma from the solid residue as soon as they are formed. *Batch melting* is the name given to the type of melting in which the liquid accumulates in equilibrium with the residue until the time comes for it to go on its way. The incorporation of trace elements from the source material into the liquids will depend on which type of melting predominates. The variation in trace element concentration of the liquids resulting from partial melting is not simply the reverse of that seen in fractional or equilibrium crystallization. For a discussion of the theory of trace element behaviour during partial melting the reader is referred to the textbooks by Cox, Bell and Pankhurst (1979) and by Rollinson (1993).

During partial melting, the initial melts (i.e. those formed at the lowest degrees of partial melting) will be enriched in trace elements which have low distribution coefficients for the solid phases of the source material. In the melting of mantle rocks (or in the crystallization of basic magmas), such elements would include K, Rb, Cs, Ba, Sr and U. These are known as '*incompatible*' elements.

There are two complications in particular which affect the trace element contents of magmas immediately following their formation by partial melting. One is that the products of low and high degrees of partial melting probably mix in the region of magma collection as the liquids find their way through the interstices of the partially melted source rocks. The other complication is zone melting. In this process the ascending magma undergoes continuous crystallization while partly melting the rocks overlying the magma chamber. This process can result in very great enrichment of the incompatible elements, since they are extracted from a total volume of source material much greater than the actual volume of magma might suggest.

Contamination

Sedimentary rocks, taken en masse, have compositions which are intermediate between those of basalts and granites. It can be seen from Table 24 that the contamination of either a basalt or a granite magma by argillaceous sediments, which are the commonest and more easily assimilated type of sediment, would result in a trend for most elements similar to that which might result from magma mixing or differentiation of a basic parent magma. As a result, it is rarely possible to discriminate between differentiation and contamination in a magma series by using the trace elements alone.

Table 24. The average concentrations of trace elements in argillaceous rocks compared with basalt and granite (in parts per million)

	Basalt	*Granite*	*Shale*
Be	1	5	3
Ce	32	120	90
Co	43	8	23
Cr	208	24	103
Cu	102	31	82
F	273	1366	945
Ga	19	22	20
La	16	61	45
Nb	12	27	16
Nd	15	52	39
Ni	101	11	69
Pb	7	39	24
Rb	21	215	160
Sc	35	6	16
Th	4	37	13
V	284	34	199
Zr	109	246	190
As	3	1	18
B	11	12	102
Cl	135	193	6228
Cs	1	7	9
S	96	118	11280

There is a small group of elements of which the abundance in sediments is not intermediate between those of basalts and granites. These elements, for example As, B, N and Se, are all enriched in sediments compared with any type of igneous rock. Because these elements mostly occur at rather low concentrations they are not often measured in igneous rocks, but they are potentially useful as tracers, for example in recognizing a contribution from subducted oceanic sediment in magmas erupted above a subduction zone.

Crust and mantle heterogeneity

Trace elements may be used to indicate heterogeneity in the magmatic source region. It is obvious that the crust is very heterogeneous, and there is also some heterogeneity in the mantle, but not all the variations in trace element content of magmas can be attributed to this cause, nor will a heterogeneous source region necessarily give rise to heterogeneous magmas.

If the solid residuum during partial melting contains minerals into which a particular trace element is strongly partitioned, then it will not be abundant in the magma regardless of its abundance in the source region. It is the incompatible elements, those that are strongly partitioned into the liquid, that reveal heterogeneity in the source region, particularly by variations in their ratio to one another.

Crustal heterogeneity is important to economic geologists, who recognize metallogenetic provinces in which particular trace elements are concentrated in igneous rocks as well as more obviously in ore deposits. For example, the granites of Cornwall contain an average of 407 ppm of B and 251 ppm of Li, compared with worldwide averages of only 12 ppm and 48 ppm respectively, as well as being unusually rich in Sn (Hall 1990), and this seems to reflect a regional enrichment of B, Li and Sn in the crust. Mantle heterogeneity is more difficult to recognize, and is usually identified by isotopic rather than by trace element evidence; the trace element and other evidence for mantle heterogeneity is discussed more fully in Chapter 8.

CHAPTER 5

Melting and crystallization

Experiments in the laboratory have helped to determine the conditions under which magmas originate and crystallize.

Two types of experimental study can be carried out. In the first, mixtures of artificial substances are melted and the course of their crystallization studied. In the second, actual rocks are melted, or their melts crystallized. Fortunately, most of the pioneering laboratory studies were on synthetic systems, whose phase relations reveal the principles underlying phase equilibria. These synthetic systems were mostly investigated at atmospheric pressure, but subsequent studies on natural rocks have been carried out at pressures comparable to those prevailing in the lower crust and upper mantle.

TEMPERATURES AND PRESSURES

TEMPERATURES

The temperature range up to 1500 °C encompasses all the igneous phenomena likely to occur near the Earth's surface. Measurements of lavas in the course of eruption (Table 5) show that magmas reach the surface at temperatures between 800 °C and 1200 °C. Melting and crystallization temperatures of igneous rocks in the laboratory vary between 950 °C and 1250 °C. Natural rocks have considerably lower melting temperatures than their constituent minerals, and they do not have melting points but melting ranges. Typically there is an interval of 100 °C or 200 °C between the temperature at which a rock starts to melt and the temperature at which melting is complete.

The temperature in the Earth increases from the surface downwards. The rate of increase varies from place to place, an average value in continental areas being about 30 °C/kilometre. Lower thermal gradients are found under ancient shield areas and higher gradients in recent fold belts and volcanic

Figure 119. Typical continental and oceanic geothermal gradients.

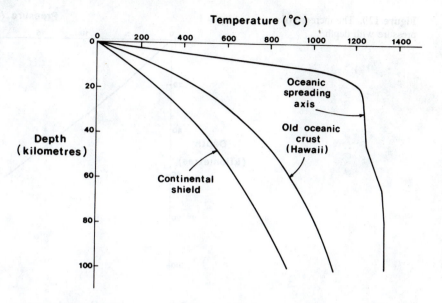

regions. More than 90% of observed geothermal gradients fall between 10 °C and 50 °C/kilometre. The variation in geothermal gradient does not exactly correspond to the variation in the heat flow towards the surface. The thermal gradient observed at the surface is partly determined by the thermal conductivities of the rocks immediately below the surface, and convection of groundwater (or magma in volcanic areas) also contributes to locally high thermal gradients.

The increase in temperature with depth which is observed at the surface does not continue uniformly with depth. The thermal conductivity of rocks decreases with temperature, and as water is expelled from pore spaces. For instance, the thermal conductivity of quartz is more than halved between 0 °C and 400 °C. This would have the result of increasing the thermal gradient with depth. On the other hand, the rocks present in the lower crust and mantle are not the same as those present in the upper crust, and in particular the concentration of radioactive heat-producing elements (K, U, Th) is less in the lower crust and mantle than in the upper crust.

In oceanic areas the geothermal gradients are more variable. At oceanic spreading centres, where new magma is being produced in large quantities, high temperatures are present at a very shallow depth. As oceanic crust ages and cools, and the oceanic lithosphere thickens, the geothermal gradient falls to a near-continental condition. Figure 119 shows estimates of the temperature gradients under typical continental and oceanic areas.

PRESSURES

Pressure in the Earth increases downwards from the surface. The SI unit of pressure is the *pascal*, which is an inconveniently small unit for petrological purposes, so petrologists use instead the *bar*, which is approximately atmospheric pressure. These units are related as follows:

Figure 120. The increase of pressure with depth.

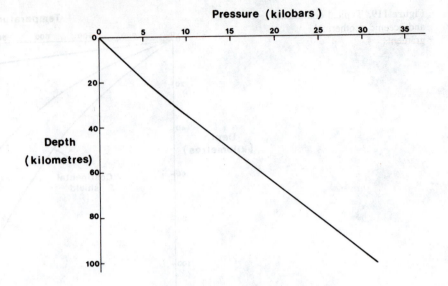

1 pascal (Pa)	$= 1$ newton/metre^2
1 gigapascal (GPa)	$= 10^9$ N/m^2
1 bar	$= 10^5$ N/m$^2 = 10^6$ dyne/cm^2
1 kilobar	$= 10^8$ N/m^2
1 atmosphere	$= 1.01325 \times 10^5$ N/m^2 (the 'standard atmosphere')

Pressures within the Earth are due mainly to the weight of overlying rock. The density of crustal rocks averages about 2.8 g/cm^3, so pressure rises with depth by 0.28 kilobars/kilometre (Fig. 120). Pressures attained in experimental studies may therefore be related to depths in the crust by assuming a depth of 3.6 kilometres per kilobar. The density of upper mantle rocks is about 3.3 g/cm^3, and pressure in the upper mantle rises by 0.33 kilobars/kilometre, corresponding to 3.0 kilometres per kilobar.

Near the surface, the pressure in pore spaces or intergranular fluid may theoretically be greater or less than the lithostatic pressure (i.e. the hydrostatic pressure of overlying rocks). However, this is only possible in a body of rock with sufficient mechanical strength to resist deformation, otherwise the rock would deform in such a way as to reduce (or increase) the internal pressures to equal lithostatic pressure. In igneous petrology, where the presence of a melt is under consideration, we do not encounter materials which could sustain a pressure significantly greater than the lithostatic pressure.

PHASE RELATIONS OF SILICATES AND SILICATE MELTS

PHASE EQUILIBRIA

The physical states in which material can exist are solid, liquid and gas. If a body of material is uniform throughout, in both its physical state and chemical composition, it is said to consist of only one *phase*. Many materials consist of more than one phase, distinct from one another and separated by boundaries.

A quartz crystal consists of one phase: solid SiO_2. A molten quartzite consists of a single phase: liquid SiO_2. Most rocks consist of more than one phase; we call the solid phases minerals. Basalt usually consists of two or three phases: solid feldspar, solid pyroxene, and (sometimes) solid olivine. A completely molten basalt consists of one phase: a basalt liquid. Natural basalt lavas contain extra phases, e.g. one or more minerals in suspension, and perhaps also an exsolved gas phase.

All gases are completely miscible; if put together they mix and become a single phase. Many liquids are also miscible, although two coexisting melts are found in some silicate systems and liquid immiscibility is common in carbonate–silicate or sulphide–silicate mixtures. The majority of solid rock-forming minerals are not miscible, although there are important exceptions such as the albite–anorthite or forsterite–fayalite solid–solution series.

The *Phase Rule* enables us to classify the types of equilibrium that can exist between the phases in any particular system. It can be stated as:

$$P + F = C + 2$$

where P = the number of phases, C = the number of components, and F = the number of degrees of freedom. The number of *components* is the minimum number of chemical constituents that are necessary and sufficient to describe the composition of all the phases in the system. The number of *degrees of freedom* is the number of variables (temperature, pressure, concentration of the components) that can be independently varied without altering the number of phases present.

For instance, consider the one-component system of SiO_2 (Fig. 121). At point X on the diagram, there is only one component ($C = 1$), there is only one phase present, i.e. high quartz ($P = 1$), and without affecting the number of phases present, it is possible to change independently the two variables, temperature and pressure ($F = 2$). Applying the Phase Rule:

$$P + F = C + 2 \text{ can be written } 1 + 2 = 1 + 2$$

At point Y on the diagram, two phases coexist. Using the Phase Rule:

$$P + F = C + 2 \text{ becomes } 2 + F = 1 + 2$$

and therefore $F = 1$, i.e. the Phase Rule predicts that we can only change one of the variables independently of the other without altering the number of phases present. Thus if we altered the temperature we could still find a pressure at which high-quartz and melt coexisted, but if we changed both temperature and pressure independently we could not expect both phases to remain. The boundary between the melt region and the high-quartz region of the diagram is known as a *univariant line*, because all the points on it have only one degree of freedom. The other boundaries in the diagram are also univariant lines.

At point Z in the diagram, three phases coexist. Using the Phase Rule again:

$$P + F = C + 2 \text{ becomes } 3 + F = 1 + 2$$

Figure 121. The system SiO$_2$.

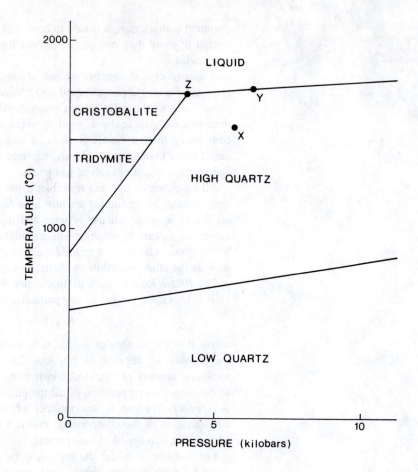

and therefore $F = 0$. This means that it is not possible to alter either of the variables (temperature or pressure) without changing the number of phases present. The system has no degree of freedom and there is no other temperature and pressure at which these three phases can be present together. This point is described as an *invariant point*.

In the system SiO$_2$, there is no way that the number of phases coexisting in equilibrium could be more than three. In more complex systems, considerations of this kind may help to decide which phase assemblages are possible and which are not, under a given set of conditions.

BINARY SYSTEMS

A binary system is one with two components. The melting relations depend on whether there is any solid solution (isomorphous substitution) between the two components, whether any intermediate compounds can be formed, and whether all the compounds (including intermediate compounds) melt congruently.

Figure 122. The system albite–anorthite (after Bowen 1913).

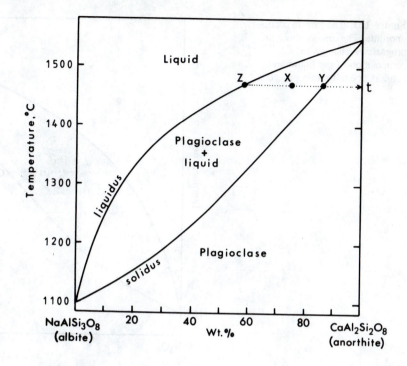

Binary system showing complete solid solution

The system albite–anorthite (Fig. 122) is an example. The phase diagram is divided into three fields. Each field is labelled to show what phases would be present under equilibrium conditions. At temperatures above the curve labelled '*liquidus*' all compositions exist as liquid, i.e. molten plagioclase. At temperatures below the curve labelled '*solidus*' all compositions exist as solid crystals, i.e. plagioclase feldspar. Any composition at a temperature between the liquidus and solidus exists as a mixture of crystals and liquid, whose respective compositions lie on the solidus and liquidus. For instance, in Fig. 122, composition X at temperature *t* should consist of a mixture of crystals of composition Y and melt of composition Z.

Imagine the crystallization of a liquid whose initial composition is shown as P in Fig. 123. At a temperature above the liquidus, it would be a melt. On cooling, it would start to crystallize when the liquidus temperature was reached, and the first crystals to form would have the composition defined by the solidus at that temperature, i.e. composition Q, which is relatively rich in anorthite. Removal of these crystals from the melt results in the remaining liquid changing composition in the other direction, i.e. towards albite. Obviously the melt cannot continue to grow crystals of composition Q because the final solid product must have a bulk composition identical to the initial melt, and there are two ways in which crystallization can proceed: by equilibrium crystallization or by fractional crystallization.

Equilibrium crystallization means that as the melt changes in composition, becoming more albite-rich, the crystals that have already separated from the melt react with it and also become more albite-rich. Thus, at each

Figure 123. The system albite–anorthite. The arrows show the progressive changes in composition of the liquid and solid as crystallization proceeds.

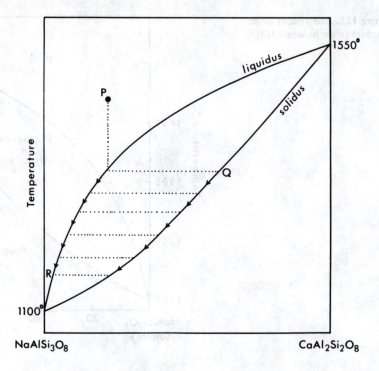

successively lower temperature there is a liquid–crystal pair whose compositions are joined by a horizontal line from the liquidus to the solidus, as shown in Fig. 123. Crystallization is completed when the liquid reaches composition R, by which time it is in equilibrium with crystals of composition P, and all the melt has been used up in bringing the more calcic early formed crystals to this composition.

In *fractional crystallization*, the melt does not react with the early formed crystals to bring them into equilibrium as its composition changes. As the crystals grow, successive layers of crystal are progressively enriched in the albite constituent, i.e. the crystal is zoned. To make up for the calcic composition of the early formed core of the crystal, the outer layers must be more sodic than the composition of the original melt. During the course of fractional crystallization, the liquid composition approaches further towards the albite end-member than in equilibrium crystallization.

The common occurrence of zoning in the plagioclase crystals of actual rocks shows that fractional crystallization is of common occurrence. As would be expected from the phase relations, a typical plagioclase crystal is zoned from a calcic core to a sodic outer shell ('normal zoning'), but 'reversed zoning' or 'oscillatory zoning' can also be found, and these may be related to changes in magma composition or water pressure during crystallization. There are obviously various intermediate possibilities between perfect equilibrium crystallization and perfect fractional crystallization, depending on the extent of reaction between the liquid and the early formed crystals. In general, rapid cooling is most conducive to fractional crystallization, and this is why the plagioclase in volcanic rocks is more strongly zoned than that in plutonic igneous rocks.

Figure 124. The system gehlenite–åkermanite (after Osborn and Schairer 1941).

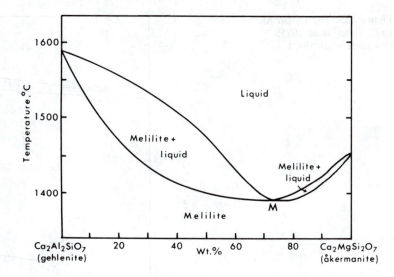

Binary system containing a solid solution series with an intermediate minimum

In several solid solution series, the composition with the lowest melting temperature is not one of the end-members but an intermediate composition. An example is the system gehlenite–åkermanite, the end-members of the melilite group (Fig. 124). As in the case of the plagioclase feldspars, the change in liquid composition during crystallization is in the downward direction of the liquidus. For melts rich in the gehlenite component, the liquid becomes more åkermanite-rich and for melts rich in åkermanite it becomes more gehlenite-rich, trending towards the minimum composition M in either case. A zoned crystal of melilite produced by fractional crystallization would have a relatively aluminous core if it crystallized from a gehlenite-rich liquid or a relatively magnesian core if it crystallized from an åkermanite-rich liquid.

Another isomorphous series which shows an intermediate minimum is the system orthoclase–albite (see Fig. 146 below), although with a number of additional complications. The opposite possibility of a solid solution series with an intermediate maximum is not shown by any of the more important silicate systems.

Binary system with complete solid solution at the melting temperatures but not at lower temperatures

An example of this type of system is Al_2O_3–Cr_2O_3 (corundum–eskolaite), which is shown in Fig. 125. At high temperatures, these two minerals form a complete solid solution series and their melting relations are similar to those of the series albite–anorthite. Below 945 °C there is a miscibility gap, and corundum having some Cr_2O_3 in solid solution coexists with eskolaite having some Al_2O_3 in solid solution. The amount of solid solution that is possible

Figure 125. The system Al$_2$O$_3$–Cr$_2$O$_3$ (after Muan 1975; Chatterjee *et al*. 1982).

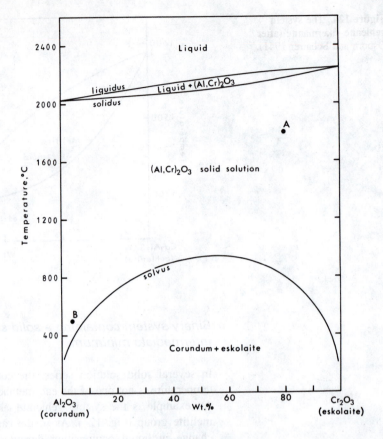

decreases with temperature. The curve delimiting the extent of solid solution is called the *solvus*.

If a homogeneous crystal of composition A were to cool from just below the solidus, it would reach the solvus at about 800 °C. At this temperature eskolaite can contain only about 20% Al$_2$O$_3$ in solid solution, and excess Al$_2$O$_3$ would start to separate as a distinct phase. It would not be pure Al$_2$O$_3$, but Al$_2$O$_3$ with considerable Cr$_2$O$_3$ in solid solution, as defined by the width of the solvus at this temperature. This process is called *exsolution*. In natural minerals, exsolved phases are often recognizable because they form an intimate intergrowth. Further cooling beyond the temperature at which the solvus was first reached would result in additional exsolution, because the miscibility gap widens with falling temperature.

The solvus shows how much solid solution is possible at any temperature. Corundum could have less Cr$_2$O$_3$ in solid solution than the maximum possible amount, if the rock in which it occurred was deficient in chromium. Composition B at 500 °C (Fig. 125) consists of Cr-poor corundum on its own, because there is not enough chromium present to saturate the corundum or to produce coexisting eskolaite.

In igneous rocks, the most important solvus is that between sodium and potassium feldspar. The system KAlSi$_3$O$_8$–NaAlSi$_3$O$_8$ is not truly binary, because KAlSi$_3$O$_8$ melts incongruently to give a compound (leucite) which is not intermediate between the two end members (Fig. 146), but the subsolidus relations are similar to those shown in Fig. 125. The exsolution intergrowths formed at the alkali feldspar solvus are known as perthites, and are commonly seen in granitic rocks. In some rapidly cooled lavas, exsolution may not have taken place, or may have taken place on such a fine scale that it is not visible under a microscope, and the feldspar present in the rock has a composition lying in the middle of the binary composition range, i.e. crystallization has taken place too rapidly for equilibrium to be completely attained.

Binary system with a eutectic point (no solid solution)

The system anorthite–silica (Fig. 126) is one in which there is no solid solution between end members. All compositions are completely molten above the liquidus and completely solid below the solidus. Between the liquidus and solidus, crystals of either anorthite or silica coexist with liquid.

Consider point A in Fig. 126, which lies at a temperature between the liquidus and solidus in the 'silica + liquid' field. At this temperature, this composition will exist as a mixture of solid silica and a melt. The silica will be in the form of cristobalite above 1470 °C or tridymite below that temperature. The crystals will be pure SiO_2 (composition B) and the coexisting liquid will lie on the liquidus for that temperature (composition C).

If a melt of composition D were to be cooled down, it would start to crystallize when it reached the liquidus. It would then lie on the edge of the 'silica + liquid' field and the crystals separating from the melt would be a silica mineral (cristobalite or tridymite, depending on the temperature). Removal of silica would leave the remaining melt enriched in the con-

Figure 126. The system anorthite–silica (after Schairer and Bowen 1947).

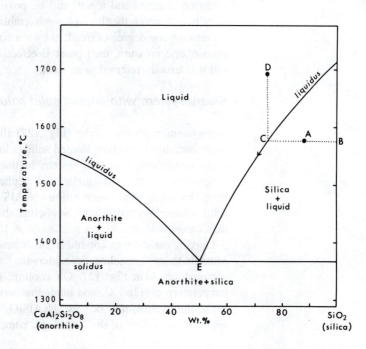

stituents of anorthite, and as it continued to cool it would move along the liquidus in the direction shown by the arrow. When, after continuing crystallization, the melt reached the composition and temperature marked E it would be at the lowest temperature at which liquid could still exist in this system. At this point the melt would complete its crystallization as a mixture of tridymite and anorthite crystals. Point E is called a *eutectic point*.

The sequence of crystallization of liquid D is therefore: (1) separation of a silica mineral on its own, followed by (2) crystallization of anorthite and silica together. If the initial composition had been chosen to lie on the anorthite side of the eutectic point, anorthite would have been the first mineral to crystallize (the 'liquidus phase') and the melt would again have moved towards the eutectic point, where anorthite and tridymite would crystallize together. Only if the initial melt had the eutectic composition would the two minerals crystallize together throughout the period of crystallization.

In natural systems containing a eutectic composition it may sometimes be possible to tell from petrographic evidence whether an actual magma had a eutectic composition when it started to crystallize, or whether it lay on one side or another of the eutectic point. One would look for evidence of the order of crystallization, as shown by inclusions of one mineral in another, or glomerophyric texture, or perhaps porphyritic texture.

Notice how the phase rule applies to this system. At the eutectic point E, three phases coexist in equilibrium (anorthite + silica + melt). Applying the Phase Rule:

$$P + F = C + 2 \text{ becomes } 3 + F = 2 + 2$$

There is therefore one degree of freedom. The phase assemblage at E is univariant. One of the three variables – temperature, pressure or composition – can be changed and it will still be possible to find a combination of the other two at which the three-phase assemblage is in equilibrium. However, if we remove one degree of freedom by not considering any pressure other than atmospheric pressure, then point E effectively becomes an invariant point, and it is usually referred to as such.

Binary system with limited solid solution

The system nepheline–albite (Fig. 127) illustrates the melting relations of two minerals which show limited solid solution with one another. There is a slight complication in this system in that $NaAlSiO_4$ has high- and low-temperature forms, carnegieite and nepheline. Carnegieite and nepheline have the same ideal composition ($NaAlSiO_4$), but in practice show some solid solution towards albite. Nepheline shows more solid solution of albite than carnegieite, reaching a maximum at 1068 °C.

Leaving aside the complications caused by the transformation from carnegieite to nepheline, consider the crystallization of a liquid with composition A in Fig. 127. On cooling, it would reach the liquidus at a temperature of 1160 °C, and nepheline would then start to crystallize. The nepheline would not be the $NaAlSiO_4$ end member but would have approximately 25% of the $NaAlSi_3O_8$ component in solid solution. Removal

Figure 127. The system
nepheline–albite (after Greig
and Barth 1938).

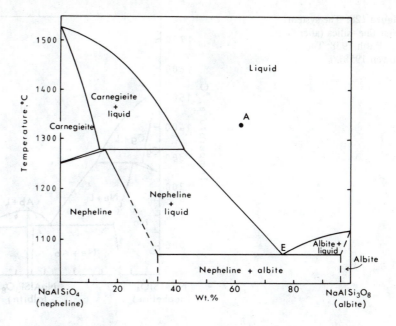

of these crystals would cause the remaining liquid to move down the liquidus
surface in the direction of albite, crystallizing more nepheline on the way.
When the liquid reached the eutectic point (E), there would be simultaneous
crystallization of nepheline crystals (with 34% of $NaAlSi_3O_8$ component in
solid solution) and albite crystals (with 4% of $NaAlSiO_4$ component in solid
solution) until all the liquid was used up.

In some phase diagrams, the subscript 'ss' (solid solution) is used to
denote minerals which are not pure end-members but are solid solutions of
varying composition. Thus in Fig. 127, the field labelled 'nepheline +
liquid' could be labelled 'nepheline$_{ss}$ + liquid'.

Binary system containing an intermediate compound

If the two components in a binary system are capable of forming an intermediate compound, as in the system nepheline–silica (Fig. 128), then it may
be possible to read the phase diagram just as though it were two separate
simple binary systems side-by-side (nepheline–albite, and albite–silica). This
is only possible if the intermediate compound melts congruently, i.e. to a
melt of its own composition. If the intermediate compound melts incongruently, the melting relations are more complicated.

Binary system containing an incongruently melting compound

A compound is said to melt congruently if it changes directly from a solid to
a melt of the same composition at the melting temperature. Incongruent
melting involves the formation of a melt with a different composition from
that of the solid.

Figure 128. The system nepheline–silica (after Greig and Barth 1938; Tuttle and Bowen 1958).

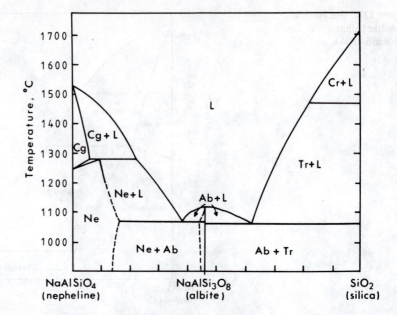

Orthoclase is an example of a compound that melts incongruently. If heated to 1150 °C, it changes to a mixture of leucite crystals and a melt whose composition is intermediate between $KAlSi_3O_8$ and SiO_2. If heated further, an increasing proportion of the leucite dissolves in the melt, but it is not until a temperature of about 1500 °C is reached that the last of the leucite disappears and the material is completely molten. By then of course the melt is of the same composition as the starting material, i.e. it is molten orthoclase.

Figure 129 shows the system leucite–silica. We could consider orthoclase–

Figure 129. The system leucite–silica (after Schairer and Bowen 1947).

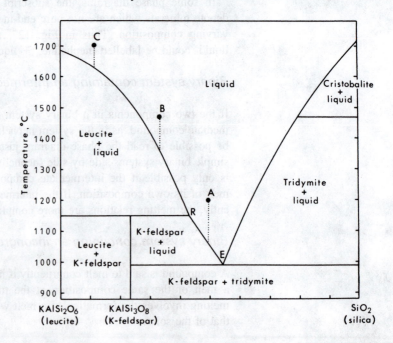

silica as a binary system in which one of the end-member compounds is incongruently melting, or leucite–silica as a binary system containing an incongruently melting intermediate compound. It can be seen that a composition existing as solid orthoclase at a temperature below 1150 °C would exist as leucite + liquid between 1150 °C and 1550 °C, and as a melt of orthoclase composition above 1550 °C.

Consider the crystallization of liquid A in Fig. 129. On cooling down to the liquidus it reaches the K-feldspar + liquid field, and therefore starts to crystallize K-feldspar. Removal of the K-feldspar crystals leaves the melt rather richer in silica, and as it cools it moves down the liquidus towards the eutectic point E. At the eutectic point, K-feldspar and tridymite crystallize together until all the melt is used up.

Liquid B has a more complicated course of crystallization. When it cools to the liquidus it reaches the leucite + liquid field, so leucite starts to crystallize. Again, this leaves the melt richer in silica, and the melt moves down the liquidus in the direction of R. By inspection of the phase diagram it can be seen that the composition of B should be represented by K-feldspar + liquid between 1150 °C and 990 °C, and by K-feldspar + tridymite below 990 °C. This provides a clue as to what happens next. When the melt reaches point R, having crystallized a substantial amount of leucite, a reaction takes place between the already formed leucite crystals and the melt, as a result of which all the leucite is converted into K-feldspar. The melt that is left continues to cool, but now separates only K-feldspar, and its composition progresses towards the eutectic point, where eventually the remaining melt crystallizes completely as a mixture of K-feldspar and tridymite.

The point R, at which the leucite reacted with the melt to be converted into K-feldspar, is an example of a *reaction point* (or *peritectic point*).

Now consider liquid C, which is much poorer in silica than liquid B and must end up below 1150 °C as a mixture of leucite + K-feldspar. On cooling, it first crystallizes leucite, and the melt migrates along the liquidus towards the reaction point, R. At the reaction point the early formed leucite reacts with the melt, being converted into K-feldspar. Because the initial composition was so poor in silica, there is not enough of the silica-rich melt left to convert all the leucite into K-feldspar, and the melt is used up before the reaction can be completed. The product is therefore a crystalline mixture of unreacted leucite and K-feldspar.

The foregoing discussion assumes that crystallization was of equilibrium type, but the possibility of fractional crystallization must also be considered. If on reaching the reaction point R, the residual melt from liquid B had failed to react completely with the early formed leucite crystals, the final crystalline product could contain the assemblage leucite + K-feldspar + tridymite, which under equilibrium conditions can not all occur together. The same result might arise if not all the early formed leucite crystals from liquid C reacted with the residual melt at R, and some liquid survived to reach the eutectic point E. Conditions that might lead to fractional as opposed to equilibrium crystallization could include: (1) rapid crystallization; (2) segregation of early formed leucite crystals from the residual melt by flotation; or (3) shielding of the early formed leucite crystals from the melt by an overgrowth of the K-feldspar reaction product.

Figure 130. The system forsterite–silica (after Bowen and Andersen 1914; Greig 1927).

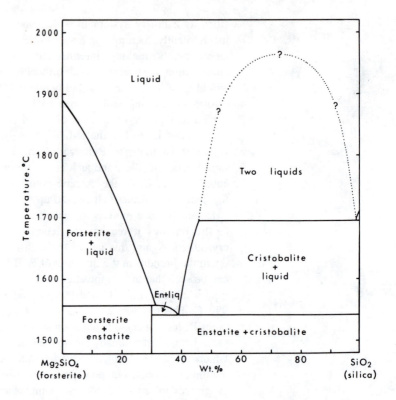

Binary system showing liquid immiscibility

In most silicate systems approximating to the compositions of actual rocks, the melts are found to be homogeneous, but examples are known of silicate liquids which do not mix. Figure 130 shows an example in the system forsterite–silica. The miscibility gap is greatest at temperatures just above the liquidus, but less at higher temperatures, and it is believed that there would be complete miscibility at temperatures above 2000 °C. Apart from the liquid immiscibility, this system resembles the system leucite–silica (Fig. 129) in that it contains an incongruently melting intermediate compound, in this case enstatite.

Pseudo-binary systems

Figure 131 shows melting relations in the system forsterite–anorthite. It will be seen that in this system there appears a mineral, spinel ($MgAl_2O_4$), whose composition is not intermediate between those of forsterite and anorthite. The system is not truly binary because the phase assemblages are determined by more than two components. The system forsterite–anorthite can be regarded as a section of the larger four-component system $CaO–MgO–Al_2O_3–SiO_2$.

The system forsterite–anorthite can be described as a pseudo-binary system or join. At the pseudo-binary 'eutectic' point P, it would appear that

Figure 131. The pseudo-binary join forsterite–anorthite. (After Osborn and Tait 1952).

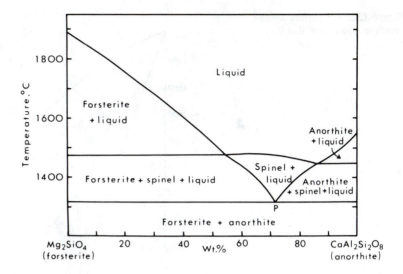

anorthite, forsterite, spinel and liquid could coexist, whereas in a true binary system only three coexisting phases are permitted by the Phase Rule.

TERNARY SYSTEMS

A ternary system is one with three components. Phase relations in ternary systems are more difficult to draw than those in binary systems. Figure 132 shows the system albite–anorthite–silica in perspective. It is represented as a triangular prism, in which temperatures are measured along the vertical axis and compositions can be projected from the basal triangle. Each face of the prism corresponds to a binary system, including the binary systems albite–anorthite and anorthite–silica which were shown in Figs 122 and 126. The main emphasis is placed on showing the liquidus temperatures. The eutectic form of the liquidus can be seen on the anorthite–silica face of the prism, and there is also a binary eutectic point between albite and silica. The complete liquidus surface for the ternary system shows that a eutectic relationship holds between silica and all the intermediate compositions of the albite–anorthite solid solution series.

Ternary diagrams of this kind are difficult to draw accurately, and it is usual to show only a projection down the temperature axis, as in Fig. 133. Besides showing the liquidus temperatures, the diagram also names the minerals which coexist with the melt at temperatures immediately below the liquidus. From this information, the course of crystallization can be worked out.

Ternary system containing a solid solution series

The system albite–anorthite–silica contains the plagioclase solid solution series. Imagine a melt of composition X in Fig. 133. On cooling, this liquid would start to crystallize at a temperature of about 1430 °C, when it reached

Figure 132. The system albite–anorthite–silica (see also Fig. 133).

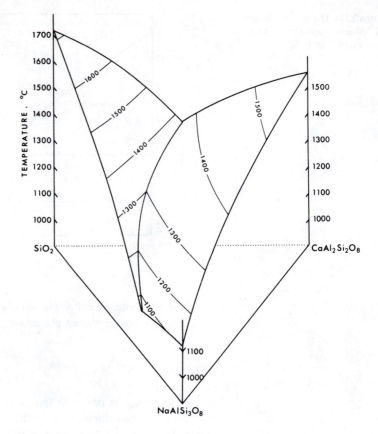

Figure 133. The system albite–anorthite–silica (after Schairer 1957; Yoder 1968).

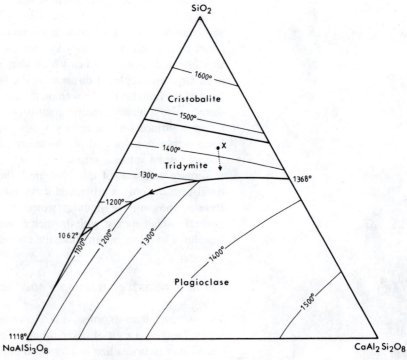

the liquidus. Since its composition lies in the silica field, the first crystals to form would be silica (tridymite at this particular temperature), and their removal from the melt would alter its composition in a direction away from that of SiO_2. The liquid would cool, separating silica, until its composition reached the silica–plagioclase boundary. On this boundary, silica and plagioclase crystallize together from the melt. This type of boundary is known as a cotectic curve. The composition of the SiO_2 crystals is fixed, as tridymite and cristobalite do not show solid solution with the other components, but the composition of the plagioclase crystals will be more calcic than that of the liquid; just how much more calcic cannot be seen by looking at the liquidus diagram alone.

Having reached the silica–plagioclase boundary, the melt changes in composition along the cotectic curve, following the liquidus temperature contours in a downward direction. Its crystallization may be of either equilibrium or fractional type, as in the system albite–anorthite, and how far the liquid moves down the cotectic curve before being completely crystallized will depend on which type of crystallization predominates. Fractional crystallization would result in residual melts approaching closer to the low-temperature end of the cotectic curve.

Ternary system with no solid solution

The system albite–fayalite–silica (Fig. 134) shows no solid solution and all the minerals melt congruently. The diagram shows the fields in which each of the minerals is the first to crystallize at the liquidus temperature, and the

Figure 134. The system albite–fayalite–silica (after Bowen and Schairer 1938).

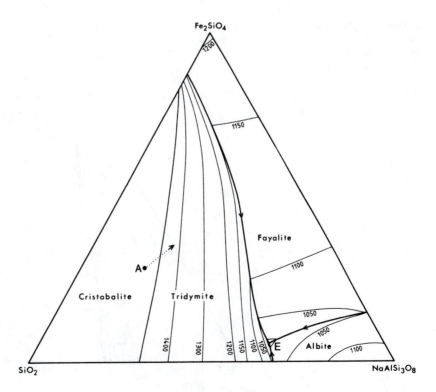

contours on part of the liquidus surface. Arrows on the cotectic lines show the direction in which crystallizing liquids change composition as they cool down on the liquidus surface. Imagine a melt of composition A. It lies in the cristobalite field and so would crystallize cristobalite if cooled down to reach the liquidus temperature. The separation of cristobalite is followed by that of tridymite when the temperature falls below 1470 °C. As silica crystallizes, the liquid changes composition in the direction shown by the dotted line, i.e. away from SiO_2. In due course, the liquid reaches the silica–fayalite boundary (cotectic curve), and fayalite and silica start to crystallize together. The melt then moves along the fayalite–silica boundary in the direction shown by the arrow. Eventually the composition of the melt reaches the point E, which is a ternary eutectic point. This is the lowest temperature point on the liquidus surface and all the remaining melt crystallizes here as a mixture of albite, fayalite and silica.

Any other initial melt composition in this system would also move to the ternary eutectic point during the course of its crystallization.

Ternary system with an intermediate compound

The system fayalite–nepheline–silica contains an intermediate compound, albite, which melts congruently. The phase diagram for this system (Fig. 135) shows two ternary eutectic points, E_1 and E_2. Crystallization of any initial melt will eventually lead to a residual liquid having the composition of one or other of these eutectic points. In this system there is a small field in which the liquidus phase is a spinel (He = hercynite $FeAl_2O_4$), whose

Figure 135. The system fayalite–nepheline–silica (after Bowen and Schairer 1938). Ca = carnegieite ($NaAlSiO_4$), Ct = cristobalite (SiO_2), He = hercynite ($FeAl_2O_4$), Wu = wüstite (FeO).

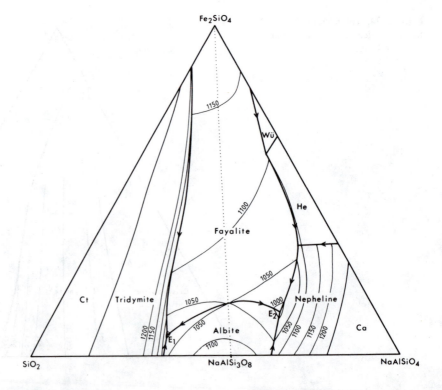

composition cannot be expressed in terms of the three components fayalite, nepheline and silica. The system is therefore actually pseudo-ternary.

In Fig. 135, the arrows on the cotectic curves show the downward slope of the liquidus surface. It can be seen that during the course of crystallization, no liquid can cross the dotted line between Fe_2SiO_4 and $NaAlSi_3O_8$. Any liquid crystallizing in the fayalite field must move away from fayalite and any liquid in the albite field must move away from albite, so all liquids with compositions close to the dotted line will move away from it as they crystallize. This line is like a watershed on the liquidus surface, and is described as a *thermal divide*.

Ternary system with an incongruently melting compound

The system anorthite–leucite–silica which is shown in Fig. 136 contains the incongruently melting compound K-feldspar. The binary section between leucite and silica was shown in Fig. 129, and may help in predicting the course of crystallization in the ternary system.

Consider the crystallization of a liquid with composition C in Fig. 136. It lies in the anorthite field and starts to crystallize anorthite on cooling to the liquidus. Removal of anorthite moves the melt in the direction shown by the broken line until it reaches the anorthite–leucite boundary. Anorthite and leucite then crystallize together while the liquid moves down the boundary towards the reaction point R, where the melt reacts with the already-formed leucite crystals to form K-feldspar. All the remaining melt is used up in this reaction, and the final product is anorthite + leucite + K-feldspar.

Figure 136. The system anorthite–leucite–silica (after Schairer and Bowen 1947).

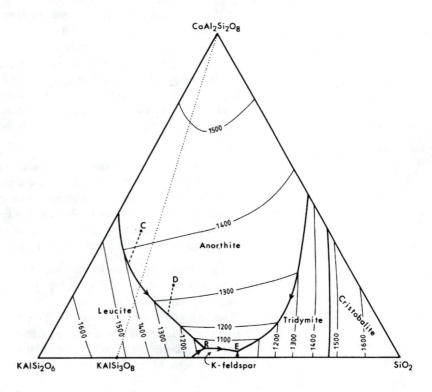

The liquid D would initially crystallize in the same way, but on reaching point R, some liquid would be left after converting all the leucite to K-feldspar. It would then continue to crystallize K-feldspar and anorthite while moving along the anorthite–K-feldspar boundary to the eutectic point E, where anorthite, K-feldspar and silica (tridymite) would crystallize together.

During equilibrium crystallization, the dotted line shown in Fig. 122 joining $CaAl_2Si_2O_8$ and $KAlSi_3O_8$ separates two types of liquid. Those liquids initially lying on the leucite side of the K-feldspar–anorthite line will end up as leucite + anorthite + K-feldspar. Liquids lying on the silica side of the K-feldspar–anorthite line will end up as anorthite + K-feldspar + silica. As in the binary system leucite–silica, fractional crystallization may result in some leucite being preserved in compositions on the silica side of the K-feldspar–anorthite line.

The short boundary curve between the fields of leucite and K-feldspar is a projection into the ternary system of the reaction point in the binary system leucite–silica, and terminates at the invariant point R. Along this boundary, reaction takes place between leucite and liquid. Along all the other boundaries in the ternary system there is simple precipitation of new crystals from the liquid. Thus we see that the primary phase boundaries can be of two types: *reaction curves* (e.g. the leucite–K-feldspar boundary), where crystals react with melt, and *subtraction curves* (e.g. the leucite–anorthite boundary), where crystals are subtracted from the melt.

Ternary system with several intermediate compounds

In the more complex ternary systems there may be several intermediate compounds and a large number of boundaries. It may be difficult to show the detailed form of the liquidus surface and to identify which boundaries are subtraction curves and which are reaction curves.

The interpretation of such diagrams is facilitated by the use of compatibility lines. A *compatibility line* (Alkemade line) is a line joining the compositions of two phases whose fields share a common boundary in the phase diagram, i.e. they can coexist with the melt. Two rules govern their use:

1. The liquidus temperature on a boundary curve tends to a maximum at its intersection with the corresponding compatibility line.
2. A boundary will be a subtraction curve if it (or its extension) intersects the corresponding compatibility line between the compositions of the two phases, but a reaction curve if it (or its extension) does not intersect the compatibility line between the compositions of the two phases.

The application of these rules can be seen in Fig. 137. The anorthite–K-feldspar boundary is shown in Fig. 137A: the liquidus temperature on this boundary tends to a maximum at the point where an extension of the boundary would meet the anorthite–K-feldspar compatibility line, and so the direction of falling temperature is as marked by the arrow. The boundary can be extrapolated backwards to meet the compatibility line between the compositions of anorthite and K-feldspar, and is therefore a subtraction

Figure 137. The system anorthite–leucite–silica, showing compatibility relations between: (A) anorthite and K-feldspar; and (B) leucite and K-feldspar.

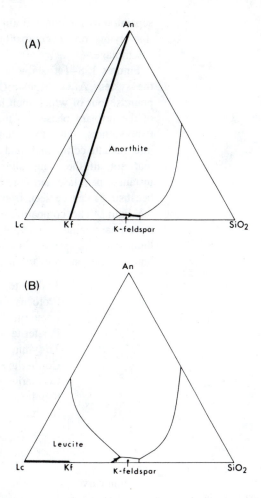

curve. The leucite–K-feldspar boundary is shown in Fig. 137B: again the temperature of the boundary falls in a direction away from the compatibility line, but because the boundary does not meet the compatibility line between the compositions of leucite and K-feldspar, it is a reaction curve and not a subtraction curve.

In the examples quoted, the boundaries are fairly short and straight, but in some cases a boundary may be sufficiently curved that one part of it points towards the compatibility line between the two phases and another part does not. Part of the boundary would then be a subtraction curve and part a reaction curve. In such cases it is necessary to draw tangents to the boundary at various points to see which of them intersects the compatibility line between the compositions of the phases.

The compatibility lines on a ternary phase diagram outline compatibility triangles. Each compatibility triangle defines the sub-solidus phase assemblage which will eventually be produced by crystallization. Thus, in the system anorthite–leucite–silica there are two three-phase assemblages to which equilibrium crystallization can lead: either anorthite + leucite + K-feldspar, or anorthite + K-feldspar + silica. The two assemblages are separated by the compatibility line anorthite–K-feldspar. An initial composition lying on a compatibility line would crystallize to a two-phase assemblage.

Figures 138–140 show how compatibility relationships can be applied to the system Al_2O_3–MgO–SiO$_2$, in which there are six intermediate compounds, four of which melt incongruently. Figure 138 shows only the fields of the primary phases on the liquidus, with no indications of the liquidus temperatures. The first step is to mark the compositions of all the intermediate compounds and draw all the compatibility lines (Fig. 139). Note that not all the compositions are connected by compatibility lines; for instance, periclase and cordierite are not connected by a compatibility line because, as can be seen from Fig. 138, they never crystallize together from the liquid. The compatibility triangles show that there are nine different three-phase mineral assemblages that can result from the crystallization of liquids in this system (in addition to one- or two-phase assemblages from liquids lying on compatibility lines):

Tridymite + protoenstatite + cordierite
Protoenstatite + forsterite + cordierite
Forsterite + periclase + spinel
Forsterite + cordierite + spinel
Tridymite + cordierite + mullite
Cordierite + sapphirine + mullite
Cordierite + sapphirine + spinel
Sapphirine + mullite + spinel
Mullite + spinel + corundum

Figure 138. The system MgO–Al_2O_3–SiO$_2$ showing the primary phase fields on the liquidus, i.e. the identities of the first solid phase to crystallize from the liquid.

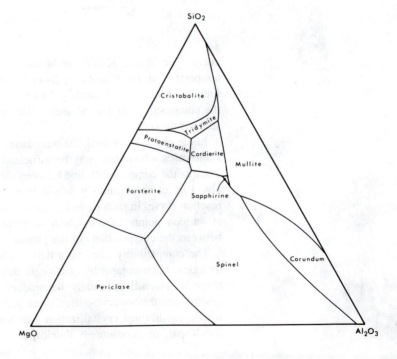

Figure 139. The system MgO–Al₂O₃–SiO₂ showing the compatibility lines. The three sides of the diagram are also compatibility lines. Each of the minerals is marked as having a fixed composition; small variations due to solid solution are ignored.

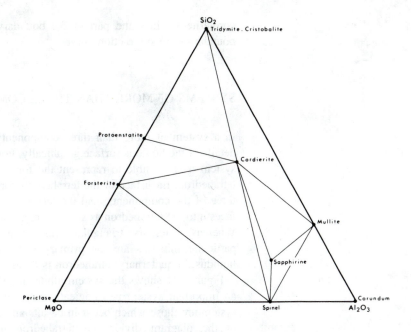

In Fig. 140, the compatibility lines have been used to find the downward directions on the liquidus surface, indicated by the arrows on the boundaries. Altogether there are nine invariant points, one for each of the three-phase sub-solidus assemblages. Three of them are eutectic points (E₁–E₃) as can be seen from the directions of the arrows. Some of the boundaries are reaction curves (protoenstatite–forsterite, sapphirine–spinel, sapphirine–mullite,

Figure 140. The system MgO–Al₂O₃–SiO₂ showing downward directions of the liquidus surface along the phase boundaries. Eutectic points are marked E₁–E₃.

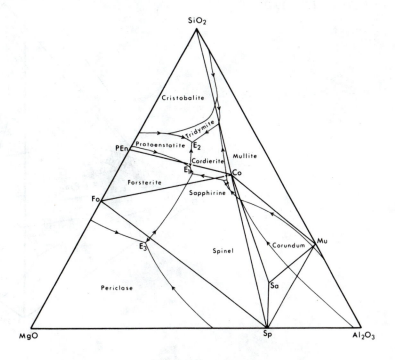

cordierite–mullite, and part of the boundary mullite–corundum); the other boundaries are subtraction curves.

SYSTEMS OF MORE THAN THREE COMPONENTS

In a system of more than three components it is difficult to show all the details of the liquidus surface graphically. For a four-component (quaternary) system it is possible to represent the four components by the corners of a tetrahedron. Each face of the tetrahedron corresponds to a ternary system of three of the components, and the ternary boundaries are projected from the faces into the tetrahedron as surfaces representing the quaternary boundaries. Whereas a ternary triangular diagram is divided into areas in which particular minerals are the primary phase (the first to crystallize on the liquidus), a quaternary tetrahedron is divided into primary-phase volumes.

Figure 141 shows the system albite–anorthite–diopside–forsterite. This is an important system petrologically because the phases crystallizing in the system are those which occur in basalts and gabbros. The boundaries shown in the diagram divide the tetrahedron into phase volumes in which plagioclase, diopside, forsterite and spinel are the liquidus phases. The presence of spinel shows that this is a pseudo-quaternary system, since the

Figure 141. The pseudo-quaternary system albite–anorthite–diopside–forsterite (after Yoder and Tilley 1962).

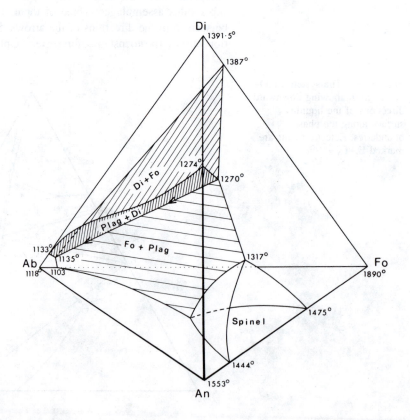

composition of spinel itself does not lie within the tetrahedron. Leaving aside spinel, most liquids in the system would initially lie within the phase volume of either plagioclase, diopside or forsterite, and on crystallizing the appropriate mineral would migrate to one of the three-phase surfaces (two minerals in equilibrium with liquid). Two minerals would then crystallize together until the liquid reached the four-phase curve (three minerals in equilibrium with liquid). This curve is shown on the diagram as falling from 1270 °C to 1135 °C. The liquid composition would migrate down this curve as the crystallization of calcic plagioclase (+ diopside + forsterite) left the liquid enriched in the constituents of sodic plagioclase.

A tetrahedron with internal detail is rather difficult to draw, and in practice it is usually necessary either to project some of the details on to one particular face of the tetrahedron, or to show cross-sections through the tetrahedron. A triangular cross-section may look like an ordinary ternary diagram but it will usually contain quaternary compounds whose compositions do not lie on the plane of the section, and is thus a pseudo-ternary section, or ternary join. A four-phase univariant line in a quaternary system, corresponding to three solids in equilibrium with a liquid, may have the appearance of a ternary invariant point when intersected by a pseudo-ternary join; such an intersection is called a 'piercing point'.

HIGH PRESSURES

The phase diagrams discussed up to now have represented equilibria at atmospheric pressure. Silicate phase relations are affected by pressure with regard to both the melting temperatures and the nature of the solid phases. The effect of pressure on the melting temperatures is shown by albite and diopside, whose melting points rise with pressure as shown in Fig. 142. Most silicate melting temperatures rise with increasing pressure, so long as there is no water present.

A system which shows several of the effects of increasing pressure is $KAlSi_3O_8$–SiO_2 (Fig. 143). In this system, there are minerals with the following densities (in g/cm^3):

Tridymite	2.27
Cristobalite	2.33
Leucite	2.47
Sanidine	2.56
Quartz	2.65

As might be expected, the stability fields of the low-density phases disappear as the pressure is increased. With the elimination of the leucite field, K-feldspar melts congruently at 20 kilobars. For most compositions in the system, including the eutectic composition, melting temperatures increase as the pressure increases, although this is not true for the low-density phase, leucite. The eutectic composition shifts appreciably, from 42% SiO_2 to 28% SiO_2, as the pressure rises from 0 to 20 kilobars. A liquid which acquired a eutectic composition by fractional crystallization at depth (high pressure) might no longer have a eutectic composition if moved to a higher level (low pressure).

Figure 142. The effect of pressure on the solidus temperatures of albite and diopside (after Boettcher *et al.* 1982).

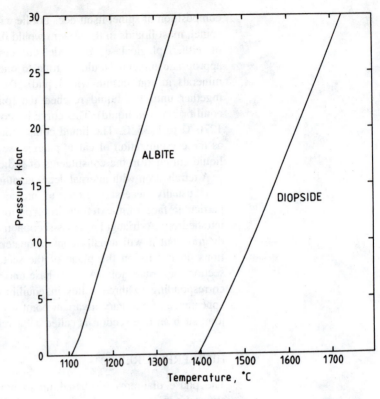

Figure 143. The system orthoclase–silica at various pressures (after Luth 1968).

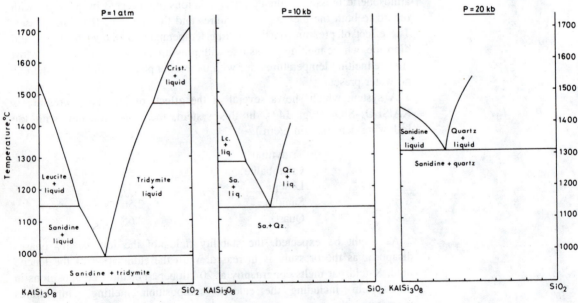

SYSTEMS CONTAINING WATER

Water as a component in silicate systems has a profound effect on the phase relations. Because pure water has much lower melting and boiling points than any silicate, we think of H_2O as a volatile component. This means that its miscibility with silicate melts, its reactions with silicate solids, and its effect on silicate melting temperatures all depend on the confining pressure. For instance, if a bucket of water is thrown on to a lava flow, it will rapidly turn to steam and be blown away. It may chill a portion of the magma, but it will not alter the actual phase relations or change the mineralogy of the rock. Only if the water is held in contact with the magma by a confining pressure, as it might be under plutonic conditions, will it participate in the crystallization and influence the phase relations.

The effective concentration of water in a system is therefore measured by the pressure under which it is held in the system. In a melt which is saturated with water, so that there is free water present, the pressure of water (P_{H_2O}) will equal the total confining pressure (P_{total}). If there is insufficient water to saturate the melt, P_{H_2O} may be less than P_{total}. In experimental studies it is much easier to saturate the experimental charge with water than to adjust the P_{H_2O} value to differ from P_{total}, so most of the phase equilibria that have been studied are of H_2O-saturated compositions. In natural magmas, it is likely that the H_2O content is often high enough to affect the phase equilibria without being great enough for P_{H_2O} to equal P_{total}, and this must be allowed for in applying experimental phase equilibria to actual problems.

The solubility of H_2O in silicate melts of common compositions is shown in Fig. 144. In general, the solubility increases with pressure, but decreases with temperature.

The principal result of high water pressures is to lower melting temperatures. The melting points of albite in the presence and absence of water are:

0 bars	1120 °C
5000 bars, dry	1180 °C
5000 bars, water-saturated	750 °C

It can be seen that the melting temperature is altered much more by water under pressure than by pressure alone, and in the opposite direction.

In describing hydrous phase equilibria, systems containing three components including H_2O are usually represented by binary diagrams for the other

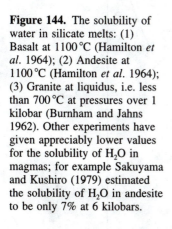

Figure 144. The solubility of water in silicate melts: (1) Basalt at 1100 °C (Hamilton *et al.* 1964); (2) Andesite at 1100 °C (Hamilton *et al.* 1964); (3) Granite at liquidus, i.e. less than 700 °C at pressures over 1 kilobar (Burnham and Jahns 1962). Other experiments have given appreciably lower values for the solubility of H_2O in magmas; for example Sakuyama and Kushiro (1979) estimated the solubility of H_2O in andesite to be only 7% at 6 kilobars.

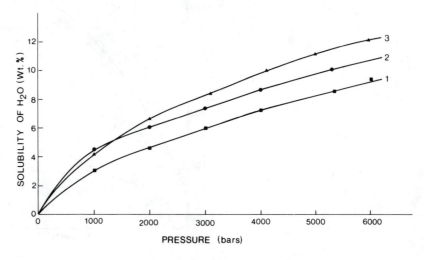

two components at fixed values of P_{H_2O}, and similarly four-component systems are represented by ternary diagrams for specified values of P_{H_2O}. The system $NaAlSi_3O_8$–$CaAl_2Si_2O_8$–H_2O (Fig. 145) is an example of a three-component system in which one of the components is water.

The effect of variations in the water pressure during crystallization can be inferred from the plagioclase phase diagram (Fig. 145). A melt with the composition $An_{50}Ab_{50}$ at a temperature of 1100 °C would be completely molten at $P_{H_2O} = 6$ kilobars, but at $P_{H_2O} = 5$ kilobars it would be in equilibrium with crystals of composition An_{90}, and at $P_{H_2O} =$ zero it would quench completely to crystals with a bulk composition of An_{50}. If the water pressure in a body of magma were to fluctuate during crystallization, perhaps as a result of intermittent eruption of an overlying volcano, the plagioclase crystals might show the alternating effects of fractional crystallization, equilibrium crystallization, and resorption as the water pressure went up and down. Such variations are the probable cause of the oscillatory zoning shown by many plagioclase crystals in volcanic rocks.

Figure 145. The system albite–anorthite–H_2O at various water pressures (after Johannes 1978).

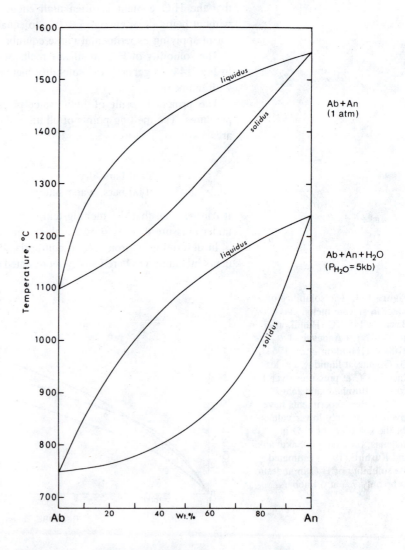

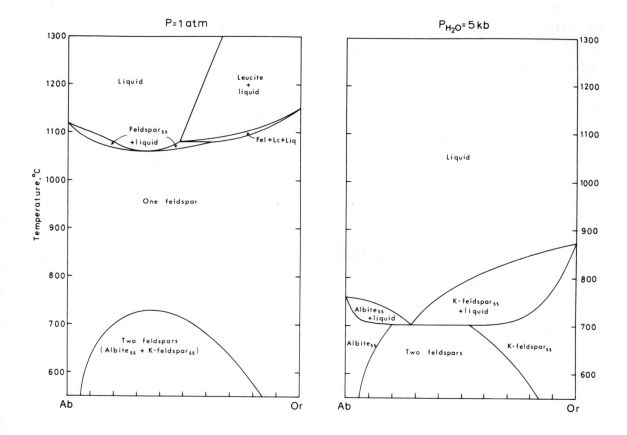

Figure 146. The system albite–orthoclase–H₂O at two different water pressures (after Tuttle and Bowen 1958; Morse 1970).

The effect of high water pressure on the alkali feldspars is shown in Fig. 146. The disappearance of the leucite field in the 5-kilobar phase diagram is partly due to the increase in pressure alone, because leucite is a low-density phase. The most important consequence of the presence of water is the lowering of the melting temperatures. As a result, at 5000 bars the solidus intersects the solvus, and instead of a single homogeneous feldspar crystallizing from the melt it is now possible for two separate alkali feldspars to crystallize together, one rich in the $KAlSi_3O_8$ end-member and the other rich in the $NaAlSi_3O_8$ end-member.

As well as lowering the melting temperature, a high water pressure may also change the composition of eutectic points or liquidus minima. This is illustrated by the system $KAlSi_3O_8$–$NaAlSi_3O_8$–SiO_2–H_2O (Fig. 147). At a low water pressure, there is a minimum on the liquidus surface which lies half-way along the quartz–feldspar cotectic curve. Crystallizing liquids which have reached the quartz–feldspar boundary migrate towards this minimum from either side (in contrast to the system Ab–An–SiO_2 shown in Fig. 133, where the minimum is at one end of the cotectic curve). At a water pressure of 10 kilobars, the minimum is found to be a eutectic point, because the lowest region of the liquidus is now below the temperature of the alkali feldspar solvus (see Fig. 146), and albite and K-feldspar crystallize as separate minerals from the melt. Furthermore, the eutectic composition is appreciably poorer in SiO_2 than the low-pressure minimum.

Figure 147. The system albite–orthoclase–silica–H$_2$O at two different water pressures (after Tuttle and Bowen 1958; Luth *et al.* 1964).

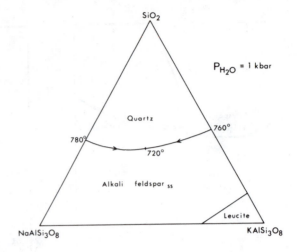

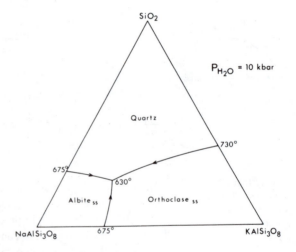

Two additional roles of water are important in magmas. One is that water decreases the viscosity of magmas. The other is that water acts as a catalyst in bringing about reactions that might otherwise be very slow. In particular, the crystallization of glassy melts is much more rapid at high water pressure than in the dry state.

Magmas may contain volatile constituents other than water, but on the evidence of volcanic gases and fluid inclusions in minerals the only one which may be present in petrologically significant amounts is carbon dioxide. It is more abundant in basic magmas than in acid magmas, and especially in alkaline basic magmas. The maximum amount of CO_2 which can be dissolved in magma at the liquidus temperature under a pressure of 10 kilobars is about 0.2% for tholeiite, and 0.6% for andesite and for olivine melilitite (Spera and Bergman 1980). The presence of CO_2 depresses the liquidus temperatures of magmas, but only by a slight amount compared with the effect of H$_2$O (Fig. 148).

Figure 148. The effect of H_2O and CO_2 on the solidus of albite (after Bohlen *et al.* 1982).

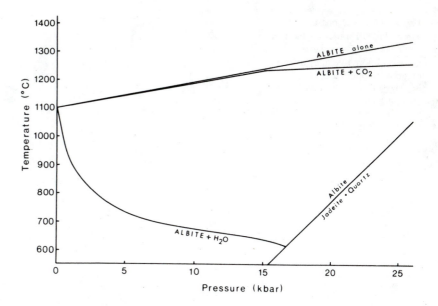

MELTING AND CRYSTALLIZATION OF NATURAL ROCKS

Much useful information can be gained from the phase relations of synthetic systems, but actual rocks are too complex to be perfectly represented by synthetic systems of only three or four components. Over the course of time there has been an increasing emphasis on experiments on the melting relationships of actual rock samples.

Various kinds of information may be sought in such a study, for example the liquidus temperature, the solidus temperature, the identity of the first solid phase to crystallize at the liquidus temperature, the identity of the crystalline phases present at various temperatures below the liquidus, and the composition of the residual liquid remaining at various degrees of crystallization. The latter can be found by actually analysing the quenched glass from a high temperature experiment using an electron probe microanalyser.

Figure 149 shows the phase diagram of a plagioclase–garnet–pyroxenite determined experimentally over a wide range of temperatures and pressures. The liquidus is readily identifiable as the line separating the field of liquid only from the fields containing liquid + crystals. The solidus is the line separating the fields of crystals + liquid from the fields of crystals without liquid. It can be seen how both liquidus and solidus temperatures increase with pressure, and how the solidus phase assemblage (i.e. the mineralogy of the rock) changes with increasing pressure.

Figure 150 shows the melting behaviour of a peridotite. The diagram shows the phase assemblage present at various temperatures, plotted against the degree of melting. For this particular rock, the experimenters did not determine the exact solidus and liquidus temperatures (about 1400 °C and

Figure 149. The phase assemblage of a rock of basaltic composition (plagioclase–garnet–clinopyroxenite) over a range of temperatures and pressures (after Obata and Dickey 1976).

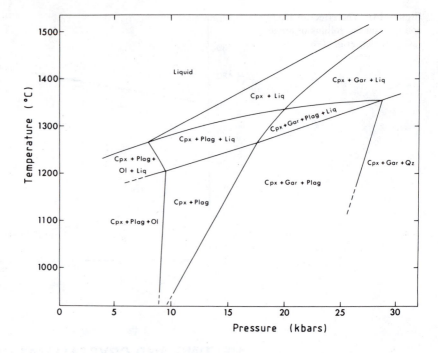

Figure 150. The melting behaviour of a peridotite with and without the presence of water (after Mysen and Kushiro 1976).

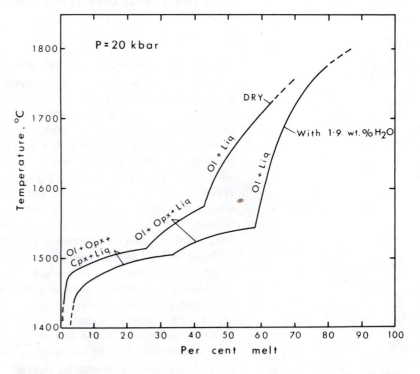

1800 °C respectively) as they were more interested in the varying composition of the melts formed at different degrees of partial melting.

Figure 151 shows the melting behaviour of a basalt at various pressures in the presence of excess water. The curve marked 'Beginning of melting' is

Figure 151. The melting relations of an olivine tholeiite in the presence of excess water (after Yoder and Tilley 1962).

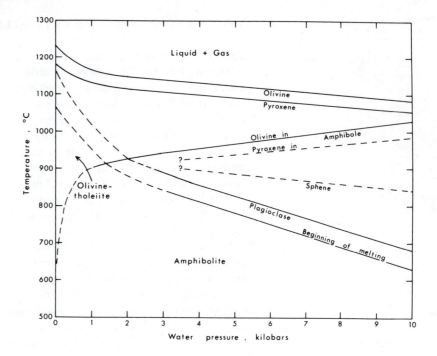

the solidus. Below this temperature, material of this composition would be completely solid; its mineral assemblage would be that of an olivine tholeiite at low water pressures or an amphibolite at high water pressures. If the temperature were raised above the solidus, melting would start, and an increasing proportion of the rock would be melted until at the liquidus temperature it was completely liquid. For example, at $P_{H_2O} = 5$ kilobars the solidus temperature is 780 °C, and the temperature has to be raised to 1130 °C before all the silicate minerals are melted.

If this basalt melt were cooled from 1300 °C down to 700 °C at 5 kilobars, the sequence of crystallization would be:

1. Olivine crystals start to appear at 1120 °C.
2. Pyroxene crystals start to appear at 1090 °C.
3. Amphibole starts to form at 965 °C by reaction of olivine and pyroxene with the melt. At this temperature all the olivine is used up and by 940 °C all the pyroxene is also used up.
4. Amphibole continues to crystallize from the melt until the temperature reaches 890 °C, when it is joined by sphene.
5. At 825 °C, plagioclase starts to crystallize, and by 780 °C all the melt is used up. The final solid assemblage is amphibole + plagioclase + sphene.

On heating up such a rock from 700 °C to 1300 °C, a similar sequence of changes would take place in the reverse order, with plagioclase being the first solid phase to be completely melted.

MELTING COMPARED WITH CRYSTALLIZATION

To a certain extent, melting behaviour is the reverse of crystallization behaviour, but there are important differences. The rocks in the Earth's mantle and crust that may undergo melting are very varied, but most of them have a large difference between their solidus and liquidus temperatures. In an episode of melting, it will be normal for them to be heated above their solidus temperature, but not as far as their liquidus temperature, so they will not melt completely. An episode of partial melting rather than complete melting is therefore the ultimate source of nearly all magmas. The most important thing to realize about partial melting is that it nearly always results in a melt which is different in composition from the starting material as a whole. It contains the low-melting constituents of the starting material, but the high-melting constituents remain in the unmelted residuum.

Just as the crystallization of a melt may take place in either an equilibrium manner or a fractional manner, so partial melting may take place in either an equilibrium manner or a fractional manner. Equilibrium melting is a process in which the liquid produced on heating remains continuously in equilibrium with the crystalline residue. Fractional melting is a process in which the liquid is separated from the crystalline residue and is unable to react with it and does not remain in equilibrium with it. In the natural situation, one can envisage a range of circumstances between perfect equilibrium fusion and perfect fractional fusion.

The differences between equilibrium and fractional fusion have been analysed by Presnall (1969) and Morse (1980). Equilibrium fusion is simply the reverse of equilibrium crystallization, but fractional fusion is not the reverse of fractional crystallization. During fractional crystallization, the composition of

Figure 152. The system diopside (Di)–forsterite (Fo)– pyrope (Py) at a pressure of 40 kilobars (after Davis and Schairer 1965; Yoder 1976).

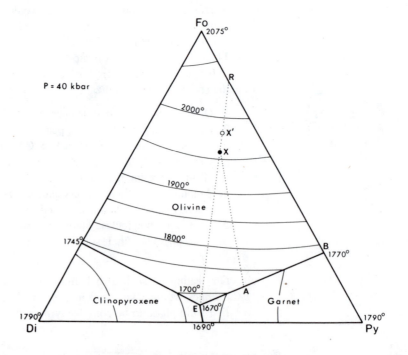

the liquid changes continuously as it cools, but during fractional fusion the successive liquids show compositional gaps. These correspond to temperature intervals in which addition of heat causes no additional melting. Consider the fractional melting of composition X in Fig. 152. Melting would start at 1670 °C and give a liquid of composition E. If this liquid is removed as it is formed, the composition of the residue will move from X to X'. Continued melting yields further batches of melt with the composition E and this does not change until the residue has the composition R, when there is no diopside component left in the residue. Melting then ceases and cannot resume until the temperature is raised to 1770 °C, when the liquid produced will have composition B. In contrast, during fractional crystallization of a liquid with initial composition X, the change in liquid composition is continuous along the path X to A and then A to E.

OXIDATION AND REDUCTION

The oxidation state of magmas has an important bearing on their course of crystallization and on the crystallization products. Iron is the most important element that occurs geologically in more than one oxidation state, although other elements such as manganese and chromium are also affected by oxidation and reduction. Iron occurs in three oxidation states:

1. Fe^0 – as in native iron. This is abundant in meteorites and present in lunar rocks, and metallic iron in a molten condition is probably the main constituent of the Earth's core.
2. Fe^{II} – as in fayalite (Fe_2SiO_4) and other ferromagnesian minerals.
3. Fe^{III} – as in hematite (Fe_2O_3) or aegirine ($NaFeSi_2O_6$).

Most ferromagnesian minerals actually contain both ferrous and ferric iron, as does magnetite.

The concentration of oxygen in magmas may be expressed as its fugacity, f_{O_2}. The fugacity is a measure of the 'escaping tendency' of a component held in solution. It may be regarded as an idealized partial pressure, and it takes account of deviations from ideal gas behaviour at high pressures. The fugacity is related to the partial pressure by the 'fugacity coefficient', γ:

$$f_{O_2} = \gamma P_{O_2}$$

The difference between f_{O_2} and P_{O_2} is small in relation to the large variations in f_{O_2} encountered in magmas, and negligible at very low pressures.

In experiments concerning the effect of oxidation on phase equilibria, the oxygen fugacity is kept at a fixed value by the use of oxidation buffers. The buffer mixtures are made up of the participants in the following reactions, which proceed to the right at increasing oxygen pressures:

$$(1) \quad 2Fe + O_2 + SiO_2 \rightleftharpoons Fe_2SiO_4$$
$$(2) \quad 2Fe + O_2 \rightleftharpoons 2FeO$$
$$(3) \quad 6FeO + O_2 \rightleftharpoons 2Fe_3O_4$$
$$(4) \quad 3Fe_2SiO_4 + O_2 \rightleftharpoons 2Fe_3O_4 + 3SiO_2$$
$$(5) \quad 4Fe_3O_4 + O_2 \rightleftharpoons 6Fe_2O_3$$

Figure 153. Plot of log f_{O_2} against temperature for assemblages used as oxygen buffers (calculated from the data of Myers and Eugster 1983).

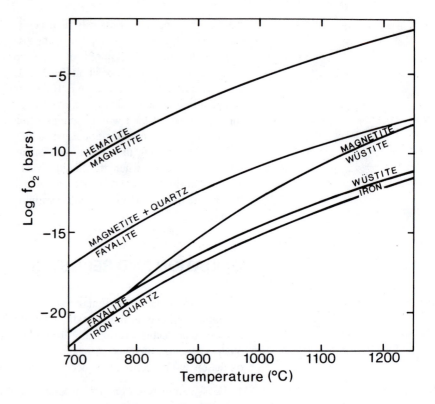

The fugacities corresponding to these reactions are shown in Fig. 153. The buffer mixtures do not necessarily have to contain iron; other buffer mixtures which can be used are Ni–NiO and MnO–Mn_3O_4.

In general, oxygen fugacities during the crystallization of igneous rocks are not sufficiently low for the formation of native iron or wüstite, and not sufficiently high for the formation of hematite. Exceptionally reducing conditions are encountered in a number of special circumstances, for example in magmas contaminated by carbonaceous sediment (Pedersen and Rønsbo 1987) or in peridotites undergoing serpentinization (Frost 1985). Exceptionally oxidizing conditions occur in some near-surface environments, for example in fumaroles, or during the cooling of some lava lakes (Sato and Wright 1966).

In normal igneous rocks the iron is present mainly in ferro-magnesian silicate minerals, and an increase in oxygen fugacity would be marked by the formation of magnetite at the expense of the iron-bearing silicates. The actual ratio of Fe_2O_3 to FeO in a crystallized igneous rock is not a direct measure of the oxygen fugacity because it also depends on the bulk composition, including constituents other than iron and oxygen (Dickenson and Hess 1986).

The first method to be developed for measuring the oxygen fugacity during crystallization of igneous rocks was based on the compositions of coexisting magnetite and ilmenite. Both minerals show isomorphous substitution:

$$Fe^{II} (Fe^{III}Fe^{III})O_4 \; — \; Fe^{II} (Fe^{II}Ti^{IV})O_4$$
$$\text{(magnetite)} \qquad\qquad \text{(ulvöspinel)}$$

$$Fe^{II}Ti^{IV}O_3 \; - \; Fe^{III}Fe^{III}O_3$$
$$\text{(ilmenite)} \quad \text{(hematite)}$$

As conditions become more oxidizing, the amount of hematite in solid solution in ilmenite increases, whereas the amount of ulvöspinel in solid solution in magnetite decreases. These solid solutions depend on temperature as well as oxygen fugacity, but Buddington and Lindsley (1964) showed how the compositions of coexisting magnetite and ilmenite could be used to determine the oxygen fugacity during crystallization, and some results which have been obtained for volcanic rocks are shown in Fig. 154. The calculated values are mostly a little more oxidizing than the fayalite–magnetite–quartz (FMQ) buffer, i.e. sufficiently high for magnetite to be often present in volcanic rocks, but not high enough for hematite.

The Fe–Ti oxide method of oxygen barometry has been considerably refined by Ghiorso and Sack (1991), and additional methods have been developed based on the composition of biotite coexisting with Fe–Ti oxides (Patiño Douce 1993), the composition of the spinel in spinel–peridotites (O'Neill and Wall 1987; Mattioli *et al.* 1989; Ballhaus *et al.* 1991), and the composition of garnet in garnet–peridotites (Luth *et al.* 1990). Direct measurements of intrinsic oxygen fugacity have also been made using oxygen-specific ZrO_2 solid electrolyte sensors (Kersting *et al.* 1989).

Oxygen fugacities determined for mid-ocean ridge basalts and upper mantle peridotites are mostly close to or slightly below the FMQ buffer, but there is some evidence that the mantle becomes more reduced with depth (Daniels and

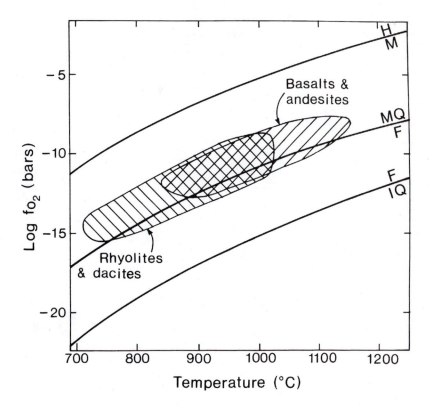

Figure 154. Equilibration temperatures and oxygen fugacities for acid and basic lavas measured from Fe–Ti oxide compositions (data compiled by Carmichael *et al.* 1974; Gill 1981).

Gurney 1991). The redox state of the mantle has many implications for the geochemistry of H, C and S. For carbon to be stored in the mantle as graphite or diamond, conditions need to be appreciably more reducing than FMQ; under relatively oxidizing conditions it would be present as CO_2 or carbonate, and thus redox conditions in the source region control the amount of CO_2 in basalt magmas. Similarly, the oxygen fugacity determines whether S is present in magmas as sulphide or sulphate, with consequences for the behaviour of chalcophile elements in magmas and their potential for giving rise to mineralization.

The apparently more oxidized state of volcanic rocks compared with the upper mantle has not yet been fully explained. One possibility is that magmas undergo auto-oxidation by some such reaction as:

$$2FeO + H_2O \rightarrow Fe_2O_3 + H_2$$

with the oxygen fugacity being controlled by the difference in permeability of the cover rock for H_2O and H_2, but this explanation may not be entirely adequate (Carmichael 1991).

CHAPTER 6

Isotopic composition

There are just over 100 chemical elements, of which 93 have now been found to occur in nature. Each element has a fixed number of protons in the atom; this number is the *atomic number* (Z). Most elements consist of more than one isotope, differing in the number of neutrons (N) present in the atom. Isotopes are denoted by their *mass number* (A), which is the total number of protons and neutrons in an atom:

A (mass number) = Z (number of protons) + N (number of neutrons)

Protons and neutrons are of roughly the same mass, and the mass number approximates to the atomic weight.

As an example, magnesium has three naturally occurring isotopes:

$$^{24}\text{Mg} \quad (78.7\%)$$
$$^{25}\text{Mg} \quad (10.1\%)$$
$$^{26}\text{Mg} \quad (11.2\%)$$

The atomic number (Z) is 12, and the isotopes contain 12, 13 and 14 neutrons respectively. Each is denoted by its mass number, written as a superscript. The atomic weight of naturally occurring magnesium is 24.3, which reflects the relative abundance of the three isotopes (indicated in brackets).

Some isotopes are stable. Others are unstable, and over the course of time break down into isotopes of other elements by radioactive decay. All the isotopes of magnesium are stable, but all the isotopes of uranium are unstable. Some elements are a mixture of stable and radioactive isotopes, e.g. potassium, which is a mixture of two stable isotopes and one unstable isotope:

$$^{39}\text{K} \quad (93.08\%) \text{ stable}$$
$$^{40}\text{K} \quad (0.01\%) \text{ radioactive}$$
$$^{41}\text{K} \quad (6.91\%) \text{ stable}$$

The different isotopes of an element have the same chemical properties and behave in the same way during magmatic crystallization and other geological processes. Therefore the magnesium in a basalt might be expected to contain the same isotopes in the same proportions as the magnesium in a granite, or in a sedimentary or metamorphic rock. In general this is true, but it is possible to detect small differences in the isotopic ratios of a number of elements.

Variations in isotopic composition arise for two reasons. One is that the individual isotopes of an element differ very slightly in their physical properties, which can lead to them becoming partially separated: this is called isotopic fractionation. The other is that high concentrations of radioactive elements lead to high concentrations of the stable isotopes which are their decay products: this is enrichment in radiogenic isotopes.

FRACTIONATION OF STABLE ISOTOPES

Isotopic fractionation can be seen in the isotopes of hydrogen, which has two main isotopes, ^1H and ^2H (the latter is also called heavy hydrogen or deuterium, D). Because atoms of ^2H are heavier than atoms of ^1H, molecules of ^2H_2O are heavier than molecules of ^1H_2O. This causes a slight difference in their physical properties:

	^1H_2O	^2H_2O
Boiling point	100.0 °C	101.4 °C
Freezing point	0.0 °C	3.8 °C

Light water molecules (^1H_2O) evaporate more readily than heavy water molecules (^2H_2O). They therefore concentrate in the atmosphere and hence in rain and river water, leaving the oceans enriched in ^2H_2O.

Isotopic fractionation of this kind can occur with the isotopes of any element, but is greatest where there is a large relative difference in mass between the atoms of the two isotopes. It occurs to a measurable extent in the isotopic compositions of hydrogen, carbon, oxygen and other elements whose isotopes have low mass numbers, but is negligible for the elements whose atoms have high mass numbers, such as silicon or iron. It can arise during the operation of many different geological processes, such as melting, solution, crystallization from melt or solution, diffusion, or ion-exchange. Elements suffer the most severe isotopic fractionation under sedimentary conditions, because fractionation is greatest at low temperatures and because of the many opportunities for fractionation during biochemical reactions. In igneous petrology, the most important elements from the point of view of stable isotope fractionation are oxygen, hydrogen and sulphur.

OXYGEN AND HYDROGEN ISOTOPES

The proportions of the oxygen and hydrogen isotopes in nature are approximately $^{16}O : ^{17}O : ^{18}O = 99.76 : 0.04 : 0.20$, and $^1H : ^2H = 99.98 : 0.02$. The major variations in these ratios occur as a result of the evaporation and condensation of the Earth's surface waters, and there is thus a close relationship between the isotopic distributions of the two elements. This is why they are usually considered together.

For oxygen, the greatest fractionation is between the heaviest and lightest isotopes, ^{18}O and ^{16}O, and is expressed as:

$$\delta^{18}O = \left[\frac{(^{18}O/^{16}O)_{sample}}{(^{18}O/^{16}O)_{standard}} - 1 \right] \times 1000$$

For hydrogen, the fractionation between ^{1}H and ^{2}H is expressed as:

$$\delta D = \left[\frac{(^{2}H/^{1}H)_{sample}}{(^{2}H/^{1}H)_{standard}} - 1 \right] \times 1000$$

The values of $\delta^{18}O$ ('delta-18-O') and δD are the deviations in parts per thousand (‰, 'per mil') of the sample from an internationally agreed standard (standard mean ocean water, SMOW).

There are six possible molecules of water containing ^{16}O or ^{18}O, whose masses range from 18 to 22 (in units of atomic weight): $^{1}H^{1}H^{16}O$, $^{1}H^{2}H^{16}O$, $^{2}H^{2}H^{16}O$, $^{1}H^{1}H^{18}O$, $^{1}H^{2}H^{18}O$, $^{2}H^{2}H^{18}O$. When seawater evaporates, the vapour is enriched in the lighter molecules and depleted in the heavier molecules. Atmospheric moisture therefore has negative values of $\delta^{18}O$ and δD. As a result, fresh water (rain, lakes, rivers) also has negative $\delta^{18}O$ and δD, even allowing for complications due to partial condensation, evaporation, freezing and thawing.

Figure 155. The relationship between mean annual $\delta^{18}O$ of rain or snow and the mean annual air temperature (after Faure 1986, based on analyses by W. Dansgaard).

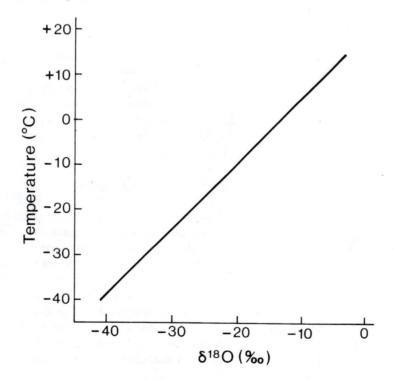

Figure 156. Average δD values of meteoric surface waters of North America, showing the variation with latitude and elevation (after Taylor 1979).

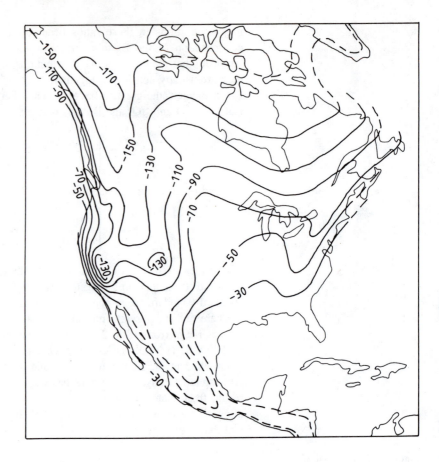

Isotopic fractionation is temperature-dependent, and is greatest at low temperatures. The isotopic composition of meteoric water shows a geographical variation which is closely related to mean annual temperatures, with the most extreme fractionation being observed at high (polar) latitudes. Meteoric water has $\delta^{18}O$ values ranging from 0 for tropical rainfall to -50 for South Polar ice (Fig. 155), and there is a similar but larger variation in δD (Fig. 156).

The oxygen isotope geothermometer

The equilibration of oxygen isotopes between individual minerals and water has been measured experimentally over a wide range of temperatures (Fig. 157). The experimental results cannot be applied directly to estimating temperatures of magmatic processes because a sample of the magma is not usually available. It is, however, possible to make use of pairs of minerals to estimate temperatures. When two coexisting minerals have equilibrated oxygen with a common reservoir at a particular temperature, the difference in their $\delta^{18}O$ values is a measure of that temperature (Fig. 158). Mineral pairs that have been used for this purpose include plagioclase–magnetite, plagioclase–pyroxene, quartz–plagioclase, quartz–muscovite, and several others.

Figure 157. Oxygen isotopic fractionation between various minerals and water as a function of temperature (after Matsuhisa *et al.* 1979). The fractionation factor, $\alpha = (^{18}O/^{16}O)_{mineral}/(^{18}O/^{16}O)_{water}$.

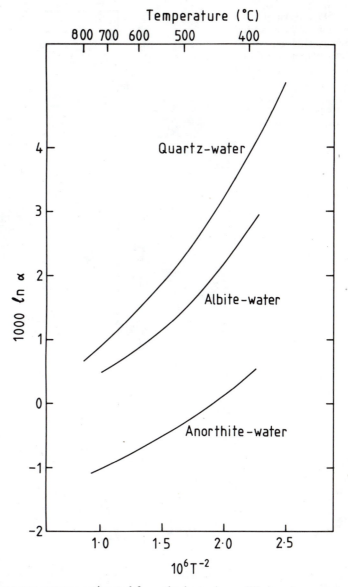

The temperatures estimated from the isotopic equilibrium between various pairs of coexisting minerals in volcanic rocks are in good agreement with each other, and with the results of independent methods of measurement, such as direct observation on lavas or the laboratory measurement of solidus temperatures. The assumptions implicit in the use of oxygen isotopes for geothermometry are that the coexisting minerals reached isotopic equilibrium with one another and with the common source of oxygen, and that the isotopic compositions have not subsequently been altered. The second assumption is not always valid. In plutonic rocks, cooling may be slow enough to permit some re-equilibration of isotopes during cooling, which leads to temperature estimates lower than those of the initial crystallization. Anderson *et al.* (1971) found that temperatures of crystallization measured by the equilibria between plagioclase, magnetite and clinopyroxene were

Figure 158. Oxygen isotopic equilibration temperatures for the assemblage plagioclase–pyroxene–magnetite in mafic igneous rocks (after Anderson *et al*. 1971). $\Delta_{plagioclase-magnetite} = \delta^{18}O_{plagioclase} - \delta^{18}O_{magnetite}$. Open circles are plutonic rocks; solid circles are volcanic rocks.

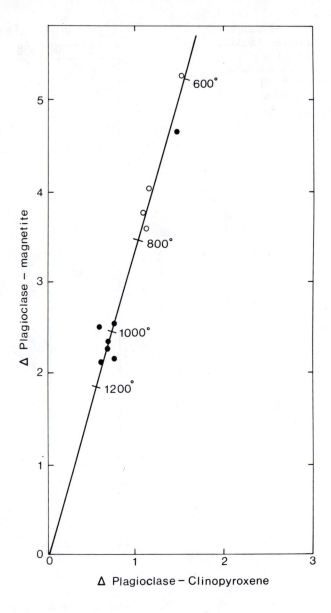

around 1000–1100 °C for basalts but below 800 °C for gabbros (Fig. 158). In both volcanic and plutonic rocks, post-magmatic hydrothermal alteration frequently results in modification of the isotopic ratios.

Characterization of magmas by oxygen isotopes

At magmatic temperatures, isotopic fractionation is much less pronounced than at atmospheric temperatures. Each mineral has a different propensity to concentrate ^{18}O. The minerals of igneous rocks have been listed by Epstein and Taylor (1967) in order of their increasing tendency to concentrate ^{18}O: magnetite (least), ilmenite, biotite, olivine, hornblende, pyroxene, anorthite,

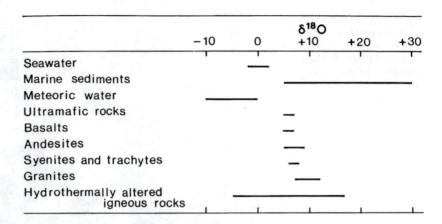

Figure 159. Approximate range of $\delta^{18}O$ in igneous and other materials.

muscovite, intermediate plagioclase, calcite, alkali feldspar, quartz (greatest). The minerals of acid igneous rocks have a greater tendency to concentrate ^{18}O than the minerals of basic igneous rocks, and therefore it is not surprising to find that acid igneous rocks generally have higher $\delta^{18}O$ values than basic igneous rocks (Fig. 159).

Normal $\delta^{18}O$ values are $+7$ to $+12$ for granites and $+5$ to $+7$ for basalts. The $\delta^{18}O$ values are controlled by a number of factors, including the $\delta^{18}O$ of the original magma, the effects of fractional crystallization or contamination, and the effects of low temperature hydrothermal activity. High values of $\delta^{18}O$ ($> +10$ for granites or $> +7$ for basalts) are almost always indicative of derivation from, or contamination by, sedimentary rocks, many of which have $\delta^{18}O$ values above the normal range for either acid or basic igneous rock types. Fractional crystallization has only a slight effect on the distribution of ^{18}O in igneous rocks, but hydrothermal alteration can have a large effect.

The effect of hydrothermal alteration on O and H isotopes

One of the principal applications of oxygen and hydrogen isotopic studies to igneous rocks is in the recognition of hydrothermal alteration, especially in granites, low values of $\delta^{18}O$ being indicative of alteration. Normal unaltered granites have $\delta^{18}O$ in the range of $+7$ to $+12$, and δD in the range -50 to -100. Unaltered magmatic compositions are also indicated by the $\delta^{18}O$ values of the individual minerals increasing in the equilibrium sequence: magnetite–biotite–hornblende–muscovite–plagioclase–alkali feldspar–quartz.

Igneous complexes that show extremely low $\delta^{18}O$ values typically show other features indicative of hydrothermal activity, such as alteration of feldspars and ferromagnesian minerals, and veining with quartz, feldspar, epidote, chlorite or sulphides. The feldspars are usually depleted in ^{18}O to a greater degree than quartz because of their greater susceptibility to alteration.

Figure 160 shows the depletion in ^{18}O in the basaltic rocks of the Skye volcanic centre in Scotland. Forester and Taylor (1977) found that the

Figure 160. Contours of $\delta^{18}O$ in basaltic rocks (lavas, dykes, tuffs, agglomerates) of the Skye igneous complex, Scotland (after Forester and Taylor 1977).

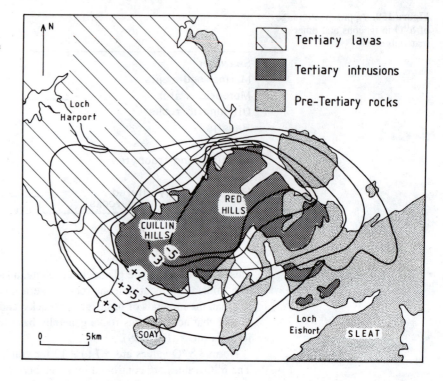

low-^{18}O character of the igneous rocks of Skye was brought about by alteration by heated groundwater during or soon after the period of igneous activity in the early Tertiary. The δD values of the hydroxyl-bearing minerals in these rocks are also much lower than normal, and depletion in ^{18}O and D is also found in the other Scottish Tertiary igneous complexes of Mull and Ardnamurchan.

In addition to recognizing that hydrothermal alteration has occurred, the depletion in ^{18}O and D may actually be used to calculate the relative amounts of water and rock involved in the hydrothermal system. For Skye, Taylor (1977) estimated the ratio of water oxygen to rock oxygen to be approximately 1:1 (atomic ratio), corresponding to a water:rock ratio of 0.5:1 by weight. The effect of various water:rock ratios on the isotopic composition of a hypothetical granodiorite is shown in Fig. 161.

There are two important but atypical igneous provinces that contain low-$\delta^{18}O$ igneous rocks that are not apparently due to postmagmatic hydrothermal alteration, i.e. some of the rhyolites of Yellowstone Park and some of the basalts and rhyolites in Iceland. According to Taylor (1987), these can be explained by assimilation or remelting of hydrothermally altered roof-rocks by a body of isotopically normal magma.

Whereas igneous rocks contain a large amount of oxygen even without the introduction of additional oxygen from meteoric sources, the hydrogen contents of magmas are low, and introduced meteoric water should have a correspondingly greater influence on the final δD than on $\delta^{18}O$, as indicated in Fig. 161. The Mesozoic and Tertiary Cordilleran batholiths of western North America show a progressive depletion in D from south to north in the

Figure 161. The calculated values of $\delta D_{biotite}$ and $\delta^{18}O_{feldspar}$ that would result from the meteoric-hydrothermal alteration of a typical granodiorite at various water/rock ratios, according to Taylor (1977). The initial composition of the meteoric waters was assumed to be $\delta D = -120$, $\delta^{18}O = -16$.

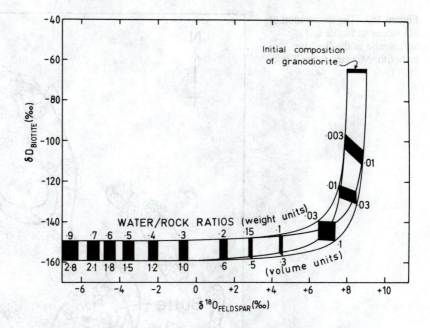

Figure 162. The distribution of δD values in biotite and hornblende in Cordilleran batholiths at different latitudes (after Taylor 1977).

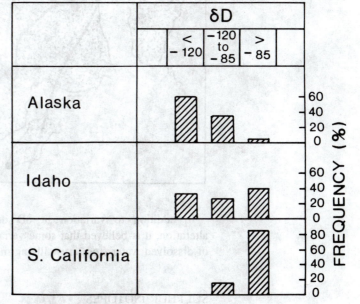

same way as present-day meteoric water (Fig. 162). This may be taken as evidence for the wholesale interaction of meteoric water with plutonic rocks throughout this large area.

Figure 163 shows the depletion in deuterium associated with hydrothermal activity during the cooling of the Boulder batholith, Montana. One of the areas of lowest δD, i.e. of most intense hydrothermal alteration, is around the important mineral deposits of Butte. As a general rule, unaltered igneous rocks have δD values in the range of -50 to -100, and values below -100 are an indication of hydrothermal alteration.

Figure 163. The distribution of δD values in biotite and hornblende in the Boulder batholith, Montana (after Taylor 1977).

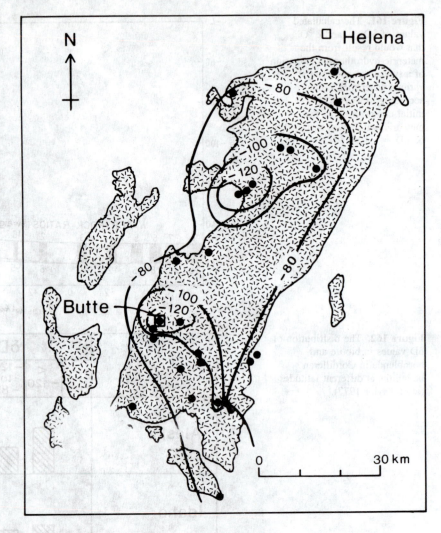

In addition to variation in δD due to post-magmatic hydrothermal alteration, it is believed that some variation may also result from exsolution of dissolved water from crystallizing magma (Nabelek *et al.* 1983).

SULPHUR ISOTOPES

Sulphur has four stable isotopes: ^{32}S (95.02%), ^{33}S (0.75%), ^{34}S (4.21%) and ^{36}S (0.02%). Isotopic variations are measured as:

$$\delta^{34}S = \left[\frac{(^{34}S/^{32}S)_{sample}}{(^{34}S/^{32}S)_{standard}} - 1 \right] \times 1000$$

The standard used for sulphur isotopes is the sulphur in the Cañon Diablo meteorite.

Variation in the proportions of sulphur isotopes is mainly brought about

Figure 164. The range of sulphur isotopic compositions observed in nature.

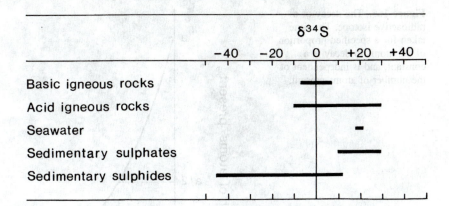

by oxidation and reduction, especially the biological reduction of sea-water sulphate to sulphide. Slight fractionation is also observed in volcanic gases, the oxidized sulphur (SO_2) being richer in ^{34}S than the reduced sulphur (H_2S).

Figure 164 shows the range of $\delta^{34}S$ in nature. Basic igneous rocks have a narrow range of $\delta^{34}S$, similar to meteorites. Granites are much more variable, and their variability suggests at least a partly sedimentary source for their sulphur, through either partial melting or contamination (Poulson *et al.* 1991).

RADIOGENIC ISOTOPES

In a sample containing a radioactive isotope, a fixed proportion of the atoms present decays in any given interval of time. This proportion is expressed by the decay constant, λ. For example the decay constant for the decay of ^{87}Rb (parent isotope) to ^{87}Sr (daughter isotope) is 0.142×10^{-10} per year, which means that out of every gram of ^{87}Rb present in a body of rock 0.142×10^{-10} grams will be transformed to ^{87}Sr during a single year, regardless of whether the ^{87}Rb is present in high or low concentration, and regardless of how much of the ^{87}Rb originally present has decayed already.

The time taken for half of any sample of a radioactive isotope to decay is called the half-life ($T_{1/2}$), which is related to the decay constant by:

$$T_{1/2} = \frac{0.6931}{\lambda}$$

As a radioactive isotope undergoes decay, 50% of the atoms originally present are left after one half-life has elapsed, 25% are left after two half-lives, 12.5% after three half-lives, and so on (Fig. 165). In the case of ^{87}Rb, the half-life is 48.8×10^9 years, and a long period of geological time has to pass before a measurable amount of the decay product ^{87}Sr can accumulate.

Figure 165. The decay of a radioactive isotope. The time taken for a specified proportion of the atoms to decay is a constant, and is independent of the number of atoms present.

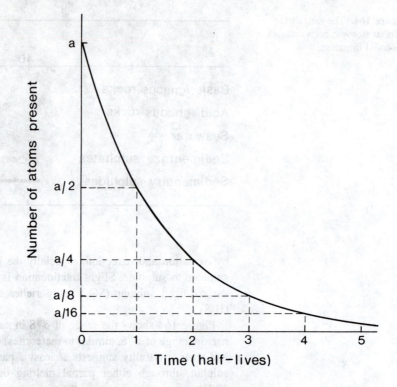

In principle, one can work out the age of an igneous rock by measuring the relative concentrations of the parent and daughter isotopes, making allowance for any daughter isotope that was present initially. Any radioactive isotope could be used for this purpose, but in practice some radioactive isotopes are too rare to be conveniently measured, while others have half-lives which are too long or too short. The principal radioactive isotopes occurring in igneous rocks are listed in Table 25. Several other naturally occurring radioactive isotopes are known, but are not widely used in the study of igneous rocks, either because their half-lives are too great or too small or because their abundance is too low.

Table 25. Decay constants and half-lives of naturally occurring radioactive isotopes

Parent isotope	Decay constant	Daughter isotope	Half-life (years)
K-40	5.543×10^{-10}/yr	Ar-40 and Ca-40	1.25×10^9
Rb-87	1.42×10^{-11}/yr	Sr-87	48.8×10^9
Sm-147	6.54×10^{-12}/yr	Nd-143	106×10^9
Th-232	4.948×10^{-11}/yr	Pb-208	14.0×10^9
U-235	9.848×10^{-10}/yr	Pb-207	0.70×10^9
U-238	1.551×10^{-10}/yr	Pb-206	4.47×10^9
Lu-176	1.94×10^{-11}/yr	Hf-176	35.7×10^9
Re-187	1.64×10^{-11}/yr	Os-187	42.3×10^9

POTASSIUM–ARGON

Potassium has three isotopes: ^{39}K (93.26%), ^{40}K (0.01%) and ^{41}K (6.73%). The isotope ^{40}K is radioactive, and gives rise to two daughter isotopes.

Approximately 11% of ^{40}K atoms change to ^{40}Ar by electron capture, and 89% change to ^{40}Ca by β-decay. The ^{40}K–^{40}Ar parent–daughter pair is invariably used for age determination because of the difficulty of measuring radiogenic ^{40}Ca in the presence of the large amounts of ordinary ^{40}Ca that nearly all rocks contain. The non-radiogenic ^{36}Ar is measured at the same time to correct for contamination of the sample by atmospheric ^{40}Ar. The half-life of ^{40}K is 1250 million years, and to accumulate a measurable amount of radiogenic argon takes many thousands of years, thus setting a lower limit to the ages that can be measured by the method. Ages as low as 5000 years have been measured but are not very precise because of the proportionately large correction for atmospheric argon.

The method has usually been applied to the dating of minerals separated from rocks, rather than to the rocks themselves. It has been found that minerals vary in their ability to retain argon. When a potassium atom changes to argon, the argon atom occupies a space in the crystal structure formerly occupied by potassium. It is not held in place by chemical bonds to the neighbouring anions and is easily lost if there is any recrystallization. In practice, the feldspars of igneous rocks are much more prone to recrystallization than the micas, and always tend to give lower K/Ar ages. The loss of argon from any mineral is greater at high than at lower temperatures, and argon is invariably lost during regional or contact metamorphism. A large igneous intrusion may take several million years to cool to say 300 °C, losing argon all the time. The accumulation of radiogenic argon cannot start until the rocks have cooled below the temperature at which argon is able to diffuse away (the 'blocking temperature'). This temperature is not a fixed value, and varies from mineral to mineral. In general, one can say that the K/Ar method dates the most recent time that a rock has cooled down to a low temperature. Potassium–argon dates on volcanic rocks are more satisfactory than those on plutonic rocks because cooling is much faster, and the K/Ar date is not significantly different from the date of crystallization.

Two refinements of the K/Ar method are the $^{40}Ar/^{39}Ar$ technique and the stepwise heating technique. In the $^{40}Ar/^{39}Ar$ technique, the sample is subjected to neutron bombardment in a nuclear reactor, which converts ^{39}K to ^{39}Ar. The ratio $^{40}Ar/^{39}Ar$ is then measured by mass spectrometry, the ^{39}Ar being a measure of the amount of undecayed potassium. This technique is more precise than carrying out separate ^{40}K and ^{40}Ar measurements on different portions of the sample being dated. Stepwise heating involves measuring the $^{40}Ar/^{39}Ar$ ratios of the argon fractions released at successive temperatures, and discriminates between argon held in crystal structures of different retentivity. It gives rather more information on the thermal history of the sample than could be obtained from the bulk $^{40}K/^{40}Ar$ ratio alone.

The main application of the K/Ar method in igneous petrology is in the dating of young (i.e. Tertiary and Quaternary) volcanic rocks. Older rocks are more likely to have lost argon and can be more reliably dated by the Rb/Sr method. Potassium–argon dates are particularly helpful in indicating the duration and frequency of magmatic events in particular igneous provinces, in revealing the migration of active volcanicity in linear volcanic chains such as the Hawaiian Islands, and in measuring the rate of magma generation at oceanic spreading axes.

RUBIDIUM–STRONTIUM

Rubidium-87 decays to strontium-87 with a half-life of 48.8×10^9 years. This half-life has proved very difficult to measure accurately, and for a long time different values of between 47 and 50×10^9 years were used by various workers while awaiting a more accurate determination of the decay constant. This should be taken into account when comparing older published Rb/Sr dates with one another or with the results of other dating methods.

Strontium in nature has four isotopes: ^{88}Sr (82.5%), ^{87}Sr (~7.0%), ^{86}Sr (9.9%) and ^{84}Sr (0.6%). A mineral contains a fixed proportion of Sr isotopes when it crystallizes initially, but the decay of ^{87}Rb causes ^{87}Sr to increase over time relative to the other Sr isotopes. The increase in ^{87}Sr content with time is given by the equation:

$$^{87}Sr = {}^{87}Sr_{initial} + {}^{87}Rb(e^{\lambda t} - 1)$$

The amount of enrichment in ^{87}Sr can be specified by comparing it with a reference isotope (^{86}Sr) which is not affected by the presence of Rb.

Each ^{87}Rb atom decays to form one ^{87}Sr atom, so the $^{87}Sr/^{86}Sr$ ratio rises at the same rate as the $^{87}Rb/^{86}Sr$ ratio falls (Fig. 166). It takes 48,800 million years for the $^{87}Rb/^{86}Sr$ ratio to fall to half of its original value (i.e. one half-life of ^{87}Rb), so even in very old rocks the actual changes in the isotope ratios are small. For example, in a typical granite, the ratios might be as follows:

	Initially	After 100 million years
Rb (ppm)	215	215
^{87}Rb (ppm)	59.8	59.7
Sr (ppm)	256	256
^{87}Sr (ppm)	17.9	18.0
^{86}Sr (ppm)	25.3	25.3
$^{87}Sr/^{86}Sr$	0.708	0.711
$^{87}Rb/^{86}Sr$	2.363	2.360

Figure 166. The change in isotopic ratios which results from the decay of ^{87}Rb to ^{87}Sr. The ratio $^{87}Sr/^{86}Sr$ increases as the ratio $^{87}Rb/^{86}Sr$ falls. The greater the Rb content of the rock, the greater the increase in $^{87}Sr/^{86}Sr$ in a given amount of time.

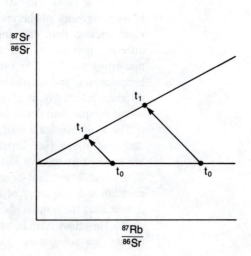

Figure 167. Sr isochron diagram for the coexisting minerals in an igneous rock. The ratio $^{87}Rb/^{86}Sr$ indicates the relative proportions of Rb and Sr in the different minerals when the rock formed. The ratio $^{87}Sr/^{86}Sr$ is a measure of the amount of enrichment in radiogenic strontium.

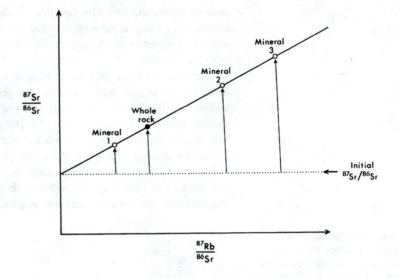

To measure ages by the Rb/Sr method it is necessary to measure the isotopic composition of several minerals from the same rock, or several rocks from the same magma source. The isotopic compositions are plotted on a diagram such as that shown in Fig. 167. It may be assumed in the first instance that the samples were initially identical in their $^{87}Sr/^{86}Sr$ ratio since their strontium was drawn from the same pool of magma. Differences in their present-day $^{87}Sr/^{86}Sr$ therefore result from their differing amounts of rubidium; the more rubidium, the greater the enrichment in ^{87}Sr since the

Figure 168. Sr isochron diagram for a set of related igneous rocks and their constituent minerals.

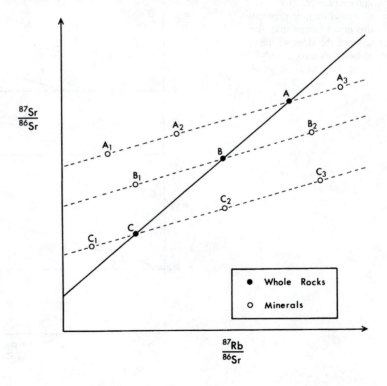

time of crystallization. The samples lie along a line whose slope becomes increasingly steep as time passes. The line is called an isochron, because its slope is a measure of the age of the rock. The whole diagram is called an isochron diagram.

When an isochron diagram is plotted for several whole-rock samples it usually gives an age consistent with other geological evidence, but for mineral samples the isochron sometimes gives a low result, as shown in Fig. 168. The mineral isochrons of each rock in Fig. 168 give the same age, but it is lower than that measured by the whole-rock isochron. This is the result of reheating, i.e. metamorphism since the original time of magmatic crystallization.

Figure 169 shows how the mineral isochrons are reset by an episode of reheating. The reheating redistributes the strontium isotopes between the minerals but does not cause any loss or gain of strontium isotopes by the rock

Figure 169. Resetting of strontium isotope ratios. (A) The initial $^{87}Sr/^{86}Sr$ ratio of the rock and its constituent minerals is R_0. After time t_1, the average $^{87}Sr/^{86}Sr$ ratio of the rock has increased to R_1, but is higher in mineral 2 (e.g. mica, which is Rb-rich) and lower in mineral 1 (e.g. plagioclase, which is Rb-poor). (B) A heating event at time t_1 causes redistribution of Sr isotopes between the minerals so as to bring them to the common ratio R_1. (C) Isotopic evolution then proceeds from this new baseline until the time t_2, when the slope of the mineral isochron records only the time elapsed between t_1 and t_2.

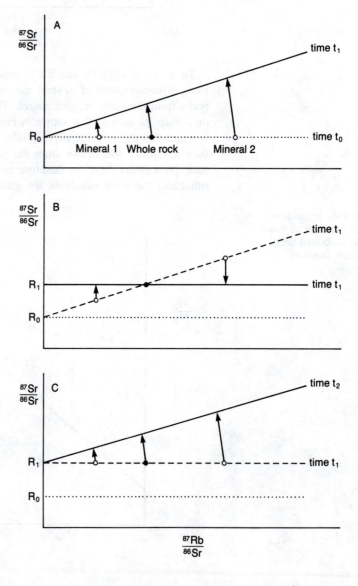

as a whole. Study of both the minerals and the whole rocks in an igneous complex therefore enables one to measure both the time of initial crystallization (from the whole rock isochron) and the time of any subsequent metamorphism (from the mineral isochron).

Strontium isotope studies give much more information than just the age of a rock. A whole-rock isochron intersects the $^{87}Sr/^{86}Sr$ axis at a value corresponding to the initial $^{87}Sr/^{86}Sr$ ratio at the time of crystallization. If rocks which are known to be of the same age do not plot on a straight-line isochron (Fig. 170), then their initial $^{87}Sr/^{86}Sr$ ratios must have differed, and they cannot have come from the same magma source.

The initial $^{87}Sr/^{86}Sr$ ratio of a magma is dependent on the amount of rubidium in the source region from which it is derived. Rubidium is very non-uniformly distributed in different parts of the Earth and in the solar system. Approximate abundances (in ppm) of rubidium are:

Ultramafic rocks	1
Basalt	30
Granite	150
Shale	140

Meteorites, in which Rb is almost absent, have an initial $^{87}Sr/^{86}Sr$ ratio of 0.699, lower than any terrestrial rocks. The Earth's primordial Sr is assumed to have had this ratio. The mantle contains a small amount of Rb, and as a result its $^{87}Sr/^{86}Sr$ ratio has grown from 0.699 to about 0.705 over the course of geological time, but it is not entirely homogeneous and contains both higher and lower ratios. Oceanic basalts, which come from the mantle, have initial $^{87}Sr/^{86}Sr$ ratios in the range 0.702–0.706, the higher values indicating the presence of a slightly larger amount of Rb in the mantle source. Sedimentary rocks have much higher $^{87}Sr/^{86}Sr$ ratios, since much of the Earth's rubidium is concentrated in the crust. The average $^{87}Sr/^{86}Sr$ of the Earth's crust has been growing throughout geological time, reaching an average value of around

Figure 170. Strontium isochron diagram for a set of rocks known from independent evidence to be of the same age. If one or more of the rocks do not lie on the same isochron as the rest, they must have started from a different initial $^{87}Sr/^{86}Sr$ ratio, and therefore from a different magma source.

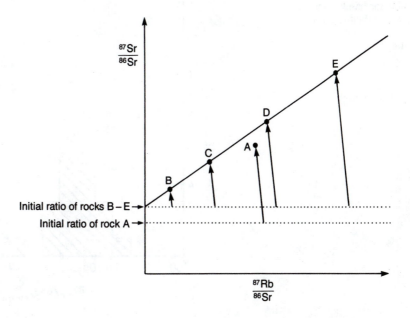

0.725 at the present day, although the crust is very heterogeneous. Measurement of initial $^{87}Sr/^{86}Sr$ therefore provides a way of deciding whether a particular magma derived its strontium from a mantle or a crustal source.

The difference between carbonatites and limestones provides a simple illustration of the different $^{87}Sr/^{86}Sr$ ratios of crust- and mantle-derived materials. Carbonatites are igneous carbonate rocks, although they have not always been recognized as such. Initial $^{87}Sr/^{86}Sr$ ratios of carbonatites have been measured, and lie around 0.703, which implies a mantle origin. Limestones derive their strontium from sea water, and so have higher $^{87}Sr/^{86}Sr$ ratios, averaging 0.709 (this is not as high as most continental sediments because sea water has interacted with sea-floor basaltic rocks, but it is high enough to indicate a substantial crustal contribution).

Granites are very diverse in their initial $^{87}Sr/^{86}Sr$ ratios, ranging from 0.700 to 0.740. At the lowest end of the range they overlap basalts, which obviously have a mantle origin, but many of their initial ratios are over 0.706, which is sufficiently high to indicate a crustal source for at least a proportion of the magma. Unfortunately this conclusion cannot be put on a more quantitative basis because crustal rocks are so variable in their $^{87}Sr/^{86}Sr$ ratio. Although the average crustal material may now have a $^{87}Sr/^{86}Sr$ ratio of around 0.725, there are volcaniclastic sediments (e.g. some greywackes) with initial $^{87}Sr/^{86}Sr$ ratios lower than 0.705, i.e. no greater than those of the andesites and other lavas from which they were derived.

In a notable early study using Sr isotopes, Moorbath and Bell (1965) examined the granites and basalts of the Skye igneous centre in Scotland, and showed that whereas the basalts had initial $^{87}Sr/^{86}Sr$ ratios of about 0.706, the granites had an average initial ratio of 0.712 (Fig. 171). This finding ruled out the possibility that the acid magmas were differentiates from the basic magma, and strongly indicated a crustal origin.

Figure 171. Initial $^{87}Sr/^{86}Sr$ ratios for basaltic and granitic rocks of the Skye Tertiary igneous complex, Scotland (after Moorbath and Bell 1965).

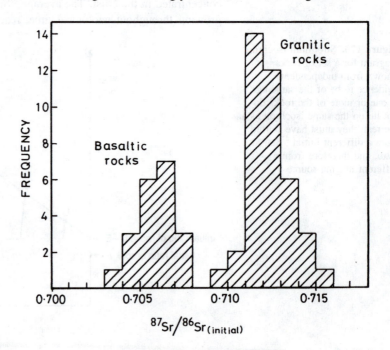

Results such as these show the value of $^{87}Sr/^{86}Sr$ measurements in igneous petrology. However, there are also several potential pitfalls in their petrogenetic application, namely: (1) heterogeneity in the magmatic source region; (2) contamination of the magma; and (3) subsequent hydrothermal alteration.

It has been found that different lava flows on the same present-day volcano sometimes have different $^{87}Sr/^{86}Sr$ ratios, even though there has been no time for radiogenic ^{87}Sr to have accumulated since the rocks crystallized. For example the lavas of Vesuvius have varied in $^{87}Sr/^{86}Sr$ from 0.703 to 0.707 within the last 250 years (see Fig. 330 in Chapter 12). If there were this much variation among ancient igneous rocks, it would be very difficult to obtain isochrons for age determination. On the other hand the present-day variations in $^{87}Sr/^{86}Sr$ may provide useful information on the composition of magmatic source regions. Specialists in the study of basalt use small variations in initial $^{87}Sr/^{86}Sr$ to distinguish between different mantle sources for basalt (see Chapter 8). It is assumed in these studies that the Sr isotopic composition of magmas reflects that of their source material, although there is a slight possibility that magmas of different $^{87}Sr/^{86}Sr$ ratio could be formed by partial melting of a homogeneous source in which different minerals had different $^{87}Sr/^{86}Sr$ ratios (disequilibrium melting).

A more serious problem, and one which has not been sufficiently taken into account, is that of magmatic contamination. In contaminated plutonic rocks, one cannot assume that all the Sr of the contaminants has been uniformly mixed into the magma, and hence the assumption that all rocks initially had the same $^{87}Sr/^{86}Sr$ ratio is not justified. Contamination is one of the main sources of petrographic variation in granitic rocks, and all the main types of likely contaminant have compositions which would lead to an underestimation of the initial $^{87}Sr/^{86}Sr$ ratio of the magma. In fact the necessity of obtaining samples with a spread of $^{87}Rb/^{86}Sr$ ratios in order to calculate the slope of an isochron may actually cause isotopic investigators to deliberately sample highly contaminated rocks in order to extend the Rb/Sr range. It is impossible to estimate how much error in the estimated $^{87}Sr/^{86}Sr$ ratios of magmas is introduced in this way, but the general effect in granitic rocks is to lead to an underestimation of the initial $^{87}Sr/^{86}Sr$ of the magma, and hence to an understatement of the case for a crustal origin.

A further problem in Sr isotope studies of igneous rocks is that Sr is readily mobilized during low grade hydrothermal alteration, so that for example sea-floor basalts with initial $^{87}Sr/^{86}Sr$ ratios of 0.702 have been found to take up relatively radiogenic strontium from sea water thereby raising their $^{87}Sr/^{86}Sr$ ratios towards that of sea water (~0.708) by an amount depending on the extent of alteration (Satake and Matsuda 1979). Hydrothermal alteration of an igneous intrusion by fluids containing extraneous strontium can also invalidate the assumption of initial isotopic homogeneity which is normally relied upon in plotting isochrons (Juteau *et al.* 1984).

SAMARIUM–NEODYMIUM

Samarium-147 decays to neodymium-143 with a half-life of 106×10^9 years. The enrichment in ^{143}Nd can be measured by reference to the non-

Figure 172. Samarium–neodymium isochron diagram for a gabbro from the Stillwater intrusion, Montana (after De Paolo and Wasserburg 1979b). This isochron corresponds to an age of 2701 million years.

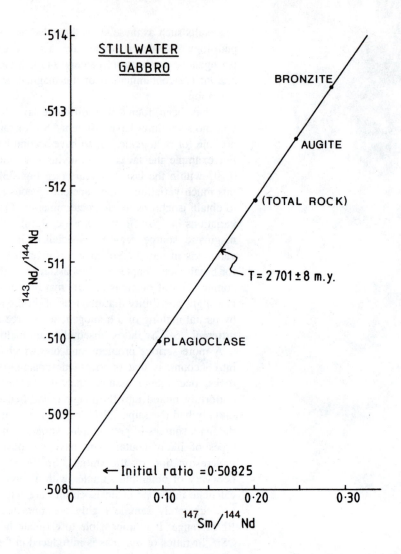

radiogenic isotopes ^{144}Nd or ^{146}Nd on isochron diagrams analogous to those used for interpreting strontium isotopes (Fig. 172). Samarium is less abundant than rubidium, and the ^{143}Nd/^{144}Nd ratios do not increase very far above the initial value because of the slow decay of ^{147}Sm, so the isotopic ratios need to be measured with great precision to obtain usable results. For age determination the technique is most suitable for very old rocks, and it is particularly valuable because Sm and Nd are relatively unaffected by weak metamorphism that can cause redistribution of Rb, Sr, U, Th and Pb (McCulloch *et al.* 1980).

Variations in the initial ^{143}Nd/^{144}Nd ratio of younger rocks are also of value in revealing differences in the source regions of magmas. The ^{143}Nd/^{144}Nd ratio in the Earth as a whole has increased from approximately 0.507 to 0.5126 during the course of geological time. Unlike the ^{87}Sr/^{86}Sr ratio, the ^{143}Nd/^{144}Nd ratio is lower in crustal rocks (~0.512) than in those derived from the mantle (~0.513) because of the relatively low Sm/Nd ratio of the crust.

Relatively large variations in $^{143}Nd/^{144}Nd$ ratio are found in Recent basalts, indicative of large scale heterogeneity in their mantle source regions.

Some isotope geochemists express initial Nd isotopic compositions by the measure ϵ_{Nd}, where:

$$\epsilon_{Nd} = 10^4 \times \left[\frac{^{143}Nd/^{144}Nd_{(sample)} - {}^{143}Nd/^{144}Nd_{(chondritic)}}{^{143}Nd/^{144}Nd_{(chondritic)}} \right]$$

the ratios being those at the time the rock was formed. The chondritic initial ratio must be calculated from the chondritic present-day ratio by the equation:

$$^{143}Nd/^{144}Nd_{(initial\ chondritic)} = {}^{143}Nd/^{144}Nd_{(present\text{-}day\ chondritic)} - {}^{147}Sm/^{144}Nd_{(present\text{-}day\ chondritic)} \times (e^{\lambda t} - 1)$$

where t is the age of the rock (De Paolo and Wasserburg 1979a; Jacobsen and Wasserburg 1980). Present-day reference values are:

$$^{143}Nd/^{144}Nd_{(present\text{-}day\ chondritic)} = 0.511836$$
$$^{147}Sm/^{144}Nd_{(present\text{-}day\ chondritic)} = 0.1967$$
$$\lambda = 6.54 \times 10^{-12}/year\ (the\ decay\ constant\ of\ {}^{147}Sm)$$

A value of ϵ_{Nd} of $+10$ would correspond to a $^{143}Nd/^{144}Nd$ ratio of 0.5131 (a typical value for mid-ocean ridge basalt). A value of ϵ_{Nd} of -20 would correspond to a $^{143}Nd/^{144}Nd$ ratio of 0.5111 (a typical value for Pre-Cambrian metasediments).

RHENIUM–OSMIUM

The decay of ^{187}Re to ^{187}Os has not been used much in age determination, mainly because of the extremely low abundance of Re except in certain sulphide minerals (e.g. molybdenite). The degree of radiogenic enrichment in Os is expressed as the ratio $^{187}Os/^{186}Os$. The variation of $^{187}Os/^{186}Os$ in igneous rocks is of special interest because unlike the other parent–daughter isotope pairs both Re and Os have strong chalcophile and siderophile character. The main feature of this isotopic system is the very strong fractionation between Re and Os during partial melting in the mantle, the Re being concentrated in basaltic melt while the Os is concentrated in residual peridotite (Pegram and Allègre 1992).

URANIUM–THORIUM–LEAD

Natural uranium consists of three isotopes, occurring in the ratio $^{238}U : {}^{235}U : {}^{234}U = 99.27 : 0.72 : 0.01$. Natural thorium consists almost entirely of ^{232}Th. These isotopes decay through a series of short-lived intermediate products until finally a stable isotope of lead is produced:

$$^{238}U \rightarrow {}^{206}Pb$$
$$^{235}U \rightarrow {}^{207}Pb$$
$$^{232}Th \rightarrow {}^{208}Pb$$

Two types of study can be carried out: (1) measurement of the relative amounts of U and Th and their decay products with a view to age determination; or (2) examination of the lead isotopes in rocks (irrespective of whether U and Th are present), with a view to deducing the source of the lead.

Age determination

The ratio of each U or Th isotope to its Pb daughter isotope is a measure of the age, but most U- and Th-bearing minerals contain a small amount of Pb when they first crystallize, and allowance must therefore be made for any lead present before the decay of the U and Th commenced. This is easily done because natural lead also contains the isotope ^{204}Pb which is not radiogenic, and serves as a measure of the original lead content. One method is to plot an isochron diagram such as is shown in Fig. 173, in which the enrichment in ^{208}Pb is proportional to the amount of its parent isotope ^{232}Th, and the slope of the isochron is then a measure of the age. Similar isochron diagrams can be plotted of $^{206}Pb/^{204}Pb$ against $^{238}U/^{204}Pb$, and of $^{207}Pb/^{204}Pb$ against $^{235}U/^{204}Pb$, but in these systems straight isochrons are rarely obtained because uranium is extremely susceptible to removal during even the slightest hydrothermal alteration or weathering. This is not so much of a problem with the $^{208}Pb/^{204}Pb$ isochron, because Th is less mobile than uranium.

Because U and Pb are so easily leached from rocks, most U/Pb and Th/Pb dating of igneous rocks has been done on separated U- or Th-rich minerals such as zircon or monazite. It is possible to measure each of the three parent–daughter pairs in the same sample and compare the results. If the

Figure 173. Thorium–lead isochron diagram for rocks of the Granite Mountains, Wyoming (after Rosholt *et al.* 1973). The isochron corresponds to an age of 2.79 billion years.

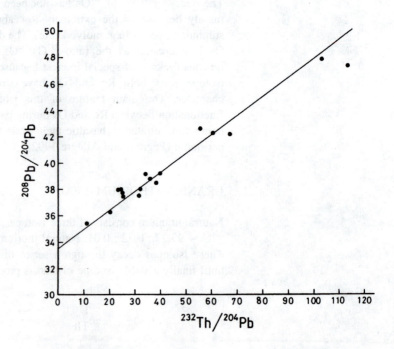

ages obtained from the various pairs of isotopes agree the results are said to be concordant, but if not they are said to be discordant.

Discordant ages are very often obtained. For example, a zircon from a granite in Colorado gave the following results (Faure 1986):

$$^{238}U-^{206}Pb \qquad \text{apparent age} = 1404 \text{ million years}$$
$$^{235}U-^{207}Pb \qquad \text{apparent age} = 1523 \text{ million years}$$
$$^{232}Th-^{208}Pb \qquad \text{apparent age} = 1284 \text{ million years}$$

The most probable explanation of the discordant ages is that some radiogenic lead has been lost from the zircon, either by diffusion while the rock was still at a high temperature or by a later episode of metamorphism.

Uranium–lead age determination is facilitated by a plot of $^{206}Pb/^{238}U$ against $^{207}Pb/^{235}U$. This is called a concordia diagram (Fig. 174). The curvature of the concordia line is due to the much faster decay of ^{235}U than of ^{238}U (the half-lives are 700 and 4470 million years respectively). Thus:

Time elapsed	*Amount of ^{235}U left*	$^{207}Pb/^{235}U$
700 million years	50%	1.0
1400 million years	25%	3.0
2100 million years	12.5%	7.0

As the remaining small amount of ^{235}U decays, the $^{207}Pb/^{235}U$ ratio rises

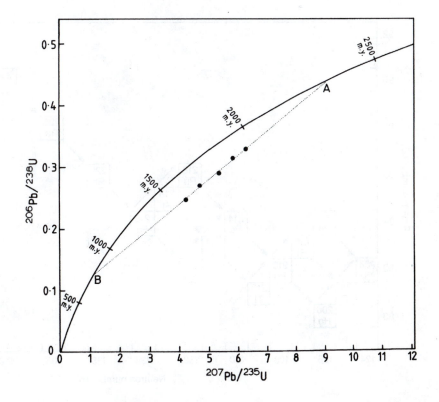

Figure 174. A concordia diagram.

rapidly, while the $^{206}Pb/^{238}U$ ratio increases at a steadier rate as there is plenty of ^{238}U left.

When discordant ages are plotted on a concordia diagram they commonly fall along a straight line, as shown in Fig. 174. One possible interpretation of such a plot is that the minerals formed at time A, but lost part of their lead at time B. Another interpretation is that there has been a continuous loss of lead over a long period of time, in which case point B may not actually correspond to the date of any particular event.

It is sometimes found that the discordant zircons in granites have a lower concordia intersection corresponding to the known age of the rock in which they occur, whereas the upper concordia intersection is much older. This could be due to preservation of old zircon carried up by the magma from its source region, or to incorporation of older zircon crystals from xenoliths assimilated by the magma. Zircon with an age greater than the time of intrusion of the host rock is known as 'inherited zircon' (Bickford *et al.* 1981, Harrison *et al.* 1987). More often, the zircon in granites has an upper concordia intersection corresponding to the age of emplacement and a lower concordia intersection denoting an episode of recrystallization.

Uranium series dating

Figure 175. The ^{238}U decay series.

Additional information on the age and petrogenesis of igneous rocks may be obtained by measuring the intermediate members of the $^{238}U-^{206}Pb$ decay series (Fig. 175), such as ^{230}Th (half-life = 75,200 years) and ^{226}Ra (half-life

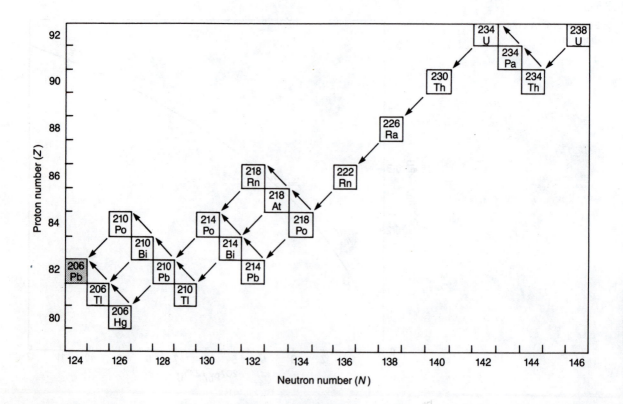

= 1600 years). In an undisturbed uranium-bearing mineral, the proportion of intermediate members of the decay series reaches an equilibrium which depends on their relative decay constants, i.e. the short-lived isotopes are present at lower concentrations than the long-lived ones. If the link between the parent and daughter isotopes is broken by partial melting or crystallization, the daughter isotopes may become separated from the uranium and enter different minerals. There are numerous examples of young volcanic rocks which show disequilibrium between the intermediate members of the ^{238}U decay series, and studies of U-series disequilibrium have been used to infer the time interval between magma formation and eruption in active volcanoes (Gill and Condomines 1992).

Radiogenic helium

The decay of U and Th gives rise not only to radiogenic lead but also to radiogenic helium, since the α-particles emitted from the radioactive isotopes are in fact nuclei of ^{4}He. The decay schemes can be summarized as follows:

$$^{238}U \rightarrow {}^{206}Pb + 8\,^4He + 6\beta^-$$
$$^{235}U \rightarrow {}^{207}Pb + 7\,^4He + 4\beta^-$$
$$^{232}Th \rightarrow {}^{208}Pb + 6\,^4He + 4\beta^-$$

Natural helium is a mixture of two stable isotopes, ^{3}He and ^{4}He, which occur in the atmosphere in the approximate ratio $^3He/^4He = 1.38 \times 10^{-6}$. The ratio in magmas and natural waters varies widely and can be used to indicate the relative contributions of U- and Th-rich sources and of atmospherically derived helium (Lupton 1983; Mamyrin and Tolstikhin 1984). The isotopic composition of helium can be expressed as percentage deviations from a standard (the atmosphere), thus:

$$\delta^3 He = \left[\frac{(^3He/^4He)_{sample}}{(^3He/^4He)_{atmospheric}} - 1 \right] \times 100$$

Basalts have $^3He/^4He$ ratios about $10 \times$ atmospheric, indicating that the mantle is much less rich in radiogenic helium than the atmosphere. The ^{3}He in basalts must be primordial, i.e. present in the Earth since it was formed, and it is rather surprising that there is still some left in the mantle after all this time. The role of basaltic volcanism in allowing the primordial ^{3}He to leak away from the mantle is graphically shown by the plume of ^{3}He rising from the crest of the East Pacific Rise (Fig. 176). Whereas mid-ocean ridge basalts have $^3He/^4He$ of $7-11 \times$ atmospheric, oceanic island basalts have ratios which are much more variable and often higher, e.g. Tristan da Cunha 5–6, Iceland 4–26, Hawaii 8–32 $\times$ atmospheric (Kurz *et al.* 1982, 1983; Condomines *et al.* 1983). The high ratios cannot be due to a deficiency of radioactive elements in the source region because the basalts themselves are comparatively rich in U and Th. Instead, the volcanism of these islands must for some reason be more effective in collecting the primordial ^{3}He from the mantle and allowing it to escape. In contrast to the high $^3He/^4He$ ratios of oceanic basalts, continental margin basalts have low $^3He/^4He$ ratios, e.g. from

Figure 176. Contours of δ^3He in the waters of the Pacific Ocean over the East Pacific Rise at 15°S, showing a plume of ^{3}He-rich water emanating from the crest of the ridge (after Lupton and Craig 1981).

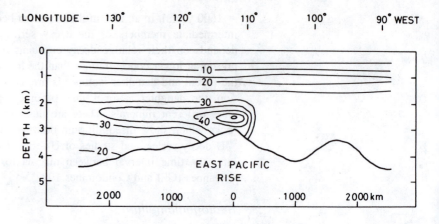

<1 to 8 × atmospheric for Indonesian volcanoes (Hilton *et al.* 1992), indicating a large contribution of radiogenic ^{4}He from the continental crust.

Common lead

The lead in the Earth is of two kinds: *primeval lead*, present throughout the Earth's history; and *radiogenic lead*, produced by the radioactive decay of U and Th. *Common lead*, i.e. lead as it actually occurs in rocks, is a mixture of primeval and radiogenic lead. Over the course of geological time the Earth's stock of common lead has gradually been changing by the addition of the radiogenic isotopes, but present-day lead is not a well-homogenized mixture of the primeval and radiogenic components. The latter are concentrated in rocks with high U and Th contents, or in magmas derived from a source containing plenty of U and Th.

The isotopic composition of primeval lead is known from the study of meteorites. Many meteorites have Rb/Sr ages which indicate that they crystallized approximately 4600 million years ago, and have remained chemically isolated from one another ever since. Of these, the iron meteorites contain very little U or Th, so that their original Pb has not been modified by the addition of any radiogenic component. The most accurate measurements are made on the troilite (pyrrhotite, FeS) in these meteorites. The troilite in the Canyon Diablo meteorite has the least radiogenic lead which has yet been discovered, and its composition is generally taken to be that of the primeval lead of the Earth and Moon as well as of the meteorites. Its composition is defined by the following ratios (Tatsumoto *et al.* 1973):

$$^{206}\text{Pb}/^{204}\text{Pb} = 9.307$$
$$^{207}\text{Pb}/^{204}\text{Pb} = 10.294$$
$$^{208}\text{Pb}/^{204}\text{Pb} = 29.476$$

Holmes and Houtermans have developed a model of the evolution of common lead with which observed isotope ratios can be compared. Their model assumes that ever since some initial time in the past (t_0), lead having the primeval composition (a_0, b_0) has been subjected to enrichment in the decay products of U and Th in one or more closed systems. The Pb isotope ratios would consequently develop along one or more growth curves, as

Figure 177. The Holmes–
Houtermans model of lead
isotope evolution. Primeval lead
has the isotopic ratios a_0 and b_0.
Curves A and B represent the
change in lead isotopic
composition of systems with
high and low U/Pb ratios
respectively.

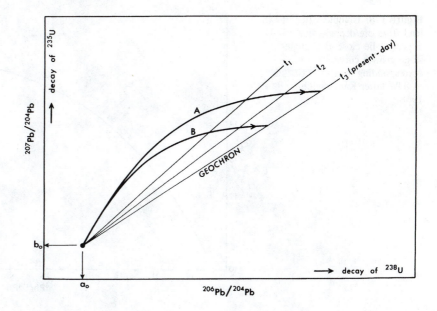

shown in Fig. 177. Growth curve A represents the change in lead isotopic
composition in a closed system with a high U/Pb ratio, and growth curve B
represents the change in a system with a low U/Pb ratio.

At successive times, t_1, t_2, etc., the isotopic composition of the lead in
different closed systems would have reached the sloping lines (isochrons) in
Fig. 177. For closed-system lead isotope evolution the present-day isochron
is called the Geochron. Thus the isotopic composition of a lead mineral
containing no uranium since its crystallization, or a lead-bearing rock
corrected for the decay of any uranium it has contained since its crystallization, can be used to infer a 'model age', i.e. the actual age if all the
assumptions of the model were valid.

Stony meteorites contain varying amounts of uranium and thorium, so
their lead is enriched to varying extents in radiogenic components. When
plotted on the Holmes–Houtermans diagram it is found that they lie on
isochrons corresponding to ages between 4.52 and 4.57 billion years. Lunar
dust lies on an isochron corresponding to an age of 4.56 to 4.58 billion years.
Recent oceanic sediments lie close enough to a 4.57 billion year isochron to
be reasonably confident that the Earth, Moon and meteorites separated from
one another at essentially the same time, and that the Earth's primeval lead
isotopic composition was the same as that of the iron meteorites.

Some sedimentary lead ores lie roughly along a growth curve corresponding to a single U/Pb ratio ($\mu = {}^{238}U/{}^{204}Pb = 8.99$), implying their derivation
from a reservoir (presumably the Earth's crust and oceans) which has been
fairly homogeneously mixed with respect to U and Pb during geological time
(Fig. 178). Such leads have been called ordinary, normal or conformable,
because they conform to a single-stage growth model, and some of them
yield model ages close to their actual geological ages.

The lead in igneous rocks and in most ore deposits is anomalous, giving
model ages which are either too high or too low, so that it cannot have

Figure 178. Isotopic ratios of lead from ore deposits that appear to lie close to a single-stage growth curve, corresponding to $\mu = {}^{238}U/{}^{204}Pb = 8.99$ (after Kanasewich 1968).

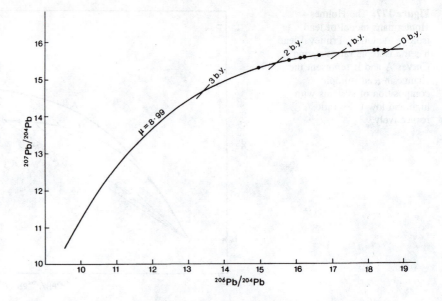

evolved along a single growth curve. In many geologically young rocks the lead is so anomalous that it yields a model age which is actually negative, i.e. it lies to the right of the zero isochron or geochron. Actually, since most sedimentary lead comes ultimately from the weathering of igneous rocks in which lead isotopic compositions are anomalous, the conformable nature of sedimentary lead is more apparent than real. It is not likely that any of the Earth's lead has evolved along a single growth curve, and the U/Pb ratios of different regions of the crust and mantle must have varied during geological time.

Lead which has not evolved in a system of constant U/Pb ratio is called multistage or anomalous lead. At each stage in its development it has gained radiogenic lead rapidly or slowly according to the amount of uranium in its environment, and its present isotopic composition is the cumulative result of the various growth stages. Figure 179 illustrates two-stage growth. Lead is shown as evolving along a single growth curve until it reached point A, at some time during geological history. A development then took place which partly separated the uranium and lead. This could have been something like a partial melting event or an episode of metasomatism. Assuming that no extraneous lead was introduced (which is a further possibility), the subsequent development of the lead isotopic composition would be along a family of growth curves diverging from point A. By the present day, the Pb evolving in the highest U/Pb milieu would have evolved to point B_1 and the Pb evolving in the lowest U/Pb milieu would have reached point B_2. These points (B_1 and B_2) lie along a line passing through A, whose slope depends on the time elapsed between the U-Pb separation event at A and the present day. It is an isochron and can be referred to as a 'secondary isochron'. Those rocks that underwent the second stage of their development in an environment of very low U content would have an eventual Pb isotopic composition (B_2) not much more radiogenic than that reached at the separation event (A).

Figure 179. A two-stage model of lead isotope evolution.

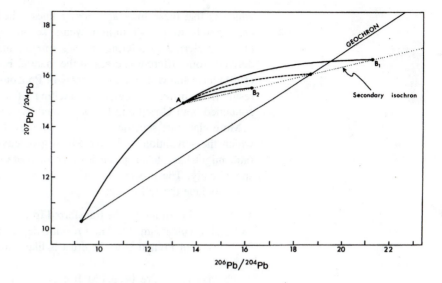

If interpreted in terms of the Holmes–Houtermans single-stage model they would yield positive model ages, i.e. lie to the left of the Geochron. The rocks that evolved in a U-rich system would become much richer in radiogenic isotopes and eventually lie to the right of the Geochron (at B_1), yielding negative model ages. The period that has elapsed since the separation event (A) has been one in which the $^{238}U/^{235}U$ ratio has been high, because most of the primeval ^{235}U had already decayed, so the second stage of evolution was characterized by particularly rapid increase in the $^{206}Pb/^{204}Pb$ ratio in those rocks which were U-rich.

Many igneous rock suites lie on lines in the $^{207}Pb/^{204}Pb$ against $^{206}Pb/^{204}Pb$ diagram which can be interpreted as secondary isochrons, for example the Hawaiian basalts shown in Fig. 180. There is an alternative explanation,

Figure 180. Lead isotope ratios for Hawaiian basalts (after Tatsumoto 1978).

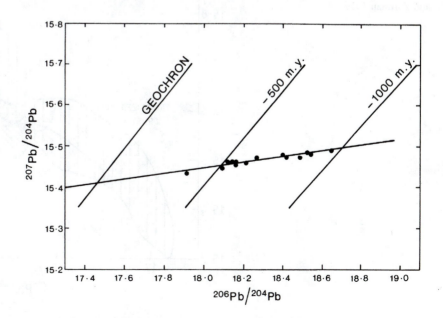

which is that these lines are mixing lines. The Hawaiian trend in Fig. 180 corresponds to a 940 million years secondary isochron, but Tatsumoto (1978) preferred to interpret it as a mixing line between two components derived from different sources in the mantle. For granites, Hogan and Sinha (1991) have shown that a linear array of Pb isotopic compositions could even result from different degrees of melting of a single source material if it contained both U-rich and U-poor minerals.

More elaborate schemes of lead isotope evolution might be envisaged, in which the separation of U and Pb to give environments of different U/Pb ratio might have taken place in three or more episodes, or continuously or sporadically. The processes controlling the distribution of U and Pb in the Earth include the following:

1. Pb and U both tend to be transferred from the mantle into the crust during basaltic volcanism, leaving a residue depleted in both elements.
2. Transfer of Pb and U to the crust is likely to have taken place at unequal rates.
3. Pb may also have been lost from the mantle to the core because of its partly siderophile nature.
4. There has been separation of Pb from U within the crust, sometimes a complete separation as in the case of Pb or U mineralization.
5. Pb that has evolved in the presence of a particular amount of U may have its isotopic composition sharply displaced by an influx of lead from a source with a completely different evolutionary history.

Doe and Zartman (1979) reviewed the probable distribution of U and Pb in different parts of the Earth during geological time, and distinguished three

Figure 181. Lead isotopic composition of major divisions of the Earth's crust (after Doe and Zartman 1979).

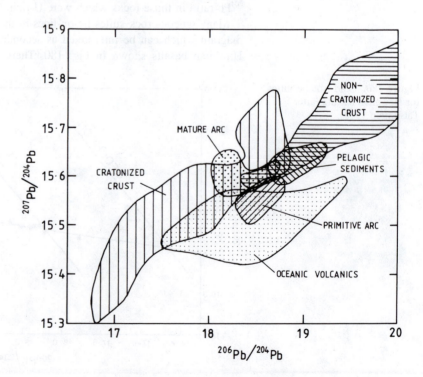

Table 26. Estimated Pb, U and Th contents of the present-day crust and mantle, after Doe and Zartman (1979)

	Mass (in 10^{24} g)	Pb (ppm)	U (ppm)	Th (ppm)	U/Pb
Upper crust	7.0	22.8	4.18	13.8	0.18
Lower crust	17.8	6.5	0.62	3.6	0.10
Mantle	775.2	0.40	0.055	0.19	0.14

major reservoirs of U and Pb: the mantle, the upper continental crust, and the lower continental crust. The average Pb, U and Th contents of these reservoirs are shown in Table 26. These contents are reflected in the isotopic compositions of the Pb from different environments (Fig. 181). The upper continental crust has the highest U/Pb ratio, so upper crustal rocks have the most radiogenic lead (highest $^{207}Pb/^{204}Pb$ and $^{206}Pb/^{204}Pb$), as do sediments derived from the weathering of the upper crust. Doe and Zartman suggested that the continental crust has grown by the periodic extraction of new material from the mantle combined with recycling of older continental crust, resulting in a progressive increase in the U/Pb ratio of the upper crust over the course of time.

CHAPTER 7

Magmatic evolution

Any igneous province, or even a single volcano, may display a range of igneous rock types. This variation may be *primary* (magmas of different composition were created at source) or *secondary* (different magmas evolved subsequently from a common parent).

Magmas originate by partial melting in the originally solid crust or mantle, and the main causes of primary variation are differences in:

1. the materials being melted in the magmatic source region
2. the degree of melting
3. the conditions under which melting took place.

At one time, very little was known about the conditions under which magmas originated, and it was customary to attribute most of the variation in igneous rocks to secondary processes, but it is now realized that much igneous variation must be primary.

The origin of primary magmas will be considered in the following chapters. Here we will be concerned with the development of secondary variation. The possibilities for secondary variation are broadly as follows:

Magmatic differentiation An originally homogeneous magma changes its composition or becomes heterogeneous. The mechanisms which have been proposed are crystal fractionation, liquid immiscibility, and liquid fractionation. Of these, the first is generally thought to be the most important.

Contamination A magma melts, reacts with, or incorporates material from its wall rocks so that the final product is a rock differing in composition from the original magma. The words assimilation and contamination are both used to describe this process.

Zone melting A body of magma changes composition by simultaneously crystallizing its coolest part while melting its wall rocks elsewhere.

Mixing of magmas Two or more magmas become mixed to give various intermediate products.

In individual plutonic intrusions, one can often see evidence of differentiation, sometimes of contamination, and rarely of mixing. In volcanic rocks, the operation of these processes is more difficult to detect, but may be revealed by detailed study of the chemical, mineralogical or isotopic composition.

CRYSTAL FRACTIONATION

Crystallization of magma is not instantaneous, but takes place over a range of temperature and over a period of time. In a partially crystallized magma, the coexisting crystals and liquid rarely have exactly the same composition, and so if the crystals are separated from a magma the remaining melt will differ from the original magma. The result is a progressive change in the composition of the magma, described as crystallization differentiation or crystal fractionation. The end-products of crystal fractionation are solid cumulates (accumulations of crystals) and residual liquids.

In describing examples of crystal fractionation, the following terms are useful. A *primitive* magma is one which is close to its original composition, i.e. it has not undergone fractionation. An *evolved* magma is one which has undergone fractionation and differs from its original composition. The *liquid line of descent* is the series of liquid compositions leading from the most primitive magma to the most evolved magma in a fractionation series. In general, primitive compositions are characterized by being rich in constituents with high melting temperatures (e.g. ferromagnesian minerals and calcic plagioclase) and have high Mg/Fe ratios and high liquidus and solidus temperatures.

The phase relations of simple synthetic systems show some of the ways in which the composition of a magma might be altered by the removal of crystals. The system albite–anorthite (Fig. 122) shows that early formed plagioclase crystals will be more calcic than the melt from which they crystallize, and the residual liquid will be more sodic. The system forsterite–fayalite (Fig. 182) shows that early formed olivine crystals are more magnesium-rich than the melt, and their residual liquid will be enriched in iron. The systems albite–fayalite–silica (Fig. 134) and albite–anorthite–diopside–forsterite (Fig. 141) show that crystallization would commonly involve the removal of ferromagnesian minerals and enrichment of the liquid in felsic constituents.

Crystallization of magma may be of either the equilibrium type or the fractional type (Chapter 5), but the change in liquid composition is greater if crystallization is fractional. Moreover, the operation of any mechanism that separates crystals from the melt will in itself be conducive to fractional crystallization. So, although it is possible to envisage crystal sorting under conditions of perfect equilibrium crystallization, for example by gravitational settling from a melt of eutectic composition, crystallization differentiation nearly always involves fractional crystallization and is often referred to simply as 'fractionation'.

Fractional crystallization is particularly effective in changing the melt

Figure 182. The system forsterite–fayalite (after Bowen and Schairer 1935). Crystals of composition A are in equilibrium with liquid of composition B at 1660 °C.

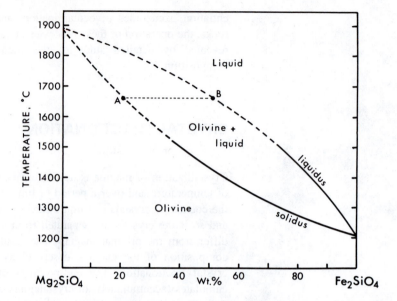

composition in systems which involve a solid solution series (e.g. Fig. 182) or an incongruently melting compound (e.g. Fig. 183). Consider the system forsterite–silica, shown in Fig. 183. Liquid X would initially start to crystallize olivine on cooling, but these olivine crystals should react with the melt when its composition reached the reaction point Y. If by reason of their high density the olivine crystals sank to the bottom of the magma chamber, they would not be available to react with the melt when it reached reaction point Y, even though crystallization might have been slow enough for equilibrium to have been attained otherwise. Thus the modification of the

Figure 183. The system forsterite–silica (after Bowen and Andersen 1914; Greig 1927).

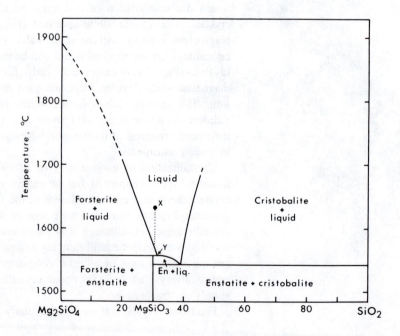

original liquid is accentuated by the operation of the crystal separation mechanism.

CRYSTAL–LIQUID SEPARATION MECHANISMS

The principal requirement of crystallization differentiation is the operation of a mechanism to remove crystals from the magma. The separation mechanisms that have been proposed are: (1) gravitational settling of heavy crystals, or flotation of light crystals; (2) flow differentiation; (3) flow crystallization; (4) filter pressing; (5) selective nucleation; (6) gas streaming; and (7) gravitational liquid separation. These mechanisms may result in separation of all the early formed crystals from the residual liquid, or only some of them, or they may involve separation of crystals from one another.

Gravitational settling

This is the most commonly invoked mechanism of crystal separation. In general, magmas are about 10% lighter than their equivalent solidified rock at the same temperature, and so more minerals are likely to sink in a crystallizing magma than are likely to float. The densities of the common rock-forming minerals (at low temperature) are:

$<2.5\,\mathrm{g/cm^3}$ – analcime, sodalite, leucite
$2.5–3.0\,\mathrm{g/cm^3}$ – quartz, feldspars, nepheline, muscovite
$>3.0\,\mathrm{g/cm^3}$ – ferromagnesian minerals, iron oxides, apatite, zircon

The potentiality for floating is greatest for the three minerals with a density of $<2.5\,\mathrm{g/cm^3}$. Floating of leucite is shown by the nearly monomineralic leucite ('italite') blocks of some Italian and East African volcanoes, made up largely of leucite phenocrysts in a sparse groundmass, and flotation of sodalite has been invoked by Ferguson (1964) to explain the formation of naujaite (sodalite–syenite) in the Ilimaussaq alkaline intrusion in Greenland.

Settling of heavy crystals is more common than flotation of light crystals. It has been held responsible for the rhythmic layering which is seen in many gabbroic intrusions. Numerous examples of rhythmic layering are described and illustrated in the monograph by Wager and Brown (1968) on layered igneous rocks. Typical examples of small-scale layering show layers a few centimetres thick with a well-defined base, a lower part rich in heavy, dark, ferromagnesian minerals, grading into an upper part rich in less dense, light-coloured feldspar.

Density is not the only factor which determines the rate of settling and the order of deposition of crystals from a magma. The size of the crystals, the viscosity of the magma, the degree of crystallization, the degree of supercooling, and the presence or absence of convection currents are also important (Bartlett 1969; Marsh and Maxey 1985).

The rate of sinking of crystals can be estimated approximately from Stokes's Law:

$$V = \frac{2gr^2(d_S - d_L)}{9\eta}$$

Table 27. Settling rates of crystals from basic magma of density 2.58 g/cm^3 and viscosity 3000 poises (300 Pa.s), from Wager and Brown (1968)

	Radius (mm)	Settling velocity (m/yr)	(cm/hr)
Plagioclase (2.68 g/cm^3)	0.5	6	0.06
	1.0	23	0.3
	2.0	92	1.0
	4.0	368	4.2
Augite (3.28 g/cm^3)	0.5	40	0.5
	1.0	160	1.8
	2.0	640	7.3
Olivine (3.70 g/cm^3)	0.5	64	0.7
	1.0	256	2.9
	2.0	1024	11.7
Magnetite (4.92 g/cm^3)	0.25	3	0.03
	0.5	134	1.5
	1.0	535	6.1

where V is the terminal velocity of a solid sphere of radius r falling through liquid of viscosity η under gravitational acceleration g, and d_S and d_L are the densities of the solid and liquid respectively. Table 27 gives some settling velocities calculated by Wager and Brown (1968) for crystallizing basaltic magma. It can be seen that large silicate crystals sink more rapidly than small magnetite crystals, despite the greater density of the latter, and that settling velocities are slow enough for settling to be easily disrupted by convection in the magma. A few examples have actually been described of rhythmic layering in which small heavy crystals overlie larger light crystals, and the same result was obtained in centrifuge experiments on basaltic liquids by Campbell *et al.* (1978).

Shaw (1965) calculated the settling velocities for crystals in granitic magmas, and showed that they are very much less than for basaltic magmas. Assuming a viscosity of 10^6 poise (10^5 Pa.s) and a magma density of 2.3 g/cm^3, crystals of feldspars, biotite and hornblende 1 mm in diameter would have a settling velocity of less than 0.1 m/yr.

In recent years, serious doubt has been cast on the role of crystal settling in differentiation. Much of the debate has centred on the behaviour of plagioclase feldspar, which may be either lighter or heavier than basic and intermediate magmas, depending on its composition and that of the magma. Bottinga and Weill (1970) calculated that the iron-rich fractionated gabbroic magma of the Skaergaard layered intrusion would actually have been heavier than the plagioclase crystals which have apparently settled from it, and a similar relationship has been found for other ferrogabbros. In many accumulative rocks, crystals occur together whose settling rates could not have been the same, taking into account both specific gravity and crystal size, and in a few layered igneous rocks the heaviest crystals are actually concentrated at the top of the layers (Parsons 1979).

Furthermore, magmas have a finite yield strength (see Chapter 2), and this must be exceeded before a crystal can begin to sink or rise, even if it is

Figure 184. The variation of density with pressure of a basaltic melt (Kilauea olivine tholeiite) and plagioclase crystals of composition An_{90} and An_{65} (after Kushiro 1980a).

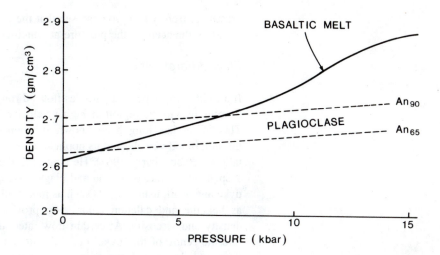

lighter or heavier than the liquid. According to McBirney and Noyes (1979) this effect may explain why the observed settling rates of olivine crystals in Hawaiian lava lakes are much less than would be predicted by Stokes's Law. The effect of yield strength on the settling of large xenoliths is illustrated in Fig. 52 in Chapter 3.

At high pressure, the relative densities of crystals and magmas are not the same as at low pressure. This is illustrated by Fig. 184, which shows how the relative densities of basaltic melt and plagioclase crystals vary with pressure. Crystals that sink at low pressure may float at high pressure. Figure 185 shows estimated settling rates of crystals in basaltic magma at different

Figure 185. Rates of sinking or floating of crystals (2 mm diameter) in olivine tholeiite melt, as calculated by Kushiro (1980a).

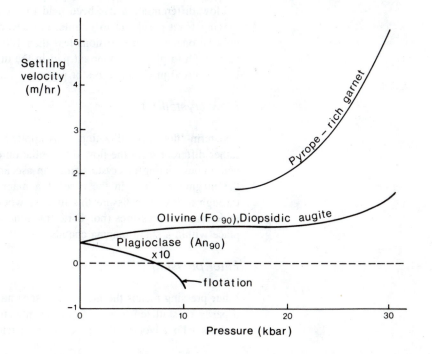

pressures, from which you can see that the course of fractionation would be greatly influenced by the pressure at which crystal settling occurs.

Flow differentiation

If a fluid with suspended particles flows through a conduit, the particles tend to migrate into the region of higher velocity flow, i.e. away from the walls. This effect has long been known in connection with the flow of blood through blood vessels, and its application to magmas was studied experimentally by Bhattacharji (1967). He simulated the flow of a basaltic liquid with suspended olivine, pyroxene and plagioclase crystals between the walls of a dyke and a sill, using liquids such as motor oil and turpentine and solids such as bakelite and calcium chloride to provide appropriate combinations of density and viscosity. At certain flow rates, the particles were concentrated into the centre of the 'dyke' or 'sill', but showed a tendency to settle as the flow rate dropped. From these experiments it can be predicted that flow differentiation could produce enrichment of either heavy crystals (for example olivine) or light crystals (for example plagioclase) in the centre of a sill or dyke, depending on what phenocrysts were present at the time of intrusion or crystallized during the period of flow.

Flow differentiation has been invoked to account for picritic sills which show a concentration of olivine in the centre, rather than at the base, as might be expected from gravitational settling, and for the picritic dykes. Of course, the chilled margin of a dyke or sill represents the first magma to enter the fissure, whereas the central part may have been flushed through by a large volume of magma before final consolidation, so a concentration of olivine in the centre could be due to the tapping of a magma reservoir which had already been differentiated by crystal settling.

Flow differentiation has been held to account for variation in intrusions ranging from peridotite to granite, but according to Barrière (1976) it can only be operative in intrusions less than 100 metres across. This could still be enough to play an important part in the differentiation of extrusive rocks if it occurred in feeder-dykes or pipes.

Flow crystallization

The term 'flow crystallization' was applied by Irving (1980) to a process rather different from the flow differentiation described above. He envisaged continuous plating of crystals from an ascending magma on to the walls of the magma conduit. In the case of a magma ascending from the mantle through a dyke-like fissure this process would involve fractionation over a wide range of pressures (polybaric fractionation), unlike fractionation taking place within a small magma chamber.

Filter pressing

Filter pressing means the mechanical separation of the still-liquid portion of a partly crystallized magma from the interstices of a mass of crystals. The authors who advocate this process are often rather vague about the exact mechanism which is envisaged. In the industrial process of filter pressing, a liquid is separated from a solid by squeezing the suspension against a porous surface or filter, as in a wine press. An analogous situation never arises in a magmatic environment, according to Propach (1976), who reviewed the various ways in which filter press action might operate. Nearly all the possible models of filter pressing are open to serious objections, not least of which is the relative incompressibility of the crystal framework once the crystals are in contact.

It has yet to be shown that filter pressing actually occurs in nature on any significant scale, but a little interstitial liquid may be displaced from crystal cumulates by compaction, or by seismic shaking, or by convection of the interstitial liquid. Anderson *et al.* (1984) proposed that a possible mechanism for driving the residual liquid from a crystalline matrix is gas pressure from H_2O exsolved from the magma in the final stages of crystallization.

Some segregations of residual liquid into small patches in lava flows and large sills may result from interstitial liquid oozing through a crystalline matrix. The best examples of segregation veins are those that have been found in consolidated Hawaiian lava lakes (Helz 1980; Helz *et al.* 1989). In recent years, several large lava lakes have formed on Kilauea volcano by the ponding of lava flows in pre-existing pit craters (Fig. 186). These lava lakes have then crystallized over a period of several years and some of them have been explored by drilling. The segregation veins form dyke- or sill-like bodies distinguishable from the surrounding basalt by their coarser grain size and more felsic composition. Their compositions correspond approximately to the liquid that would remain after about 50% crystallization of the host basalt. Many of these veins are only a centimetre or so thick, but the biggest ones are over a metre thick and extend for over 100 metres.

Bea *et al.* (1994) described a horizontal sheet-like body of granite (the Pedrobernardo pluton in Spain) in which the compaction of crystal mush at an advanced stage of crystallization has caused residual melt to be expelled upwards. The resulting layered structure shows 30 m of muscovite leuco-granite grading down into 300 m of muscovite–biotite granite, grading down into 500 m of biotite–muscovite granite. This is one of the very few convincing examples of *in situ* fractional crystallization in a body of granitic magma.

Figure 186. The summit region of Kilauea volcano, Hawaii, showing the main caldera and neighbouring pit craters. The ones which are named on the map are those which have contained lava lakes at various times since 1950.

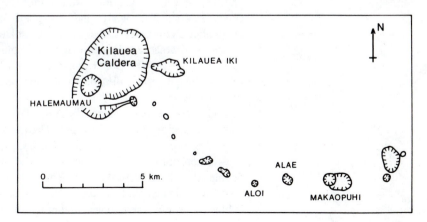

Selective nucleation

Under perfect equilibrium conditions, every magma would have completed crystallization by the time it cooled to its solidus temperature, and would then consist of an assemblage of minerals whose crystallization sequence was determined by the equilibrium phase relationships. Actually, crystallization does not take place so readily. Some volcanic rocks (obsidians) cool right down to atmospheric temperature without crystallizing at all, and many more have at least a glassy groundmass. Plutonic rocks have more time available for crystallization, but it is still unsafe to assume that crystallization is instantaneously completed as soon as the temperature falls below the solidus. If supercooling of magmas is a commonplace occurrence, then the sequence of crystallization could be determined by the nucleation behaviour of the crystals rather than by their equilibrium phase relations.

A textural feature particularly indicative of selective nucleation in supercooled magma is the crescumulate texture (= harristic structure, comb layering), which is seen in some layered basic intrusions. This consists of a development of olivine-rich layers in which the olivine crystals have an elongated branching morphology and are orientated perpendicularly to the plane of the layers. Donaldson (1977) was able to reproduce this structure experimentally in picritic melts which were supercooled by 30–50 °C. An example of an intrusion in which the inversely graded layers could not have formed by crystal settling is the Klokken syenite, in which Parsons (1979) believed that the layering was controlled by the order of nucleation and rates of crystal growth.

Gas streaming

The term 'gas streaming' describes an upward flow of gas bubbles in a magma. Bubbles might contribute to fractionation by attaching themselves to crystals or by helping to extract an interstitial liquid from a crystal cumulate.

There has been a certain amount of confusion about the role of volatiles in magmatic differentiation. Gas streaming is not to be confused with gas transfer, in which material is transported in a gas phase with or without crystallization. The word 'gas' is itself misleading, because in most magmatic environments the common volatiles are above their critical pressures. A volatile-saturated magma in a plutonic environment might be at a temperature of 700 °C and pressure of 3 kilobars; under these conditions free water and carbon dioxide would have densities of 0.62 and 0.76 g/cm^3 respectively.

This does not really matter, because the mechanism is just as valid with bubbles of dense fluid as with a gas. The movement of crystals by bubbles is a familiar process in the mining industry, where froth flotation is used to separate ore minerals from gangue. A possible example of flotation by gas bubbles has been described by Rose *et al.* (1978): basalts from Fuego volcano in Guatemala have plagioclase phenocrysts containing zones of gas bubble inclusions; the bubbles are not themselves large enough to have lifted the plagioclase crystals, but they do indicate that larger gas bubbles may have been adhering to the crystal surfaces during crystallization. The process of flotation by bubbles has been observed experimentally by Campbell *et al.* (1978).

A specialized role for bubbles is that of extracting interstitial liquid from a cumulate. By creating movement in the crystal–liquid mush, bubbles could help to pack the cumulate crystals closer together and reduce the space available for interstitial liquid. This might be described as fluid stirring. The source of the fluid could be volatiles either exsolved from the crystallizing magma or driven from adjacent country rocks by contact metamorphism.

Gravitational liquid separation

During the crystallization of a magma, the density of the remaining liquid changes, sometimes becoming greater and sometimes less, depending upon whether the crystallization results in iron-enrichment or depletion. McBirney *et al.* (1985) described how crystallization at the walls of a magma body might result in either an upward or downward current of differentiated liquid which could then accumulate at the top or bottom of the magma chamber. A somewhat similar mechanism was invoked by Helz *et al.* (1989) to explain the differentiation of Kilauea Iki lava lake. They envisaged that a low-density interstitial melt rose from within the crystal mush at the base of the magma body and passed up through overlying denser uncrystallized melt without mixing. They called this process 'diapiric melt transfer'.

CRYSTALLIZATION DIFFERENTIATION OBSERVED

The products of crystal fractionation are a variety of crystal concentrates, and a series of evolved liquids. It is relatively easy to recognize the rocks produced by crystal accumulation, such as rhythmically layered gabbros or phenocryst-enriched lavas, but the recognition of liquid differentiates is more difficult.

Accumulations of crystals which have separated out from magma are known as *cumulates*, and they have various distinctive textures which enable their mode of crystallization to be deduced. The simplest distinction that can be made is between *orthocumulates*, in which the settled ('cumulus') crystals are enclosed by material that has crystallized from their interstitial melt, and *adcumulates*, in which the interstitial liquid has been displaced by outgrowth from the cumulus crystals (Fig. 187). Some adcumulates are completely monomineralic. Many descriptions of accumulative rocks are given in the monograph on layered igneous rocks by Wager and Brown (1968).

One method of examining fractionated liquids is to measure the composition of the glassy groundmass in partially crystallized lavas. Another is to identify segregations of residual liquid formed during the crystallization of lava flows or minor intrusions. The largest volumes of fractionated liquid that are identifiable with absolute certainty are those which constitute the later fractions of the large differentiated intrusions such as Skaergaard.

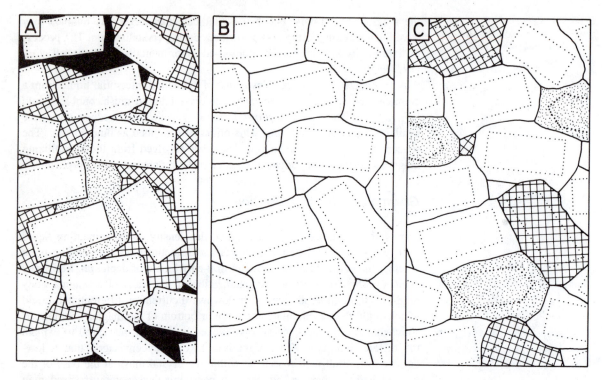

Figure 187. Diagrammatic representation of different types of cumulate that might be formed from a gabbroic magma:

A : Plagioclase–orthocumulate
B : Plagioclase–adcumulate
C : Plagioclase–olivine–pyroxene–adcumulate

Plagioclase – unshaded, olivine – dotted, pyroxene – square grid, iron oxide – black, interstitial quartz and feldspar – dashes. The cumulus crystals are outlined by dotted lines (after Wager 1963).

Glassy groundmasses

The composition of the liquid that would be left by the removal of early formed crystals can be estimated from the composition of the glassy groundmass in those lavas that are only partly crystalline. Analysis of groundmass glasses can be done by using an electron microprobe, although with difficulty because of heterogeneity and the presence of crystalline inclusions.

Table 28 shows the composition of the glassy groundmass in three different types of basaltic rock. In the tholeiitic basalt the residual glass is rhyodacitic. The alkali olivine basalt has a trachytic glass. The nepheline basanite has a glass of phonolitic composition. Other analysed rocks have given similar results, although there are relatively few measurements of alkaline basalts because they less often contain a glassy matrix than tholeiites.

Hawaiian lava lakes

Makaopuhi crater is situated on the eastern rift zone of Kilauea volcano, Hawaii (Fig. 186). In prehistoric times it contained a lava lake up to 1000 metres across and 150 metres deep. During the 1960s a 70-metre high cliff (subsequently covered by new lava) exposed a complete section through the crystallized lava lake, enabling its crystallization history to be studied. The lava, which is tholeiitic, underwent fractionation as it crystallized (Moore and Evans 1967; Evans and Moore 1968).

The main process of fractionation was the settling of olivine crystals,

Table 28. Chemical composition of the glass phase in some partially crystallized basaltic rocks

	1		2		3	
	Rock	*Glass*	*Rock*	*Glass*	*Rock*	*Glass*
SiO_2	49.44	76.13	45.60	52.84	42.63	52.00
TiO_2	2.09	0.43	2.42	1.05	2.11	0.80
Al_2O_3	14.96	13.65	15.36	20.63	13.07	19.24
Fe_2O_3	—	—	2.44	1.24	2.02	0.92
FeO	12.47	1.52	8.89	2.74	10.78	3.06
MnO	0.23	0.02	0.15	0.04	0.20	0.11
MgO	6.42	0.02	7.36	0.51	10.19	0.36
CaO	11.42	0.70	8.99	2.56	10.97	1.92
Na_2O	2.47	3.72	3.13	5.08	3.35	7.80
K_2O	0.32	3.85	1.68	5.07	0.93	6.09
P_2O_5	0.24	0.02	0.80	0.89	1.00	0.45
H_2O+	—	—	3.27	6.99	1.77	7.10

1. Tholeiite (94% crystalline), Ogurnes, Iceland (Meyer and Sigurdsson 1978).
2. Alkali olivine basalt (78% crystalline), Guyra, New South Wales (Wilkinson 1966).
3. Nepheline basanite (87% crystalline), Inverell, New South Wales (Wilkinson 1966).

which are concentrated in the lower part of the body although not at the base (Fig. 188). Moore and Evans estimated that it took 30 years for the centre of the lava lake to become 90% solid, and that the settling rate of olivine phenocrysts averaged 4×10^{-6} cm/s. The end result of fractional crystallization was a glassy groundmass of rhyolitic composition (SiO_2 75.5%, K_2O 5.7%, Na_2O 3.1%) which is most abundant in the centre of the lake, which presumably was the last part to crystallize. In the upper parts of the lake there are also a series of 'micropegmatite' segregation veins. These vary from less than 1 up to 25 centimetres thick, and taper out towards the edge of the lake.

Figure 188. Modal variation with depth of the basalt of the Makaopuhi lava lake, Hawaii (after Evans and Moore 1968).

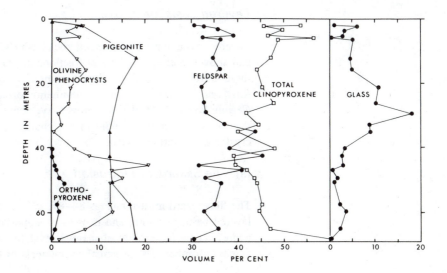

An outstanding experimental study of differentiation in place is that carried out on the solidification of the *Alae* lava lake (Wright and Peck 1978). This lake formed during the eruption of Kilauea in August 1963, and was about 15 metres deep. Six days after the eruption, when the solidified crust was less than a metre thick, an 87-centimetre long hole was drilled into it. Another 11 holes were drilled during the following 15 months as the whole lava lake solidified. Samples were collected of the newly crystallized basalt, and of glasses representing interstitial melt that had oozed from the walls of the drill holes. Chemical analyses of the bulk rock and of the successive liquid residues showed the trend of liquid composition brought about by fractional crystallization. The initial magma had an olivine–tholeiite composition, but the final melt composition was rhyolitic, with the following approximate normative composition: quartz 33.5%, feldspars ($Or_{50}Ab_{47}An_3$) 63.3%, hedenbergite 1.2%, magnetite 1.3%, ilmenite 0.7%.

The most thoroughly studied of the Hawaiian lava lakes is *Kilauea Iki*. This was filled with lava in the eruption of November–December 1959. The lava was a picrite, i.e. a basaltic liquid with about 20% of suspended olivine crystals. The crust of the cooling lava lake was drilled in 1960–62 by staff of the Hawaiian Volcano Observatory, and more holes have been drilled subsequently, enabling many samples to be collected for analysis (Helz 1987; Helz *et al.* 1989). Some of the drill core was partially molten right up to the point at which it was quenched by the water used to cool the drill bits.

Figure 189 shows the compositions of the bulk samples and interstitial glasses from the Kilauea Iki drill cores. Again, the final melt composition was rhyolitic. It is noticeable that several of the constituents show a non-linear variation, and the abrupt changes in the compositional trend can be related to changes in the crystallizing mineral assemblage. Olivine was the only phase crystallizing at 1200 °C; it was joined by augite at about 1160° (CaO in the liquid then starts to fall), plagioclase at about 1150° (Al_2O_3 starts to fall), ilmenite at 1110° (TiO_2 starts to fall), and apatite at 1065° (P_2O_5 starts to fall).

Layered basic intrusions

Layered basic intrusions contain rocks which are obviously accumulative, and show the effects of crystal fractionation *in situ*. The layering shown by these intrusions is of two types: rhythmic layering and cryptic layering. *Rhythmic layering* is the visible alternation of layers of different mineralogy. *Cryptic layering* consists of a progressive change in chemical composition of the rocks and minerals indicating a change in the magma composition as crystallization progressed.

(a) *The Skaergaard intrusion*

The Skaergaard intrusion (Figs 84–87, 190–192) is the classic example of a layered basic intrusion, and it owes its importance to the fact that it results from a single large body of magma crystallizing to completion in a closed magma chamber. The principal components of the intrusion are:

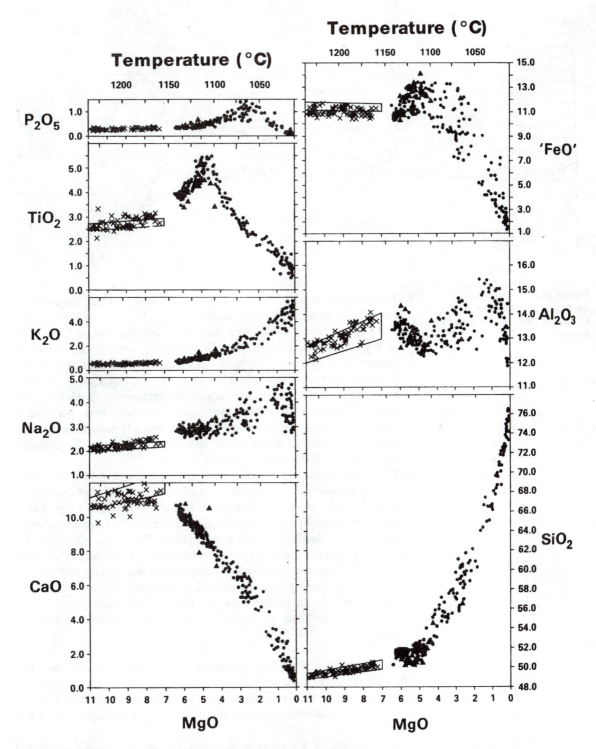

Figure 189. Variation diagrams for the bulk samples and interstitial glasses of Kilauea Iki lava lake, Hawaii (after Helz 1987). The samples with more than 7% MgO are olivine–phyric rocks. The temperature scale at the top gives the estimated quenching temperatures of the glasses.

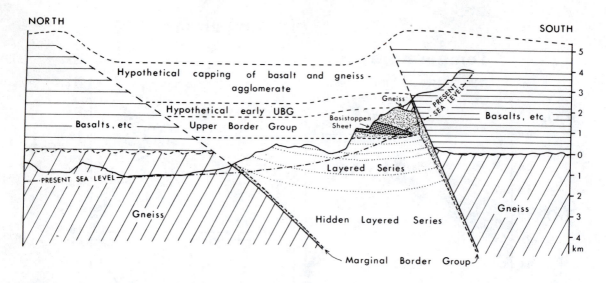

Figure 190. Cross-section of the Skaergaard intrusion, east Greenland (after Wager and Brown 1968).

1. The Marginal Border Series, chilled near the outer contact, composed of fine-grained tholeiitic olivine gabbro.
2. The Layered Series, constituting the bulk of the intrusion, and consisting of rhythmically layered gabbro. The rhythmic layering consists of an alternation of light and dark bands, in which the dark bands are enriched in olivine, pyroxene and magnetite, and the light bands are enriched in plagioclase. The bands are typically a few centimetres in thickness and may be repeated over and over again, or they may occur only sporadically, separated by unlayered gabbro. Notable features are igneous lamination (preferred orientation of tabular crystals, especially feldspars) and various well-developed cumulate textures.
3. The Upper Border Series. Coarse-grained gabbros with poorly developed horizontal layering; they probably crystallized from the roof downwards throughout the period of development of the main layered series. Fallen blocks from the Upper Border Series occur within the Layered Series.
4. Granophyres. Small bodies of granophyre are found near the top of the intrusion: (a) melanogranophyres (hedenbergite granophyres) which occur among the Upper Border Series – they do not have sharp contacts and their exact relationship to the gabbros is uncertain; (b) the Tinden acid granophyre – a sill intruded into the Upper Border Series; and (c) the Sydtoppen transitional granophyre – also intruded into the Upper Border Series.

The fractionation of the Layered Series is revealed by its cryptic variation. Although the rocks look similar, their bulk composition (taking light and dark bands together) progressively changes throughout the height of the intrusion, from gabbro to ferrodiorite, as shown by the examples in Table 29.

Because most of the rocks in the intrusion are the products of crystal accumulation, analyses of individual hand specimens are not necessarily identical to the compositions of the magmas from which they crystallized. The compositions of the individual minerals are a better guide to the change

Table 29. Representative chemical and modal compositions of Skaergaard rocks (from Wager and Brown 1968)

	1	*2*	*3*	*4*	*5*
Structural height (m)	0	1295	2150	2540	2540
FeO (wt %)	9.3	14.9	22.7	15.1	9.1
MgO	11.6	6.4	1.7	0.1	0.5
CaO	10.5	9.2	8.7	7.8	5.1
$Na_2O + K_2O$	2.3	2.8	3.3	4.9	5.9
Quartz (%)	—	—	1	7	33
Plagioclase	56	37	45	43	40
Olivine	11	—	17	4	5
Clinopyroxene	29	51	28	34	13
Orthopyroxene	3	—	—	—	—
Opaques	1	12	6	11	7

1. Olivine gabbro (Lower zone, no. 4087)
2. Gabbro (Middle zone, no. 3661)
3. Ferrodiorite (Upper zone, no. 4145)
4. Quartz–ferrodiorite (Sandwich horizon, no. 4330)
5. Hedenbergite granophyre (no. 4332)

in composition of the magma. The appearance and disappearance of each mineral phase, and their changes in composition, reveal the progress of differentiation as the whole body of magma crystallized. Figure 191 shows the compositions of the cumulus minerals. Plagioclase was a cumulus phase throughout; pyroxenes were not present as cumulus phases initially although interstitial pyroxene is always present.

The Skaergaard intrusion is uniquely valuable as a source of information because it appears to represent the complete crystallization of a single body of magma which was not tapped or replenished, and its initial composition is

Figure 191. Mineralogy of the Skaergaard intrusion: the compositions of the cumulus minerals in the Layered Series are shown in relation to height above the lowest exposed level (after Wager and Brown 1968).

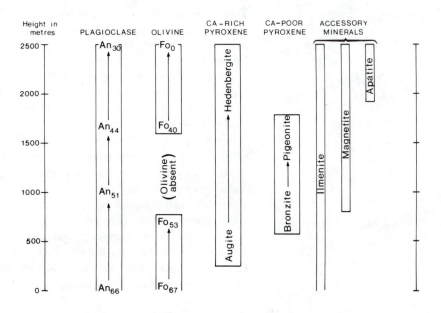

that of the commonest type of basaltic magma. There are two main problems for its complete interpretation: (1) clarification of the precise mechanism of fractionation; and (2) the nature of the granophyres – do they represent the final melt fraction?

There is no doubt that the layered gabbros are the products of crystal accumulation, but what is uncertain is the degree to which this was brought about by gravitational settling. There are many features in the layered gabbros reminiscent of depositional structures in sediments, such as graded layering and cross-bedding. Wager (1963) emphasized the role of movements in the magma, and distinguished between two types of current: relatively slow convection currents, depositing uniform gabbro; and rapid currents heavily loaded with suspended crystals, depositing layers of crystals graded by size or density. Bottinga and Weill (1970) predicted, and Murase and McBirney (1973) confirmed experimentally, that the more Fe-rich magmas of Skaergaard would have had a higher density than the coexisting plagioclase. Since crystallization was predominantly from the bottom upwards, this means that gravitational settling may not have been the dominant crystal separation mechanism. It is possible that the minerals accumulated as they crystallized, with little or no sorting.

McBirney and Noyes (1979) radically reinterpreted the crystallization of the Skaergaard gabbros, further questioning the role of gravitational separation. They attributed cumulate textures, layering and preferred orientation to *in situ* crystallization. In their opinion, layering was controlled by the relative rates of chemical and thermal diffusion in the magma during cooling. The combined effect of density and temperature gradients on convection in a magma has been described by the term 'double diffusion' (McBirney 1985).

Of the granophyres, the Sydoppen and Tinden granophyres have initial $^{87}Sr/^{86}Sr$ ratios much higher than those of the main layered sequence (Leeman and Dasch 1978) and cannot be differentiates of the gabbro. It seems most likely that they are partial melt fractions of the country rocks. The melanogranophyres do have initial $^{87}Sr/^{86}Sr$ ratios similar to the gabbros and may represent the last liquid fraction in the differentiation sequence. The calculated trend of compositions of liquids during fractionation of the intrusion as a whole is shown in Fig. 192.

It was discovered by McBirney and Nakamura (1974) that the composition reached by the fractionating magma during the crystallization of the upper zone of the layered gabbros was one in which separation of immiscible liquids may have occurred. When mixtures intermediate in composition between the upper zone ferrodiorites and the granophyres were held experimentally at a temperature just above the liquidus at an appropriate oxygen fugacity, separation of immiscible globules was observed. The melanogranophyres may therefore have been concentrated from an immiscible melt fraction coexisting with the ferrodiorite magma, rather than being a liquid residue of the ferrodiorite.

The trend in composition of the evolving liquids during crystallization was calculated by Wager and Brown (1969) as shown in Fig. 192. However, since the layered Skaergaard rocks are all cumulates it is very difficult to estimate the liquid composition, and Hunter and Sparks (1987) believed the

Figure 192. FMA diagram for (a) the actual rocks, and (b) the calculated liquids of the Skaergaard intrusion (after Wager and Brown 1968).

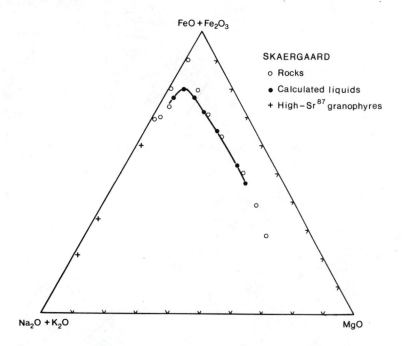

trend of liquid composition was quite different from that shown in Fig. 192. Wager and Brown's calculated trend is one of extreme iron enrichment, but Hunter and Sparks believed the evolved liquids trended towards basaltic andesite, icelandite and rhyolite, more in line with the association of magmas observed in many ordinary volcanoes. They also questioned whether the magma chamber was really closed, and suggested that some andesitic and rhyolitic magma could have been lost by eruption. Their interpretations have been strongly challenged by McBirney *et al.* (1990), while Brooks *et al.* (1991) have pointed to examples of iron-rich dykes and lavas in East Greenland as proof that fractionation did indeed lead towards a ferrobasaltic liquid.

Subsequent isotopic work (Stewart and De Paolo 1990) showed that the Skaergaard magma might have undergone replenishment by up to 20 or 30% of its total mass during fractionation, and that it may have assimilated between 2 and 4% of gneissic country rock.

(b) *The Kiglapait intrusion*

The Kiglapait intrusion is on the coast of Labrador in eastern Canada, and has been very thoroughly described by Morse (1969a, 1979, 1981). It is roughly circular in plan (Fig. 193), and shows both layering and cryptic variation. The layering dips inwards at angles which decrease from about 40° near the margin to about 20° in the centre. On the assumption that the layering is parallel to the floor of the intrusion, the shape can be envisaged as a lopolith (Fig. 194), and such a structure is also consistent with geophysical (gravity) evidence (Stephenson and Thomas 1979).

About 94% of the volume of the intrusion is occupied by a Layered Group, which grades upwards from troctolite through olivine gabbro to

Figure 193. Sketch map of the Kiglapait intrusion, Labrador (after Morse 1979).

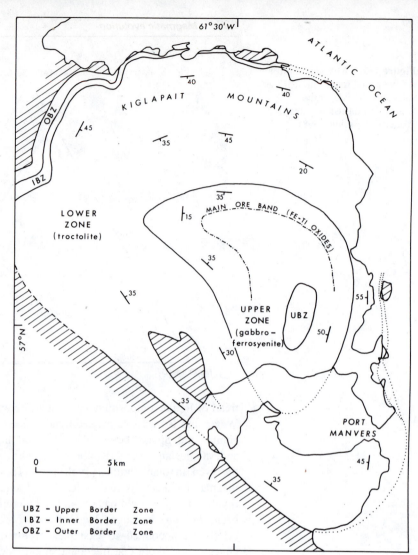

Figure 194. Cross-section of the Kiglapait intrusion, Labrador (after Morse 1979).

UBZ – Upper Border Zone
IBZ – Inner Border Zone
OBZ – Outer Border Zone

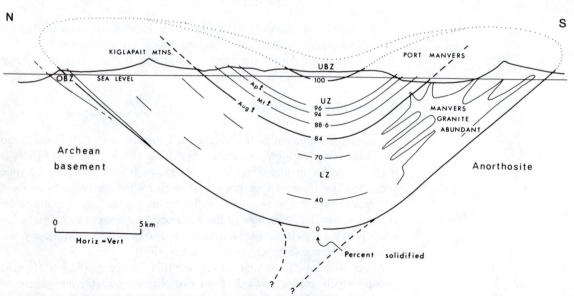

ferrosyenite. It shows strong rhythmic layering, with examples of igneous lamination, graded layering, and slump and scour structures, but there are also large areas of homogeneous unlayered rock. The rocks of the main Layered Group are underlain by a zone of gabbro (Inner Border Zone) and overlain by a zone of troctolite, gabbro and ferrodiorite (Upper Border Zone), of which the latter shows an inverted cryptic variation and may be a roof-grown equivalent of part of the main layered group. An Outer Border Zone originally mapped by Morse contains a mixture of gabbroic and metamorphic rocks whose relationship to the Kiglapait magma is doubtful.

According to Morse, the magma of the main Layered Group crystallized completely in a closed system and without contamination, although De Paolo (1985) has subsequently found isotopic evidence for both magma replenishment (in the Lower Zone) and contamination (in the Upper Zone). The original magma was estimated from the bulk composition and mineralogy to have been a dry troctolitic high-alumina basalt, somewhat more alkaline than that of the Skaergaard intrusion. The average modal composition of the whole intrusion is feldspar 67%, olivine 23%, augite 7%. Accumulating crystals from the floor upwards, the magma is estimated to have taken about 1 million years to crystallize. The fractionation sequence is more completely preserved than in any other layered intrusion, and Fig. 195 shows the trend of liquid variation. It is estimated that the last ferrosyenites represent the final 0.01% liquid fraction of the initial volume. This final differentiate is represented by rocks containing alkali feldspar, fayalite and ferroaugite, neither strongly oversaturated nor undersaturated with silica. Extreme cryptic variation is shown by the compositions of the main minerals in the intrusion: plagioclase from An_{67} to An_{10}, olivine from Fo_{69} to Fo_0, and augite from En_{73} to En_0 (Fig. 196).

Figure 195. FMA diagram for (a) actual rocks and (b) calculated liquids of the Kiglapait intrusion (after Morse 1981).

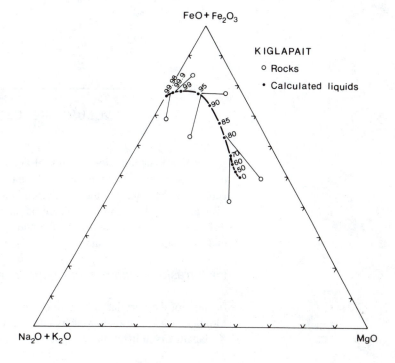

Figure 196. Mineralogy of the Kiglapait intrusion: the variation in composition of olivine, augite and plagioclase in relation to degree of solidification of the original magma (after Morse 1979). The appearance of cumulus augite at 84% solidified is used to separate the Lower Zone (troctolites) from the Upper Zone (gabbros).

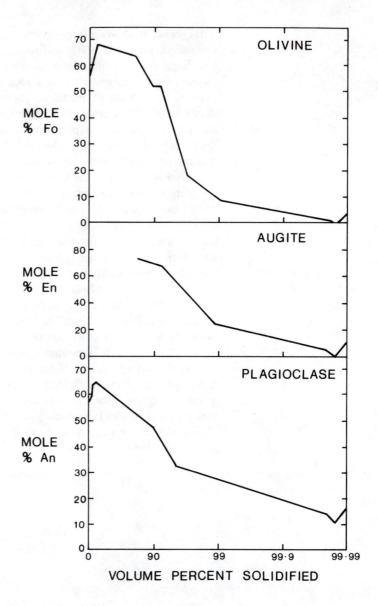

The Kiglapait intrusion is unusual among the larger differentiated gabbro intrusions in the syenitic nature of the most differentiated rocks. Most of the gabbros which show extreme differentiation are tholeiitic and gave rise to oversaturated differentiates. There is no good example of a completely differentiated alkali gabbro body, and hence no example with a silica-undersaturated end product.

(c) *Intrusions with macro-rhythmic layering*

In some of the very large basic intrusions, such as the *Stillwater* intrusion of Montana, fractionation has taken a different course from that of Skaergaard or Kiglapait. Crystallization has apparently been dominated by the accumulation of only one or two cumulus phases for long periods, thus giving rise to thick layers of extreme composition, such as peridotite, pyroxenite, and anorthosite. There are even completely monomineralic layers of dunite and chromitite. The Stillwater intrusion is described in more detail in Chapter 14.

The *Great Dyke of Zimbabwe* (Fig. 91) consists mainly of layered dunites and pyroxenites, overlain by gabbro and norite at four centres. The ultramafic rocks are cumulates of olivine + bronzite + chromite, with several layers of chromitite. One pyroxenite band about 200 metres thick can be followed for 200 kilometres. The mafic rocks are unlayered gabbros and norites, which are not rhythmically layered but show strong cryptic variation. The immense volume of ultramafic rocks, combined with their very magnesian composition (olivine Fo_{94}, pyroxene En_{93} in the lowest cumulates) have given rise to the suspicion that the parent magma of the Great Dyke might have been ultramafic rather than basaltic. No chilled margin has been found in the Great Dyke itself, but some neighbouring and possibly related dykes have chilled margins with MgO contents between 15 and 20%, and fractionation of a parent liquid with about 15% MgO would account well for the compositional variation in the complex (Wilson 1982).

The very large-scale layering shown by the Stillwater and Great Dyke intrusions is known as *macro-rhythmic layering*, or *cyclical layering*. In contrast to Skaergaard, in which most of the rock could be described as gabbro, all these intrusions contain very thick bands of ultramafic rock. Typical macro-rhythmic layering consists of units 30–300 metres thick with olivine cumulates at the base followed by orthopyroxene or plagioclase cumulates towards the top. The sudden reappearance of olivine (a high-temperature phase) at the base of each unit, together with a reversal of the chemical trend of cryptic variation, suggests that each cyclical unit represents the entry of a new pulse of magma into the magma chamber (Campbell and Turner 1989).

Many other large basic intrusions show rhythmic layering and cryptic variation. These two features do not always go hand-in-hand. The gabbros of the Great Dyke are not layered even though there is substantial cryptic variation. Conversely, the Rhum intrusion in Scotland is strongly layered, with accumulative rocks ranging from peridotite to anorthosite, but a relative lack of cryptic variation shows that the residual liquids were not strongly differentiated (Wager and Brown 1968). Well-described intrusions which show a high degree of differentiation are the Bushveld complex in South Africa, the Muskox intrusion in Canada, and the Jimberlana intrusion in Australia. Each of these intrusions resembles Skaergaard and Kiglapait in their extreme range of liquid compositions, but each resembles Stillwater and Great Dyke in the presence of thick monomineralic layers, and in all three the differentiation sequence is complicated by replenishment of the magma reservoir during fractionation.

ROLE OF CRYSTALLIZATION DIFFERENTIATION

The importance of crystal fractionation was advocated so effectively by Bowen (1928) that for many years it was taken almost for granted that all igneous rocks were related by this process. It was assumed that in any varied suite of igneous rocks, one magma type was parental and the rest were derived from it by differentiation. This assumption is no longer reasonable, because the experimental studies of recent years have shown that nearly all the common magma compositions can equally well be produced directly by the partial melting of suitable source materials under appropriate conditions of melting. So although crystallization differentiation obviously does take place, it is by no means certain that it plays an essential role in the development of the most voluminous types of magma.

The observed examples of crystal fractionation show what kind of magma variation would result from this process. Basalt is nearly always assumed to be the parental magma in possible examples of fractionation series. Three main magma series are widely recognized by petrographers:

1. A tholeiitic series: tholeiitic basalt–andesite–dacite–rhyolite
2. An alkali rock series: alkali basalt–trachybasalt–trachyte–phonolite
3. A calc-alkaline series: basalt–andesite–dacite–rhyolite

These series are based primarily on observed rock associations, but advocates of crystallization differentiation view them also as probable fractionation series. Alkali basalts are assumed to give rise to trachytic and phonolitic differentiates, and tholeiitic basalts to rhyolitic differentiates. The tholeiitic and calc-alkaline series differ in that the tholeiitic series is said to be characterized by greater iron-enrichment (increase in Fe/Mg ratio) during fractionation than the calc-alkaline series, and the latter series is characterized by a greater abundance of magmas of intermediate (andesitic) composition.

It should be emphasized that the arrangement of rocks into 'magma series' does not automatically prove that their magmas must be related by fractional crystallization, since there are several other processes by which a series of magma compositions can be related.

The trend of liquid composition in all the definitely observed examples of differentiation of basic magmas can best be represented by the FMA diagrams (Figs 192, 195 and 197). In every case, we see an initial increase in Fe/Mg followed by enrichment in alkalis. These features are readily predicted from simple phase equilibria. In systems containing anhydrous ferromagnesian minerals, such as Fo–Fa (Fig. 182), the lower liquidus temperatures of the iron-bearing compositions are bound to result in the residual liquids being enriched in iron. In systems containing alkalis along with magnesium or iron, such as the system Fo–Di–An–Ab (Fig. 141), the lowest liquidus temperatures are found for alkali-rich compositions, so that differentiation must lead to enrichment in alkalis.

In actual volcanic rock series the degree of iron-enrichment is not as great as that shown by the Skaergaard and Kiglapait intrusions. Tholeiitic rock series sometimes contain fairly iron-rich 'ferrobasalts' and 'ferroandesites', for example in the Craters of the Moon district (Leeman *et al.* 1976) or the sea-floor rocks of the Galapagos spreading centre (Byerly 1980). A feature of the Craters of the Moon lavas which is suggestive of fractionation is the alignment of rock-groundmass pairs along the same trend as the lava series as a whole (Fig. 198). A smaller degree of iron-enrichment is

Figure 197. FMA diagram showing the trend of liquid variation in Alae lava lake compared with that in the Skaergaard intrusion.

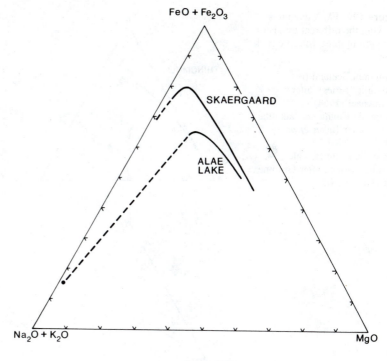

Figure 198. FMA diagram for the lavas of the Craters of the Moon district, Snake River Plain, Idaho (after Leeman *et al.* 1976).

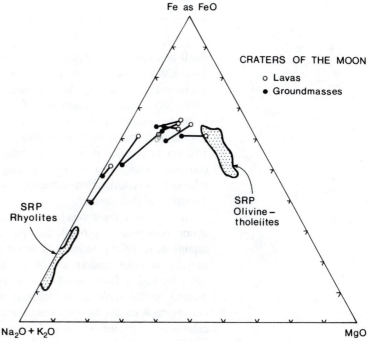

typical of alkalic rock series such as that of Tristan da Cunha (Fig. 199).

The 'calc-alkaline series' is said to encompass the basalt–andesite–rhyolite association typical of orogenic belts. When such rocks are plotted on an FMA diagram, they typically show a high degree of scatter and only a

Figure 199. FMA diagrams showing the different trends of variation in three volcanic rock series:

Thingmuli, Iceland (a 'tholeiitic' series), after Carmichael (1964).
Tristan da Cunha (an 'alkaline' series), after Baker *et al.* (1964).
Lorne, Scotland (a 'calc-alkaline' series), after Groome and Hall (1974).

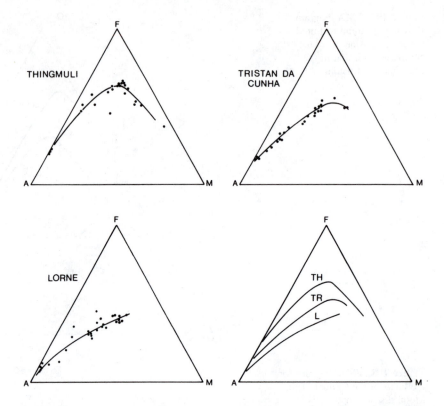

small degree of iron-enrichment, for example the Lorne lavas in Fig. 199. This FMA trend is not shared by any of the observed examples of differentiation, such as layered basic intrusions, and the abundance of intermediate and acid magmas in this association is far greater than is seen in proven examples of differentiation. There are mechanisms to explain the calc-alkaline trend by fractionation at high oxygen fugacity (see Chapter 10), but in practice it is very doubtful whether this range of rock types is consanguineous (derived from a single parent magma) at all, let alone related by crystallization differentiation or by any of the other processes discussed in this chapter.

The strongest objections to the hypothesis of fractional crystallization as a major petrogenetic process are the difficulty of extracting fractionated liquids in sufficient quantity to account for the observed volumes of felsic magma, and the inadequacy of the known fractionation mechanisms. In order for highly fractionated lavas to be extruded, a differentiating body of magma would have to be tapped, but this in turn would prevent the development of the extreme liquid compositions. Wager and Brown (1968) calculated that the original Skaergaard magma was more than 98% solidified by the time the residual liquid composition reached the sharp bend in the FMA curve shown in Fig. 192, at which point the magma was still very basic with less than 50% SiO_2.

A way in which differentiation might take place without extreme iron enrichment is for the differentiating magma to be replenished with more of the original magma, as seems to have happened periodically in some large differentiated intrusions such as Bushveld and Muskox, but this would make the generation of a felsic melt fraction even more difficult to achieve. Replenishment is certainly an important factor in magmatic evolution. It has been suggested that repeated injection of new magma into high-level magma reservoirs has prevented the diversification of ocean-floor basalts that might have taken place if differentiation had been able to continue (Dungan and Rhodes 1978; Rhodes *et al.* 1979; Huppert and Sparks 1980). The consequences of replenishment in a body of fractionating magma have been discussed at length by O'Hara and Mathews (1981).

The mechanisms of crystallization differentiation are largely controlled by physical conditions, such as the densities of magmas and crystals, and the viscosity and rate of movement of magmas. The importance of density in gravitational crystal separation has already been emphasized. The extent of fractionation is equally strongly influenced by viscosity, since any mechanism of crystal–liquid separation is likely to be less effective in high-viscosity magmas. Basaltic magmas are much less viscous than acid magmas, and it is in intrusions of basaltic magma that the most obvious examples of crystal fractionation are found.

The rate of magma flow is also important. Magmas that are viscous enough to support, or rise rapidly enough to carry, large solid xenoliths can hardly be depositing small crystals at the same time. For alkali basalts ascending from the mantle, Maaløe (1973) calculated that unless the velocity of upward movement is at least 1 centimetre/second, magma could not reach the surface without consolidating by heat loss to its surroundings, and suggested that this rate of ascent is too great for crystal settling to occur (see Table 27). Indeed alkali basalts often manage to carry large ultramafic nodules, centimetres across, some of which have almost certainly come up from the mantle.

For fractionation to take place, it may be necessary for magma to be stored in a reservoir or magma chamber for some period of time between the generation of the parent magma and the eruption of its fractionated derivatives. This in turn may require particular tectonic conditions to permit a body of magma to form and remain undisturbed while fractionation takes place. This may be why large bodies of trachyte and phonolite magma only occur in regions of extensional tectonics, and are erupted in large volume even though magmas of this composition are not particularly common.

An objection which has been raised to several magma series having been related by fractionation is that they contain gaps, for example between andesite and rhyolite in continental orogenic lava suites or between basalt and trachyte in some oceanic volcanoes. It is difficult to be sure that such a gap is not due to sampling problems, but in any case a gap is easily explained by a shallow-sloping liquidus surface (Grove and Donnelly-Nolan 1986).

LIQUID IMMISCIBILITY

Liquid immiscibility involves the separation of an originally homogeneous magma into two coexisting fractions. In silicate melts, experimental studies have identified a very limited extent of liquid immiscibility, and petrographic evidence for immiscibility is rather sparse. In melts of silicate–carbonate or silicate–sulphide composition, liquid immiscibility is more extensive and the process is likely to have operated on a large scale.

SILICATE–SILICATE IMMISCIBILITY

Among simple binary silicate systems, liquid immiscibility has been found in the systems SiO–CaO, SiO_2–MgO, SiO–FeO and SiO–Fe_2O_3. It does not occur in the systems SiO_2–Al_2O_3, SiO_2–Na_2O or SiO_2–K_2O. In the systems in which it does occur, the addition of a small amount of alumina or alkalis brings about a homogeneous melt. In view of the limited extent of immiscibility in these simple systems, most igneous rock compositions might be expected to give rise to homogeneous melts above the liquidus temperature. More complex synthetic systems have generally confirmed this prediction, but there is one notable exception, the system leucite–fayalite–silica (Roedder 1951).

In the system leucite–fayalite–silica (Fig. 200) there are two fields of liquid immiscibility. One adjoins the silica–fayalite tieline and is not geologically relevant. The other lies along the fayalite–tridymite boundary, and has a much more interesting composition range. There are igneous rocks whose compositions are not far from this region of liquid immiscibility, namely the fayalite–granites and rhyolites. The system is also relevant to the melting of pelitic rocks, in which SiO_2, Al_2O_3, Fe_2O_3 and K_2O are the major involatile constituents. The related systems fayalite–nepheline–silica (including Na_2O), forsterite–leucite–silica (including MgO), and leucite–anorthite–silica (including CaO) have been investigated, but do not show any corresponding region of liquid immiscibility.

Figure 200. The system leucite–fayalite–silica, showing the two regions of liquid immiscibility (after Roedder 1951).

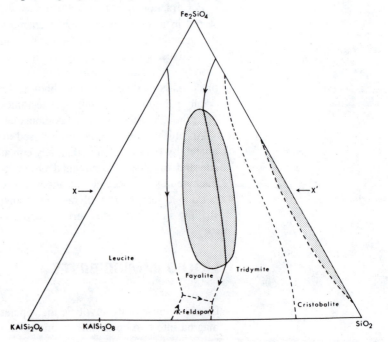

At first glance it might appear that the extent of liquid immiscibility in the system Lc–Fa–SiO$_2$ is too small to be of much practical relevance. This is not altogether so. Firstly, the range of liquid immiscibility has been found to vary with pressure and with oxygen fugacity (Watson and Naslund 1977; Visser and Koster van Groos 1979c). Secondly, the extent of liquid immiscibility in this system is increased by the presence of P and Ti in the liquids (Visser and Koster van Groos 1979b). Thirdly, supercooling of the magma could extend the range of compositions in which immiscibility could occur, since the extent of liquid immiscibility increases with falling temperature.

In the system leucite–fayalite–silica, the upper unmixing surfaces of the liquid immiscibility fields are connected below the liquidus of tridymite and cristobalite by a wide region in which metastable liquid immiscibility might occur (Fig. 201). It is possible that some other systems that have no stable region of liquid immiscibility might contain a potential field of metastable liquid immiscibility lurking not far beneath the liquidus. On the other hand, the role of metastable liquid immiscibility would be limited by the very rapid increase in viscosity of melts cooled metastably below the liquidus, which would prevent the unmixed liquids from separating on a large scale.

In a small number of volcanic glasses, liquid immiscibility is actually observed. The first recorded examples were in the lunar basalts studied by Roedder and Weiblen (1970). Small amounts of melt were trapped in the crystals and are preserved as glass inclusions. During most of the course of crystallization the melts were apparently homogeneous, but when the magmas were more than 90% crystalline immiscibility occurred, and the late stage crystals contain inclusions of two coexisting glasses. Subsequently,

Figure 201. Cross-section through the liquidus surface of the system leucite–fayalite–silica along the 30% FeO isopleth (Fig. 200), showing the extent of liquid immiscibility. A dotted line indicates the metastable extension of the 2-liquid field (after Visser and Koster van Groos 1979a).

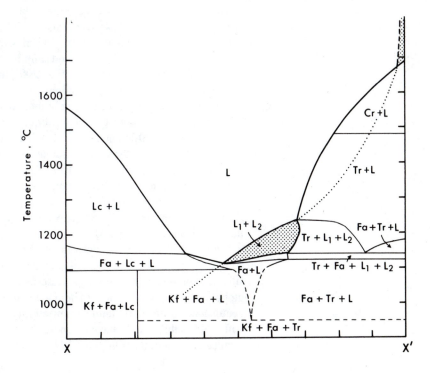

immiscibility was discovered by De (1974) in residual glass of the Deccan basalts in India, and by Fujii *et al.* (1980) and Philpotts (1981, 1982) in a number of other basalts and andesites.

Table 30. Compositions of coexisting liquids in an olivine-basalt of Fuji volcano, Japan (averaged from Fujii *et al.* 1980)

	Dark globules	*Light matrix glass*
SiO_2	30.4	76.0
TiO_2	9.6	0.6
Al_2O_3	0.8	11.8
Fe as FeO	40.2	2.5
MgO	1.2	0.1
CaO	8.2	1.0
Na_2O	0.3	3.3
K_2O	0.2	3.0
P_2O_5	4.3	0.2

As can be seen from the analyses given by Fujii *et al.* (Table 30), unmixing is into an Fe-rich liquid and an Si-rich liquid, the latter being similar to rhyodacite but the former not corresponding to any familiar magma type. These glasses have not of course unmixed from the basaltic magma but from its fractionated inter-crystalline residual liquid. The Fe-rich droplets coexist with a larger volume of Si-rich liquid, so the melts in which immiscibility occurs are relatively siliceous, and also Mg-poor judging by the paucity of MgO in both the liquid compositions. Most of the basalts in which liquid immiscibility has been discovered are tholeiitic, but a small number are alkaline, and Philpotts (1982) found that the Si-rich glass in these examples is trachytic or phonolitic.

There is indirect evidence that immiscibility may be more widespread than is suggested by these few occurrences. Melting experiments by McBirney and Nakamura (1974) showed that the Skaergaard Upper Zone gabbros lie on the border of a field of liquid immiscibility comparable to the low temperature unmixing region of Fig. 200, and they would not have been miscible with the overlying granophyres. In fact a mixture of Upper Zone gabbro with granophyre actually split into two liquids when it was melted in the laboratory. The densities of the liquids were sufficiently different, and their viscosities sufficiently low, that separation of the immiscible melts would have been possible. Immiscible liquids were also produced experimentally by Dixon and Rutherford (1979) by extreme fractional crystallization of a basaltic melt. Their compositions approximated to those of a ferrobasalt and a plagiogranite, the latter being rather similar to the plagiogranites of some ophiolite complexes. The compositions of the immiscible liquids are shown in Fig. 202.

Liquid immiscibility has been invoked to explain a variety of rocks which consist of inclusions of one silicate composition in a matrix of another silicate composition, variously described as having variolitic, ocellar, globular, spherulitic or amygdaloidal texture. Some of them are probably genuine occurrences of liquid immiscibility and some are not. Basic rocks with feldspathic ocelli have been melted experimentally to produce two immiscible fractions (Philpotts 1971; Ferguson and Currie 1971). On the

Figure 202. Liquid immiscibility as revealed by fractionation in the laboratory of a liquid of oceanic tholeiite composition. The open circle shows the original composition; the dots show successive liquid fractions; and fields A and B represent the final immiscible fractions (after Dixon and Rutherford 1979). The Skaergaard liquid trend (S) is shown for comparison.

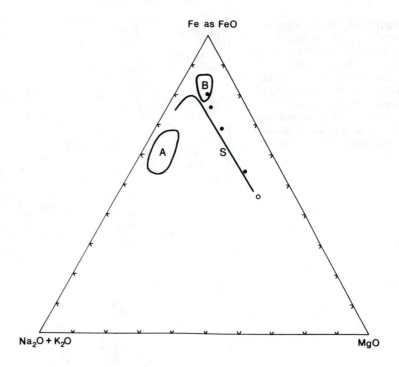

other hand some postulated examples of liquid immiscibility may be segregation vesicles, formed by the flow of interstitial residual melt into gas cavities without immiscibility having occurred. Roedder (1979) has reviewed the many possible occurrences of silicate liquid immiscibility.

SILICATE–OXIDE AND SILICATE–PHOSPHATE IMMISCIBILITY

Philpotts (1967) demonstrated experimentally that magnetite–apatite-rich liquids can coexist immiscibly with silicate liquids, and Naslund (1976) showed that an iron oxide melt can coexist with a silicate liquid in the system $KAlSi_3O_8$–FeO-Fe_2O_3–SiO_2 at high oxygen fugacities (Fig. 203).

There is abundant field evidence for the existence of iron oxide-rich magmas, including some volcanic examples (Nyström and Henriquez 1994). The magnetite flows of El Laco volcano in northern Chile are composed of magnetite and hematite with accessory apatite and silica. This occurrence is at a very high altitude (~5000 metres above sea level), and contains a total of 500 million tons of iron ore in the form of flows, dykes and tuffs. The even larger Proterozoic iron ore deposit of Kiruna in Sweden may also be volcanic; this deposit is mainly magnetite, with a high apatite content and minor actinolite and diopside. In both cases the iron oxide-rich magmas are assumed to have separated from the accompanying acid and intermediate magmas.

Several authors have attributed the formation of apatite-bearing magnetite layers in the Bushveld complex to liquid immiscibility, although the majority of magnetite-rich layers in the Bushveld appear to be due to simple crystal accumulation (Reynolds 1985). Badham and Morton (1976) described

Figure 203. The influence of oxygen fugacity on the extent of immiscibility between silicate and oxide liquids in the compositional range $KAlSi_3O_8$–SiO_2–FeO. The fields of liquid immiscibility are shaded; tielines show the compositions of coexisting liquids at liquidus temperatures (after Naslund 1976).

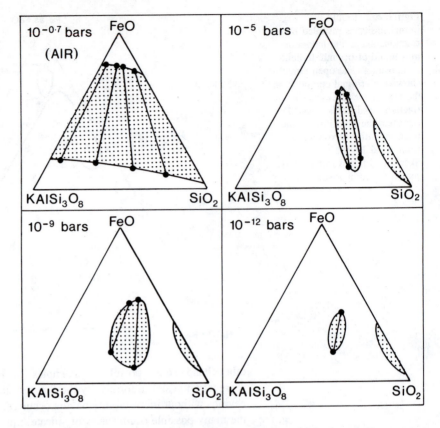

magnetite–apatite intrusions in northern Canada which they believe separated by immiscibility from a silicate magma, and Bergstøl (1972) proposed a similar origin for a dyke of 'jacupirangite' (magnetite 37%, apatite 24%, pyroxene 25%) in southern Norway. Magnetite–apatite rocks occupy quite large areas in some carbonatite complexes, for example at Phalaborwa in South Africa (see Fig. 301 in Chapter 11).

In addition to the intrusive magnetite–apatite rocks there are also ilmenite bodies which are thought by many authors to be magmatic. Several of these form substantial ore deposits, notably the Allard Lake ilmenite deposits in Quebec, which occur as dykes, sills and lenses in anorthosite country rock. In the Roseland anorthosite in Virginia, there are dykes of ilmenite–apatite, magnetite–apatite, and rutile–apatite rock which have all been interpreted as products of liquid immiscibility (Kolker 1982).

SILICATE–CARBONATE IMMISCIBILITY

Koster van Groos and Wyllie (1966, 1968) studied the join $NaAlSi_3O_8$–Na_2CO_3 in the presence of CO_2 and H_2O, and found large miscibility gaps between silicate-rich melts and carbonate-rich melts (Fig. 204). Immiscibility between carbonate and silicate liquids has subsequently been found in several other simple systems, and in compositions intermediate between ijolite and carbonatite and between phonolite and carbonatite (Freestone and

Figure 204. Liquid immiscibility along the join $NaAlSi_3O_8$–Na_2CO_3 in the presence of excess CO_2 at 1 kilobar pressure (after Koster van Groos and Wyllie 1966). L_1 = silicate-rich liquid, L_2 = carbonate-rich liquid, V = vapour; the solidus is not shown.

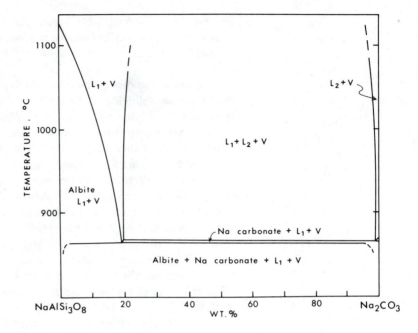

Hamilton 1980; Kjarsgaard and Hamilton 1989). As a result of these discoveries it is now widely believed that liquid immiscibility is involved in the formation of carbonatite magmas. One possibility is that a silicate magma with some dissolved carbonate might become carbonate-enriched by crystallization of silicate minerals, leading to the carbonate content reaching the limit of miscibility, whereupon a separate carbonate melt would separate. The development of carbonatite magmas is discussed further in Chapter 11.

There are several direct observations of carbonate immiscibility. The carbonatite lava erupted from Oldoinyo Lengai in June 1993 contains abundant spherical silicate inclusions composed of nepheline, pyroxene and garnet, which themselves contain rounded inclusions of carbonatite, and this was interpreted by Dawson *et al.* (1994) as an example of coexisting carbonate and silicate liquids. In alkaline plutonic rocks, immiscibility is sometimes shown by the occurrence of carbonate-rich ocelli in a silicate matrix, or by microscopic inclusions of coexisting silicate and carbonate melt within crystals.

SILICATE–SULPHIDE IMMISCIBILITY

It has long been known from the smelting of ores that molten silicates and sulphides are immiscible. Their immiscibility in magmas was directly observed following the August 1963 eruption of Kilauea. During the cooling of basaltic lava in the Alae crater, the residual siliceous fraction became saturated with sulphide when its sulphur content reached 0.038% (Skinner and Peck 1969). There then separated a copper-rich pyrrhotite and an immiscible sulphide melt with an estimated composition near Fe 61%, Cu 4%, S 31%, O 4%. This quenched to a mixture of pyrrhotite, chalcopyrite and magnetite. Several other basalts have been described which contain immiscible sulphide inclusions, the sulphide phase normally being rich in Fe, Ni and Cu (Czamanske and Moore 1977; Pedersen 1979; Distler *et al.* 1983).

Experimental studies of the system $FeS-FeO-Fe_3O_4-SiO_2$ (Maclean 1969) and of sulphur solubility in silicate melts (Haughton *et al.* 1974; Shima and Naldrett 1975; Mysen and Popp 1978; Wendlandt 1982) have clarified the likely extent of immiscibility and the conditions under which exsolution of a separate sulphide melt might occur. The solubility of sulphide in silicate melts has been found to increase with temperature and with FeO content of the melt, but to decrease with pressure, oxygen fugacity and SiO_2 content.

Many plutonic igneous rocks contain sulphides in sufficient amount that a separate sulphide melt may have been present during at least the later stages of their crystallization. The former presence of immiscible droplets of molten sulphide is sometimes revealed by globular sulphide inclusions in silicate rock, and by globular silicate inclusions in sulphide rock (Hawley 1965). Sulphide immiscibility is believed to have occurred during crystallization of many of the larger gabbroic intrusions, such as those of Skaergaard (Wager *et al.* 1957), Sudbury (Hawley 1965), and Bushveld (Liebenberg 1970).

The density of molten sulphides is considerably higher than that of molten silicates, and droplets of any size would be susceptible to concentration by gravity. This may account for some of the sulphide ore deposits associated with large basic intrusions such as the Sudbury complex (Naldrett and Kullerud 1967).

A rare occurrence in modern volcanoes is that of natural molten sulphur. Flows or accumulations of molten sulphur have been described from several volcanoes in Japan, the Kuril Islands, Hawaii and New Zealand (Macdonald 1972). Although such a melt would presumably be immiscible with silicate magma, these liquids did not necessarily separate from a normal magma by immiscibility but are more likely derived from the melting of fumarolic sulphur.

SILICATE–WATER IMMISCIBILITY

Water has a finite solubility in silicate magmas, and if for any reason it is present in a larger amount than can be dissolved in the silicate melt it will form a separate immiscible phase. This will be an aqueous solution containing a small proportion of dissolved silicates. In contact with the magma, water would be above its critical temperature, and strictly speaking would be a supercritical fluid rather than a true liquid, but under all except near-surface pressures the density of supercritical water is high enough that it behaves very much like a liquid. The coexistence of silicate melt and hydrothermal fluid is therefore regarded here as another example of liquid immiscibility.

Some experimental data on the solubility of water in silicate melts are shown in Fig. 205. The solubility increases with pressure, and acid magmas can hold as much as 10% H_2O at pressures above 5 kilobars, but under the range of pressures encountered in the crust there is still a large miscibility

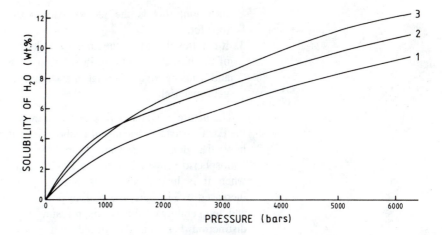

Figure 205. The solubility of water in silicate melts at different pressures: (1) basalt at 1100°C; (2) andesite at 1100°C; (3) granite at its liquidus. For additional details see Fig. 144.

gap between water and the common silicate magma compositions. Burnham (1967) has reviewed the possibility that the gap might be closed by the presence of large amounts of F, Cl, or CO_2, or if the magma were peralkaline or peraluminous, but the experimental evidence is that none of these factors could bring about continuous miscibility between a silicate melt and a hydrothermal solution.

Bearing in mind the increasing solubility of water in silicate melts with increasing pressure, it follows that decrease in pressure on a melt containing dissolved water may bring the dissolved water content of the magma to saturation point. A further decrease in pressure would then cause water to be exsolved from the magma. If the rising magma were crystallizing hydrous minerals such as amphiboles and micas, sufficient water might be used up in forming these minerals to prevent exsolution from happening. Water might also be lost from the magma into wall rocks and xenoliths. Any water that is not used up in these ways must inevitably be exsolved as crystallization of the magma proceeds to completion.

Exsolution of water can therefore be expected to be a common but not universal occurrence. Basic magmas generally have lower water contents than acid magmas, and have enough Fe and Mg that they can use up all the available water in making hornblende and biotite. Acid magmas can only crystallize as much biotite and muscovite as the available Fe, Mg and Al will permit, so they are more likely to end up with a surplus of dissolved water. Much depends on the source of the magma. If it formed by the melting of rocks containing such hydrous minerals as amphibole or mica, it may contain just enough water to permit the eventual crystallization of these minerals from the magma without any being left over.

Under volcanic conditions, exsolved water is lost to the atmosphere and does not influence the composition of the remaining magma. Under plutonic conditions, exsolved water has three main effects:

1. By remaining in contact with the silicate melt it can bring about textural modifications such as those seen in pegmatites and aplites.
2. In migrating to another part of the magma it can transfer dissolved material, altering the composition of the magma in different parts of the

intrusion: this is the phenomenon known as gas-transfer or volatile-transfer.

3. It can travel off into the country rocks or into an already crystallized part of the intrusion, carrying in solution material extracted from the magma, and causing hydrothermal metasomatism and depositing hydrothermal vein material.

The nature of water is very different under plutonic conditions from that at the Earth's surface, because it is above its critical pressure. Figure 206 shows how the density of water varies with temperature and pressure. At atmospheric pressure, water (a liquid) changes suddenly to steam (a gas) when it is heated above its boiling point (100 °C). As the pressure is increased to 218 bars, the boiling point rises to 374 °C (the critical pressure and temperature). Above this pressure and temperature, there is no sharp distinction between the liquid and gaseous states, and under plutonic conditions water must be considered as a supercritical fluid rather than as a liquid or gas. In its physical properties, such as density and viscosity, it resembles low temperature water more than it resembles steam. The term 'volatile-transfer' might therefore be preferable to 'gas-transfer' to describe the transport of material via the supercritical H_2O phase.

Figure 206. The density of H_2O in g/cm^3 at different temperatures and pressures. The critical point of H_2O is at 374 °C, 218 bars. Measurements of the properties of H_2O at high temperatures and pressures are listed by Burnham *et al.* (1969).

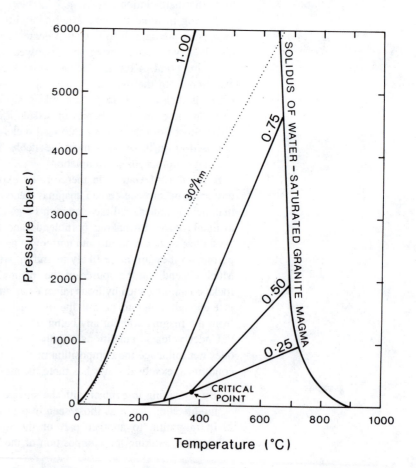

Figure 207. The isotopic composition of hydrothermal fluids from the Cornubian orefield, south-west England, compared with the magmatic water of the Cornubian granites and with present-day meteoric waters (after Alderton and Harmon 1991).

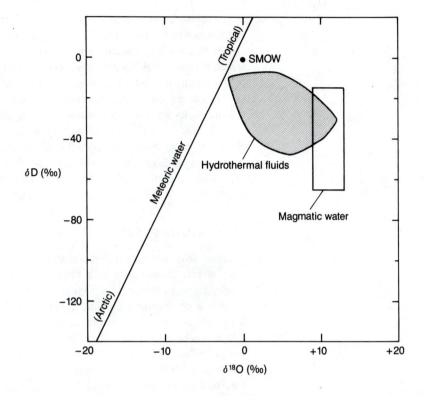

Stable isotope studies have been widely applied in an attempt to discover what proportion of the hydrothermal phenomena associated with the crystallization and cooling of igneous intrusions can be attributed to exsolved magmatic water and what proportion to meteoric water. The isotopic composition of magmatic water is difficult to establish, but can be estimated on the assumption that magmatic water would have been in equilibrium with the various igneous minerals at about 700–1100 °C. Because of the relatively limited range of $\delta^{18}O$ and δD in most igneous rocks their magmatic waters must be similarly limited, but the isotopic compositions of hydrothermal fluids (e.g. the fluid inclusions in ore minerals) are much more varied, as indicated in Fig. 207. Taylor (1979) reviewed the isotopic evidence and showed that most of the hydrothermal phenomena associated with igneous intrusions are due to meteoric rather than magmatic water.

LIQUID FRACTIONATION

Several processes are known by which an initially homogeneous liquid magma might become heterogeneous without crystallizing and without splitting into immiscible fractions. Diffusion in a thermal gradient can be demonstrated experimentally, and the process is well known to physical chemists. Diffusion due to a pressure gradient or by the separation of light or heavy atoms by gravity is a possibility in a vertically extended magma body, but has not been studied much experimentally.

Petrologists have taken some interest in diffusion as a mechanism by which magmas may become homogenized rather than differentiated, so there is a certain amount of information available on diffusion rates. Bowen (1921) carried out some experiments which showed that diffusion rates in silicate melts are extremely slow, implying that most magmas would have cooled and crystallized before significant diffusion could take place. Bowen's experiments temporarily discredited differentiation mechanisms dependent on diffusion, but subsequently Shaw (1974) showed that rates of diffusion increase with temperature and water content, and that the effectiveness of diffusion at an igneous contact is enhanced by convection.

THERMAL DIFFUSION

Thermal diffusion means that constituents migrate through a medium from a zone at one temperature to a zone at a different temperature. It does not simply mean diffusion at a high temperature, which will happen anyway if there is a composition gradient. Thermal diffusion can occur in a perfectly homogeneous medium, making it heterogeneous.

If a homogeneous gas, liquid or solid is placed between a hot surface and a cold surface, certain constituents migrate towards the high temperature and others towards the low temperature. In liquids, this phenomenon is known as the Soret effect (Tyrrell 1961). The experience of industrial application (e.g. in the separation of uranium isotopes) has shown that the efficiency of separation by thermal diffusion can be greatly increased by convection. Thermal diffusion has been demonstrated experimentally in silicate liquids (Reuther and Hinz 1980; Walker and De Long 1982; Lesher and Walker 1986), and it has been invoked from time to time as a mechanism of magmatic differentiation, particularly by Wahl (1946), Hildreth (1981) and Schott (1983). Thermal diffusion may also be important in the differentiation of diagenetic and hydrothermal fluids (Dandurand *et al.* 1972), of the Earth's core, and of the solar system as a whole (Mason *et al.* 1966).

A detailed study of thermal diffusion in basaltic liquid was carried out by Walker and De Long (1982). They studied the compositional variation which developed in a basalt held above its liquidus for up to 219 hours in a temperature gradient of 265 °C over a distance of 8 millimetres. The amount of liquid fractionation is illustrated by the analyses in Table 31. The hot end of the charge became enriched in Si, Na and K, and the cold end became enriched in Ti, Fe, Mg and Ca. The effect of liquid fractionation was comparable in magnitude to the degree of differentiation that could be brought about by crystal fractionation. The end products were also somewhat reminiscent of the products of silicate liquid immiscibility, although fractionated to a lesser degree (see Table 30). However, it remains to be seen whether the Soret effect could occur on a petrologically significant scale in a large body of magma, and some authors believe that it could not (Carrigan and Cygan 1986).

A very significant aspect of thermal diffusion is that it can bring about fractionation of stable isotopes, even heavy isotopes that are not normally

Table 31. Thermal diffusion: analyses of a Soret-fractionated basaltic liquid (from Walker and De Long 1982)

	Cold base (1215 °C)	Hot top (1480 °C)	Bulk composition
SiO_2	44.9	56.7	53.1
TiO_2	2.81	1.78	2.08
Al_2O_3	13.6	16.2	15.2
Cr_2O_3	0.10	0.03	0.04
FeO	16.4	6.38	9.70
MnO	0.26	0.16	0.18
MgO	7.39	4.27	5.24
CaO	12.3	9.96	10.8
Na_2O	2.22	3.91	3.44
K_2O	0.18	0.57	0.45

considered to be susceptible to fractionation in nature. Lodding and Ott (1966) experimentally fractionated the isotopes of Rb in liquid metal: they used temperature gradients of 100–200 °C and found that the light isotope was concentrated in the hottest part of the liquid during a period of a few days. If isotopic fractionation of Rb can be so readily achieved in the laboratory, then fractionation of strontium (i.e. variation of the $^{87}Sr/^{86}Sr$ ratio) by this mechanism must be a serious possibility. Fractionation of the isotopes of H, S, K and Rb by a combination of thermal and gravitational diffusion has also been shown experimentally by Costesèque and Schott (1976).

GRAVITATIONAL DIFFUSION

Gravitational diffusion involves the separation of heavy and light elements in a completely molten magma by gravity. Atoms are always in motion, each atom has a finite mass, and movement of any mass in a gravitational field involves work: consequently there must be a tendency for the atoms even in a homogeneous liquid to rearrange themselves by gravity. No one knows how fast such a rearrangement might be, or how rapidly a concentration gradient set up by this process would be counteracted by cohesive forces and by convective motion in the magma. Centrifugal accelerations of several thousand *g* are required by biochemists to separate heavy molecules from aqueous solution in an ultracentrifuge, and by chemical engineers to separate uranium isotopes in UF_6 gas, so it seems improbable that the natural gravitational acceleration of 1*g* would cause any significant separation in a viscous silicate liquid.

The nature of the changes that could result from gravitational diffusion in magmas can be estimated from the ratios of molecular mass to partial molar volume listed in Table 10. The constituents enriched in the upper part of a magma body would be SiO_2, Na_2O, K_2O and H_2O, and at the bottom of the magma body there would be enrichment in FeO, MgO, CaO and TiO_2. This corresponds to a differentiation into the major constituents of the common felsic and mafic magma types, and would lead to a concentration of felsic material at the top of the magma resembling that seen in some differentiated basic intrusions. On the other hand, gravitational diffusion could not account for the trend of Fe/Mg enrichment shown by such intrusions as Skaergaard.

Magnesium and ferrous iron are of similar ionic size and probably occupy similar sites in a silicate melt structure, but iron atoms are much heavier than magnesium. Mutual displacement would cause the iron to become concentrated downwards in preference to magnesium, leading to the lowest Fe/Mg ratios in the most felsic melt fraction. This does not correspond to the relationship observed in natural igneous rock series.

Water is the magma constituent most likely to be concentrated by gravitational diffusion. Some estimates of the potential scale of gravitational diffusion of water have been made by Verhoogen (1949). He calculated, as an example, that an albite–water melt at 900 °C and 500 bars containing 4% of water might be in equilibrium with a melt 2900 metres below it containing only 2.8% of water.

Gravitational diffusion was studied experimentally by Ivanov *et al.* (1975). They took basalts melted at 1400 °C and subjected them to an acceleration of about 1000 g in a powerful centrifuge for 30 minutes. They then chilled the melts and analysed them by electron microprobe. Appreciable fractionation was said to have been found, but it did not correspond to any familiar type of magmatic variation. SiO_2 migrated to the top and Al_2O_3 to the bottom of the melt, but Fe, Mg, Ca and Ti remained uniformly distributed. In Walker and De Long's (1982) experiments on thermal diffusion they studied vertically placed experimental charges with the hot end at the top and others with the hot end at the bottom, and found that the Soret separations were almost the same in both cases, indicating that gravitational diffusion was insignificant compared with thermal diffusion. The diffusion of Al_2O_3 was small, but in both cases it was enriched at the top end of the charge, the opposite of what was claimed by Ivanov and his collaborators.

Gravitational diffusion has been invoked to explain the variation in many igneous intrusions of varied type. Hamilton (1965) applied the mechanism to some of the large differentiated tholeiite sills in Antarctica, and reviewed other possible examples. Boone (1962) used the mechanism to account for the development of a syenitic facies in a predominantly diorite–granodiorite intrusion. Wilshire (1967) suggested that gravitational diffusion was responsible for part of the variation in the Prospect alkaline dolerite sheet in New South Wales. He believed that there had been upward diffusion of volatiles, alkali elements and possibly calcium, explaining the concentration of the least mafic rock varieties under structural highs in the undulating roof of the intrusion. He also pointed to other consequences of the upward diffusion of volatiles, namely retarded crystallization of the volatile-enriched magma and a higher oxidation state. Saether (1950) attached an important role to gravitational upward diffusion of CO_2 in alkaline magma, and suggested that the carbonate ions concentrated in the upper part of the magma would also attract the cations of the strongly electropositive elements so that they too would be enriched towards the top of the magma.

THE ROLE OF LIQUID FRACTIONATION

The case for liquid fractionation has been presented comprehensively by Hildreth (1981). He reviewed the evidence for magmatic heterogeneity, especially as shown by the products of large rhyolitic eruptions. He said that the proportion of crystals (phenocrysts) to glass in some rhyolites is so small that the compositional variations in the liquid cannot have resulted from fractionation of crystals, and he also discounted other causes such as primary heterogeneity or magma mixing.

Diffusion of water in a magma, by whatever mechanism, would be of particular importance because of the tendency of certain cations, particularly of the alkali elements, to be coordinated by water molecules. Kennedy (1955) suggested that the water dissolved in a magma would tend to migrate in such a way as to attain a uniform chemical potential throughout the magma body. In order for P_{H_2O} to be constant, the water content would have to be greater at shallow depths, where P_{total} is low, than at greater depth, where P_{total} is high. Hildreth (1981) listed mineralogical features of ash flow tuffs consistent with just such an upward enrichment in water within the magma body.

Arguments against a significant role for liquid fractionation of the type advocated by Hildreth have been put by Baker and McBirney (1985). It should be noted that Baker and McBirney have confusingly used the term 'liquid fractionation' to describe separation of differentiated liquids in a crystallizing magma (i.e. crystal fractionation), although also addressing themselves to liquid fractionation in the sense understood here.

CONTAMINATION

Magmas very often melt, react with, or mechanically incorporate material from the surrounding rocks and from xenoliths. The nomenclature of this process is somewhat confused. The magma is said to be contaminated, and the country rock assimilated, but authors disagree as to whether the use of these terms requires that the added material should have been completely liquefied. It is preferable not to be too restrictive, as melting, reaction and mechanical incorporation commonly operate together. I will call the process as a whole contamination, regardless of whether the assimilated material ever became completely melted. The term hybridization will not be used, because although sometimes applied to this process the word hybrid is also used to describe rocks produced by the mixing of magmas.

There are two different mechanisms by which contaminating material can be incorporated into an igneous body. The first is by melting of some of the contaminant and mixing of the melt fraction with the main body of magma. The second is by chemical reaction and mechanical incorporation not involving melting. The second process is by far the most important.

ASSIMILATION BY MELTING

For a magma to be able to melt its country rocks it must be able to supply heat equivalent to the latent heat of fusion of the country rock material. It can provide this heat in two ways: (1) if it is above its liquidus temperature it can lose heat by cooling without undergoing crystallization (this is called 'superheat'); (2) below its liquidus temperature it loses heat by crystallization (latent heat of fusion), as well as by further cooling.

It was once thought that the superheat contribution would be very small (Bowen 1928), because a magma formed by partial melting or fractional crystallization must initially be at its liquidus. However, several ways are now known in which a magma can attain a temperature well above the liquidus. The most important is due to the dependence of the liquidus on pressure. As can be seen from Fig. 208, a basalt magma rising rapidly from a depth of 100 kilometres (35 kbar) could easily be 200 °C above its liquidus temperature when it reached the surface. Some heat would of course be lost to the surroundings on the way up, but some would also be gained by the release of gravitational energy (Harris 1962). In the case of an H_2O-bearing magma such as granite, the liquidus and solidus temperatures can also be depressed by an increase in the water content of the magma. Clear evidence that many magmas arrive at the surface at super-liquidus temperatures is the common eruption of lavas which are completely liquid.

The ability of magmas to melt solid silicates may be estimated from their heat capacities (specific heats) and enthalpies of fusion (latent heats of melting):

Figure 208. The variation with pressure of the liquidus and solidus temperatures of basalt (after Green 1982).

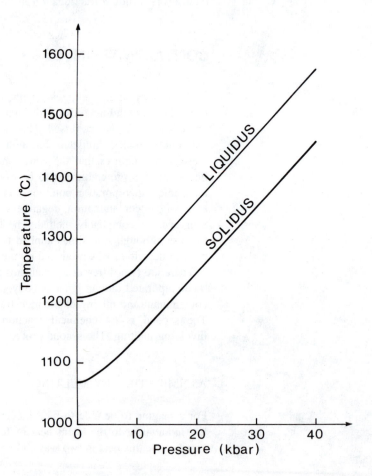

1. For a melt of albite composition, the heat capacity at constant pressure, C_p = 358.7 J/mole/°C (Carmichael *et al.* 1977). One mole (262 grams) of albite liquid cooling from 100 °C above its liquidus would release 35,870 joules of heat.

2. For solid albite at 1120 °C, the enthalpy of melting, ΔH = 64,500 J/mole (Lange and Carmichael 1990). Therefore the amount of heat required to melt one mole (262 grams) of solid albite is 64,500 joules.

So, an albitic liquid 200 °C above its liquidus could melt more than its own weight of albite crystals. This calculation is very approximate, because the latent heats of fusion of silicates are very poorly known. Furthermore, both the latent heat of fusion and the heat capacity of the liquid would be somewhat different if water were present (Rosenhauer 1976; Kesler and Heath 1968).

After any initial superheat has been used up, the amount of heat available for melting of foreign material is limited to what can be provided by crystallization of the magma itself. On thermodynamic grounds it is estimated that the mass crystallizing would have to be about twice the mass being melted (Nicholls and Stout 1982). This limits the power of the magma to melt large volumes of country rock, because the loss of much heat would effectively consolidate the outer margins of the magma body. Convective movement in the magma or sinking of stoped blocks from a roof zone could help by carrying xenoliths into the deeper parts of the magma body. Magma flow also speeds up assimilation by dispersing melted xenolithic material into the magma, and by causing dissolution of crystals even below their solidus temperatures (Sachs and Strange 1993).

The main limitation on the melting of country rocks by magma is not the inability of the magma to provide heat for melting, but the need for the country rocks to be raised to their melting temperatures first. In the deeper parts of the crust, the country rocks may already be at temperatures approaching their solidus, and the greatest scope for contamination by melting would be where crustal rocks were already in a state of incipient melting even before the arrival of the magma.

If a magma has its composition changed by the addition of new melt from xenoliths or by diffusion of constituents from solid xenoliths, there may be important consequences for the crystallization behaviour of the magma. For example it can be seen from Fig. 209 that the addition of granitic melt to a nepheline–syenite magma, or vice versa, would not so much alter its composition as cause it to crystallize, because the intermediate compositions have higher liquidus temperatures than the granite and nepheline–syenite minima. Similarly, there are thermal barriers between tholeiitic and alkali basalt liquids (Yoder and Tilley 1962), so that tholeiitic magma could not easily be made alkaline by contamination or vice versa.

ASSIMILATION WITHOUT MELTING

Assimilation without melting takes place in two stages: (1) the constituents of the xenoliths are brought into chemical equilibrium with the melt; and (2)

Figure 209. The system KAlSiO₄–NaAlSiO₄–SiO₂ (after Schairer 1957). In the two compositional ranges indicated by shading the liquidus surface lies below 1050 °C. These two minima, one on the tridymite–feldspar boundary and the other around the feldspar–nepheline–leucite eutectic point, correspond approximately to the compositions of rhyolite and phonolite respectively.

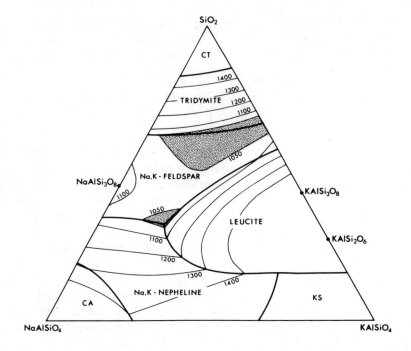

the equilibrated xenocrysts are mechanically dispersed through the magma, losing their separate identity. As the magma is normally crystallizing while this is taking place it is common for equilibration and dispersal to be incomplete, and the study of contaminated intrusions reveals many examples of incomplete assimilation where the processes have been arrested in varying degrees of completion.

Equilibration of xenoliths

The reaction between a xenolith and the surrounding melt can often be predicted from the phase relations of the cooling liquid. A xenolithic mineral may be:

1. A phase which the liquid could have been precipitating earlier in its cooling history.
2. A phase which the liquid is currently precipitating.
3. A phase which the liquid might eventually precipitate at a lower temperature.

The systems forsterite–silica and albite–anorthite illustrate some of the theoretical possibilities. Imagine liquid A in Fig. 210, crystallizing enstatite. A xenocryst of forsterite would be an example of the first case – a crystal which could have been precipitating earlier in the cooling history. It would react with the melt and be converted to enstatite. A xenocryst of enstatite would be an example of the second case – it is currently precipitating and would not react since it is already in equilibrium with the melt. A xenocryst of cristobalite (or quartz) would be an example of the third case – a phase which could crystallize at a lower temperature; it would dissolve in the melt, even though the temperature is well below the melting point of cristobalite.

Figure 210. The systems forsterite–silica and albite–anorthite (see also Figs 122 and 130).

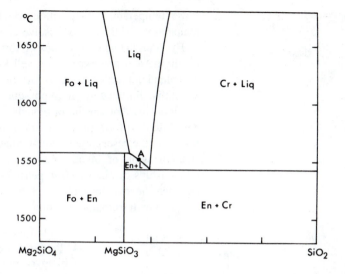

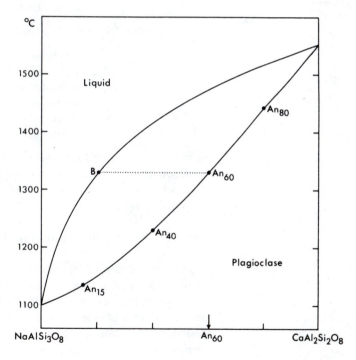

Similarly consider liquid B in Fig. 210, crystallizing labradorite (An_{60}). A xenocryst of bytownite (An_{80}, a higher temperature composition) would react with the liquid and be converted to labradorite. A xenocryst of labradorite (An_{60}) would be in equilibrium with the liquid and would not react. A xenocryst of andesine (An_{40}, a lower temperature composition) would selectively dissolve in the magma while being brought up to the temperature of B, and leave a residue of An_{60}. A xenocryst of oligoclase (An_{15}, an even lower temperature composition which would lie above the liquidus at the temperature of B) would dissolve completely in the melt.

The liquidus phase relations of some of the simpler igneous minerals such as olivine and plagioclase are well known, but there is less information on hornblende, biotite, or most of the metamorphic minerals that occur in xenoliths. Fortunately, it is possible to infer quite a lot from the reaction relationships that can be observed petrographically. The reaction of olivine with liquid to form pyroxene gives rise to the well-known corona structure. Likewise the resorption of low-temperature phases is indicated by rounding and corrosion of alkali feldspar xenocrysts in basalts. Pictures of these relationships can be found in petrographic textbooks. Bowen (1928) brought together the evidence of this kind, and summarized it into two 'reaction series'. For ferromagnesian minerals, the common reaction sequence is:

> Olivine (highest temperature)
> Orthopyroxene
> Clinopyroxene
> Amphibole
> Biotite (lowest temperature)

For plagioclase feldspars, the sequence is:

> Anorthite (highest temperature)
> Bytownite
> Labradorite
> Andesine
> Oligoclase
> Albite (lowest temperature)

It must be emphasized that these reaction series are empirical statements of common liquidus phase relationships encountered along the crystallization paths of the main magma types (basalt, andesite, rhyolite), and should not be regarded as rigid rules or applied to unusual magma compositions.

Incorporation of the equilibrated foreign matter

A small proportion of the material introduced to the magma by contamination is introduced as melt, and may be incorporated in the body of magma by diffusion and magma flowage. If mixing of a partial melt phase does occur the magma may be selectively contaminated by the constituents that have the fastest diffusion rates (Watson 1982). The bulk of the contaminating material remains in a solid condition. In the absence of any movement it remains in position at the igneous contact as altered wall rock or xenoliths, but for contamination to be effective it must be dispersed into the igneous body. This happens in two ways: by magma flowage, and by disaggregation due to grain boundary melting.

Consider an equilibrated xenolith of simple mineralogy, such as that shown in Fig. 211. The quartz–orthoclase–albite assemblage is such as might be present in a xenolith of originally arkosic or granitic composition. Its composition is amenable to partial melting if the surrounding magma is at a high enough temperature, and the melting starts not on the surface of the

Figure 211. Grain boundary melting: white = crystals, shaded area = melt. Progressive partial melting starts at the triple junctions where the solidus temperature is lowest, and proceeds along grain boundaries (see text). Mehnert *et al.* (1973) and Dodge and Calk (1978) have published photographs showing actual melting along grain boundaries in experimental and natural examples respectively.

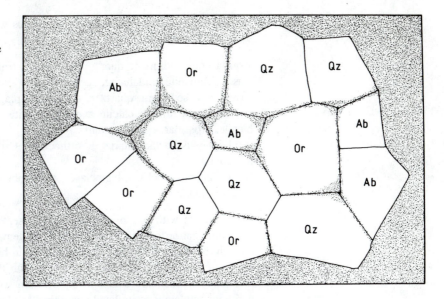

xenolith, but at whichever points within the xenolith have the lowest melting-temperature composition. Tuttle and Bowen (1958) give the lowest melting temperatures (in °C) for the following compositions at $P_{H_2O} = 1$ kilobar as follows:

Quartz	1170
Albite	905
Orthoclase	990
Quartz + albite	790
Quartz + orthoclase	760
Albite + orthoclase	870
Quartz + albite + orthoclase	720

Consequently, melting starts at the triple points where quartz, albite and orthoclase crystals are in contact, and spreads thereafter along the grain boundaries. The insides of the individual crystals remain unmelted. It is obvious that by this process a xenolith can be completely disaggregated by intergranular melting even though no more than say 20% of the bulk composition has become liquid. With the loss of mechanical coherence, the crystals can be dispersed throughout the enclosing magma and lose their identity. The result is a rock of igneous mineralogy and texture in which many of the crystals are non-magmatic.

The incorporation of melt fraction from the xenoliths into the host magma takes place alongside the incorporation of xenocrysts, and some energy may need to be used up in effecting the mixing. The heat of mixing of silicate melts may be either positive or negative, but is usually small compared with the heats of fusion of crystals (Navrotsky *et al.* 1983).

OBSERVED REACTIONS OF MAGMA AND CONTAMINANT

Under plutonic conditions, a typical contaminant has been subjected to regional as well as contact metamorphism before coming into contact with the magma. The sort of mineral assemblages which have to be equilibrated with the magma are:

Argillaceous sediment: quartz + biotite ± muscovite, orthoclase, plagioclase, sillimanite, andalusite, cordierite.
Arenaceous sediment: quartz ± orthoclase, plagioclase, muscovite, biotite.
Calcareous sediment: calcite and/or dolomite ± diopside, tremolite, hornblende, grossular, epidote, quartz.
Acid igneous rocks: quartz + orthoclase + plagioclase + biotite and/or muscovite.
Basic igneous rocks: plagioclase ± hornblende, hypersthene, diopside, epidote, quartz.

In reaction with one of these mineral assemblages, the magma can melt any combination of minerals with a lower temperature range of melting, if enough heat is available. Minerals with a higher temperature range of melting cannot be melted, but may react with the magma and be converted into phases which are stable in equilibrium with the magma. As a result, many incompletely assimilated xenoliths have been so transformed as to bear little resemblance to their original composition. Many granites contain dark, fine-grained inclusions which might be either modified hornfelses or igneous inclusions more mafic than the host magma. This identity problem has been discussed at length by Grout (1937) and Didier (1973).

(a) Granite magma reacts with pelitic material

Of the minerals in a pelitic assemblage, quartz, orthoclase, muscovite, plagioclase and biotite are all ones which crystallize from granitic magma. If these minerals are crystallizing from the magma they cannot be melted from the xenoliths at the same time. On the other hand, those that show isomorphous variation, such as plagioclase and biotite, must change their compositions to those of the same minerals crystallizing from the melt. The Fe/Mg and Na/Ca ratios of sediments are rather variable, so the biotite and plagioclase in the xenoliths may or may not be in equilibrium with the magma already; the plagioclase for instance may become either more calcic or less calcic when it reacts with the magma, depending on its initial composition. Equilibration will rarely be complete, and the effects of incomplete reaction are often seen in contaminated granites as patchy or irregular extinction in plagioclase.

Two of the minor constituents of pelitic material, andalusite and hornblende, are minerals which can crystallize from granitic magma under particular conditions, namely low P_{H_2O} for andalusite (Haslam 1971) and a bulk composition rich in CaO and low in Al_2O_3 for hornblende. If the appropriate conditions prevail, these minerals may survive the reaction with the magma; otherwise they will be converted to muscovite or biotite respectively. Other pelitic minerals, such as sillimanite, kyanite, staurolite, cordierite and almandine, all react with the magma and are converted into the actual minerals of a granitic rock; they only survive if equilibration is very incomplete.

(b) *Granite magma reacts with psammitic material*

Very little reaction can take place between a granite magma crystallizing quartz and a sandstone mainly composed of quartz, but arkoses and greywackes are more promising because they contain an appreciable proportion of low-melting constituents. The Crossdoney granite in Ireland provides an excellent example of the interaction between granite magma and greywacke country rocks, although Skiba's (1952) account of the relationship is written in terms of granitization. The greywacke was converted first by contact metamorphism into a quartz–feldspar–biotite hornfels, effectively requiring no further reaction to bring it into equilibrium with the magma. Near the contact the hornfelses contain pods, lenses and veins of tonalitic composition and igneous appearance, which may be interpreted as partial melt segregations. The hornfelses, with their igneous-looking patches, pass gradationally into xenolithic banded diorite, quartz–diorite and granodiorite. The uncontaminated (or least contaminated) magma is represented by a small area of adamellite in the centre of the intrusion.

(c) *Granite magma reacts with calcareous material*

The principal minerals of metamorphosed calcareous sediments are not stable in contact with granitic magma. Since they are all calcium-bearing they cannot be melted, because an increase in the CaO content of the magma would raise its liquidus. The most common initial reaction is for calcite and dolomite to be converted to calc-silicates, depleting the magma in silica. Further equilibration converts these to hornblende and calcic plagioclase. Although the xenoliths may be greatly altered in appearance by these changes, the magma is much less affected, since not much of the solid material enters the melt. Depletion of the melt of the contact zone in silica and other constituents of the calc-silicate minerals produces a quartz-poor reaction zone, and enough calcium enters the melt for a calcic pyroxene to crystallize instead of biotite.

Knopf (1957) described the effects of limestone contamination in the Boulder batholith of Montana. Large calcareous xenoliths have been converted to skarns (tactites) with garnet, clinopyroxene, wollastonite and vesuvianite, while the host granodiorite shows a contact zone of augite–granodiorite instead of the normal hornblende–biotite–granodiorite. Very

Table 32. Chemical analyses of quartz–monzonite and its contaminated derivatives from the Terre Neuve stock, Haiti (after Kesler 1968)

	Quartz-monzonite (least contaminated)	Granodiorite	Syenodiorite (most contaminated)
SiO_2	65.0	59.2	53.1
TiO_2	0.7	1.0	1.4
Al_2O_3	14.3	14.3	14.0
Fe as Fe_2O_3	5.6	8.4	11.3
MgO	1.7	3.4	6.5
CaO	3.9	5.5	8.5
Na_2O	3.8	3.8	3.7
K_2O	3.6	2.8	1.3

severe modification of a granitic magma was described by Kesler (1968) from a granite–limestone contact in Haiti. The least contaminated granitic rock is a quartz–monzonite, which grades into granodiorite and then into syenodiorite as the contact is approached (Table 32). At the contact there are small bodies of syenite, some of which are nepheline-bearing. In a more complex example at Crestmore, California, Burnham (1959) has described the conversion of quartz–monzonite porphyry into diorites and monzonites with increasing pyroxene and decreasing quartz. In all these examples the contamination is very localized, occurring within 100 metres or so of the igneous contacts.

(d) *Granite magma reacts with basic igneous rocks*

The principal minerals of metamorphosed basic igneous rocks are hornblende and plagioclase, and both of these can crystallize from granite magma, although some reaction is necessary to bring their individual compositions into equilibrium with regard to the Mg/Fe ratio of hornblende and Ca/Na ratio of plagioclase. It is more usual for granite magmas to be crystallizing biotite than hornblende, and there may well be conversion of some of the xenolithic hornblende to biotite, resulting in a loss of potassium and water from the magma, with raising of the liquidus and increased crystallization due to the lowering of P_{H_2O}. Other minerals of metamorphosed basic rocks, such as epidote and diopside, are not stable in contact with granitic magma and react with it to produce plagioclase and hornblende respectively.

A small-scale contact exemplifying some of these features occurs at Curran Hill in the Thorr granodiorite of Donegal (Hall 1965). A large mass of metamorphosed dolerite, consisting essentially of hornblende and plagioclase, is in contact with biotite–granodiorite. Near its contact with the granodiorite the hornblende has been replaced by biotite. In being converted to the biotite-rich rock, the metadolerite has gained potassium from the magma and lost calcium. The granodiorite close to the metadolerite has suffered the opposite changes, losing potassium and gaining calcium, thereby being converted into a trondhjemite (quartz–plagioclase rock).

A more advanced stage of contamination has been described by Wells and Wooldridge (1931) from Jersey. Gabbro xenoliths at a granite contact have undergone equilibration, resulting in the conversion of their pyroxene to hornblende and biotite, and dispersal has taken place by streaking out of the xenoliths in their granitic matrix until they merge into the surrounding granite (Fig. 212). The product is a 'hybrid' rock of quartz–dioritic composition.

Very extensive assimilation of gabbroic xenoliths by granitic magma is shown by the Bonsall tonalite in southern California (Hurlbut 1935). The tonalite is rich in dark rounded xenoliths of hornblende–biotite–plagioclase rock, which make up as much as 30% of the volume in some outcrops and are rarely completely absent. They are usually elongated and often streaked-out. The tonalite is heterogeneous, with many clusters of mafic minerals, and its average content of hornblende and biotite is about 20%. Hurlbut identified the xenoliths as having been derived from the neighbouring San

Figure 212. Sketch of gabbroic xenoliths in granite at Ronez, Jersey, from Wells and Wooldridge (1931). Xenolithic streaks in the granite clearly originated by disaggregation of the gabbro. The area shown is about 4 metres across.

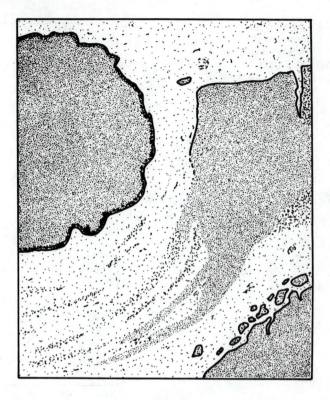

Marcos gabbro. The xenoliths have undergone equilibration with the acid magma but contain relict minerals which indicate their parentage. Traces of augite and hypersthene were found, altering to hornblende. Cores of corroded bytownite (An_{90}) are found in the centres of andesine crystals. The contamination here extends throughout an outcrop 15 kilometres across. Equilibration is very advanced, and a considerable amount of the equilibrated material has already been incorporated into the tonalite.

(e) *Basic magma reacts with pelitic material*

The lowest-melting fraction of a pelitic rock will normally have a granitic composition, and as can be seen from Fig. 213 granitic melts can exist at temperatures well below the solidus of basalt. Basaltic magmas are therefore capable of melting out the low-melting fraction of a pelitic rock, the extent of melting depending partly on the availability of water. Pelitic xenoliths included in a basaltic magma may be severely modified by removal of melt. The residue is enriched in mafic minerals, and often desilicated to the extent that corundum and spinel take the place of aluminosilicates.

The reaction of pelitic xenoliths with basic magma has a more subtle effect on the mineralogy of the contaminated rock. Olivine is converted by the additional silica into orthopyroxene, and plagioclase crystallizes in place of the diopsidic component of augite. These changes may be represented in a simplified form by the reactions:

$$Mg_2SiO_4 + SiO_2 \rightarrow 2MgSiO_3$$
$$CaMgSi_2O_6 + Al_2SiO_5 \rightarrow CaAl_2Si_2O_8 + MgSiO_3$$

Figure 213. Solidi of granite and basalt, dry and in the presence of excess water (from Yoder and Tilley 1962; Robertson and Wyllie 1971).

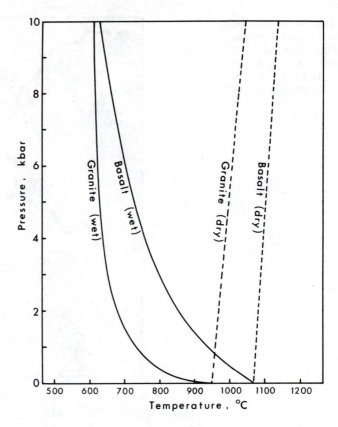

The products of both reactions are quite normal minerals in basic igneous rocks, so a large amount of contamination will simply cause an increase in the volume of crystals without giving the crystallized rock an abnormal appearance. In fact, if the second of the above reactions predominates there may not even be any xenoliths left behind to tell the tale. Potassium and associated trace elements entering the magma will merely give the contaminated rock the geochemical appearance of being rather more fractionated than it really is. Contamination must therefore be suspected in any basic rocks that are rich in orthopyroxene, i.e. norites.

The norite of Haddo House in north-east Scotland is a classic example (Fig. 214). It is very heavily contaminated by metasedimentary country rocks, predominantly schists and gneisses. These contain andalusite, cordierite, biotite, muscovite, quartz, and alkali feldspar. In the innumerable small rounded xenoliths which occur in the eastern part of the intrusion, these rocks have lost their quartz, alkali feldspar, andalusite and biotite, and have been enriched in various combinations of spinel, corundum, sillimanite, cordierite and plagioclase. In the xenolith-bearing areas of the intrusion the normal rather variable quartz–norite passes into a very heterogeneous quartz–biotite–norite. Olivine–norite occurs in two small areas of the centre of the intrusion; it is homogeneous and free of xenoliths and is presumed to be the least contaminated type of rock. The compositions of typical rock types are given in Table 33.

Figure 214. The Haddo House intrusion, Scotland (after Read 1935; Gribble 1967).

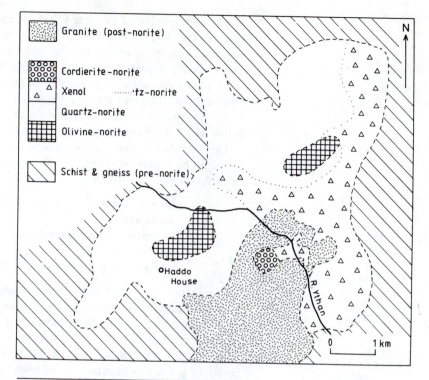

Table 33. Chemical analyses of representative rocks from the Haddo House intrusion, north-east Scotland (after Read 1935)

	Olivine–norite	Quartz–norite	Cordierite–norite
SiO_2	47.0	56.2	59.6
TiO_2	0.6	2.9	2.6
Al_2O_3	18.0	15.9	15.8
Fe_2O_3	1.1	0.6	1.4
FeO	6.9	9.0	6.9
MgO	12.6	3.5	3.0
CaO	10.2	6.2	3.6
Na_2O	1.9	2.6	2.1
K_2O	0.6	2.1	2.8
Normative composition			
Q	—	10.9	20.6
Or	3.9	11.7	17.2
Ab	16.2	22.0	17.8
An	38.4	25.8	18.1
C	—	—	2.6
Di	9.2	4.2	—
Hy	4.3	17.9	14.6
Ol	24.0	—	—
Mt	1.6	0.9	2.1
Il	1.2	5.5	5.0

In addition to xenolithic norites, this intrusion also shows marginal xenolith-rich patches in which the igneous matrix is cordierite-bearing. Rocks with mineral assemblages such as plagioclase–quartz–biotite–cordierite–hypersthene ('cordierite–norite') and plagioclase–cordierite–quartz–

garnet were interpreted by Read (1935) as extreme products of contamination of the originally gabbroic magma, but it has subsequently been suggested that the cordierite-bearing rocks and their xenoliths are the melt and residuum of country rocks fused during contact metamorphism, and are not modifications of the original gabbroic magma (Gribble and O'Hara 1967; Gribble 1968). This idea is very difficult to credit since the cordierite-bearing rocks are much more like partial melting residues of pelites than initial melts, which one would expect to be granitic.

The strontium isotopic compositions of the Haddo House rocks were measured by Pankhurst (1969), and the initial $^{87}Sr/^{86}Sr$ ratios were found to be as follows:

Olivine–norites	0.706
Quartz–norites	0.709–0.715
Cordierite-bearing igneous rocks	0.720–0.728
Xenoliths	0.724–0.731
Aureole schists and gneisses	0.718–0.719

The high ratios of the Haddo House quartz–norites are shared by other noritic intrusions of north-east Scotland, even those which are relatively free of xenoliths, and some authors have chosen to appeal to large scale Sr isotope exchange by influx of water rather than admit the contaminated origin of the norites (Busrewil *et al.* 1973). The initial $^{87}Sr/^{86}Sr$ ratios of the cordierite-bearing rocks are in accord with Gribble and O'Hara's belief that they are partially melted country rocks rather than extreme products of contamination.

In the Loch Scridain district of Mull, in the west of Scotland, there are numerous sills and sheets of tholeiite and andesitic pitchstone (Thomas 1922; Bailey and Anderson 1925). Many of the sills, especially the andesitic ones, are very rich in xenoliths which are predominantly argillaceous. Many of the xenoliths have melted to form buchite, and contain sapphire (corundum), cordierite, anorthite or mullite (named after this occurrence). Although Thomas did not believe that the magma had been modified by reaction with these xenoliths, assemblages such as anorthite–corundum–spinel in fully crystalline xenoliths are so deficient in low-melting constituents as to imply a loss of melt to the surrounding magma. Andesite is a rare composition among the predominantly basic igneous rocks of the Mull complex, and the region in which there are sills of andesitic composition approximately corresponds to the region in which the xenoliths occur, so it seems likely that the andesites of this area are products of contamination.

Pelitic xenoliths in the Cashel–Lough Wheelaun basic–ultrabasic intrusion in western Ireland have been studied chemically by Leake and Skirrow (1960) and compared with their equivalents in the surrounding aureole (Table 34). It was calculated that they had lost silica and alkalis, and this is exactly what would be expected if a partial melt fraction had been extracted.

One of the most remarkable examples of contamination of basic magma by pelitic sediments occurs on Disko Island in west Greenland. Here the contaminating sediments were carbonaceous, and their assimilation has resulted in severe reduction of the contaminated magma. The most notable

Table 34. Composition of desilicated xenoliths in the Cashel–Lough Wheelaun intrusion (after Leake and Skirrow 1960)

	Average unhornfelsed pelite	*Average pelitic xenolith*	*Material removed (calculated)*
SiO_2	46.2	15.3	63.4
TiO_2	1.6	3.8	0.5
Al_2O_3	25.0	32.6	19.8
Fe_2O_3	3.1	15.0	0.0
FeO	8.9	17.8	0.9
MgO	3.4	8.2	0.7
CaO	1.1	0.7	1.3
Na_2O	1.4	0.2	2.0
K_2O	4.8	0.5	7.4
H_2O	4.2	5.0	3.7

result is the presence of native iron and graphite, but the contaminated lavas show many other abnormal mineralogical features (Ulff-Møller 1990). Native iron has also been found in basalts contaminated by coal (Medenbach and El Goresy 1982).

(f) *Basic magma reacts with psammitic material*

Reaction of basic magma with psammitic rocks is only likely to take place if the latter are sufficiently impure or arkosic to contain some fusible constituents. As with pelitic rocks, the addition of silica to the melt will cause it to crystallize orthopyroxene in place of olivine, but the xenoliths will not be so desilicated as to contain spinel or corundum.

There are many examples of vitrification of arkose by basalt magma on a small scale (Butler 1961), but the acid melt has usually been too viscous to mix with the basic magma on an appreciable scale. Ackermann and Walker (1960) described an example from South Africa of a 15-metre thick sill of dolerite which fused the overlying arkose to a dark-coloured buchite containing up to 70% glass, and in which hybridization of the basic and acid melts took place in a zone a few centimetres wide.

(g) *Basic magma reacts with calcareous material*

Limestones do not contain a significant proportion of fusible material, and their influence on basic magmas is mainly confined to removal of silica from the magma at the contact, associated with the conversion of carbonates to calcium silicates. The removal of some silica has the tendency to make the magma undersaturated or more undersaturated with silica, but as mentioned already there is a thermal barrier between tholeiitic and alkali basalt liquids which prevents large-scale desilication of tholeiitic magmas.

Contaminated gabbros may contain melilite, wollastonite, nepheline and alkali pyroxenes very locally near limestone contacts, a good example being that of Camas Mor in the Island of Muck (Tilley 1947, 1952). The occurrence of nepheline and melilite in such gabbros was once thought to be significant in relation to the origin of alkaline igneous rocks as a whole, but both nepheline and melilite are now regarded as normal minerals of uncontaminated basic rocks.

(h) *Basic magma reacts with acid igneous rocks*

In principle, basalt magma can extensively melt acid igneous rocks, as will be obvious from Fig. 213. Under volcanic conditions the process is ineffective, because of the high viscosity of the acid melts. Even rhyolite magma which is already molten has difficulty in mixing with basaltic magma, as can be seen from their coexistence in some composite lava flows. Plutonic conditions may be more favourable for mixing of melted acid material with basic magma because their viscosities are lowered by the combination of water and pressure. The temperature interval of melting of acid igneous rocks is lower than that of any sedimentary material, and there is a good chance of granitic inclusions in basic magma being melted completely, so that they do not leave behind partially reacted xenoliths to record what has happened.

Philpotts and Asher (1993) described the melting of gneissic wall rocks at the margins of a basic dyke in Connecticut. Melting started at grain boundaries in the gneiss, and some of the melts that were produced are preserved as felsic wisps in the borders of the dyke and as granophyric veins. Chemical profiles across the dyke suggest that a significant amount of acid melt has been completely incorporated into the basalt, changing its composition from olivine-normative to quartz-normative.

PETROGRAPHIC AND GEOCHEMICAL CHARACTERISTICS OF CONTAMINATED ROCKS

A body of contaminated igneous rock contains the following components:

1. High-melting crystals which crystallized from the original magma during its reaction with the foreign material. These may be represented by cumulates, perhaps in greater abundance than if the magma had not been contaminated.
2. The low-melting fraction of the original magma and additional assimilated low-melting material.
3. Equilibrated xenocrysts dispersed in the igneous body without having melted.
4. Undigested xenoliths and wall rocks.

The first three of these components do not necessarily differ in appearance from the normal crystallization products of uncontaminated magma. The xenoliths are the most conspicuous indication of contamination, but their abundance is not a reliable guide to the amount of contamination that has taken place. Furthermore, not all xenoliths are relics of partially assimilated country rocks. Some may be cognate, i.e. inclusions of igneous rocks related in some way to their host, or have crystallized from a different but coexisting magma. Some xenoliths may even be melting residues which have come all the way from the magmatic source region.

The most difficult task petrographically is to recognize xenocrysts after they have been dispersed in the magma by intergranular melting. One class of undoubted xenocryst which can be positively identified is 'inherited zircon', i.e. crystals of zircon with Pb isotopic ages greater than the age of crystallization of the enclosing rock. Xenocrysts are also relatively conspicuous if they are of minerals which do not normally crystallize from the magma, such as cordierite or garnet, but there is still a possibility that such crystals could be residual crystals carried from the magmatic source region. Calcic plagioclase and quartz may have had either of these origins, in addition to crystallizing from the magma itself.

The chemical evidence for assimilation is equally ambiguous. The chemical changes involved are not simply additive. The composition of a contaminated igneous rock is made up of:

the original magma
+ completely assimilated xenoliths
+ the melt fraction of partially melted xenoliths
+ constituents gained from xenoliths during equilibration
  other than by melting
− constituents lost to the wall rocks or xenoliths during equilibration

The most obvious examples of contamination are those in which the assimilation is incomplete and many xenoliths are modified but undigested, and in these examples the end-product is most unlikely to have the composition of a simple mixture. Tables 34 and 35 give an indication of how residual xenoliths may differ from their original compositions.

Table 35. Average compositions of various rock-types before and after reaction with granitic magma (from Hall 1966a)

	Argillaceous rocks		Limestones		Basic igneous rocks	
	A	B	A	B	A	B
SiO_2	62.6	56.9	8.9	50.8	48.4	50.5
TiO_2	0.8	1.1	0.1	1.2	1.8	1.5
Al_2O_3	17.2	20.3	1.4	13.9	15.5	15.0
Fe_2O_3	2.6	1.1	0.6	2.2	2.8	2.6
FeO	4.0	6.3	0.3	5.4	8.1	8.6
MgO	2.6	2.9	13.5	5.2	8.6	7.1
CaO	1.8	3.0	73.0	14.4	10.7	8.0
Na_2O	1.9	2.6	0.1	2.7	2.3	2.3
K_2O	3.5	3.7	0.6	1.7	0.7	2.0
P_2O_5	—	0.4	0.1	0.3	0.3	0.2
H_2O+	3.5	1.3	1.0	1.0	0.7	1.2

A – unmodified rock types
B – xenoliths in granite
The compositions have been recalculated free of CO_2.

The contaminants offering most scope for modification of the magma are those which are rich in low-melting constituents, such as argillaceous rocks, arkoses and greywackes, but since their low-melting constituents give rise to a melt of granitic composition the contaminated rock may be very similar to

Table 36. The average composition of the continental crust compared with the compositions of basalt and granite

	Basalt	Continental crust	Granite
SiO_2	50.1	61.5	72.0
TiO_2	1.9	0.7	0.3
Al_2O_3	16.0	15.1	14.4
Fe as Fe_2O_3	12.2	6.3	3.1
MgO	7.0	3.7	0.7
CaO	9.7	5.5	1.8
Na_2O	3.0	3.2	3.7
K_2O	1.1	2.4	4.1

The data are from Le Maitre (1976) and Wedepohl (1994).

igneous rocks which have not been contaminated. The average composition of continental crust lies between the composition of granite and basalt (Table 36), so even large-scale contamination by a mixture of crustal rocks might still lead to a contaminated rock of perfectly normal igneous composition.

In some cases isotopic studies may be helpful. The contamination of basaltic magma by crustal materials should be indicated by initial $^{87}Sr/^{86}Sr$ ratios above the normal values for basalts, and in fact many high $^{87}Sr/^{86}Sr$ ratios in basic rocks have been ascribed to contamination. However, the degree of contamination needed to account for basalts with high $^{87}Sr/^{86}Sr$ ratios is not always compatible with their major and trace element geochemistry. Some continental flood basalts have $^{87}Sr/^{86}Sr$ in excess of 0.710 without any other sign of having been contaminated, leading to the suspicion that they may be derived directly from a mantle source with a high $^{87}Sr/^{86}Sr$ ratio.

An alternative interpretation of high $^{87}Sr/^{86}Sr$ in otherwise normal basalts is that they have been selectively contaminated by crustal Sr without the incorporation in bulk of other crustal constituents. One way in which this might happen is by disequilibrium melting of xenoliths in which a small melt fraction is highly enriched in ^{87}Sr. This has actually been observed by Pushkar and Stoeser (1975), who found that the glass in a partially fused quartz–monzonite xenolith in basalt had a $^{87}Sr/^{86}Sr$ ratio of 0.723, compared with 0.706 for the unmelted portion.

A clear example of contamination has been described by Carden and Laughlin (1974) from a large basalt flow in New Mexico. The flow varies in $^{87}Sr/^{86}Sr$ from 0.7040 to 0.7084 along its 50 kilometre length, and shows corresponding chemical variations correlated with the presence of partly digested sandstone xenoliths.

LARGE-SCALE CONTAMINATION

In the examples quoted so far, the reactions that are observed are between magma and country rocks at the level of emplacement, where the magma as a whole became crystalline and where assimilation was arrested before completion. They give the impression that contamination is a localized phenomenon at igneous contacts. To be of major significance in magmatic evolution, contamination has to have occurred on a large scale before reaching the level of emplacement. Under these circumstances, a high proportion of the assimilated material will have lost its identity by dispersion and homogenization in the magma, and its recognition is difficult.

Contaminated granites

The most convincing examples of large-scale contamination are of granitic rocks in which some xenolithic material remains undigested. A common feature of such intrusions is that in addition to discrete, sharply bounded xenoliths, there are dark patches, streaks or wisps which are in more advanced stages of digestion. This is exemplified by the *Thorr* granodiorite (Fig. 50), in which inclusions of recognizable country rocks constitute up to 20% of the outcrop. Where they are abundant the igneous rock between them is a quartz–diorite, but when followed to the north-west the rock grades imperceptibly into an alkali-granite as the proportion of xenoliths becomes less. Quantitive modal analysis (Fig. 215) indicates the degree to which the rock has been modified by assimilation over a large area, assuming the most leucocratic variety of granite to be nearest in composition to the original magma.

Figure 215. The Thorr granodiorite pluton, Donegal, Ireland. Left: the distribution of xenoliths (after Pitcher and Berger 1972; see also Fig. 50 for a more detailed map of the southern part of the intrusion). Right: the variation in colour index of the granodiorites, i.e. the volume percentage of dark minerals (after Whitten 1962).

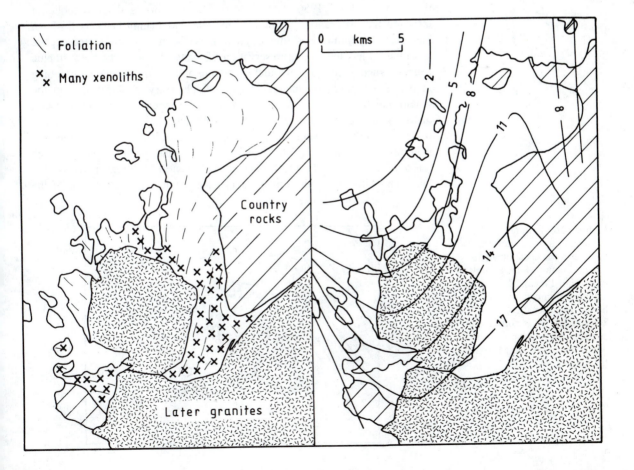

Another large intrusion in which granite is believed to have been severely modified by contamination is the *Dartmoor* granite in south-west England. In this intrusion the proportion of remaining xenoliths is small, because of a greater degree of dispersal into the magma. In the words of Brammall and Harwood (1932): 'Small relics of the inclusions fade almost imperceptibly into the granite'. About xenoliths they wrote: 'The hybrid marginal zone is liable to tail away and be dispersed by magma flowage. Where preserved it is often discernible only on polished blocks. The fact that a nodular inclusion is only the residuum of a larger mass must be taken into account ...'. Small amounts of accessory cordierite and garnet are further indications of the contaminated nature of this granite. The relatively homogeneous nature of the granite as it is now exposed implies that much of the contamination took place below the level of emplacement, enabling the foreign material to be substantially assimilated before the intrusion finally consolidated. The over-all chemical effect of assimilation has been a slight basifaction of the granite, even though the predominant country rock is shale or slate. The potassium content is also high, so much so that the composition of the granite is appreciably richer in normative orthoclase than the minimum of the system Q–Or–Ab–H$_2$O, and as in many other contaminated granites porphyritic K-feldspars are particularly large and abundant.

The *Ardara* pluton in Donegal (Figs 63 and 64) is contaminated throughout (Akaad 1956; Hall 1966a). Because of its diapiric mode of intrusion the dispersal of assimilated material has mainly been by flowage, but many xenoliths and mafic clots have been preserved. Incomplete assimilation has led to the preservation of many characteristic petrographic features, such as irregular undulatory zoning and poorly developed twinning in partially equilibrated plagioclase, small phenocrysts of feldspar in the granite identical to those in some xenoliths, heterogeneous amphiboles with many inclusions, and the preservation of epidote derived from the country rocks showing corrosion by the magma (Fig. 216). The contaminating material was a mixture of pelitic sediments and metamorphosed basic igneous rocks, mixed in different proportions in different parts of the intrusion. Although both contaminants have resulted in basification of the

Figure 216. Textural relationships of xenocrystic epidote in the contaminated Ardara granite, Donegal, Ireland (after Hall 1966a). The epidote is euhedral in contact with biotite which is also xenocrystic, but is corroded in contact with feldspar or quartz, which have crystallized from the magma.

Table 37. Compositions of contaminated granites from the Ardara Pluton, Ireland (after Akaad 1956 and Hall 1966a)

	Least contaminated granite	Contaminated outer granite	Contaminated inner granite
Modal compositions			
Quartz	24.4	14.0	14.5
Potassium feldspar	16.8	21.5	14.9
Plagioclase	49.8	46.1	47.8
Biotite	7.7	13.6	12.9
Hornblende	—	2.6	6.0
Sphene	0.1	0.7	0.5
Epidote	1.1	1.0	2.8
Chemical compositions			
SiO_2	71.7	62.4	62.6
MgO	1.0	2.0	2.4
K_2O	3.2	4.2	3.5

original granite magma, there are trace element differences (for example Rb is high in pelites, but low in basic igneous rocks) which enable the nature of the contaminant to be recognized even though much of the assimilated material has been completely incorporated into the granite. It has been estimated that the magma has assimilated at least its own weight of xenolithic material and perhaps several times as much. The magma must have been nearly a solid mass of crystals by the time it came to a halt. Table 37 shows the composition of the more and less contaminated parts of the intrusion. In the outer unit of the intrusion, a high proportion of pelitic contaminant is revealed by high K-feldspar and biotite, and by high K_2O. In the inner unit, a greater proportion of basic contaminant is shown by high hornblende and epidote, and by high MgO.

Contaminated gabbros

The role of contamination in basic magmas is more debatable. Examples such as the pitchstone sills of Mull suggest that contamination of basaltic magma by pelitic sediments would lead to rocks of andesitic composition, but the low $^{87}Sr/^{86}Sr$ initial ratios of most andesites and the scarcity of dioritic facies in large intrusions suggest that this type of contamination does not occur on a significant scale. The modification of gabbro to produce norite is a different matter. Almost all the large basic intrusions in which hypersthene–gabbro or norite is present show signs of contamination, and as Bowen (1928) wrote: 'It may very well be that if there were no argillaceous sediments, norites would be of much rarer occurrence than they are.'

Extensive evidence of contamination is found in the *Bushveld* complex of South Africa (Hall 1932; Willemse and Viljoen 1970). This is the largest basic intrusion in the world; norite is the predominant basic rock type, and in the layered succession there are layers of orthopyroxenite as well as anorthosite. Xenoliths of sedimentary rock are found in various stages of modification; one spectacular inclusion is a 25-centimetre wide layer of ruby-coloured corundum, spinel and sillimanite, surrounded by anorthosite in an alternating succession of norite, chromitite and anorthosite. Poorly defined patches of plagioclase–quartz–orthopyroxene–cordierite–biotite rock in the basal part of the norite represent xenoliths in an advanced stage of assimilation, and biotite-bearing norite is widespread near the base of the intrusion. In general, the norite does not have the high K_2O content that would be expected from a magma contaminated by pelitic sediments, but of course many of the rocks of this layered basic intrusion are accumulative, so the rock compositions do not accurately indicate the composition of the magmas from which they were deposited. Incidentally, the presence of cumulus textures does not mean that the norites are uncontaminated. On the contrary, contamination may well have controlled the nature of the cumulus assemblages.

Irvine (1975) postulated contamination in the roof zone of the *Muskox* layered basic intrusion (Figs 89 and 90). The granophyre near the top of this intrusion contains many inclusions of quartzite which have been rounded and corroded by the magma. In the north-eastern corner of the outcrop the granophyre is transitional into an extensive contact breccia consisting of closely packed blocks of quartzite in a 10–20% matrix of granophyre. The schists of the roof are locally permeated by small granitic veins. The granophyre is interpreted as formed by the melting of the roof rocks (schist, quartzite, granite), of which the relatively refractory quartzite forms the residual xenoliths. High initial $^{87}Sr/^{86}Sr$ ratios for the ultramafic and gabbroic rocks of the intrusion suggest that some of the granophyric melt has become mixed with the basic magma.

COMBINED ASSIMILATION AND FRACTIONATION

De Paolo (1981b) pointed out that assimilation and fractional crystallization are often likely to occur together, the heat required for assimilation being partly provided by the latent heat of crystallization of the magma. He presented equations for modelling their combined effects on the trace element and isotopic compositions of a magma. This approach has been widely copied, and AFC (assimilation–fractional crystallization) modelling has been applied to many igneous complexes. In order to evaluate the combined process quantitatively it is necessary to take into account not only its chemical budget but also the heat budget (i.e. the heats of crystallization and fusion of the various participating liquids and solids), as was done by Ghiorso and Carmichael (1985).

GENERAL SIGNIFICANCE OF CONTAMINATION

Consideration of the way in which contamination takes place leads to two general conclusions: (1) plutonic rocks are more likely to have been affected by contamination than volcanic rocks; and (2) acid magmas are more likely to be contaminated than basic magmas.

The high proportion of crystals present in a magma undergoing contamination means that crystallization is likely to be completed at or not far above the place of maximum assimilation. Magmas extruded in the volcanic environment in a mainly liquid condition are unlikely to have been appreciably contaminated. Furthermore, magmas emplaced under plutonic conditions take longer to consolidate, which is conducive to reaction and mixing, and they crystallize under higher P_{H_2O} which facilitates reaction and lowers the minimum melting temperatures of possible contaminants.

The larger volumes of acid intrusions compared to basic intrusions give a greater opportunity for heating large amounts of country rock to near their melting temperatures, and the loss of heat by the magma can be partially made good by convection. Gabbroic intrusions usually have a chilled margin but granitic intrusions do not; a chilled margin would tend to isolate the remaining magma from the country rocks. The higher water contents of acid magmas are conducive to reaction with xenolithic material. The greater viscosity of acid magmas results in more effective mechanical dispersal of xenocrysts into the igneous body. The prevalence of stoping and other subsidence mechanisms in emplacing granite intrusions facilitates incorporation of xenolithic material into the igneous mass. Indeed, we rarely see accumulations of stoped blocks in the deepest exposed levels of igneous complexes, which means that they must be at least partly assimilated into the magma in which they sink. Magmas emplaced by stoping must inevitably be heavily contaminated by the rocks through which they have been intruded. We may conclude that assimilation is a major factor in the evolution of granitic rocks, but that it plays a relatively minor role in the evolution of basalts.

ZONE MELTING

Zone melting (or zone refining) is the name of an industrial process in which an impure metal is refined by allowing a zone of melting and recrystallization to pass through it, so that the impurities are concentrated into the melt. A similar process may occur naturally in the Earth. Imagine a body of magma in the mantle, crystallizing from the bottom upwards, while at the same time melting the overlying solid rocks. During the course of time the body of melt could progress slowly upwards through the mantle, changing gradually in composition as it went.

In a large body of magma the pressure at the top would be less than the pressure at the bottom, and since the melting temperatures of silicates increase with pressure (in the absence of volatiles) there could be simultaneous melting at the top and freezing at the bottom of the magma body. Convection would be an efficient method of transferring heat upwards to the zone of melting.

The operation of zone melting need not be restricted to a large coherent body of magma. An equally plausible situation would be of a small fraction of melt travelling through the interstices of a partially melted source region, interacting continuously with its crystalline matrix (McKenzie 1984).

The extent of zone melting in the mantle is as yet unknown. It has mainly been invoked as a way of explaining the minor element abundances in basic magmas (Harris 1957; Kay 1979). The elements which would become concentrated in the melt by zone melting are those which are not essential constituents or isomorphous substituents in the minerals of mantle rocks, i.e. olivine, pyroxenes, plagioclase, spinel and garnet. Elements which do not normally enter the crystal structures of these minerals include K, Ti, P, Ba, Rb, Sr, U and Th, and they are referred to as the 'incompatible' elements. One might envisage a body of magma 1 kilometre in vertical extent travelling upwards for 100 kilometres, collecting all the incompatible elements from 100 times its own volume of mantle. Zone melting was originally proposed as a mechanism by which leucite–basalts and similar rocks become so rich in potassium, but it has also been applied to alkali basalts in general (Alibert *et al*. 1983).

Although it is an attractive idea, zone melting is not a satisfactory explanation for the formation of K-rich basic magmas. It can explain why these magmas are rich in incompatible elements but it does not explain why they have different isotopic compositions from normal basalts. The origin of the K-rich magmas is discussed more fully in Chapter 12.

Zone melting in the crust would have a rather different character from that in the mantle for two reasons. One is that the rocks most likely to form by crustal melting, i.e. granites, do not have any incompatible elements. All the major elements and trace elements occurring at the level of 10 ppm or more can be accommodated by the rock-forming or accessory minerals of granitic rocks. The other difference is that the vertical extent of the crust is small. The average crustal thickness of about 30 kilometres is only two or three times the lateral extent of most plutonic intrusions, so there is no scope for the minor constituents of a large volume of rock to be concentrated into a small volume of magma.

On the other hand, the upward movement of a body of granite magma through the crust may still involve a combination of melting at the roof and crystallization at the base. The partial melting of stoped xenoliths near the top and the accumulation of refractory residues near the bottom has the effect of changing the composition of the magma, enriching it in the low-melting constituents of its roof rocks as it moves upwards. This process could be equally well considered as large-scale contamination.

Crustal zone melting will be especially facilitated if dissolved water is able to migrate upwards within the magma, depressing the melting temperature of the rocks at the top of the magma body, and raising the crystallization temperature of the liquid at the bottom. Hamilton and Myers (1967) described the process in these words: 'Zone melting – whereby volatile components rising in response to pressure gradients within magmas lower the melting temperature of the roof while forcing crystallization low in the chamber – can cause great assimilation; indeed very little of the final high-level magma need represent material present in the initial melt.'

Barker *et al*. (1975) used the term 'reaction-melting' to describe a similar process, in which basaltic magma arriving at the base of the crust became modified by the addition of partial melt from overlying crustal rocks and subtraction of cumulates and melting residues. They envisaged upwards progress through the crust of a magma body continuously changing from basaltic through syenitic to granitic composition as zone melting proceeded and different roof rocks were encountered.

MIXING OF MAGMAS

Magmas of more than one composition occur together so often that there must be a likelihood of occasional mixing. On the other hand, differences in density and viscosity are likely to inhibit mixing when magmas of different composition are brought together, and proof of extensive mixing is difficult to find. The geochemical evidence is usually ambiguous. For example, isotopic studies have suggested that the Pb, Sr and O in many magmas is derived from more than one source, but it is often difficult to decide whether this represents the mixing of magmas or the melting of a mixed source.

Discussion of magma mixing has in the past mainly focused on the mixing of dissimilar magmas such as granite and basalt. In recent years there has been more emphasis on the mixing of similar magmas, for example, one batch of basalt magma with another, or the mixing of two granite magmas of different origins.

MIXING OF SIMILAR MAGMAS

Mixing is most likely to involve basaltic magmas, which are generated in large quantities and have low viscosities. A possible site of mixing would be a high-level magma chamber which was undergoing crystallization while at the same time being continuously fed with fresh magma from below. Even if the magma supply were of constant composition, the liquid in the magma chamber might be partly fractionated before being mixed with new magma, giving a mix-product different from both the original magma and from any normal product of closed-system fractionation. This may have happened in those large differentiated intrusions where there is evidence of magma replenishment (e.g. Bushveld, Muskox), and is especially likely to have happened in the magma chambers underlying sea-floor spreading axes. This may account for variations in the composition of ocean floor basalts (Dungan and Rhodes 1978; Rhodes *et al.* 1979; Johnson 1979; Walker *et al.* 1979; Sparks *et al.* 1980), and is also relevant to ocean-floor gabbros as represented in ophiolite complexes (Smewing 1981).

Mixing is rather less likely between acid magmas, because of their higher viscosity, but several possible examples have been suggested. Bowman *et al.* (1973) described a composite flow of rhyolitic obsidian and dacite at Borax Lake, California, in which there was such a linear variation in composition between two end-members as to suggest that mixing and not fractional crystallization was responsible.

MIXING OF DISSIMILAR MAGMAS

Simultaneous eruption of magmas of different composition has been recorded from volcanoes in many parts of the world. Good examples are the

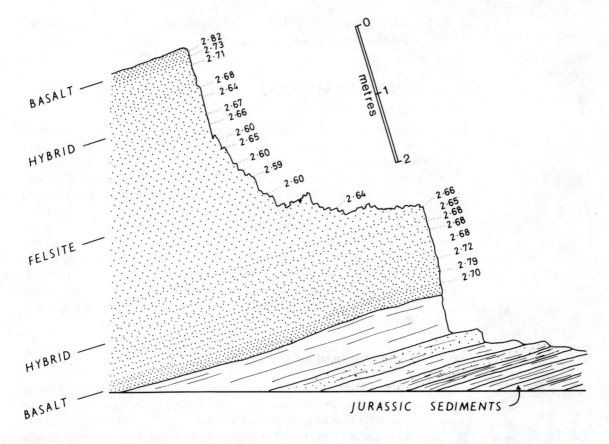

Figure 217. The composite sill of Rudh'an Eireannaich, Skye, which shows continuous gradation from basic at the margins to acid in the centre (after Harker 1904). The figures are specific gravities of specimens from different levels in the sill.

mixed lavas of the 1875 eruption of Askja, Iceland (Sigurdsson and Sparks 1981), or the mixed rhyolite–trachyte–basalt ignimbrite flow described by Freundt and Schmincke (1992) from Gran Canaria. In Iceland, Gibson and Walker (1963) described a particularly interesting example of a composite basalt–rhyolite lava fed by a composite dyke.

Many further examples from plutonic environments show that basic and acid magma have coexisted without effective mixing. Figure 217 shows one of the numerous composite sills of the British Tertiary volcanic province. This sill is 4 metres thick and consists of basalt adjacent to the top and bottom contacts and rhyolite in the centre, with gradational relationships between the two rock types. Other plutonic occurrences of coexisting magmas are in net-vein complexes (Fig. 218), where molten basalt has chilled against neighbouring acid magma (Blake 1966; Walker and Skelhorn 1966; Wiebe 1988).

All these examples show that basic and acid magmas can exist together in contact without becoming appreciably mixed. Nevertheless, Yoder (1973) confirmed experimentally that basaltic and granitic melts definitely are miscible, and can be fused together to produce a uniform glass of intermediate composition. He also found that the rate of inter-diffusion was extremely slow, and a capsule containing powdered granite at one end in contact with powdered basalt at the other end could be heated for 1 hour at 1200 °C in the presence of excess water at a pressure of 1 kilobar without

Figure 218. The relationships between acid and basic rocks characteristic of a net-vein complex. The basic rock is veined by or enclosed by acid rock, but is chilled against the acid rock.

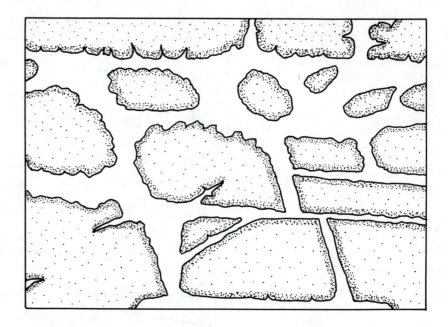

any detectable mixing of the resulting melts. In contrast, Kouchi and Sunagawa (1985) were easily able to produce homogeneous andesite by experimentally mixing basaltic and dacitic liquids.

There are other physical reasons why acid and basic magmas often do not mix. It is normal for acid magma to overlie basic magma, as when basic magma intruded into the crust partially melts the overlying crustal materials. In such a situation the superposition of light acid magma above heavier basic magma is gravitationally more stable than a homogeneous mixture of the two would be. In fact if acid and basic magmas were mixed without being immediately homogenized, gravitation would tend to separate them again.

A classic example of possible magma mixing is the Gardiner River rhyolite–basalt complex in Yellowstone Park, Wyoming. The rocks are very heterogeneous, and a complete range of compositions from basalt (50% SiO_2) to rhyolite (75% SiO_2) is present (Fig. 219). The chemical analyses show an almost linear variation between the two end-members, which Wilcox (1944, 1979) interpreted as the result of simple mixing of the basaltic and rhyolitic liquids. The dolerites and granophyres of the Slieve Gullion complex in Northern Ireland have likewise given rise to a complete range of intermediate hybrids despite for the most part having remained unmixed (Gamble 1979). The lavas of Medicine Lake volcano in California have also been identified as mixing products (Gerlach and Grove 1982) – phenocryst compositions and glass inclusions in the phenocrysts show the end-members to have been rhyolite and basalt magmas, but the mixed rocks cover the whole range from basalt through andesite and dacite to rhyolite.

In the examples mentioned above, the mixing is a relatively localized phenomenon. If basaltic and rhyolitic magmas undergo mixing on a large scale, their mixing products are presumably to be found among igneous rocks of intermediate composition. Honjo and Leeman (1987) described

Figure 219. Variation diagrams for rocks of the rhyolite–basalt complex of Gardiner River, Yellowstone Park (after Fenner 1938).

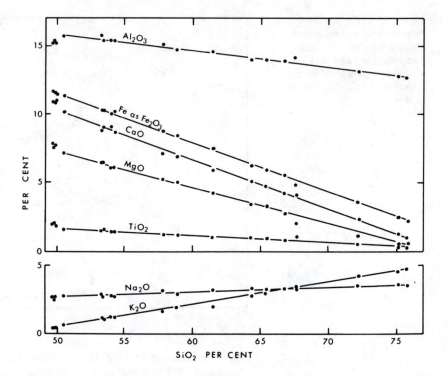

intermediate lavas (ferrolatites) from Idaho which show many features indicating that they formed by mixing of the rhyolitic and basaltic magmas that were erupted in the same area. Petrographic features indicative of mixing include the presence in the same lava of plagioclase phenocrysts of two distinct compositional ranges (An_{27-32} and An_{49-60}), partially resorbed quartz crystals, and two generations of augite and orthopyroxene. In recent years, the mixing of acid and basic magmas has been generally thought to play only a minor role in the development of intermediate rocks, but its importance has been asserted by several authors (Eichelberger 1975; Anderson 1976; Bloomfield and Arculus 1989; Oldenburg *et al.* 1989).

CHAPTER 8

Basalts

WHERE BASALTS OCCUR

Basalts occur in every tectonic environment. Figure 220 shows the main situations that can be distinguished: (1) the basalts formed at the mid-ocean ridges and making up most of the oceanic crust; (2) the basalts of oceanic islands; (3) the basalts of orogenic continental margins and island arcs; and (4) the basalts of continental volcanoes not situated above subduction zones.

Differences in deep structure and thermal regime require us to consider the possibility that the basalts of these different environments may not all have the same origin. The ocean floor and oceanic island basalts can only originate in the mantle. The continental within-plate basalts have a similar compositional range to the oceanic island basalts, and are presumably also from the mantle. For the basalts of island arcs and continental margins there is an alternative possible source, namely subducted oceanic crust. A strong reason for suspecting a different source from the other environments is that the volcanic arc basalts are intimately associated with andesites, which do not occur in the other environments.

Figure 220. The main occurrences of basalt in relation to tectonic environment. Variations in the thickness of the lithosphere are indicated diagramatically (crust – oblique shading; mantle – dotted).

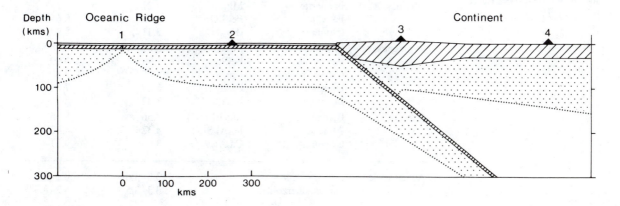

The compositional ranges of the main categories of basalt are as follows:

1. Ocean floor basalts – formed at divergent plate margins. The term 'MORB', i.e. mid-ocean ridge basalt, is used to describe these. They are uniformly tholeiitic basalts with very low contents of the 'mantle-incompatible' elements, i.e. K, Ba, P, Sr, U, Th and Zr.
2. Basalts of oceanic islands and stable continental areas – formed within plates. Those of oceanic islands vary very widely, from tholeiitic to strongly alkalic, grading into nephelinite. A similar range occurs in continental areas, including extensive tholeiitic flood basalts as well as more frequently occurring but less voluminous alkali basalts. Continental tholeiites and those of the volcanic islands are not as poor in potassium and other incompatible elements as the ocean floor tholeiites.
3. Volcanic arc basalts – formed at convergent plate margins, either island arcs or orogenic continental margins. These range from tholeiitic to alkalic. Some authors distinguish between a 'low-K tholeiite' type and a 'calc-alkali basalt' type. Strongly alkaline basalts also occur but only in relatively small quantity. All the volcanic arc basalts are high in K_2O and low in MgO and CaO compared with ocean floor basalts (Pearce 1976).

PARENT MATERIALS OF BASALT

The source rock which is melted to give basalt magma must either have a basaltic composition, or contain the constituents of basalt plus additional refractory constituents which remain as a residue when basaltic melt is extracted. A number of basic and ultrabasic rock types comply with this requirement, including peridotite, pyroxenite and hornblendite, as well as basalt itself and its metamorphic equivalents (amphibolite, eclogite).

Mantle source rock must have a seismic P-wave velocity of 8.0–8.4 km/s, which is the observed range in the upper mantle underneath stable

Table 38. Densities and seismic velocities of minerals and mantle rock types (after Anderson 1989)

			Density (g/cm^3)	Vp (km/s)
Olivines:	Forsterite	Mg_2SiO_4	3.21	8.57
	Fayalite	Fe_2SiO_4	4.39	6.64
	Olivine	(Fo_{95})	3.31	8.42
Pyroxenes:	Enstatite	$Mg_2Si_2O_6$	3.21	8.08
	Ferrosilite	$Fe_2Si_2O_6$	3.99	6.90
	Orthopyroxene	(En_{80})	3.35	7.80
	Diopside	$CaMgSi_2O_6$	3.29	7.84
	Jadeite	$NaAlSi_2O_6$	3.32	8.76
Garnets:	Pyrope	$Mg_3Al_2Si_3O_{12}$	3.56	8.96
	Almandine	$Fe_3Al_2Si_3O_{12}$	4.32	8.42
	Grossular	$Ca_3Al_2Si_3O_{12}$	3.60	9.31
Actual rocks:	Eclogite		3.5–3.6	8.2–8.6
	Garnet–lherzolite		3.3–3.5	8.2–8.3

continental areas and ocean basins. This indicates a combination of olivine, pyroxene and perhaps garnet as the most likely mineralogical constituents (see Table 38). Basalt and gabbro are disqualified because their seismic velocities are too low (6.4–7.3 km/s). Amphibolite and hornblendite are also improbable candidates because they are hydrated, and would give rise to a magma with a high water content. The rarity of residual or phenocrystic hornblende in basalts generally, and the absence of other hydrous minerals, aqueous fluid inclusions, or the explosive emplacement expected of volatile-rich magmas, indicate the unlikelihood of a hydrated source material except in orogenic environments. The remaining candidates are peridotite, pyroxenite and eclogite. These rock types have appropriate seismic velocities, can be partially melted to give basaltic magmas, and have been found as inclusions in basaltic lavas and in kimberlite pipes.

PERIDOTITE AS A SOURCE MATERIAL

The main evidence that the mantle has a peridotitic composition comes from the following sources: (1) the nature of the ultramafic xenoliths in basalts and kimberlites; (2) the mineralogy of some Alpine-type peridotites, which indicates equilibration under pressures corresponding to those prevailing in the upper mantle; and (3) melting experiments on ultramafic rocks.

Ultramafic xenoliths in basalt are most often found in alkali basalts, rarely in olivine tholeiites, and never in oversaturated tholeiites. In order of decreasing abundance, their compositions are: spinel–peridotite, garnet–peridotite, and garnet–pyroxenite (including eclogite). They have several possible origins. Some are crystal accumulates from the enclosing magma. Some have been plucked from the walls of the magma channel between the source region and the Earth's surface ('accidental xenoliths'). Others may be melting residues carried up along with their complementary melts. None of them can represent the unmodified material of the basaltic source region, but they do undoubtedly include little-modified samples of the mantle overlying the magmatic source region.

Much better evidence comes from the xenoliths in kimberlite pipes. The mantle origin of these xenoliths is very firmly established from their high-pressure mineralogy. Kimberlite itself is not likely to be a major constituent of the mantle, because its composition is much more potassic than that of most basalts, but the xenoliths in kimberlite provide a very good sample of the mantle traversed by the kimberlite on its ascent to the surface. They are a more reliable guide to mantle composition than the xenoliths in basalt because of the deeper origin of kimberlite, its more rapid ascent, and the greater erosive effect of gas-rich kimberlite magma on its walls during ascent. The commonest types of xenolith in kimberlite are lherzolites and harzburgites; eclogites are abundant at a few localities. The compositions of these xenoliths are described more fully in Chapter 12.

The mineralogy of Alpine-type peridotites provides an approximate indication of their depth of origin. Experimental studies show that in rocks of the same composition, plagioclase–peridotite, spinel–peridotite and garnet–peridotite represent stable mineralogical assemblages at successively higher

Figure 221. The stability fields of plagioclase–, spinel– and garnet–lherzolite with respect to temperature and pressure. The anhydrous solidus is after Takahashi and Kushiro (1983). The spinel–garnet facies boundary is after O'Hara *et al.* (1971), and the plagioclase–spinel boundary is after Saxena and Eriksson (1983). In the field labelled 'spinel–lherzolite' there may not be any actual spinel present if the Al_2O_3 content of the rock is small and the temperature is close to the solidus; this is because the pyroxenes are able to accommodate all the Al themselves and a separate Al-bearing phase does not then develop.

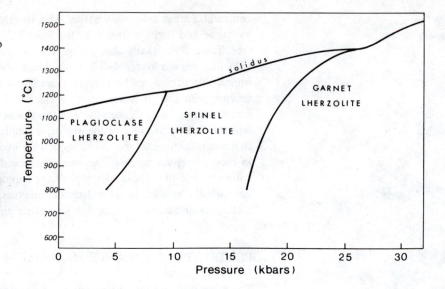

pressures (Fig. 221). Garnet–peridotites must have equilibrated at depths greater than 50–70 kilometres (depending on the geothermal gradient), and although these rocks are of relatively rare occurrence they represent samples definitely lifted from the upper mantle. Some may be unmodified mantle material and others may be residues depleted by partial melting.

In the past, attempts have been made to estimate the average composition of mantle peridotite as though it were a homogeneous material, but it is now recognized that the mantle is heterogeneous and there is no basis for such a calculation. Table 39 provides some guidelines as to the possible composi-

Table 39. Limiting and possible compositions of typical upper mantle material

	1	*2*	*3*	*4*
SiO_2	46.0	44.5	45.1	46.5
TiO_2	0.1	0.3	0.2	0.3
Al_2O_3	2.2	2.8	4.6	3.1
Cr_2O_3	0.6	0.3	0.3	0.5
Fe_2O_3	2.2	—	0.3	1.9
FeO	6.2	10.2	7.6	9.5
MgO	39.9	37.9	38.1	32.9
CaO	2.3	3.3	3.1	4.8
Na_2O	0.2	0.3	0.4	0.1
K_2O	0.00	0.14	0.02	0.01

1. An Alpine-type peridotite (spinel–lherzolite) believed to be a partial melting residuum, Ronda, Spain (Schubert 1977).
2. A garnet–lherzolite inclusion from a kimberlite pipe, Thaba Putsoa, Lesotho (Boyd and Nixon 1973).
3. Estimated composition of mantle 'pyrolite' (Ringwood 1975).
4. A quenched peridotitic komatiite believed to be the liquid product of a high degree of partial melting, Barberton, South Africa (Smith and Erlank 1982).

All compositions have been recalculated on a water-free basis; all iron is given as FeO in no. 2.

tion of an upper mantle peridotite that could yield basaltic magma on melting. Columns 1 and 4 give limiting compositions. The peridotite in column 1 is the residual portion of an Alpine-type peridotite which contains partial melt segregations; the original unmelted mantle must have been richer than this in fusible constituents. The komatiite lava in column 4 must have formed by such a high degree of partial melting that not only the basaltic fraction of its source rock but also a good deal of ferromagnesian refractory material entered the liquid; to leave any refractory residue the source material would have to have been even richer in MgO and poorer in Na_2O and K_2O.

The types of peridotite which can be partially melted to yield a basaltic liquid are those which contain clinopyroxene and either spinel or garnet (e.g. lherzolites), but most Alpine-type or stratiform peridotites are composed mainly of olivine and orthopyroxene and are too refractory to give a significant amount of basic liquid. The garnet–lherzolite of which the analysis is given in column 2 is an actual piece of upper mantle brought up in a kimberlite pipe. It is a 'fertile' or undepleted sample of mantle, meaning that it has not yet been depleted of its low-melting fraction. In fact Kushiro (1973) actually melted this sample experimentally and obtained a basaltic liquid.

Reasoning that basaltic liquids and Alpine-type peridotites might represent complementary melt and residuum from a common mantle source, Ringwood proposed that unmelted mantle might have a composition intermediate between the two. He proposed a model composition for the unmelted mantle consisting of the constituents of refractory peridotite and average basalt mixed in the ratio of 3:1, and called this mixture 'pyrolite'. Column 3 is Ringwood's (1975) estimate of a typical mantle composition based on three methods of calculation: (1) mixing peridotite and basalt compositions in various proportions; (2) calculating the bulk composition of an ophiolite complex by combining its mafic and ultramafic constituents; and (3) a composition based on ultramafic inclusions in basalt.

ECLOGITE AS A SOURCE MATERIAL

It has been pointed out by Yoder and Tilley (1962) that eclogites have the same range of chemical composition as basalts, and that for every basalt there is a chemically equivalent eclogite. Eclogite is a heavy rock consisting mainly of the high density minerals garnet and omphacite (clinopyroxene), and is in effect the high pressure metamorphic equivalent of basalt under dry conditions. The conversion of basalt to eclogite and vice versa has been studied experimentally, and the changes are illustrated by Figs 222 and 223.

The transition from basaltic to eclogitic mineralogy must be marked by a change in the density of the rocks concerned, and hence by a change in seismic velocity. It was once thought that if eclogite were the material of the upper mantle over a large area, the level of the basalt–eclogite transition must correspond to the velocity change at the crust–mantle boundary, i.e. the Mohorovičić discontinuity, but this can certainly not be the case under the

Figure 222. Mineralogy of a quartz–tholeiite in the pressure range 0–30 kilobars at 1100 °C (after Green and Ringwood 1967).

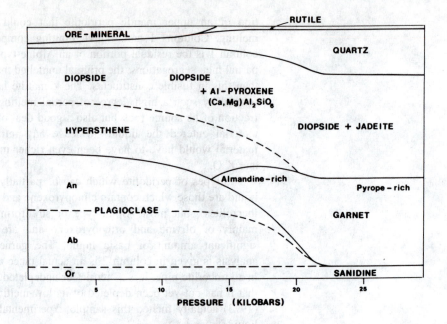

Figure 223. Mineral assemblages in an olivine–tholeiite composition between 0 and 30 kilobars, showing the transition from basalt to eclogite (after Ito and Kennedy 1971).

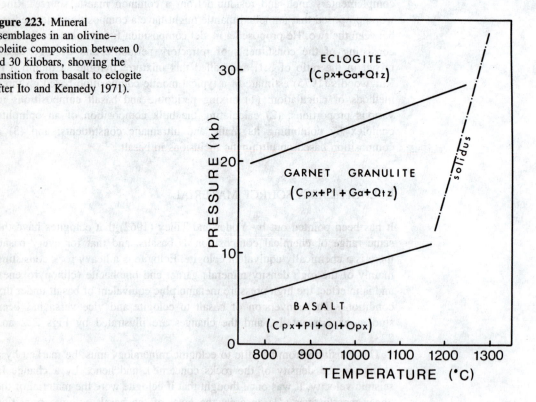

oceans, where the Mohorovičić discontinuity is at too shallow a depth. The pressure–temperature slope of the mineralogical boundaries in Fig. 223 indicates that the transition from basaltic to eclogitic mineralogy would take place at different depths in regions of different geothermal gradient, being at the greatest depth in regions where the geothermal gradient is highest. The transition from basalt to eclogite is in fact gradational, because of the overlapping stability ranges of plagioclase and garnet.

There are two main possibilities for the presence of eclogite at normal mantle depths. Firstly, it is widely assumed that eclogite is present in subduction zones, where it is formed by the transformation of basaltic oceanic crust under increasing pressure. Melting of eclogite in this situation to give basalt magma is contemplated even by some petrologists who believe that basalt magmas elsewhere are derived from peridotite. Secondly, there is a possibility that some part of the mantle outside subduction zones may also be eclogitic (Ito and Kennedy 1970; Ito 1974; Anderson 1979, 1981a, 1984).

Despite these possibilities, most petrologists assume that it is the partial melting of peridotite which gives rise to basaltic magmas, and it is only in subduction zones that melting of eclogite need be considered. The melting of eclogite will be considered further in the discussion of andesites (Chapter 10), but it must be emphasized that the generation of basalt magma by melting of eclogite is a real possibility and has been reproduced experimentally.

An argument against an eclogite source for basalts outside subduction zones is that partial melting of eclogite provides rather a high proportion of melts with intermediate composition, whereas in most volcanic provinces basalt is the dominant type of lava. There are two additional lines of evidence that indicate large parts of the mantle to have compositions more mafic than eclogite: (1) the existence of komatiites, the ultramafic lavas which occur in most of the Pre-Cambrian shield areas; and (2) the association of ultramafic cumulates with gabbro and basalt in many ophiolite complexes, which suggests derivation of both from a parent magma more mafic than basalt.

PERIDOTITE OR ECLOGITE?

Anderson (1989) reviewed the physical lines of evidence which might help to discriminate between peridotite and eclogite as possible upper mantle materials. One of the most promising is the measurement of seismic velocity anisotropy. It has been found in several parts of the world that P-wave velocities in the upper mantle vary according to the direction of travel. Near Hawaii, Morris *et al.* (1969) found an anisotropy of 0.6 km/s, the P-wave velocities varying between 8.5 km/s in a roughly E–W direction and 7.9 km/s in a roughly N–S direction. Close to oceanic spreading axes the velocities are greatest parallel to the direction of spreading. Anisotropy has also been found in upper mantle overlain by continental crust, as in Germany (Bamford and Prodehl 1977). These anisotropies are easier to explain in a peridotite mantle than one of eclogitic composition, since olivine and pyroxenes are highly anisotropic in their elastic properties and many peridotites show a strong preferred orientation of crystals. In contrast, garnet is an isotropic mineral and eclogites do not usually show a strong preferred orientation.

Ito (1974) compared various possible mantle constitutions in the light of heat flow and seismic velocities and found that a two-layer model with peridotite overlying olivine–eclogite best fitted the geophysical data, while Herzberg and O'Hara (1985) suggested on geochemical grounds that a peridotite upper mantle overlies a pyroxenite and eclogite lower mantle.

The densities of the various possible mantle materials have an important bearing on their potential distribution. Eclogite (~3.5 g/cm^3) is appreciably denser than peridotite (~3.3 g/cm^3), and so would tend to sink relative to peridotite. The density of peridotite itself would depend on its precise composition. A 'depleted' peridotite (i.e. the residue left behind after melting out of a basaltic liquid) would be lighter than a 'fertile' peridotite (i.e. one that still contained some potential basaltic constituents). This is because the liquid would have a higher Fe/Mg ratio than the residue, and because the dense phase garnet would be removed during melting. The density relationships of the various solid rock types are appreciably altered as soon as any partial melting takes place. Thus Anderson (1981b) postulated that partially molten eclogite might rise diapirically from the deep mantle to provide a potential source of basaltic magma even though completely solid eclogite would be denser than any of the other likely mantle materials.

Many geochemists have adopted a chondritic model for the bulk composition of the mantle. If we assume that the Earth grew from the accumulation of small bodies similar to meteorites in their composition, then the Earth's main silicate portion (the mantle) should have a bulk composition similar to that of the commonest type of stone meteorite (the chondrites). If this analogy is valid, then it is noteworthy that the bulk composition of the mantle would have to be richer in Al, Ca and Na than peridotite, but less rich than eclogite.

Whichever evidence is given most weight, it is certain that there are inhomogeneities in the composition of the mantle. There are obviously regions which have lost their low-melting constituents through the formation and uprise of basaltic magma during past episodes of magmatic activity. Other parts of the mantle may still be in their 'original' condition, with their low-melting constituents intact. The distribution of these depleted and fresh parts of the mantle will depend on how thoroughly the material of the mantle has been circulated by the Earth's internal movements over the last 4600 million years. Apart from variations in chemical composition there are also variations in mantle mineralogy caused by pressure-related phase changes. In the upper mantle (above a depth of 300 kilometres) these phase changes would involve familiar minerals such as olivine, pyroxenes, garnet and spinel. Below 300 kilometres high-pressure minerals may be present which are not encountered in rocks at the Earth's surface. Examples are the conversion of olivine to its β-crystalline form at about 120 kilobars, and of pyroxenes to a garnet structure at a similar pressure. Such transformations are probably responsible for the seismic discontinuities which occur between the upper and lower mantle (Ringwood 1991).

PRIMARY ORIGINS

MELTING OF PERIDOTITE AND ECLOGITE

Various natural rock types ranging in composition from peridotite to basalt (eclogite) have been melted experimentally and have yielded melts of basaltic or picritic composition. Examples of such experiments in which the melts produced have actually been analysed are those of Ito and Kennedy (1974) on eclogite, and of Takahashi and Kushiro (1983) on peridotite. The composition of the melt produced by the partial melting of a peridotite or eclogite depends on the conditions of melting and on the degree of melting.

The role of pressure is particularly important because it determines which phases are present in the parent rocks of the source region. At successively higher pressures a typical *peridotite* at a temperature below the solidus would consist of the following mineral assemblages (Fig. 221):

$P = 0–10$ kilobars   Olivine + orthopyroxene + clinopyroxene + plagioclase ('plagioclase–lherzolite')
$P = 10–25$ kilobars   Olivine + aluminous orthopyroxene + aluminous clinopyroxene ± spinel ('spinel–lherzolite')
$P > 25$ kilobars   Olivine + orthopyroxene + clinopyroxene + garnet ('garnet–lherzolite')

An *eclogite* just below the solidus would consist of the following assemblages (Fig. 223):

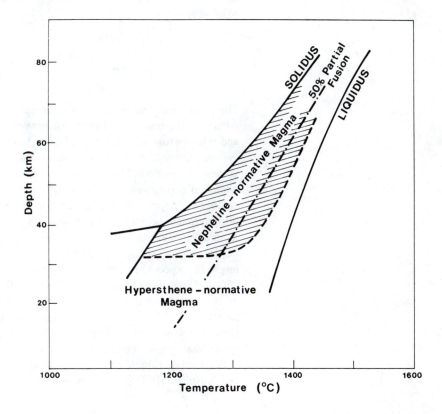

Figure 224. The relationship of silica-undersaturation to pressure and degree of partial melting in an eclogite investigated by Ito and Kennedy (1974).

$P = 0$–10 kilobars   Plagioclase + clinopyroxene ± orthopyroxene ± olivine ('gabbro')

$P = 10$–20 kilobars   Clinopyroxene + plagioclase + garnet ± orthopyroxene ± quartz ('garnet–granulite')

$P > 20$ kilobars   Sodic clinopyroxene + garnet ± orthopyroxene ± quartz ('eclogite')

The degree of partial melting is important because the source rock does not melt instantaneously, and the solid phases do not go into the melt simultaneously. Eclogites have a very limited melting range, sometimes going from almost completely crystalline to completely molten in a temperature rise of as little as 25 °C, whereas peridotites have a very wide melting range and are still partially solid at temperatures several hundred degrees above the beginning of melting.

Ito and Kennedy (1974) analysed the melts produced by subjecting an eclogite to various degrees of partial melting at various pressures. They found that by varying the conditions of melting, either a tholeiite or an alkali basalt could be produced. At high pressures or low degrees of partial melting a nepheline-normative melt was formed, and at low pressures or greater degrees of partial melting the melt was hypersthene-normative (Fig. 224). Any type of basaltic magma may of course be produced if an eclogite of the appropriate composition undergoes complete melting.

The melting behaviour of garnet–peridotite can be predicted from simple phase equilibria. One may assume a mineralogical composition similar to that of the average garnet–lherzolite nodule in kimberlite, as given by Maaløe and Aoki (1977):

Olivine	63%
Orthopyroxene	30%
Garnet	5%
Clinopyroxene	2%

The garnet and clinopyroxene play an essential role in providing the Ca and Al which are major constituents of basaltic magma. The melting of garnet- and clinopyroxene-bearing peridotite may be roughly modelled by the system forsterite–diopside–pyrope as shown in Fig. 225. Melting in this system is 'eutectic-like', because despite the richness in olivine of the unmelted peridotite the initial liquid must have an olivine-poor composition analogous to point E in Fig. 225. In the four-phase assemblage olivine–orthopyroxene–garnet–clinopyroxene the initial melt composition is rich in the constituents of garnet and clinopyroxene, being approximately $Di_{47}Py_{47}Fo_3En_3$ (Davis and Schairer 1965). The composition of the liquid might be expected to remain close to this invariant composition up to about 25% melting of the source material, i.e. until one of the minor phases (garnet or clinopyroxene) is used up. This is the most likely reason for the relative uniformity of basaltic magmas. The melting residue of such a garnet–lherzolite would consist largely of olivine and orthopyroxene, and would be termed harzburgite or dunite.

A garnet–peridotite nodule melted experimentally at 20 kilobars by Mysen and Kushiro (1977) gave the following analysed liquids (Table 40):

Figure 225. The system forsterite–diopside–pyrope at 40 kilobars (after Davis and Schairer 1965; Yoder 1976).

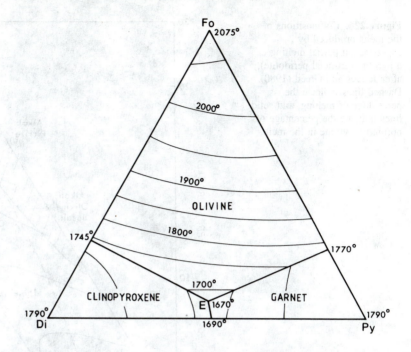

Table 40. Chemical and normative compositions of the liquids produced by partial melting of a garnet–lherzolite at 20 kilobars in the absence of water (after Mysen and Kushiro 1977)

Temp (°C): (% melting):	1450° (1%)	1475° (2%)	1500° (16%)	1600° (44%)	1700° (60%)	Starting composition
SiO_2	43.80	44.60	50.20	48.70	46.30	43.70
TiO_2	1.40	1.00	0.80	0.60	0.40	0.25
Al_2O_3	12.70	12.20	12.20	6.10	4.50	2.75
Fe as FeO	13.70	12.50	9.60	17.80	14.60	10.05
MnO	0.30	0.20	0.30	0.10	0.10	0.13
MgO	12.00	14.30	13.80	18.50	27.40	37.22
CaO	13.70	13.20	10.70	7.20	5.30	3.26
Na_2O	1.40	1.10	1.50	0.70	0.50	0.33
K_2O	0.60	0.50	0.60	0.30	0.20	0.14
Or	1.81	2.94	3.51	1.77	1.14	
Ab	—	1.94	13.32	6.26	4.33	
An	26.70	26.73	24.52	12.58	9.11	
Ne	7.57	4.73	—	—	—	
Lc	1.40	—	—	—	—	
Cpx	33.22	30.72	22.40	18.42	13.00	
Opx	—	—	18.98	35.07	23.84	
Ol	27.34	31.56	16.18	25.06	48.03	
Il	1.96	1.39	1.10	0.83	0.54	

1450 °C – alkali melabasalt   (1% melting)
1500 °C – olivine tholeiite   (16% melting)
1600 °C – pyroxenite (websterite)   (44% melting)
1700 °C – tholeiitic picrite   (60% melting)

Similarly, Jaques and Green (1980) found that melting of spinel–peridotite at 15 kilobars gave alkali olivine basalt melt at 15% melting, olivine–tholeiite

Figure 226. Compositions of the melts produced by experimental partial melting of a pyrolite (artificial peridotite), after Jaques and Green (1980). Dashed lines indicate the percentage of melting, and solid lines indicate the percentage of normative olivine in the melt.

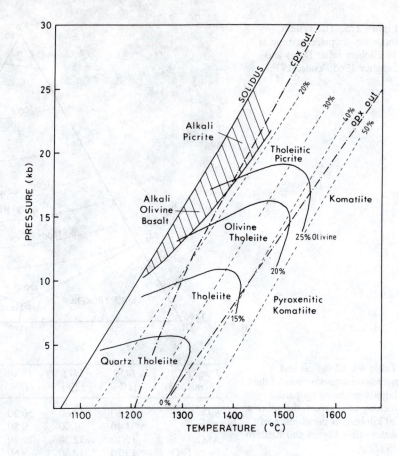

at 20–30% melting, and a picritic or komatiitic liquid at 40–60% melting (Fig. 226). At very low pressures (<5 kbar) they found that quartz-tholeiite and tholeiite were the only melts produced, becoming komatiitic at a high degree of melting. These results are in agreement with the findings of Takahashi and Kushiro (1983) that the initial melting product of a spinel–lherzolite was a tholeiitic melt at low pressures (<15 kbar) but an alkali basaltic melt at higher pressures, and that a high degree of partial melting gave rise to a liquid of picritic or komatiitic composition.

Because of the olivine-rich bulk composition of peridotites, a high degree of partial melting always leads towards picritic* liquids, whatever the initial melt composition. Some Alpine-type peridotites contain bodies of gabbro or pyroxenite which have been interpreted as partial melt segregations, and these vary from basalt to picrite in composition. Examples from the Lherz, Ronda and Beni-Bousera massifs are described in Chapter 13.

---

* The use of the term 'picritic' by experimental petrologists to describe melt compositions is rather unfortunate, because traditionally a picrite or picrite–basalt means a rock that has crystallized from a basaltic magma carrying a lot of suspended olivine crystals. A 'picritic liquid' is something essentially different, meaning a liquid with a much higher MgO content than basalt; 'komatiitic liquid' might be a better term.

MELTING OF PERIDOTITE IN THE PRESENCE OF H_2O AND CO_2

The effect of water on the melting of peridotite mineral assemblages is in general to enlarge the stability field of olivine and to produce liquids which are more siliceous than they would be in the absence of water. Mysen and Boettcher (1975) studied the hydrous melting of peridotite in detail and showed that in the presence of excess water the products of partial melting are andesitic at pressures from 7.5 to 22 kilobars.

Increasing pressure and decreasing P_{H_2O}/P_{total} leads to undersaturated melting products. Mysen and Kushiro (1977) partially melted a garnet–peridotite nodule at 20 kilobars with 1.9% of added water and obtained the following liquids:

1460 °C – alkali melabasalt   (6% melting)
1500 °C – olivine tholeiite   (34% melting)
1525 °C – tholeiitic picrite   (40% melting)

Perhaps more important than its effect on the composition of the melting products is the effect of water on the melting temperatures. As in all silicate systems these are greatly lowered by water under pressure (Fig. 227). To be realistic, one should consider melting of peridotite in the presence of only a small amount of water, because the lowering of the solidus by water would result in very extensive melting of the upper mantle under actual geothermal gradients. Water present in mantle peridotite before melting would most likely be in the form of amphibole, and the solidus of peridotite would then be closely related to the conditions under which the amphibole breaks down to release H_2O (Fig. 228).

Carbon dioxide is no less important than water as a possible volatile constituent of the mantle. Carbon dioxide is the main constituent of fluid

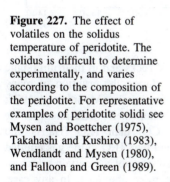

Figure 227. The effect of volatiles on the solidus temperature of peridotite. The solidus is difficult to determine experimentally, and varies according to the composition of the peridotite. For representative examples of peridotite solidi see Mysen and Boettcher (1975), Takahashi and Kushiro (1983), Wendlandt and Mysen (1980), and Falloon and Green (1989).

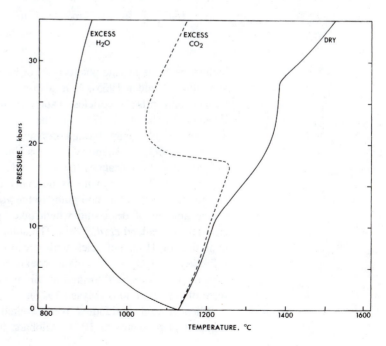

Figure 228. Melting relations of a synthetic peridotite in the presence of a small amount (~0.2 wt%) of water (after Green 1973).

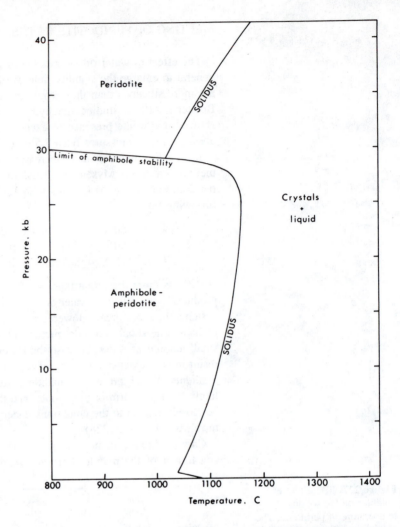

inclusions in the olivine phenocrysts of basaltic rocks and in their ultramafic xenoliths (Roedder 1965). It is also evolved during the eruption of many basalts and it fills the vesicles in some submarine basalts (Moore *et al.* 1977; Pineau and Javoy 1983). On the other hand, melt inclusions in the phenocrysts of many basalts contain less CO_2 than H_2O (Harris and Anderson 1983, 1984) and CO_2 is less abundant than H_2O in the gas emitted by some basaltic volcanoes (Gerlach 1982), so it is not yet certain which of the two volatile constituents is normally the more abundant. In any case, erupted lavas are not a good guide to the volatile content of magmas because of the amount of degassing which takes place at shallow depths prior to eruption (Greenland *et al.* 1985). The solubility of CO_2 in basic melts is less than that of H_2O, but rises with pressure. Above about 20 kilobars the solubility of CO_2 in the melts coexisting with peridotite rises abruptly (Wyllie and Huang 1976), and at this pressure the dissolved CO_2 is in the form of carbonate ions (Mysen 1977).

When natural peridotites were melted in the presence of H_2O–CO_2 mixtures at pressures of 10–15 kilobars, the liquids produced ranged from quartz-normative to nepheline–larnite–normative as the CO_2/H_2O ratio increased (Mysen and Boettcher 1975). The melts were comparable to andesites at high H_2O/CO_2 ratios, to olivine–tholeiites at intermediate H_2O/CO_2 ratios, and to olivine–nephelinite and melilitite at low H_2O/CO_2 ratios. However, Eggler (1978) claimed that if the total amounts of H_2O and CO_2 present were small, the initial product of partial melting would be nephelinitic whatever the H_2O/CO_2 ratio. Increasing pressure leads to increasingly undersaturated melt compositions (Wyllie and Huang 1976) and eventually to the formation of carbonatite melts (Wendlandt and Mysen 1980).

Because of the greater solubility of H_2O than CO_2 in silicate melts, partial melting of peridotite containing CO_2 and H_2O would result in an enrichment of the residual mantle in CO_2. Continuous removal of H_2O relative to CO_2 during an episode of basaltic magma formation could enrich the source region in CO_2 sufficiently to cause later melts to become more undersaturated, especially at pressures above 20 kilobars where CO_2 can be fixed as carbonate. Eggler (1974) suggested that this could be the reason why the basalts of the Hawaiian islands have changed from tholeiitic to alkaline in the later stages of volcanicity.

Despite the important role which volatiles can obviously play in determining basaltic compositions, it is by no means certain that this role is always brought into play. Presnall *et al.* (1979) have emphasized the low volatile contents of the most abundant (ocean-ridge) basalts, and the fact that the main compositional variations in basalts can be brought about by differences in pressure alone without recourse to high H_2O or CO_2 contents in the source rocks.

The main results of the experimental studies may be summarized as follows. All the main types of basalt magma that are erupted could be primary magmas. If the source material is peridotite the factors conducive to the formation of silica-undersaturated melts are:

(a) a low degree of partial melting,
(b) high pressure,
(c) a high CO_2/H_2O ratio.

Silica-saturated or oversaturated melts are produced by a higher degree of melting, a lower pressure, or a high H_2O/CO_2 ratio. Very high degrees of melting always yield picritic (komatiitic) magmas.

DEGREE OF PARTIAL MELTING

It is clear from the melting experiments on peridotites that as they are subjected to progressive partial melting the composition of the melt changes, and some of the variation in basalt compositions could reflect different degrees of melting in the mantle source region. The source may itself vary, and the melting products may undergo further fractionation, but different degrees of melting are certain to occur in any source which starts by being solid and proceeds to discharge large amounts of magma.

The experimental evidence that alkali basalts are products of a low degree of melting and tholeiites of a higher degree is obviously consistent with the relative abundance of these two main types of basalt. Tholeiitic basalts are produced in very large quantities at oceanic spreading centres and in island arcs, but alkali basalts are largely confined to isolated volcanoes. There is, however, a serious obstacle to comparing magma volumes and compositions, and this arises from the variation of density with composition. The higher the degree of partial melting the greater the density of the magma, since the product of a high degree of partial melting in a peridotitic source material will be rich in the constituents of olivine. It has even been suggested that the basalts erupted at the Earth's surface are not at all representative of the magmas which form at depth, but are a density-controlled selective sample. If this is true, which seems very likely, then it follows that large quantities of 'picritic' magma may be formed which either never reach the surface or do so only after fractionation.

The idea that alkaline and tholeiitic basalts are related by different degrees of melting was quantitatively examined by Gast (1968). He pointed out that alkali basalts are rich in the so-called 'incompatible' elements, i.e. those that do not readily enter the crystal structures of the major mantle phases (olivine, pyroxenes, garnet). These are just the elements which would be most enriched in an initial melt fraction. He calculated that the abyssal tholeiites and the alkali basalts might correspond to extensive (20–30%) and limited (3–7%) partial melting respectively. He also acknowledged the possibility that tholeiites might not only be the products of more extensive melting but also have had a source slightly depleted in its lowest melting fraction at an earlier date.

The partial melting experiments on peridotites show that initial melting products (i.e. products of a small degree of melting) are usually silica-undersaturated compared with the products of a higher degree of melting, and this is consistent with the relative abundance and timing of alkaline and tholeiitic magmatism in individual basaltic provinces. For example, Hoernle and Schmincke (1993) described cyclical variation in the basic magmas of Gran Canaria from nephelinite to tholeiite, with the greatest rate of magma production and the most silica-saturated lavas in the middle of each cycle, and the lowest rate of production and most SiO_2–undersaturated compositions at the beginning and end. Of course it is understandable that the lowest degrees of partial melting of the source would occur at the beginning and end of a magmatic episode, but one could argue that by the time the magmatic activity started to wane the source would already have been depleted of its initial melt fraction. However, the source region could be rejuvenated by the arrival of fresh material from a rising plume.

In each of the Hawaiian volcanoes it is observed that the eruptions become more infrequent and the lavas more alkalic as time progresses (Table 41). For Kohala volcano, the rates of eruption were calculated by Feigenson and Spera (1981) as 3000×10^{-5} cubic kilometres/year for the tholeiites, 85×10^{-5} for the transitional basalts, and 8×10^{-5} for alkalic basalts. They believed that the relationship was best interpreted in terms of the degree of partial melting.

Despite all this evidence that alkali basalts are produced by a lower degree of melting than tholeiites, it does not necessarily follow that these magmas

Eruptive stage	Rock types	Eruption rate	Volume
Pre-shield stage	Basanite Alkalic basalt Transitional basalt	Low	~3%
Shield-building stage	Tholeiitic basalt Tholeiitic picrite	High	95–98%
Post-caldera stage	Alkalic basalt Transitional basalt Ankaramite Hawaiite Mugearite Benmoreite Trachyte	Low	~1%
Post-erosional stage	Alkalic basalt Basanite Nephelinite Nepheline–melilitite	Very low	<1%

Table 41. Hawaiian eruptive products (after Clague 1987)

were produced by different degrees of melting of the same material. Isotopic evidence, to be dicussed later, shows that the source material of mid-ocean ridge tholeiites is different from that of the alkali basalts erupted on oceanic islands.

MANTLE HETEROGENEITY

The factors discussed above explain how different basalts could be produced from a uniform source material. More fundamentally, basalts may also vary because they have come from the melting of different source materials. Thus subducted oceanic crust or even continental crust may be involved in the genesis of island arc and continental basalts. Even the basalts which are unequivocally derived from the mantle, i.e. those of the ocean floor and oceanic islands, may come from different sources in the mantle.

The strongest petrological evidence for the heterogeneity of the mantle comes from isotopic ratios, because they are less subject than major or trace element contents to modification by fractionation. In modern oceanic basalts, whose actual $^{87}Sr/^{86}Sr$ ratio has not changed since crystallization, there are variations in the ratio which cannot be due to differentiation or contamination. The initial $^{87}Sr/^{86}Sr$ ratios for mid-ocean ridge basalts (MORB) are around 0.703, but for the basalts of oceanic islands the ratio varies from 0.703 up to 0.706. The basalts with high ratios must have come from a source in the mantle in which there was a relatively high concentration of rubidium prior to the episode of magma generation. Brooks *et al.* (1976) plotted initial $^{87}Sr/^{86}Sr$ against Rb/Sr for oceanic basalts, averaged by island group, and found that tholeiitic basalts lie along a line which corresponds to a 1600 (±200) million year isochron (Fig. 229), even though the analysed rocks were Recent or very young. This indicates that the magmas have come from a depth in the mantle where there are heterogeneities dating from 1500–2000 million years ago.

Figure 229. Plot of initial $^{87}Sr/^{86}Sr$ against Rb/Sr for different basaltic provinces. Closed circles = oceanic island groups; triangles = ocean floor basalts; open circles = continental provinces (after Brooks and Hart 1978).

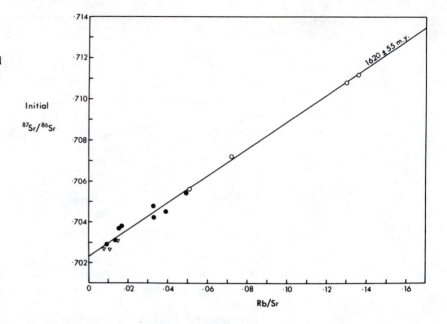

The lead isotope ratios of oceanic basalts are even more variable (Fig. 230). For a particular $^{206}Pb/^{204}Pb$ ratio, the $^{207}Pb/^{204}Pb$ ratio differs between island groups, indicating evolution of the source along different growth curves (i.e. different U/Pb ratios). In each island group there is a linear trend, aligned between MORB and a more radiogenic composition, which suggests either that there has been magma mixing, or more likely that the magma was derived from a mixed source.

Figure 230. Lead isotopic composition of oceanic basalts (after Sun 1980).

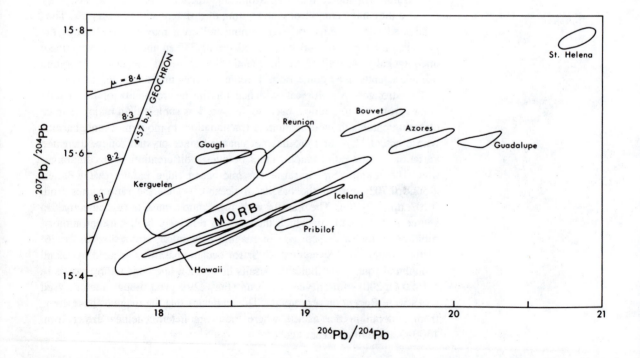

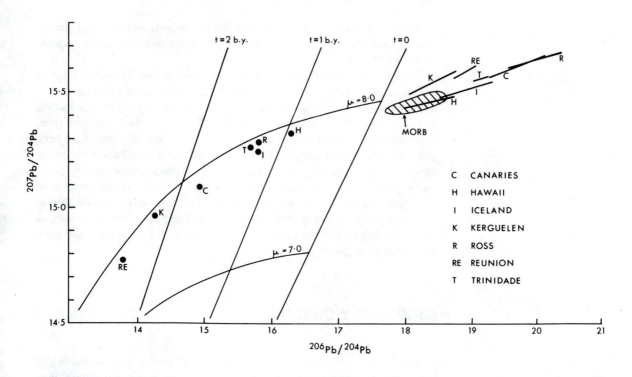

Figure 231. Lead isotopic composition of oceanic island basalts. Straight lines marked by letters indicate the range of composition in each island group. Dots indicate the primary isotopic ratios from which the secondary isochrons are derived in a two-stage model (after Chase 1981).

Figure 231 shows the linear trends of Pb isotope composition for seven island groups. Chase (1981) calculated that these trends would fit very well to a two-stage evolutionary model. According to his calculations they correspond to secondary isochrons derived from U/Pb enrichment events at different times from 2500 to 1000 million years ago. He postulated that up to that time their source regions formed part of a common mantle reservoir, whose μ value (^{238}U/^{204}Pb) was 7.91, the same as that of MORB basalts. To yield the source material of modern oceanic island basalts, this common reservoir must have been augmented by material with a higher U/Pb ratio, and Chase suggested that this was oceanic crust, returned to the mantle by subduction.

The problem has been raised (O'Nions and Pankhurst 1974) that magmas with different Sr isotopic ratios might be derived from a homogeneous source as a result of the differential melting of Rb-rich and Rb-poor mantle phases. Phlogopite in particular would be Rb-rich, although it would not be present in any quantity. Allègre *et al.* (1980) recorded ultramafic nodules with ^{87}Sr/^{86}Sr ranging from 0.7022 in diopside to 0.710 in garnet, and Pushkar and Stoeser (1975) have found a partially melted granitic xenolith with a ratio of 0.706 containing glass with a ratio of 0.723. Other authors believe that disequilibrium melting is uncommon, because rapid diffusion would enable isotopic homogenization to take place during partial melting (Sun 1980).

There are many possible reasons for the mantle being heterogeneous. For example, it may have been partially depleted in fusible constituents as a result of previous melting episodes, or it may have been locally enriched in fusible constituents by metasomatism, or it may be unevenly contaminated by subducted material. The relative roles of depleted and enriched mantle sources will be discussed in relation to the basalts of each tectonic environment.

The composition of the mantle as a whole has continually changed during the course of geological time. Uranium, lead, rubidium and strontium are all elements which are 'incompatible' in terms of the major mantle minerals, and they have been extracted from the mantle over the course of time. Uranium and rubidium, and to a lesser extent lead and strontium, have been extracted into the crust by basaltic volcanism. Lead may have been partly removed downwards by extraction into a sulphide phase or by direct exchange with the core. There is much interest in whether the removal of these elements from the mantle happened continuously, episodically or catastrophically. These matters have a great bearing on the processes of crust, mantle and core formation, and on the nature of convection and plate movement throughout geological time.

SECONDARY ORIGINS

Although the experimental evidence shows that various kinds of basalt can be produced as primary magmas from a peridotite or eclogite source, they can also be produced by the fractionation of magmas more mafic than basalt. Moreover, even if basalt magmas are primary they may still be heavily modified by fractionation or contamination before appearing at the Earth's surface.

The main lines of evidence used to decide which process has contributed predominantly to the variation of basalts in any particular area are isotopic studies, the distribution of 'compatible' elements (e.g. Mg, Fe, Ni, Cr) and the distribution of 'incompatible' elements (e.g. K, Rb, REE). The dependence of the various geochemical parameters on the major sources of compositional variation may be tabulated as follows:

	Isotopes	*Compatible elements*	*Incompatible elements*
Source heterogeneity	YES	–	YES
Degree of melting	–	YES	YES
Fractionation	–	YES	YES
Contamination	YES	–	YES

None of these geochemical criteria is able by itself to give a completely unambiguous answer to the petrogenesis of any particular group of basalts. Isotopic studies are the most valuable because the isotopic compositions are insensitive to crystal fractionation, and also to the degree of partial melting (unless isotopic disequilibrium is a more serious problem than has generally been assumed). The variation in compatible elements is mainly a measure of

Figure 232. The average Sr and Nd contents of 19 different basaltic suites (shown by dots) compared with the predicted compositions of magmas produced by melting in the mantle and crust (after Zindler *et al.* 1981). The widths of the mantle and crustal melting fields are to allow for a range of possible crystal–liquid partition coefficients; the melt compositions at specific percentages of partial melting are shown by ticks on the melting curves. Arrows show the direction in which the magmas would be altered by fractionation of olivine or plagioclase (ticked at 10% fractionation intervals).

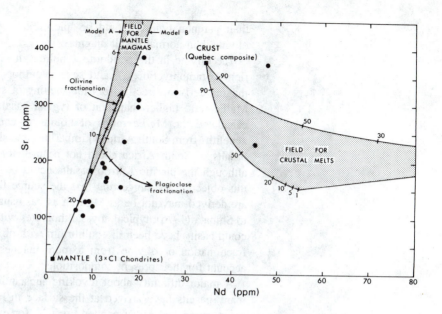

fractionation and/or the degree of melting. The variation in incompatible elements is dependent on too many factors to be decisive, although it can be of some help.

Figure 232 shows an attempt at geochemical modelling by Zindler *et al.* (1981) using the incompatible elements Sr and Nd. They made reasonable estimates of the composition and mineralogy of potential mantle and crustal source regions, assumed a range of partition coefficients for Sr and Nd in the minerals of the source materials, and calculated the compositions of the melts that would be produced by various degrees of partial melting. It can be seen that observed Sr and Nd contents of basalts could result from many different combinations of primary variation, fractionation and contamination.

FRACTIONATION

Basalt may not be a primary magma type at all. We do not know how much of the diversity of basalt magma is due to different degrees or conditions of melting or different materials in the source region, and how much is due to fractional crystallization during ascent of the magma to the surface. Petrological opinion is divided as to whether most basalts are primary magmas (Wilkinson 1982, 1991; Presnall and Hoover 1984; McKenzie and Bickle 1988) or are derivatives of a more mafic parental magma (O'Hara 1968; Yoder 1976; Stolper 1980; Hess 1992).

If ophiolite complexes are correctly identified as sections through oceanic crust formed at spreading axes (Chapter 13), the presence of thick gabbroic and ultramafic cumulates in them is an indication of fractionation in the magma chambers from which the mid-ocean ridge basalts are erupted. This analogy with ophiolites, together with the Mg-rich composition of the likely mantle source rocks, suggests that if basalts are not primary magmas then their parental liquids would be 'picritic', i.e. richer in the constituents of olivine than normal basalt magmas.

The olivine in the peridotite xenoliths in kimberlite is very magnesium-rich (commonly Fo_{90-95}), and several authors have argued that a primary magma derived directly from the melting of such peridotites would have a Mg/Fe ratio higher than that of typical basalts. For example, Cox (1980) calculated the Mg/Fe ratios of liquids in equilibrium with the olivine in xenoliths from South African kimberlites and showed that the normal Karroo basalts of South Africa could not be primary melts from such peridotites, although the picritic Karroo basalts could be. Wilkinson (1982, 1985) met this objection by suggesting that the source lherzolites from which basalts are derived may not be as Mg-rich as has usually been assumed. According to Shibata (1976), typical abyssal tholeiites with FeO/MgO ratios around 1.0 could easily have been in equilibrium with olivine of composition Fo_{90}, and fractionation of olivine from a more magnesian liquid is not required to account for their present compositions.

A major difficulty about invoking substantial crystal fractionation is that many basalts have arrived at the surface in large volumes with a uniform composition and no suspended crystals, for example the thick flows of the Columbia River province. If they are the products of crystal fractionation, the separation of crystals must have operated very effectively and then ceased before eruption.

Assuming that picrite is the parental magma of basalt, Stolper and Walker (1980) calculated how the magma would vary in density as it underwent fractionation. They showed that the melt composition which has the minimum density ($\sim 2.7 \, g/cm^3$) corresponds to the most commonly erupted composition of oceanic basalts (Fig. 233). This common basalt composition is thus better able than either a picritic parent melt or an iron-enriched derivative melt to be able to rise through the rocks of the oceanic or lower continental crust, which have densities above $2.7 \, g/cm^3$. If such a density effect does control the fractionation and eruption of basalts, then it also follows that many parts of the continental and oceanic crust may be immediately underlain by picritic rocks crystallizing from magmas that have failed to penetrate the crust, or by complementary cumulates of basaltic magmas that have done so. In the volcanic region of Central America there is a relationship between basaltic magma density, volcano height and crustal thickness which according to Carr (1984) implies that basalt magma is ponded at the base of the crust and can only get through to the surface once its density has been reduced by fractionation. Some seismic evidence to support the idea of ponding of magma at the crust–mantle boundary beneath Hawaii has been presented by Ten Brink and Brocher (1987).

Whether or not it is a primary magma, basalt certainly does undergo some fractionation, as is obvious from the variation which is found in layered basic intrusions. Comparison of the volume relations in completely fraction-ated intrusions (e.g. Skaergaard, Kiglapait), or of basalts with glassy groundmasses, suggests that the main result of fractionation is to produce a diverse range of basaltic liquids (and cumulates) rather than a large quantity of felsic derivatives. We may therefore expect that at least some of the variation seen in erupted basalts will be due to fractionation.

Figure 233. Plot of calculated liquid density against degree of fractionation, measured as molar Fe/(Fe + Mg), for mid-ocean ridge basalts. The histogram shows the actual distribution of the Fe/(Fe + Mg) ratio in a sample of 2000 mid-ocean ridge basalts (after Stolper and Walker 1980).

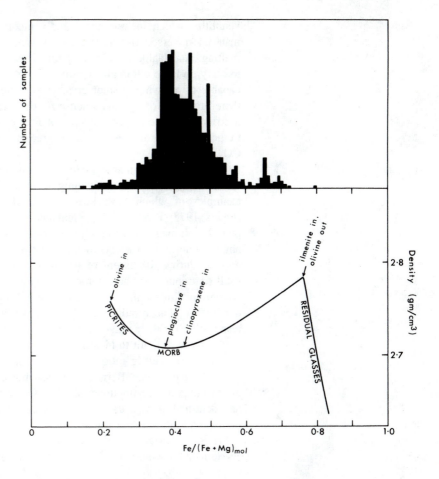

The course of fractionation would depend on the pressure at which crystallization took place, because the minerals crystallizing from a basaltic magma are different at high and low pressures (see Fig. 223). At low pressures, the fractionating minerals would be those which crystallize from basalt when it is erupted, i.e. olivine, pyroxenes and plagioclase. At high pressures (above 20 kilobars), the crystallizing phases would correspond to an eclogite mineralogy, i.e. clinopyroxene and garnet. If the parental magma were considerably more mafic than basalt then olivine might be the main fractionating phase at any pressure. Furthermore, it is quite possible that a magma would undergo fractionation over a range of pressures as it rose from its source towards the surface ('polybaric fractionation'). Phase relations in natural and synthetic compositions corresponding approximately to basalt suggest that high-pressure fractionation would cause the derivative magmas to become not only less mafic than the parental magma, but also more alkaline (O'Hara and Yoder 1967; Green and Ringwood 1967; O'Hara 1968; Gupta *et al.* 1987).

Although an alkali basalt magma could hypothetically be produced by the fractionation of a tholeiitic liquid at high pressure, it is unlikely that alkali basalts do actually originate in this way. The strongest evidence for the primary origin of alkali basalts is that they frequently carry ultramafic xenoliths, which include accidentally incorporated lumps of upper mantle material traversed by the magma as well as residues from the region of melting. As Maaløe (1973) has pointed out, a magma rising rapidly enough to carry up large ultramafic xenoliths from the source region could hardly be capable of allowing smaller crystals to settle out along the way. We must, therefore, look for a primary cause for the formation of alkalic basalts, either a different source material from tholeiites or different conditions of melting in the source region, e.g. a lower degree of partial melting or the presence of CO_2.

Basaltic magmas that we can say with certainty have undergone crystal fractionation are those containing identifiable cumulate inclusions, a few examples of which have been confidently identified by textural criteria (Lewis 1973; Baxter 1978; Blaxland and Upton 1978). Moreover, many other basalt magmas arrive at the surface carrying suspended phenocrysts, so ample opportunity for crystal fractionation obviously exists. The settling of crystals during the ascent of basalt magma to the surface is sufficient by itself to account for one common variety of basalt, i.e. picritic basalt. Picrites are enriched in olivine, and analyses of picritic basalts from Hawaii leave no doubt that concentration of olivine crystals to varying extents is solely responsible for their range of compositions (Murata and Richter 1966).

It is more difficult to identify fractionation using geochemical criteria, but a variation in Fe/Mg ratio without any variation in isotopic composition is *prima facie* evidence. Bernstein (1994) showed how trace elements might be used to identify the fractionating phases in such a series: from the Zr/Y ratio he identified garnet as a fractionating phase in Faeroe Island basalts, indicating high-pressure fractionation.

We can expect that the liquid products of more advanced fractional crystallization would have the composition of trachybasalts or ferrobasalts.

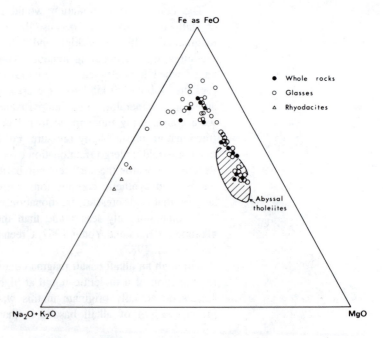

Figure 234. FMA diagram for basalts and andesites from the Galapagos and Ecuador Rifts, eastern Pacific (after Perfit and Fornari 1983). The usual field of abyssal tholeiites is also outlined (after Miyashiro *et al.* 1970).

There are some basaltic rocks whose trend of variation is very similar to that of fractionated gabbro intrusions such as Skaergaard, for example the lavas shown in Fig. 234, although such extreme iron-enrichment is rare. This may reflect the physical circumstances in which fractionation takes place. Fractionation to completion can obviously only take place in a closed reservoir, and if this happened there would be no opportunity for the derivative liquids to be erupted. It is much more likely that fractionation involves a body of magma that is being both constantly tapped and constantly replenished, so that magmas in different stages of fractionation become mixed; this is particularly true of the mid-ocean ridges where the rate of magma production and eruption is high (Dungan and Rhodes 1978).

CONTAMINATION

The contribution of crustal contamination to the variability of basalts is very uncertain. It has generally been assumed because of their local uniformity that most basalts are not contaminated to any significant extent. In contrast, basic plutonic rocks are often variable and show signs of contamination. This is particularly true of norites (see Chapter 7) and of very large basic intrusions in general. O'Hara (1980) reviewed the way in which contamination takes place, and pointed out that large-scale contamination might be very difficult to detect. In continental basalts one might imagine a reaction of the kind:

$$CaMgSi_2O_6 + Al_2SiO_5 \rightarrow CaAl_2Si_2O_8 + MgSiO_3$$
$$\text{(in magma)} \quad \text{(in pelites)} \quad \text{(in cumulates)}$$

Assimilation of pelitic material would thus cause precipitation of orthopyroxene and plagioclase in place of clinopyroxene, but apart from a change in the ratio of the phenocryst or cumulus phases the liquid would not necessarily be much altered in its major element composition. There would, however, be an enrichment in the incompatible elements including potassium. Contamination of basic magma certainly takes place in large basic intrusions, such as the Bushveld complex, and may be partly responsible for the abundance of norite in these intrusions. However, this kind of contamination is accompanied by large-scale crystallization, and gives rise to contaminated rocks rather than contaminated liquids.

Most of the evidence for crustal contamination of basalt comes from continental flood basalts. For example, Philpotts and Asher (1993) described the partial melting and assimilation of wall rocks adjacent to a feeder dyke of the Mesozoic flood basalts in Connecticut; mass balance calculations suggested that the basalt assimilated about 6% of granophyric melt during its ascent through the upper 10 km of crust. High initial $^{87}Sr/^{86}Sr$ ratios are often taken to indicate a degree of contamination. This interpretation has been placed on the high initial $^{87}Sr/^{86}Sr$ ratios in continental flood basalts from the Columbia River (0.703–0.714 – Carlson *et al.* 1981) and the Deccan plateau (0.704–0.713 – Allègre *et al.* 1982). Some flood basalts are also enriched in ^{18}O, which is another possible indication of possible contamination, or at least of the presence of crustal material in the source region (Hoefs *et al.*

1980; Fodor *et al.* 1985). In principle, high $\delta^{18}O$ values are a more sensitive measure of contamination than high $^{87}Sr/^{86}Sr$, although they are also more susceptible to postmagmatic alteration.

The amount of added crustal material required to account for the high $^{87}Sr/^{86}Sr$ ratios in some flood basalts is so large (up to 50% in some cases) that it should be very evident in the major element compositions of the basalts. In the Antarctic plateau basalts, Faure *et al.* (1974) found a strong correlation between initial $^{87}Sr/^{86}Sr$ and elements such as Si, K and Rb, but in Indian basalts with initial $^{87}Sr/^{86}Sr$ between 0.704 and 0.708 Alexander and Paul (1977) found no such correlation. As De Paolo (1983) has pointed out, large-scale melting in the crust sufficient to bring about the required degree of contamination would almost certainly result in the formation of a substantial amount of granite or rhyolite magma, but there are no rhyolites associated with the Columbia River basalts. Many isotope specialists believe that high $^{87}Sr/^{86}Sr$ ratios of basalts are better explained by high $^{87}Sr/^{86}Sr$ in their mantle source regions than by contamination.

BASALT PETROGENESIS IN RELATION TO TECTONIC ENVIRONMENT

MID-OCEAN RIDGE BASALTS (MORB)

There is a general consensus that mid-ocean ridge basalts originate by partial melting of mantle peridotite near the crest of the spreading ridges, but considerable disagreement as to whether they have undergone fractionation before being erupted. The scope for fractionation varies from place to place depending on the heat flow, the rate of spreading, and the presence or absence of fracture zones.

Figure 235. Typical geothermal gradients at an oceanic spreading axis (after Bottinga and Allègre 1978), in older oceanic crust (Hawaiian geotherm of Mercier and Carter 1975), and in a continental shield area (after Mercier and Carter 1975). The anhydrous peridotite solvus is after Takahashi and Kushiro (1983).

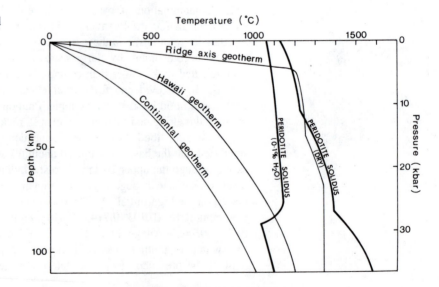

Figure 235 shows some estimates which have been made of the geothermal gradients at and away from an oceanic spreading axis. As new oceanic crust spreads away, its place is taken by mantle material ascending from depth towards the surface at the ridge axis. The solidus temperature of peridotite falls as the pressure decreases, so that during its ascent to the ridge axis the mantle peridotite starts to melt. The depth at which this happens depends mainly on the spreading rate, and widely varying estimates have been made of this depth depending on what is assumed about the physical behaviour and phase relations of the upwelling material. For example Hess (1992) envisaged the start of melting at a depth of 90–105 kilometres (30–35 kilobars), whereas McKenzie and Bickle (1988) estimated that melting would start at a depth of only 40 kilometres.

Presnall *et al.* (1979) studied the effect of the plagioclase– to spinel–lherzolite transition in the system CaO–MgO–Al$_2$O$_3$–SiO$_2$ (CMAS) and believe that it plays a key role by controlling the onset of melting. Their experiments (Fig. 236) show that there is a cusp in the solidus of plagioclase– and spinel–lherzolite corresponding to the subsolidus boundary between the stability fields of plagioclase and spinel. Hot peridotitic mantle might be expected to rise along a geotherm such as that shown for the ridge axis in Fig. 235 and intersect the solidus at the cusp. The main production of basalt magma under oceanic ridges would therefore be at this point, which is at a pressure of 9 kilobars in the simplified CMAS system. At this pressure,

Figure 236. Solidus curve for plagioclase–lherzolite and spinel–lherzolite mineral assemblages in the synthetic system CaO–MgO–Al$_2$O$_3$–SiO$_2$ (after Presnall *et al.* 1979).

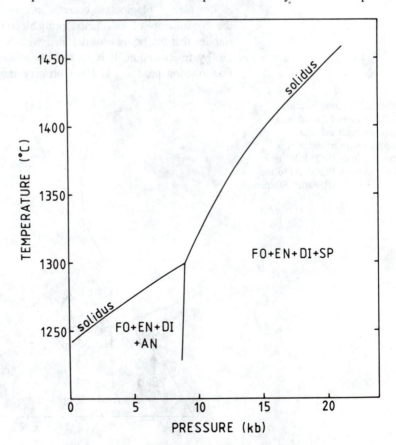

the melts would be tholeiitic in character, with the experiments yielding quartz-normative liquids at lower pressures (from plagioclase–lherzolite) and olivine-normative liquids (from spinel–lherzolite) at higher pressures. Takahashi and Kushiro (1983) determined the solidus of a natural spinel–lherzolite under dry conditions and found the corresponding cusp to be at 11 kilobars, with another at 26 kilobars where the subsolidus mineral assemblage changes from spinel– to garnet–lherzolite; liquids produced by melting close to the 11-kilobar cusp had compositions similar to MORB basalts.

How far partial melting proceeds in the upwelling mantle will determine whether the bulk of the magma produced is basaltic or ultramafic. Elthon (1979) put the case for most ocean-floor basalts being derivatives of an ultramafic parent magma. He said that if one takes into account the presence of ultramafic cumulates as well as basaltic rocks in the oceanic crust, the bulk composition of the oceanic crust must be more magnesian than basalt. From the abundance of different components of ophiolite complexes, he estimated the bulk composition of typical oceanic crust to be picritic (MgO = 18%). He did not think it possible for normal oceanic tholeiite magmas to have precipitated the volumes of ultramafic cumulates that ophiolite complexes contain, and it follows that fractionation of the postulated high-magnesian parent magma must have occurred to give oceanic tholeiites.

Fractionation is also suggested by experimental evidence. Stolper (1980) found that actual MORB compositions lie close to the 1 atmosphere olivine + diopside + plagioclase cotectic in a multicomponent system comprising the constituents of basalt and probable refractory phases (Fig. 237). This implies that the liquids equilibrated with crystalline phases at low pressures, i.e. by fractionation. It is reasonable to assume that if oceanic basalts are fractionation products and not primary magmas, then their parent magma

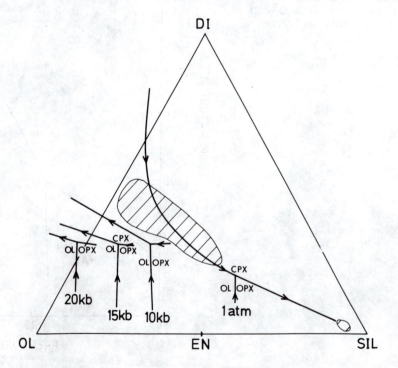

Figure 237. Compositions of MORB glasses (shaded area) compared with calculated liquidus phase boundaries at pressures from 0–20 kilobars projected from PLAG on to the plane OL–DI–SIL (after Stolper 1980).

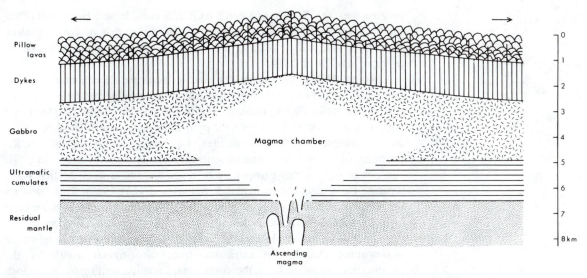

Pillow
lavas

Dykes

Gabbro

Magma chamber

Ultramafic
cumulates

Residual
mantle

Ascending
magma

Figure 238. Hypothetical cross-section through an oceanic spreading axis, showing a body of differentiating magma which crystallizes gabbro and ultramafic cumulates and feeds a basaltic dyke swarm and lava flows.

was picritic. The scarcity of erupted picrites is readily understandable in view of their high density, even if they are common parent magmas. There are some oceanic picrites which may be representative of the primary parent magma (Maaløe and Jakobsson 1980), even if others are only phenocryst-enriched basalts.

Figure 238 shows one of the many models that have been proposed for crystallization of magma beneath a spreading axis. Most such models involve the precipitation of ultramafic cumulates and crystallization of the overlying basaltic crust from the complementary differentiated melt. There are many refinements which have been introduced into such models, such as the distinction between fast and slow spreading axes, with greater opportunities for differentiation beneath the latter (Flower 1981), the possibility that the magma chamber is continuously being refilled while fractionating (O'Hara and Mathews 1981), and the possibility that the products of low and high degrees of partial melting may become mixed (Ahern and Turcotte 1979). The existence and extent of magma chambers beneath oceanic spreading axes is still a matter of controversy. There is positive evidence that magma chambers exist, from the analogy between oceanic crust and exposed ophiolites, with their gabbros and ultramafic cumulates. Seismic studies of oceanic ridges have suggested that magma chambers may be present in some areas (Reid *et al.* 1977) and absent in others (Fowler 1976). The extent of a magma chamber is likely to depend on several factors, such as the rate of spreading (a rapid rate maintains a high geothermal gradient at the axis) and the presence of a sea-floor hydrothermal convection system (which would have a cooling effect).

The geochemistry of ocean ridge basalts has been very thoroughly investigated. Their compositions show that MORB basalts are derived from a source which is relatively homogeneous compared with that of other basalts, and is relatively poor in 'mantle-incompatible' constituents, such as K, Ti, Rb, Sr, Ba, Nb, REE, P, U, Th, H_2O and CO_2. The depleted nature of the MORB basalts suggests that their source material has previously had some low-melting constituents extracted from it. This depletion is of long standing, because the $^{87}Sr/^{86}Sr$ ratio of MORB (0.703) is considerably below that of the Earth as a whole. Assuming a ratio of 0.699 (i.e. that of the achondrite meteorites) 4.55 billion years ago, the Earth as a whole should now have an average $^{87}Sr/^{86}Sr$ of 0.7045.

Although MORB basalts are depleted in incompatible elements, they are relatively less depleted in those which can be accommodated by clinopyroxene and garnet (including sodium). For this reason, Anderson (1981a, 1989) suggested an eclogite or garnet–clinopyroxenite source for MORB basalt (or olivine–eclogite for MORB picrite), but most discussions of MORB petrogenesis assume a peridotite source. There is some evidence from Hf isotopes to suggest that garnet may be a residual phase when MORB melts separate from their source material (Salters and Hart 1989). This does not necessarily indicate an eclogitic source, but if the source is peridotitic it must be in the garnet–lherzolite facies, implying a depth of at least 80 km.

It is widely believed that the MORB source owes its depletion in incompatible elements to extraction from the original mantle of the constituents needed to build the continental crust. According to McCulloch and Bennett (1994) most of the chemical and isotopic features of the continental crust can be explained by its stepwise extraction from the mantle over the course of geological time, assuming only a 'three-reservoir' model of the Earth, i.e. primitive mantle, depleted mantle, and continental crust. Figure 239 shows the complementary nature of the continental crust and the depleted MORB source, and Figure 240 shows how their Sr isotopic compositions may have developed over the course of time. Other authors feel that the recycling of crustal material back into the mantle should also be taken into account, and that a three-reservoir model does not adequately account for the enrichred source of oceanic island basalts.

Iceland occupies a special place in the oceanic ridge system because of the abnormally high rate of magma production and the consequent building of a

Figure 239. A comparison between the chemical compositions of the continental crust and the depleted MORB source (after McCulloch and Bennett 1994).

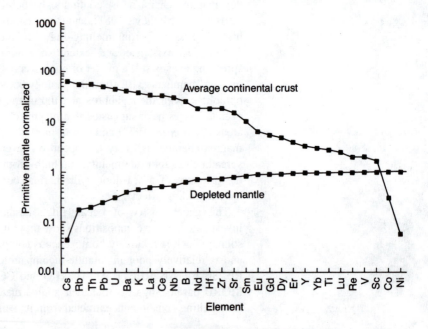

Figure 240. A model of the
evolution of $^{87}Sr/^{86}Sr$ ratios with
time, based on the progressive
extraction of continental crust
from the mantle leaving a
depleted residue (after
McCulloch and Bennett 1994).

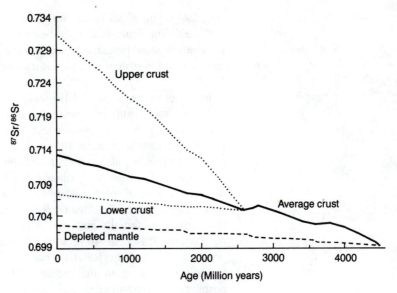

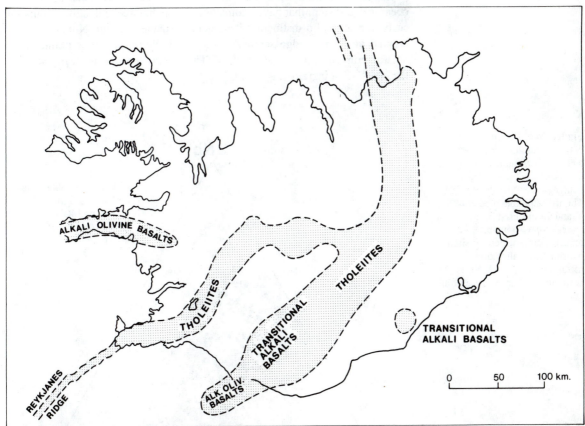

Figure 241. Map of Iceland
showing the postglacial volcanic
areas (shaded) and the nature of
the basaltic magmas erupted
(after Jakobsson 1979).

lava pile above sea level. The lavas of Iceland are about 90% basalts, and
they are predominantly erupted along a zone which is a continuation of the
Mid-Atlantic Ridge. In this zone the basalts are tholeiitic, but in the flank
zones which do not lie on the ridge axis there is alkali-basaltic volcanism

(Fig. 241). The alkali basalts are erupted in much smaller volume than the tholeiites. The crust is thicker beneath the alkalic flank zones (up to 14 kilometres) than beneath the axial zone (8–9 kilometres), so there may simply be a lower degree of partial melting under the flank zones (Jakobsson 1979).

By comparison with MORB tholeiites dredged from the Mid-Atlantic Ridge to the north and south of Iceland, the Icelandic tholeiites are anomalously rich in potassium and other 'mantle-incompatible' elements, and the association of about 10% of intermediate and acid magmas with the Icelandic tholeiites also contrasts with the absence of these magmas in the submarine parts of the Mid-Atlantic Ridge. The anomalous nature of Iceland has clearly been of long standing, because there is a rise in the sea floor at right angles to the Mid-Atlantic Ridge (the Greenland–Faeroes rise) which indicates an abnormally high rate of magma production extending back into the Tertiary. It remains to be seen whether the anomalous composition and production rate of the Icelandic basalts is due to anomalous upper mantle beneath Iceland or to the presence of a rising plume similar to those postulated for intra-plate oceanic islands. There are several other sites in the ocean-ridge system that have anomalous basalts like Iceland, although not on such a scale, and a distinction is sometimes made between N-type MORB, erupted at normal ridge segments, and E- or P- (enriched or plume) type MORB erupted in places like Iceland. This distinction can be clearly seen on the Nb–Y–Zr diagram (Fig. 242).

Figure 242. Discrimination between basalts of different tectonic environments on the basis of their trace element content (after Meschede 1986). The incompatible elements Nb, Y and Zr are used because they are relatively immobile, i.e. relatively unaffected by slight hydrothermal alteration or metamorphism. High Nb is particularly characteristic of an enriched source.

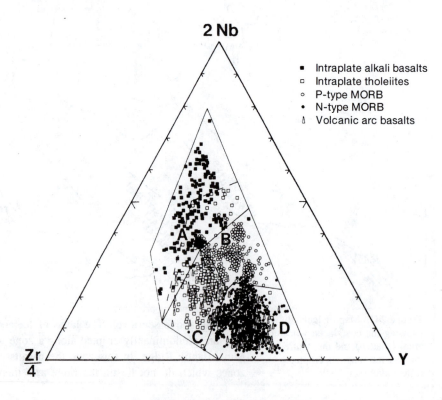

- Intraplate alkali basalts
- Intraplate tholeiites
- P-type MORB
- N-type MORB
- Volcanic arc basalts

OCEANIC ISLAND BASALTS (OIB)

Whereas the production of basaltic magma is a normal feature of oceanic spreading axes and of subduction zones, it occurs only sporadically in other tectonic regions of the Earth. The oceanic island basalts are much more varied than those of mid-ocean ridges, and have compositions ranging from tholeiite to alkali basalt and even to nephelinite. Some large oceanic islands display the whole of this range (e.g. Hawaii), but in general the smaller oceanic islands are conspicuously rich in alkali basalts.

The chemical and isotopic compositions reveal a clear difference between a 'depleted' source for MORB basalts, low in incompatible elements and with low $^{87}Sr/^{86}Sr$ and high $^{143}Nd/^{144}Nd$ (Fig. 243), and 'less depleted' or 'enriched' sources for the oceanic island and continental basalts. Anderson (1981a, 1982, 1984, 1989) suggested that the two major source materials are each of worldwide extent, and that all basalt magmas are mixtures from the two sources. The depleted source is taken to be the main source for MORB basalts. The isotopically more variable basalts of oceanic islands and continental volcanoes would correspond to varying contributions from the enriched source. Iceland, which lies across the Mid-Atlantic Ridge and

Figure 243. Plot of $^{87}Sr/^{86}Sr$ against $^{143}Nd/^{144}Nd$ for oceanic basalts (compiled from various sources). Possible compositions of the enriched mantle components EM1 and EM2 are indicated in Fig. 327.

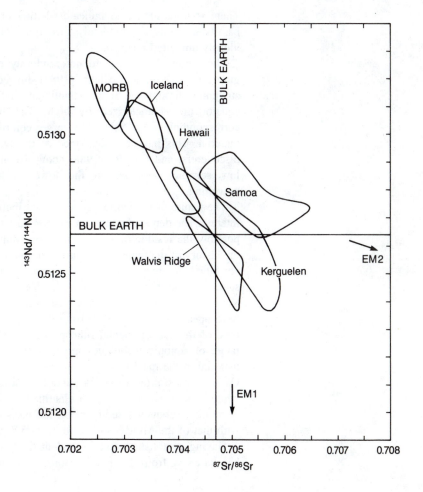

contains a very large volume of basalt, has an isotopic composition close to that of MORB, indicating a large contribution from the depleted source, whereas the basalts of the smaller oceanic islands are not only more 'enriched' in their isotopic composition, but generally more alkaline in character.

Most isotope geochemists believe that a single enriched source is insufficient to explain the variability of oceanic island basalts. Zindler and Hart (1986) reviewed the isotopic characteristics of some of the mantle components that must contribute to oceanic island basalts. They showed that some (e.g. Iceland and Hawaii, Fig. 243) incorporate a large proportion of a MORB-like depleted source component. Others deviate considerably from the MORB composition but in different ways:

1. Some (e.g. the Walvis Ridge and Tristan da Cunha) have low ^{143}Nd/^{144}Nd ratios, indicating a source component with a low Sm/Nd ratio (component 'EM1').
2. Some (e.g. Samoa) have high ^{87}Sr/^{86}Sr ratios, indicating a source component with high Rb/Sr (component 'EM2').
3. Some (e.g. St Helena, Fig. 230) have very high ^{206}Pb/^{204}Pb and ^{207}Pb/^{204}Pb, implying a source with a high U/Pb ratio (component 'HIMU').

There is no evidence to indicate whether or not the source region of any basalts contains a completely primitive component (i.e. primordial, previously unmelted mantle).

The enriched sources are not randomly distributed in the mantle geographically. Dupré and Allègre (1983) showed that the basalts in a wide belt stretching through the South Atlantic and Indian Ocean roughly 40°S of the equator are characterized by high ^{87}Sr/^{86}Sr, ^{207}Pb/^{204}Pb and ^{208}Pb/^{204}Pb, corresponding to a relatively large contribution from enriched source material. This has been described as the Dupal anomaly in honour of its discoverers, and Castillo (1988) showed that it corresponds to a region of low seismic velocities in the lower mantle, i.e. the site of mantle upwelling.

The nature of the various sources contributing to basalt magma is of great interest. The depleted source of MORB basalts is of worldwide extent and is possibly the residue of a major melting event of global proportions early in the history of the mantle. The enriched mantle components have been widely attributed to recycling of subducted oceanic crust and perhaps even sediment, and their dispersal into the deeper part of the mantle (White 1985; Weaver 1991). There is also a possibility that some enriched material may have been derived indirectly from the continental crust, or eroded from the base of the subcontinental lithospheric mantle (Class *et al.* 1993). A wide range of isotopic techniques are now being deployed to identify recycled material in the mantle source regions of basalt. Hauri and Hart (1993) used Os isotopes to infer a role for subducted oceanic crust in the HIMU source, while Woodhead *et al.* (1993) identified the EM1 and EM2 components as containing subducted sediment on the basis of their higher $\delta^{18}O$ compared with that of the MORB source ($\delta^{18}O = 5.7 \pm 0.2$).

Having established that several distinct source materials are tapped by basalts rising from the mantle, isotope geochemists have tried to go further by identifying where in the mantle these sources are located. There are broadly three types of relationship which have been proposed between the depleted MORB source and the enriched OIB source or sources (Morris and Hart 1983):

1. The *plume model*. The MORB source is assumed to lie in the upper mantle and the OIB material rises as plumes from an enriched lower mantle. On the whole, density considerations support the idea that the depleted mantle should overlie the undepleted or enriched mantle. This is because melting residues would have higher Mg/Fe ratios than their protoliths (and hence a lower density), and because undepleted mantle is more likely to be garnet-bearing (garnet is denser than olivine or pyroxene).

2. The *inverse model*. The OIB source overlies the MORB source. Upwelling of the MORB source at the oceanic ridges generates new oceanic lithosphere from the MORB source, and oceanic lithosphere spreads out over the OIB source, enclosing it within the cores of upper mantle convection cells. The inverse model has lost ground with the increasing popularity of the plume hypothesis, but Anderson (1989) has reiterated the points in its favour, for example the shallow depth of origin of the most enriched magmas (kimberlites) and the deep roots of mid-ocean ridges.

3. The *plum-pudding model*. 'Plums' of OIB source material are enclosed in a matrix of MORB source material. Variations of the plum-pudding model call for the less refractory component to be present as veins rather than as 'plums'. The plum-pudding model offers the simplest explanation of the variation in oceanic basalts. It may be assumed that the 'enriched' (less refractory) OIB source will have a lower solidus than the depleted MORB source, so where there is a low degree of partial melting (under small oceanic islands) the OIB component will predominate in the magmas generated. Where there is large-scale melting as at oceanic spreading axes, the OIB component is diluted in a larger volume of MORB material.

The uppermost part of the asthenosphere is the most obvious source of basaltic magma in intraplate regions. It is the shallowest region in which temperatures are high enough for partial melting to occur, and in many areas there is a seismic low-velocity zone at the top of the asthenosphere which may well be due to the presence of a very small proportion of interstitial melt (Ringwood 1975; Yoder 1976; Murase *et al.* 1977; Anderson 1989). However, the top of the asthenosphere is not the only possible source of basaltic magma: the material which is melted could have been brought up by a plume from a greater depth in the mantle, or the lower part of the lithosphere could undergo melting if there is an influx of heat.

The thickness of the lithosphere in any region is determined primarily by the geothermal gradient, its base being where the temperature in the mantle is too high for the mantle material to maintain its rigidity. The thickness of the oceanic lithosphere varies from <50 kilometres at spreading axes to 125 kilometres in the oldest oceanic regions where the greatest cooling has occurred. The sub-continental lithosphere varies in thickness from 50–100 kilometres in tectonically active regions to 300 kilometres under the largest Pre-Cambrian shields, such as those of Africa or eastern Canada. The low-velocity zone is not uniformly developed everywhere. It is well developed under tectonically active regions, but under some shield areas the drop in seismic velocities is slight or even undetectable. Direct evidence suggesting that the source of intra-plate magma is near the base of the lithosphere is the correlation between volcano height and plate thickness in the oceanic regions. According to Vogt (1974), volcano heights increase away from oceanic spreading axes in such a way as to maintain hydrostatic equilibrium between the magma column and the adjacent ocean water and solid lithosphere as shown in Fig. 244.

According to the plume hypothesis, melting in the sub-lithospheric mantle is triggered off by plumes of hot material ascending from the deeper part of the mantle (Fig. 245). The isotopic character of the rising plume is most plausibly associated with those magmas formed at the highest temperatures, i.e. picrites. Campbell and Griffiths (1992) pointed out that if the plume hypothesis for hotspot volcanism is correct, the magmas produced by melting at the head of mantle plumes should be at least partly derived from material brought up from the base of the plume, which they considered to originate at the core–mantle boundary, although not everyone agrees that plumes do originate at the core boundary (Ringwood 1991). Davies and Richards (1992) summarized all the evidence for convection in the mantle and concluded that subducted oceanic lithosphere eventually accumulates near the base of the mantle, from where it can be brought up by plumes.

The isotopic character of the lithosphere under a basaltic volcano may be determined from lithospheric mantle xenoliths. The lithosphere is the coldest part of the mantle, and can only contribute to magma formation if (a) the peridotite solidus is lowered by an influx of volatiles, or (b) its temperature is increased by the arrival of the head of a plume rising from below, or (c) the lithosphere is thinned, encouraging upwelling of the asthenosphere.

Stille *et al.* (1986) interpreted the isotopically variable Hawaiian basalts as

Figure 244. Diagrammatic representation of the relation between volcano heights and lithospheric plate thickness in an oceanic area (after Vogt 1974).

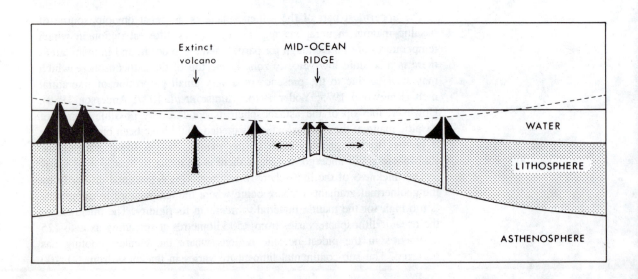

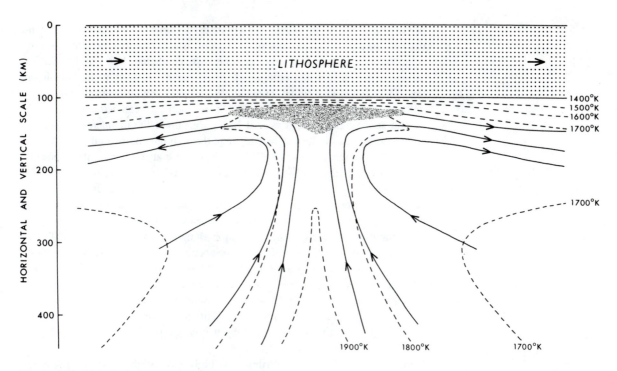

Figure 245. Hypothetical cross-section through a plume rising to the base of rigid 100 kilometre-thick lithosphere. Continuous lines are flow lines; dashed lines are isotherms. The zone of partial melting is shaded (after Parmentier *et al.* 1975).

representing different proportions of lithosphere, asthenosphere and plume sources. Much guesswork is involved in attributing particular isotopic compositions to specific depth regions of the mantle, but the task is not impossible.

CONTINENTAL FLOOD BASALTS

On the whole, continental and oceanic intra-plate volcanism are similar to one another in the scale and type of magma production. Relatively large melting events must have been involved in the production of continental flood basalts, with the corresponding likelihood that the degree of mantle melting was large enough to generate picritic liquids. Cox (1980) favoured fractionation of a picritic parent magma because he believed the Mg/Fe ratios of flood basalts to be too low for them to be primary melts from likely mantle materials. He suggested that the fractionation took place at the crust–mantle boundary, and that the resulting ultramafic cumulates form a layer the top of which is the Mohorovičić discontinuity. Wooden *et al.* (1993) interpreted the Siberian flood basalts to have been derived by fractionation of a parent magma containing more than 20% MgO. In the Tertiary flood basalts of West Greenland picrites are unusually abundant (30–50% of the whole eruptive volume), and although they include some which are the products of olivine accumulation Gill *et al.* (1992) estimated the MgO content of the parental liquids to have ranged up to 19%.

Although flood basalts give the impression that large quantities of very uniform basalt are erupted, the continental flood basalts are actually much more variable than was once thought. In some provinces alkali basalts predominate while in others tholeiites predominate. Geochemically, continental intra-plate basalts resemble oceanic island basalts more than ocean-ridge basalts, being relatively 'enriched' in character. Very high $^{87}Sr/^{86}Sr$ ratios are found in some continental flood basalts, leading to the suspicion that they have undergone considerable crustal contamination, but most workers believe that the high $^{87}Sr/^{86}Sr$ is mainly a characteristic of their mantle source. Even so, it is still likely that flood basalt magmas undergo some contamination during their passage through the continental crust. Fodor *et al.* (1985) and Petrini *et al.* (1987) suggested that the Parana flood basalts in S. America with $^{87}Sr/^{86}Sr$ in the range 0.706–0.711 have been substantially contaminated by crustal material. Thompson *et al.* (1982) described Tertiary basalts from Scotland in which a simple contamination model involving the underlying Lewisian (Pre-Cambrian) gneisses would require about 5–10% contamination to account for the major element compositions, 15–20% to account for the trace elements, 50% to account for the Sr and Nd isotopes, and 90% to account for the Pb isotopes. To some extent this must reflect the fact that crustal rocks are much richer in Pb (10–20 ppm) than basalts (~5 ppm), so that a small amount of contaminant has a disproportionate effect on the isotopic composition, but it is very difficult to assess contamination in a wholly quantitative way when it takes place at depth and the contaminants are invisible.

Many workers prefer to attribute the high and variable $^{87}Sr/^{86}Sr$ ratios of continental flood basalts to the presence of an enriched component in the mantle source (Brooks and Hart 1978; Kyle 1980; Anderson 1982; De Paolo 1983). This could possibly be ancient subducted sediment. Such an explanation has been applied to the Jurassic dolerites of Tasmania, which are considered to be intrusive counterparts of flood basalt lavas. The Tasmanian dolerites have very high initial $^{87}Sr/^{86}Sr$ ratios of 0.709–0.713 and have other isotopic and trace element characteristics more appropriate to a crustal than a mantle source, but their bulk compositions cannot easily be reconciled with contamination by the lower crust. Hergt *et al.* (1989) argued for introduction of a small amount of sediment (<3%) into a mainly depleted mantle source by subduction. If the bulk of the mantle source material was depleted, and therefore low in incompatible elements (such as Sr and Pb), then the isotopic characteristics of the magmas would be largely governed by the composition of the sedimentary component regardless of its low abundance.

BASALTS OF ISLAND ARCS AND OROGENIC CONTINENTAL MARGINS

In the island arcs and orogenic continental margins basalts are abundant, and are so intimately associated with andesites that a common origin is often assumed. Nevertheless there are few, if any, andesites associated with the basalts outside these areas, and the origin of the two magma types must be considered separately. Even so, many volcanoes erupt a continuous range of magmas from basalt to andesite and it is difficult to draw a line between the two.

There are two features which emphasize the close connection between the occurrence of orogenic basalts and the presence of a subduction zone. One is the concentration of volcanic activity at a constant distance above the inclined seismic zone, normally between 100 and 200 kilometres. The other is the relationship between the type of basalt and the depth of the subduction zone. Tholeiitic basalts occur where the subduction zone is shallowest, and above deeper levels of the subduction zone they are succeeded by high-alkali or high-alumina basalts, followed by alkali basalts where the seismic zone is deepest. These features are well illustrated by the Quaternary basalts of Japan (Fig. 246), but have been recognized in many other areas. In addition to the general increase in alkalinity of the basalts with increasing depth to the subduction zone, there is in particular a correlation between potassium content and depth (sometimes referred to in abbreviation as the K–h relationship). A similar relationship exists for Sr and perhaps for other trace elements. On a worldwide basis, these correlations are fairly well established (Fig. 247), although in any particular area they may not be apparent.

There are obviously two main possibilities for the generation of basaltic magma in subduction zones. One is by melting of the subducted oceanic crust itself, which is mainly basaltic and obviously has a suitable composition. The other is by melting of adjacent mantle, which is the presumed

Figure 246. The distribution of tholeiitic, high-alumina, and alkaline olivine basalts in the Quaternary volcanoes of Japan, in relation to the depth of the deep seismic zone (after Aramaki and Ui 1982).

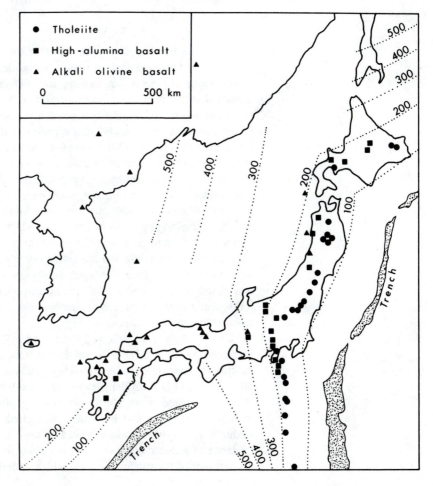

Figure 247. The relationship between the K$_2$O content of volcanic arc basalts (at SiO$_2$ ~50%) and the depth (h) of the inclined seismic zone (after Dickinson 1970).

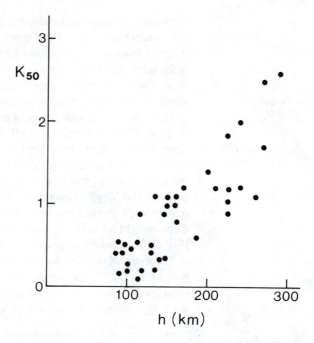

source of the basaltic magma in all other tectonic environments. Different authors have favoured each of these possibilities.

The main argument for melting in subducted oceanic crust is that it already has an appropriate composition to yield basaltic melt; the main argument against is that the subducted lithosphere may still be too cold for melting at the time it reaches the appropriate depth (i.e. beneath island arc volcanoes). Marsh (1982b) compared the composition of the commonest type of lava in the Aleutian Islands, a low-Mg, high-Al basalt (~50% SiO$_2$), with the melts produced by high-pressure partial melting of basalt (eclogite) and peridotite. He found that he could match the Aleutian basalt fairly well with the product of 60% melting of eclogite, but not with any of the anhydrous melting products of peridotite, which he considered to be too Mg-rich and Al-poor to be a possible source material (see Table 40). Similarly, Johnston (1986) investigated the anhydrous high-pressure phase relations of an island arc basalt from the South Sandwich Islands and found that olivine was absent at the liquidus at all pressures up to 30 kilobars, indicating that the source material was not olivine-bearing, although Bartels *et al.* (1991) reached the opposite conclusion.

Many workers have followed Ringwood (1974) in favouring a source in peridotite mantle overlying the subducted slab, and it is usually assumed that melting of the mantle wedge overlying a subduction zone would be triggered off by rising H$_2$O from dehydration of the subducted oceanic crust. Kushiro (1987) estimated from the geothermal gradients beneath Japan that a lherzolitic mantle would be about 2% melted in the centre of the mantle wedge overlying the subducted oceanic crust (Fig. 248), and pointed to the evidence of reduced seismic wave velocities in this region. Above the very deepest part of the subduction zone which underlies the Sea of Japan there is

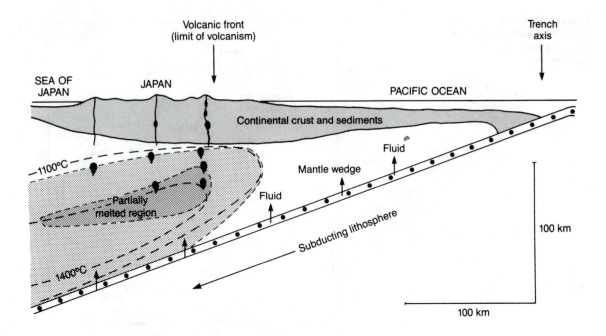

Figure 248. Schematic cross-section across the Japanese arc, showing the possible extent of partial melting in the mantle wedge above the subduction zone (after Kushiro 1987).

a region of low seismic velocity and very low Q (high attenuation) in the asthenosphere, which Suyehiro and Sacks (1983) suggested is a region of partial melting, and this could be the source of the alkali basalts which occur on this side of the volcanic arc.

Whatever the source region, water is certainly present because subduction-related basalt magmas are relatively water-rich compared with the basalts of other tectonic settings. This is shown by the H_2O contents of basaltic glass where it is trapped in phenocrysts (Sisson and Layne 1993), and by the hydrous nature of the ultramafic cumulates associated with these magmas (see Chapter 10).

Island arc basalts are more similar to oceanic island basalts (OIB) than they are to ocean ridge basalts (MORB). Compared to MORB, they are enriched in the incompatible elements and have higher $^{87}Sr/^{86}Sr$ and lower $^{143}Nd/^{144}Nd$. Compared to OIB, they have a similar range of isotopic compositions but some differences in trace element composition, e.g. low contents of 'high field strength' elements such as Ti, Nb and Ta (Fig. 249) and high boron. The abundance of high field strength elements is probably controlled by the stability of accessory minerals such as rutile and sphene during the melting process. The abundance of boron may be an indication of a sedimentary contribution to the magma, because boron is much more abundant in seawater and marine sediments than in any kind of igneous rock.

There has been much debate about whether any sedimentary material is present in the source region of arc basalts. In the island arcs it would have to come from subducted sediment in the oceanic crust; in continental margins it could be from contamination by continental crust. Barreiro (1983) showed that a contribution of a few per cent of subducted terrigenous sediment is required to account for the Pb isotopic composition of the basalts in the

Figure 249. Plot of V against Ti for the basalts of different tectonic environments (after Shervais 1982).

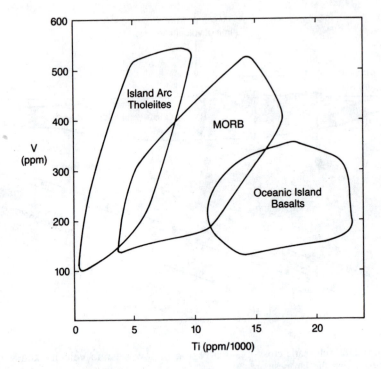

South Sandwich arc. Slight enrichment in ^{18}O and ^{34}S also indicates the presence of a small sedimentary component in the basalts of the Mariana Island arc (Woodhead *et al.* 1987). Lead isotopes are more sensitive to the presence of sedimentary material than Sr or Nd isotopes because the Pb content of oceanic sediments is much higher than that of basalts, and only a small sedimentary contribution is required to influence the isotopic ratios of the magmas. Some authors believe that the isotopic and trace element characteristics of subduction-related basalts require contributions from all three sources, i.e. subducted oceanic crust, mantle wedge, and subducted sediment (Ellam and Hawkesworth 1988).

For a further discussion of the origin of subduction-related magmas see Chapter 10.

KOMATIITES

In general, ancient volcanic rocks can be recognized as belonging to petrographic types similar to those being erupted at the present day. There is, however, one group of lavas which is almost restricted to the Pre-Cambrian and is not matched by present-day examples – the komatiites.

Komatiites were first described from the Barberton Mountains in South Africa, where a volcanic sequence 8000 metres thick contains ultramafic as well as mafic and felsic lavas. The ultramafic lavas (komatiites) are associated with normal tholeiites, and range from rocks not very different from ordinary basalts ('basaltic komatiites') to extremely ultramafic types with more than 20% MgO ('peridotitic komatiites'). An analysis of one of these rocks is given in Table 39. The high MgO and low Al_2O_3 contents reflect the extremely high and low contents of olivine and feldspar respectively. To avoid confusion, Arndt and Nisbet (1982) have suggested that in future komatiite should be defined as an ultramafic lava with >18% MgO, and that less magnesian lavas should be called komatiitic basalts.

Unlike picritic basalts, which contain well-formed olivine crystals and are the products of crystal accumulation, komatiites show a very distinctive 'spinifex' texture (elongated, skeletal, branching crystals) which is indicative of rapid quenching of liquid.

Lavas from a number of areas have subsequently been described as basaltic komatiites or komatiitic basalts, but true komatiites are virtually restricted to Archaean greenstone belts such as those of South Africa, western Australia and eastern Canada. Some of these are among the oldest parts of the Earth's crust still preserved; the Barberton lavas are intruded by granites dated at 3200–3400 million years old. The basaltic komatiites from Gorgona Island, Colombia, with 15–20% MgO are among the youngest ultramafic lavas that have been recorded, and are probably of Cretaceous age (Walker *et al.* 1991). The early Tertiary picritic basalts of West Greenland and Baffin Island may also have crystallized from highly magnesian liquids with up to 19% MgO (Gill *et al.* 1992).

Plutonic counterparts of the komatiite lavas are difficult to identify. Most ultramafic plutonic rocks are crystal cumulates or melting residues, but there are a few that show definite evidence of intrusion as ultramafic magmas, for example the Nordre Bumandsfjord pluton in Norway (Sturt *et al.* 1980). This peridotite intrusion was emplaced by stoping, and evidence for the existence of ultramafic magma includes dilational ultramafic dykes varying in width from millimetres to metres, intrusion breccias with rotated inclusions, and penetration of dunite along cracks in gabbroic xenoliths. High magma temperatures are shown by a contact metamorphic aureole up to 3 kilometres wide with granulite facies mineral assemblages, and partial melting of gabbroic xenoliths and wall rocks. The original magma is considered to be dunitic (~40% MgO), but it was extensively modified by assimilation of gabbroic and sedimentary material. Various intrusive features suggest that the magma was of unusually low viscosity, which is consistent with the low viscosity of ultrabasic melts in the laboratory. The Nordre Bumandsfjord pluton is approximately Cambrian in age, but it is interesting to note that some of the very largest Pre-Cambrian basic intrusions, such as the Bushveld and Great Dyke intrusions (2095 and 2514 million years old respectively), are also believed to have formed from parent magmas more magnesian than basalt (Hamilton 1977).

Since a predominantly peridotitic mantle is now assumed by most petrologists, all that is required to produce a komatiitic melt is a high degree of melting. This is supported by numerous experimental studies in which melts of various basaltic compositions have been produced at low degrees of partial melting with the melt composition becoming more 'picritic' (i.e. komatiitic) as the temperature was raised further above the solidus. A high degree of melting requires a correspondingly high temperature to be reached.

Although Mysen and Boettcher (1975) suggested that komatiite magma might be formed at temperatures as low as 1300 °C by melting under hydrous conditions, Green *et al.* (1975) argued that the magmas were anhydrous, and estimated the extrusion temperature of one South African komatiite as approximately 1650 °C. This requires the magma to have formed either at a very great depth, or under a higher geothermal gradient than those of the present day, or both.

A temperature of 1650 °C could hardly be attained in the upper mantle without almost the whole of the upper mantle becoming molten. This could well have been the case during the early Pre-Cambrian. One of the most intriguing possibilities is that the outer layers of the Earth underwent extensive melting due to major impacts similar to those that produced the lunar maria (Green 1972). It is now believed that the whole outer part of the Moon once melted to form a magma ocean, and there is no reason to suppose that such a process would not have happened on Earth. Nisbet and Walker (1982) suggested that komatiite might have been erupted from the remains of such a primeval worldwide magma ocean lingering beneath an overlying layer of lighter dunite and harzburgite. On the other hand, McKenzie (1984) has argued that high-temperature komatiite magmas need not have been produced by a high degree of melting if the source region were at a sufficiently great depth, perhaps 150 kilometres, and Edwards (1992) has suggested that basic magma formed at a moderate temperature could be enriched in MgO by the breakdown of magnesite in the mantle.

Campbell *et al.* (1989) interpreted komatiites in terms of the mantle plume hypothesis. They pointed out that Archaean komatiites are associated with normal basalts, and suggested that if basalts were formed by melting at the head of a plume where it mixes with cooler upper mantle material, komatiite might be produced by melting in the deeper and hotter axial region of the plume.

Whatever the exact temperature of magma formation, the virtual restriction of the most magnesian komatiites to the early Pre-Cambrian implies that the thermal regime necessary for a high degree of partial melting may no longer exist. In any case, the density of komatiitic liquids is so high that they could not easily rise through continental crust even if they did still exist, although they might be able to rise as far as the crust–mantle boundary.

CHAPTER 9

Granites

THE NATURE OF ACID MAGMAS

Rhyolites erupted at the Earth's surface bear little resemblance to granites intruded in deep metamorphic terrains. There are major differences in the viscosity, the water content, and the burden of suspended crystals of acid magmas encountered at different levels in the crust. There are also differences in intrusive style, from the permissively emplaced sub-volcanic ring complexes of high crustal levels to plutons forcefully emplaced into deeper country rocks which were undergoing deformation at the time.

Rhyolites provide the best glimpse of actual magma. Most rhyolites are erupted as ash flows, frothing expansively as they exsolve water; gas-free rhyolites are rarer, and are so viscous they can hardly flow at all. However, in both cases we can see that the magma is a liquid, with usually only a small proportion of suspended crystals and sometimes none at all. In contrast, there is every indication that granites intruded lower down in the crust are emplaced as a mixture of crystals and liquid. The clusters and streaks of mafic minerals that occur in many granites reveal the extent of crystallization as the magma flowed to rest.

On the face of it, this is a paradox, since you would expect magma to cool down as it rises, and therefore to become more crystalline. The crystals which are present at depth must either settle out, or be dissolved in the magma as it rises towards the surface. There is little evidence for settling out, but fortunately the phase relations show that dissolution is possible if the rising magma contains water.

THE WAYS IN WHICH GRANITE MAGMA CAN BE FORMED

Much evidence on the conditions under which granitic magmas can be produced has come from experimental phase equilibrium studies on the synthetic system albite–orthoclase–anorthite–silica–water (Tuttle and Bowen 1958; Luth *et al.* 1964; James and Hamilton 1969; Luth 1969; Steiner *et al.* 1975; Johannes 1984; Ebadi and Johannes 1991), and from melting experiments on natural rocks (Green and Ringwood 1968; Winkler 1974; Vielzeuf and Holloway 1988; Holtz and Johannes; 1991; Patiño Douce and Johnston 1991).

Figure 250 shows the phase relations in the system $NaAlSi_3O_8$–$KAlSi_3O_8$–

Figure 250. The system quartz–orthoclase–albite–water at $P_{H_2O} = 1$ kilobar and 5 kilobars (after Tuttle and Bowen 1958; Luth *et al.* 1964).

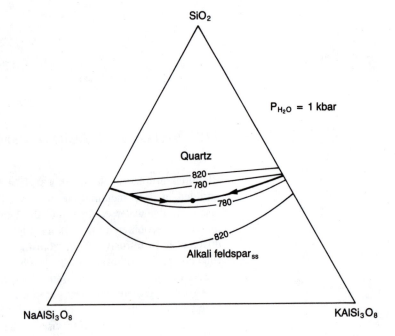

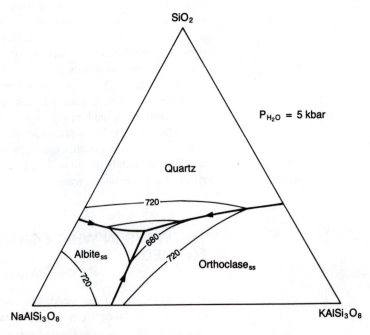

SiO₂–H₂O in the presence of excess water at high and low pressures. At 1 kilobar, the minimum point on the liquidus surface is at a composition of approximately Q:Or:Ab = 37:29:34 and a temperature of approximately 720 °C. At high water pressures the liquidus surface is lowered, and the minimum point moves gradually towards a more Ab-rich composition. As the liquidus surface is lowered, the minimum falls below the solvus of the system albite–orthoclase (Fig. 146), so that K-feldspar and Na-feldspar coexist with the melt as separate phases and not as a single $(Na,K)AlSi_3O_8$ solid solution. Thus at 5 kilobars, the diagram has the appearance of a ternary eutectic system, and the minimum resembles a ternary eutectic point. Of course, this is really a pseudoternary system because there is a fourth component, H_2O, which is not directly represented on the triangular diagram.

Figure 251 shows the range of normative compositions of granitic rocks. Granites do not contain the constituents quartz, orthoclase and albite in just any proportion, but are restricted to the vicinity of the pseudoternary minimum at moderate water pressures. The same is true of rhyolites. This relationship clearly indicates that the compositions of granites are determined by crystal–melt equilibria, and there is therefore no reason to doubt that most granitic rocks have crystallized from a melt.

Between about 1920 and 1960 there were many petrologists who were willing to entertain the idea that some granites might originate by a metasomatic process, described as granitization. It was postulated that sedimentary or other non-granitic rocks could, by the addition or removal of

Figure 251. The distribution of normative quartz, albite and orthoclase in plutonic igneous rocks with more than 80% Q + Or + Ab, based on a sample of 571 analyses. Contours and shading indicate the maximum concentration of compositions (after Tuttle and Bowen 1958).

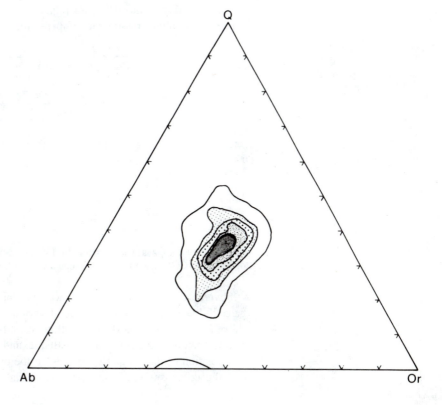

the necessary extra constituents, be transformed into granite without going through a melt stage. Various mechanisms were suggested, and field evidence adduced to support the proposition. Since the publication of Tuttle and Bowen's comprehensive experimental study of the relevant phase relations, the granitization hypothesis has been generally abandoned.

The correspondence between the compositions of granites and rhyolites and the minimum in the system Q–Or–Ab–H_2O means that granite magmas must be the products of either partial melting or crystallization differentiation. The main possibilities involve the partial melting of either sedimentary rocks and their metamorphic equivalents, or of igneous rocks of intermediate to basic composition, or the differentiation of basaltic or andesitic magma. No other source materials or parental magmas have suitable compositions or exist in sufficient quantities to produce granite magma in the required amounts.

MELTING OF BASALT OR ANDESITE

Given appropriate conditions, a melt of broadly granitic composition can be produced by the partial melting of a wide variety of rock types, both igneous and sedimentary. Table 42 gives the average compositions of some possible starting materials. Obviously, the greater the proportion of granitic constituents in the source rock, the higher the proportion of granitic melt that could be derived. Each of the starting materials has limitations as a potential source. Basic igneous rocks are generally too poor in K_2O to yield a high proportion of granitic melt, and require the addition of H_2O if they are to undergo melting at reasonable temperatures.

Table 42. Average compositions of possible source rocks for granitic magma

	Possible source materials				Granitic compositions	
	Shale	Greywacke	Andesite	Basalt	Granite	Granodiorite
SiO_2	62.8	67.8	57.9	49.2	71.3	66.1
TiO_2	1.0	0.9	0.9	1.8	0.3	0.5
Al_2O_3	18.9	14.3	17.0	15.7	14.3	15.7
$FeO + Fe_2O_3$	6.5	5.9	7.3	10.9	2.6	4.1
MgO	2.2	2.5	3.3	6.7	0.7	1.7
CaO	1.3	3.0	6.8	9.5	1.8	3.8
Na_2O	1.2	3.3	3.5	2.9	3.7	3.8
K_2O	3.7	1.7	1.6	1.1	4.1	2.7

Sedimentary compositions are from Taylor and McLennan (1985), recalculated free of H_2O and CO_2. Igneous compositions are from Le Maitre (1976).

Several experimenters have shown that acid liquids can be produced by partially melting basalt or andesite at various combinations of P_{total} and P_{H_2O} (Green and Ringwood 1968; Holloway and Burnham 1972; Helz 1976; Spulber and Rutherford; 1983, Beard and Lofgren 1991). The exact composition of these liquids depends on the pressure, the degree of melting, and the composition of the starting material, but in general they are only rhyolitic if the degree of melting is low, because of the lack of K_2O in the basic parent material. They become dacitic or trondhjemitic if generated in any quantity. The average K_2O content of tholeiitic basalt is 0.43%, compared with 2.73% in granodiorite and 4.07% in granite. Obviously a large amount of basalt, with very effective concentration of potassium into the melt fraction, is needed to produce a small amount of granitic liquid. The average andesite contains 1.62% K_2O and is a richer potential source than basalt. Many pyroclastic rocks also have suitable compositions to yield granite by partial melting.

Melting under almost dry conditions is likely to be rare because of the very high temperature required (~1000 °C), although some igneous charnockites may have had this origin (Kilpatrick and Ellis 1992). Temperatures in the crust are generally too low for melting under anhydrous conditions, so we must assume that if basic rocks are to be a source of granitic magma, either they must be in the form of amphibolites or the episode of melting must be triggered by a large influx of water. Wyllie and Wolf (1993) showed on the basis of experimental studies that if water were supplied by the dehydration of hornblende, significant melting of amphibolite in the lower crust would only take place at temperatures of 900–1000 °C, but if sufficent water were introduced for H_2O-saturated melting, then melting could take place at much lower temperatures (700–800 °C).

Melting of metabasic rocks is widely favoured as an origin for tonalites and trondhjemites, which have low K_2O/Na_2O ratios, and the relative abundance of these rocks in Archaean terrains may be attributable to the high proportion of basaltic to sedimentary rocks in the Earth's early crust. It is less certain that normal granites form in this way, but some authors believe that the whole range of igneous compositions from diorite to granite can be produced by partial melting of mafic lower crust (Tepper *et al.* 1993).

As well as being common in the continental crust, basic igneous rocks are also the main constituents of subducted oceanic crust. At first sight, melting of the thin mafic oceanic crust does not look such a promising source as melting of the thick, relatively felsic continental crust. On the other hand new oceanic crust is continually being created, and although any particular segment of oceanic crust may only be capable of generating a small amount of granitic magma, subduction of oceanic crust in any one region may continue for a very long period, perhaps as long as 100 million years. If a sufficiently large area of oceanic crust were fed into a subduction zone over a period of 100 million years, enough acid magma might be formed to produce a body of batholithic dimensions, although the accompanying volumes of more intermediate and basic melts would also be very large. It is probable that subducted oceanic crust might have a rather higher K_2O content than the average tholeiitic basalt, because the widespread alteration of submarine crust in hydrothermal convection systems results in the fixation of additional K_2O from sea water. Hart (1970) showed that as ocean-floor basalts are carried away from oceanic spreading axes their average K_2O increases from less than 0.2% to more than 0.8% within 20 million years. It is not known how far down in the oceanic crust from the sea floor this potassium metasomatism extends.

Stern and Wyllie (1981) investigated the phase relations of a granite at

Figure 252. Phase relations of a biotite–granite at 0 to 35 kilobars in the presence of excess water (after Stern and Wyllie 1981). Bi = biotite, Ct = coesite, Jd = jadeite, Ky = kyanite, Or = orthoclase, Pl = plagioclase, Qz = quartz, L = liquid, V = vapour.

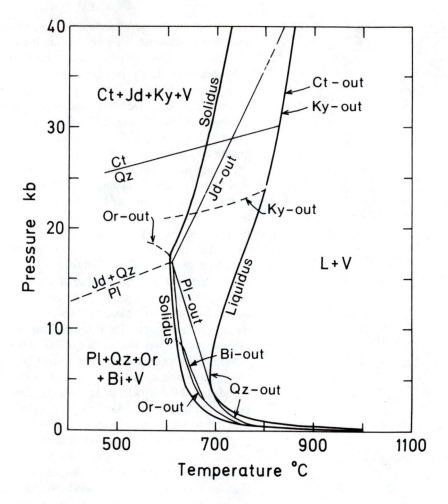

pressures up to 35 kilobars in the presence of excess water (Fig. 252). These phase relations cast doubt on the possibility of producing granite as a primary magma by the melting of subducted oceanic crust. Granite does not have such a near-eutectic composition at higher pressures as it does below 5 kilobars. There is a wide temperature interval of crystallization with a large quartz or coesite field below the liquidus, so that minimum-melting compositions at mantle pressures would be much less siliceous than granite. Furthermore, for a primary magma to be in equilibrium with the residual minerals of its parent rock, its near-liquidus minerals should be major minerals of the parent rock. It is rather unlikely that quartz or coesite would be the liquidus phase of a partial melt fraction of a basaltic parent. These conclusions apply equally to granite magmas undersaturated with water (Huang and Wyllie 1981).

MELTING OF SEDIMENTS AND METASEDIMENTS

Winkler (1974) and co-workers showed experimentally that a melt of granitic composition is produced by the partial melting of some of the most abundant sedimentary rock types such as shale and greywacke. They examined the metamorphic changes brought about when various sedimentary materials were heated in the presence of excess water at 2–4 kilobars. After passing through the various mineral assemblages of progressive regional metamorphism, the rocks started to melt at around 650–700 °C. The melts produced were broadly granitic in composition, containing the constituents of quartz, alkali feldspar and sodic plagioclase. A substantial amount of unmelted residuum was left, made up of biotite, cordierite, sillimanite, calcic plagioclase and excess quartz.

In practice, it is improbable that crustal melting actually does take place under water-saturated conditions. At lower crustal pressures granite magma can contain up to 10% of dissolved H_2O, and the average metasediment only contains 1–2% H_2O, so assuming that the sediment underwent 50% melting to leave a dehydrated residuum, there would not even be enough water available to half-saturate the melt. Under these circumstances, much higher temperatures are needed to bring about the partial melting of all the potentially low-melting constituents in a metasediment. It is most likely that the water needed for melting would come from the dehydration of hydrous minerals, which in practice means muscovite, biotite and hornblende. Figure 253 shows the temperatures at which the breakdown of various hydrous minerals could bring about dehydration melting. Muscovite breaks down at the lowest temperature, followed by biotite at a higher temperature, and hornblende at the highest temperature. In the case of the latter two minerals the actual breakdown temperature depends somewhat on their Fe/Mg ratios and on the oxidation conditions.

Formation of an H_2O-saturated magma by partial melting requires a low degree of partial melting so that all the available water is concentrated in a small amount of melt. Water-saturated granite magmas are probably not common. Any excess of H_2O in the magma over that required for the crystallization of hydrous minerals such as biotite and muscovite would be exsolved during eventual crystallization of the magma and would be revealed by high-temperature hydrothermal alteration and veining. These effects are by no means uncommon, but are extensively displayed by only a minority of granitic intrusions. Furthermore, an H_2O-saturated magma originating at a high pressure appropriate to the lower crust would not be able to rise into the upper crust because it would freeze on ascent due to the negative slope of the water-saturated melting curve (Fig. 253). In practice, one must envisage either a rather small amount of melt being formed at the water-saturated solidus with melting limited by the availability of water, or a larger amount of melting taking place only when temperatures reached much higher levels, say 800–1000 °C.

Despite the limitations imposed by the limited availability of water, the temperatures needed for at least some partial melting are certainly attainable within the crust. A crust of average thickness (30 kilometres) in an area of average geothermal gradient (30 °/kilometre for continents) would have a temperature of about 900 °C at the base. Many high-grade metamorphic rocks provide direct mineralogical evidence that such temperatures have been reached (Tarney and Windley 1977).

Figure 253. The effect of pressure on the solidus temperature of granite, either with excess water (dashed line) or with all the water for melting supplied by the breakdown of muscovite (solid line 1) or biotite (solid line 2). The shaded area indicates the conditions under which granite magma can form if all the water for melting is supplied by the breakdown of muscovite. The breakdown temperature for biotite can vary over about 100 °C depending on its composition. After Hyndman (1981).

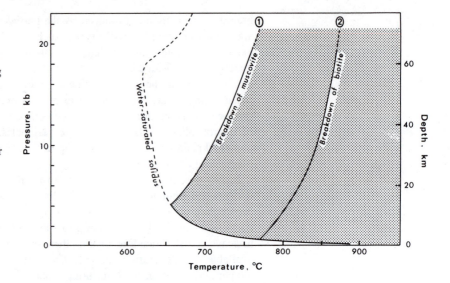

Since the lower crust does not normally contain a significant amount of melt fraction (judging from its seismic properties), we may assume that the lower crustal composition is not normally undergoing partial melting, either because of its bulk composition or because it contains insufficient water. However, it must be sufficiently near to melting that a small increase in temperature would start the process. The two most obvious circumstances that would initiate melting are: (1) changes in crustal thickness and geothermal gradient associated with orogenesis; and (2) the injection of a large amount of basic magma into or through the crust from below. Examples of the role of both heat sources will be given later in this chapter.

A striking feature of partial melting experiments on natural rocks is that a very restricted range of granitic melt compositions (close to the liquidus minimum in the system Q–Or–Ab–H_2O) is produced by the partial melting of a wide variety of sediments (from shale to greywacke). However, there is some relationship in detail between the compositions of the melts and the compositions of their sedimentary precursors. Pelitic sediments tend to yield granitic melts, and greywackes tend to yield granodioritic melts (Conrad *et al.* 1988; Patiño Douce and Johnston 1991). The composition of the source material becomes particularly important once the degree of partial melting increases beyond a low proportion. Pelitic schists are deficient in Na_2O compared with granite, and greywackes are deficient in K_2O (Table 42), so that as soon as all of these constituents have been incorporated into the melt any further melting must cause the melt composition to deviate from the minimum region of the system Q–Or–Ab–H_2O. A high degree of partial melting of pelites will give a relatively K_2O-rich melt, and of greywackes a relatively Na_2O-rich melt. In practice, any region of the crust which underwent partial melting would probably contain a wide range of rock types, some of which such as limestone or quartzite would be relatively refractory.

EVIDENCE OF CRUSTAL MELTING

Migmatites

If melting takes place in the lower crust, it should not be difficult to find localities in which lower crustal rocks are exposed that have undergone this process. This is the interpretation now generally placed upon migmatite terrains.

The term *migmatite* means 'mixed-rock'. It was introduced by Sederholm (1907), and describes a mixture of granite and metamorphic rocks so intimate that they cannot be separated on a geological map, although individual small hand specimens can be given appropriate igneous or metamorphic rock names. The granitic component typically forms veins, lenses or sheets in the associated metamorphic rock and is often conformable with the banding or foliation of the latter. Other terms which are widely used in describing partial melting are *anatexis*, which is synonymous with partial melting; and *leucosome* and *melanosome*, which are respectively the paler and darker components in a migmatite. Additional terms to describe different types of migmatite and their structures are given by Mehnert (1968). The word *protolith* refers to the original source rock from which a magma formed by partial melting.

A typical migmatite complex is that of the *Loch Coire* district in Scotland (Fig. 254), which was described by Read (1931). His mapping shows a three-fold zonation: (1) an outer zone of veins, in which many veins and sheets of granite cut the mainly psammitic country rocks, but in which the total amount of granite is much less than the amount of non-granitic material; (2) a zone of migmatites in which granites and semi-pelitic gneisses are roughly equal in amount; and (3) a zone of migmatites in which granite and pegmatite sheets are dominant over the pelitic country rocks; in this zone

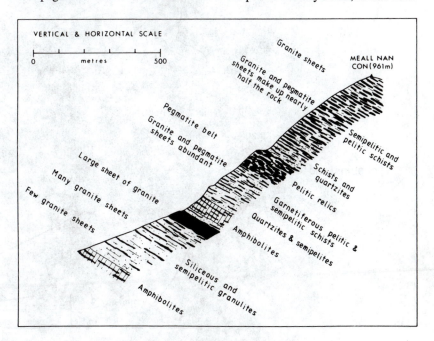

Figure 254. The Ben Klibreck section in the Loch Coire migmatite complex, Scotland (after Read 1931). Granite is shown in black.

there are several irregularly shaped bodies of homogeneous granite, the largest of which is several square kilometres in area.

There are three ways in which complexes of this type have been interpreted: (1) as the result of injection of granite magma into the metamorphic host rocks; (2) as the result of conversion of part of the metamorphic material to granite in place; or (3) as the result of partial melting of the metamorphic rocks, with the granitic veins representing the melted portion. Since the melting experiments of Winkler (1974) and his collaborators, partial melting of the metasediments has been thought the most likely origin for the granitic veins.

A more thorough-going melting history may be postulated for the migmatite complex in the Black Forest shown in Fig. 255. Within a distance of several kilometres, paragneisses and orthogneisses pass by stages into a homogeneous rock with the mineralogy and appearance of granite or granodiorite. The first stage of the transformation was described by Mehnert (1968) as 'metatexis'. The metatexites appear to have undergone incipient partial melting, with a petrographically distinguishable granitic portion and a mafic residuum ('restite'). This is followed by the more advanced stage of 'diatexis', in which the structure of the migmatite becomes streaky or nebulous and then finally homogeneous.

Figure 255. The distribution of migmatites in the area south of Kirchzarten, south-west Germany (after Mehnert 1968).

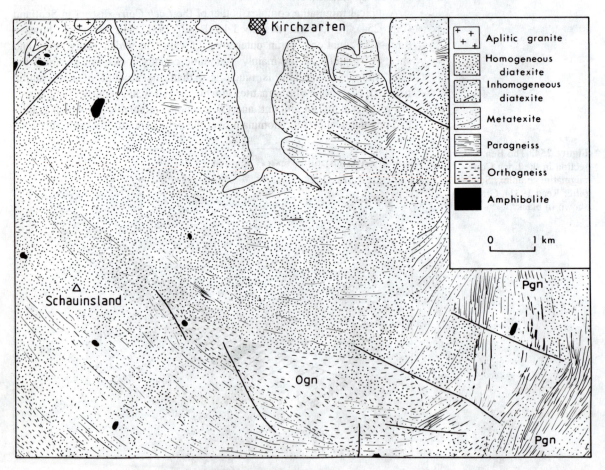

The conversion of metamorphic rock to magma and residuum can obviously take many forms, depending on the heterogeneity of the metamorphic complex, the degree of partial melting, and the coherence of the solid–melt mixture. One may imagine such varied possibilities as:

1. A homogeneous metasediment in which the mafic residuum becomes dispersed in the melt as the degree of melting increases and the melt begins to flow.
2. A heterogeneous metasedimentary succession in which relatively unmelted bands become broken up and assume the appearance of xenoliths as the less refractory layers melt around them.
3. A metasediment, homogeneous or heterogeneous, from which the melt is squeezed rather effectively, leaving most of the melting residues behind.

The separation of melt fraction from residuum, and its coalescence into large bodies of granitic magma, is controlled mainly by the buoyancy of the lighter liquid portion. A cross-section through the crust in a region where anatexis has taken place might be expected to show migmatites at deep crustal levels, passing gradationally into intrusive granite bodies and unmelted country rocks at higher levels. Large-scale geological mapping of several orogenic areas, for example by Hutchison (1970) in British Columbia, has revealed such a sequence of relationships between migmatites and intrusive granites.

Figure 256. The Cooma granite, New South Wales (after Joplin 1942, 1943). The granite and its metamorphic country rocks are of Palaeozoic age.

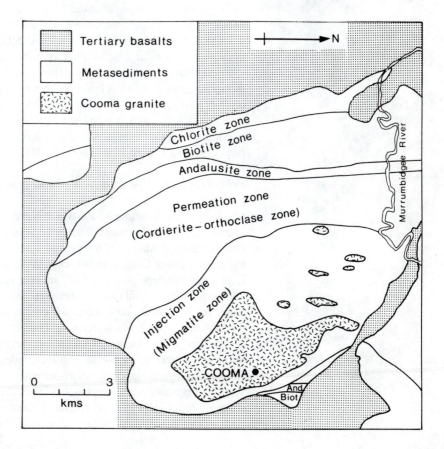

In general, melts produced by a low degree of partial melting will remain close to the source, and will tend to be found as small leucocratic bodies within migmatite complexes (Wickham 1987; Bea 1991). Figure 256 shows the example of the Cooma granite in Australia which White *et al.* (1974) believe has moved very little from its source. The Cooma granite shows sharp contacts and apophyses into the country rocks, but is crowded with xenoliths which are petrographically and isotopically indistinguishable from the adjacent country rocks.

On a larger scale, Chacko *et al.* (1994) described a granite body from northern Alberta which contains abundant metre-sized inclusions of pelitic gneiss which they interpreted as portions of the melt-depleted source material. These inclusions were rich in spinel, cordierite and garnet, and had a migmatitic texture and highly refractory bulk rock compositions. From their mineral compositions it was calculated that they crystallized at 5–7 kilobars and 830–940 °C, suggesting that the host granite may not have moved far from its source.

Various studies of migmatite complexes have been made with a view to checking whether the granite-like component of the migmatite really was a partially melted portion. The results do not always confirm this expectation. There are some migmatitic leucosomes (leucocratic granite-like portions) which do not have compositions near the minimum of the system Q–Or–Ab–H_2O. This could mean that they formed by metasomatism (Brown 1971) or by metamorphic differentiation (Amit and Eyal 1976) rather than by anatexis, but such compositions could also be magmatic if the source rock did not contain enough Na_2O or K_2O or SiO_2 for the liquid to stay at the liquidus minimum at a high degree of partial melting.

In addition to the many migmatites derived from the partial melting of metasediments, there are a few examples of migmatites apparently representing the partial melting of metabasaltic rocks. In these mafic migmatites the leucocratic veins and patches which correspond to the melt fraction are trondhjemitic in composition, i.e. they have high SiO_2 but a very low K/Na ratio (Tait and Harley 1988, Sawyer 1991).

Presence of restitic components in granites

Many granites contain zircons which yield isotopic ages greater than the known age of emplacement. The early Tertiary Pan Tak granite in Arizona contains zircons whose U:Pb ratios are discordant, and plot on a line extending from a lower Concordia intercept of 58 million years towards an upper Concordia intercept of about 1100 million years. Wright and Haxel (1982) interpreted this as a mixing line between Tertiary magmatic zircon and non-magmatic Pre-Cambrian zircon, and considered the alternative possibilities that the older zircon was xenocrystic (from contamination by country rocks) or inherited (from the magmatic source region). They favoured the latter, because the granite shows no outcrop evidence of contamination and the Jurassic country rocks contain zircons which give concordant Jurassic ages. Thus the magmatic source region was inferred to lie in Pre-Cambrian crust. In contrast, Johnson (1989) interpreted Proterozoic cores in the zircons of a Tertiary tuff as being due to high-level assimilation.

Figure 257. Sketch of a zoned zircon crystal from a Cretaceous granite in California, showing the ages of the core and rim determined by ion microprobe (after Miller *et al.* 1992).

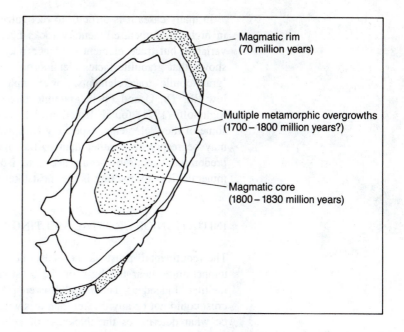

Magmatic rim
(70 million years)

Multiple metamorphic overgrowths
(1700 – 1800 million years?)

Magmatic core
(1800 – 1830 million years)

Many other examples of inherited zircon, monazite and xenotime have been discovered in granites. Recently, the use of a sensitive, high mass-resolution ion microprobe (SHRIMP) to analyse zoned zircons and other accessory minerals has permitted the dating of cores and overgrowths formed at different times. Some zircons (see Fig. 257) show a very complicated history of overgrowth, zoning, resorption and abrasion, and could potentially give a great deal of information on the pre-magmatic history of their source regions and the processes of magmatic evolution (Paterson *et al.* 1992).

Depletion of the lower crust in fusible constituents

If granitic magmas originate by melting within the crust, then parts of the lower crust should be found to be depleted in low-melting constituents. Some very high-grade metamorphic rocks have in fact been identified as melting residues from which the melts have been removed (Grant 1968; Clifford *et al.* 1975). Such rocks might be identified by their general chemical composition, for example paucity in the constituents of quartz, albite or orthoclase, bearing in mind that a melting episode could still leave a residuum containing any two of these in abundance. Melting residues may also be expected to be poor in hydrous minerals; their mafic minerals might be pyroxenes and garnet rather than micas or amphiboles. Many rocks of the granulite facies have the required characteristics. Nesbitt (1980) examined the bulk compositions of metamorphic rocks in the New Quebec area and found that extraction of varying amounts of the granitic minimum-melting composition from the amphibolite facies rocks of that area would yield the compositions of the granulite facies rocks. He found a similar relationship in the Adirondacks, where amphibolite and granulite facies rocks also occur in proximity to one another.

In many cases it is difficult to recognize whether melting has taken place in high-grade metasedimentary rocks because they are petrographically so variable, but trace element evidence may be helpful. Several studies have shown that granulite-facies metamorphic rocks are strongly depleted in the 'granitophile' trace elements. For example Heier and Thoresen (1971) found low Rb, Th, U and Pb in granulite facies rocks of northern Norway, and Sighinolfi (1971) found depletion of Li and Rb in the Brazilian shield. For some of these elements, especially U, upward migration in aqueous solution may be responsible rather than partial melting, but for Li and Rb, whose predominant mode of occurrence is as isomorphous substituents in silicate minerals, loss by melting is the most likely explanation.

INITIATION OF CRUSTAL MELTING

The continental crust is normally below its melting temperature, but temperatures near the base of the crust may approach those required for melting. Indeed as Tuttle and Bowen (1958) pointed out, the continental crust could not be any thicker than it is without starting to melt, and this may be what determines the thickness of continental crust. Therefore, a small change in the conditions at the base of the crust may be all that is required to initiate partial melting. The main possibilities are: (1) introduction of heat by basic magma intruded into the crust from below; (2) a tectonic mechanism which thickens the crust, depresses crustal rocks into a region of higher temperature, or raises the temperatures in the crust; or (3) introduction of water into the lower crust thereby depressing the solidus of rocks which would not otherwise be hot enough to melt.

Large scale introduction of basaltic magma into the crust from below, or perhaps ponding of dense basaltic magma immediately beneath the continental crust, can potentially trigger off crustal melting in any place where the lower crustal temperatures are close to the level needed for anatexis. There are many igneous provinces in which basaltic and granitic liquids have clearly been present at the same time, often with little or no evidence for the presence of intermediate liquid, and in such places the arrival of the basaltic magma may well have been responsible for the formation of the granitic magma.

Various theoretical studies have been carried out to assess the scale of crustal melting that might result from the injection of basaltic magma into the continental crust. Huppert and Sparks (1988) showed that acid magma could be generated very rapidly and in large quantities, and De Paolo *et al.* (1992) demonstrated that in such a situation the proportion of crustal to mantle melt would be a function of the ambient crustal temperatures and the rate of basalt supply.

Bearing in mind the greater density of basaltic magma, and given the difficulty of mixing large bodies of magma, it could be predicted that in many situations a pool of granitic magma formed by anatexis would be underlain by a pool of the basaltic magma responsible for the melting event. Many field relationships between acid and basic rocks have been interpreted as being the result of this juxtaposition of magmas. They include net veining of acid and basic rock types (see Fig. 218), mafic xenoliths with the appearance of chilled liquid inclusions (Larsen and Smith 1990), synplutonic dykes (see below), and the observation that granitic plutons are often marked by positive gravity anomalies (indicating the presence of basic rock at a shallow depth). Classic examples of the coexistence of basic and acid magma occur in the intrusive complex of Mount Desert Island on the coast of Maine (Wiebe 1994), where silicic magma chambers have apparently acted as traps for rising basaltic magma.

The name *synplutonic dyke* is applied to a dyke-like body of basic magma that has been intruded into a mass of granite while the latter was hot, mobile and perhaps still liquid. Such dykes show chilling against the host granite but are partly disrupted by the continuing flow of their host granite before its final consolidation. In many cases the dykes are partly dismembered into trains of partly or completely rounded bodies indistinguishable from the microgranular mafic enclaves which are present in many granites (Fig. 258).

The availability of water is as important as the availability of heat if melting is to take place in the continental crust on a large scale. The amount of water in crustal metamorphic rocks is not sufficient to permit a high degree of partial melting. Pelitic schists typically have H_2O contents in the region of 1–2%, and other lower crustal rocks have even less water. If sufficient heat is introduced, melting could progress under water-undersaturated conditions, but without an increase in temperature the degree of melting must be limited by the amount of water available to saturate the melt. Wickham (1987) considered that an influx of water from an external source would be needed for initial melt bodies to develop into kilometre-sized plutons. He reasoned that in order for small melt bodies to grow, water is needed both to promote a high degree of melting and to lower the viscosity of the melt so that convective homogenization can occur. Most

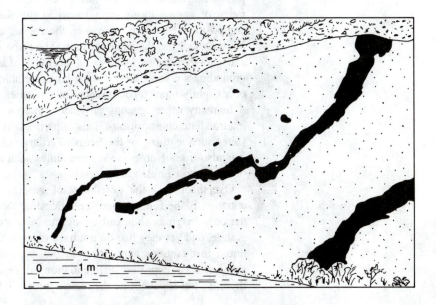

Figure 258. Synplutonic mafic dykes cutting granite in the Monte Capanne pluton, Elba (from a photograph by the author). Detached rounded blobs of the dyke material are indistinguishable from the mafic microgranular enclaves which are abundant elsewhere in the pluton.

0   1 m

granite specialists assume that crustal melting most often occurs under conditions where P_{H_2O} is high but falls well short of water saturation.

The Himalayas are a particularly good region for assessing the role of water in crustal melting. Le Fort (1981) described how the thrusting of the Tibetan crustal slab over the unmetamorphosed sediments of the Lesser Himalayas (see Fig. 276 below) caused inverted metamorphism up to kyanite grade and consequent dehydration of the Lesser Himalaya rocks. The water driven from these rocks then rose into the Tibetan slab and caused the episode of crustal melting which gave rise to the Higher Himalayan leucogranites.

DIFFERENTIATION OF BASALT OR ANDESITE

The rhyolitic or dacitic groundmass in some glassy tholeiites such as that of Alae lava lake (Chapter 7) show that it is possible for acid magmas to form by fractional crystallization. In tholeiitic lavas, and in differentiated tholeiitic intrusions, the residual melt does not attain a rhyolitic composition until crystallization is very far advanced. It is not obvious how such residual acid melts could become separated in quantities sufficient to form large bodies of magma. The viscosities of acid liquids are several orders of magnitude higher than those of basic liquids, so that gravitational separation is unlikely to be effective, and no plausible mechanism of filter pressing has yet been elaborated. Differentiation of primary andesite magma presents fewer difficulties in that an andesitic parent could yield a greater proportion of a residual acid fraction, but the existence of a realistic differentiation mechanism still remains to be demonstrated.

Many igneous provinces show a continuous range of magma compositions from basaltic through andesitic and dacitic to rhyolitic, and in these provinces a differentiation relationship has sometimes been proposed. For instance Nicholls (1971) favoured differentiation from a high-alumina basalt parent magma to yield the dacites of Santorini. Here there is a continuous range of compositions from basalt to rhyodacite, and in addition there is compositional similarity between the dacites and the glassy groundmass of basic rocks from the same volcano. The same relationship is observed in the andesite–rhyodacite lava series of Crater Lake, Oregon (Fig. 259). The groundmass composition of each lava resembles the whole-rock composition of a slightly more felsic member of the series.

In many other igneous provinces, there are breaks in the continuity of intermediate compositions. This is particularly true of non-orogenic magma suites such as those of the British Tertiary volcanic province (Table 47) but it is also true of orogenic magma suites such as the Variscan igneous rocks of south-west England (Fig. 260).

Overshadowing the possibility of generating granitic liquids by differentiation is the fundamental limitation that tholeiitic basalt and andesite are the only magmas abundant enough to be considered as parent magmas for granite, and yet they just do not have enough K_2O to yield derivative granite liquid in quantities appropriate to the huge size of granitic batholiths and rhyolitic sheet deposits. Differentiation can be a minor source of granites and rhyolites but it cannot be the main source.

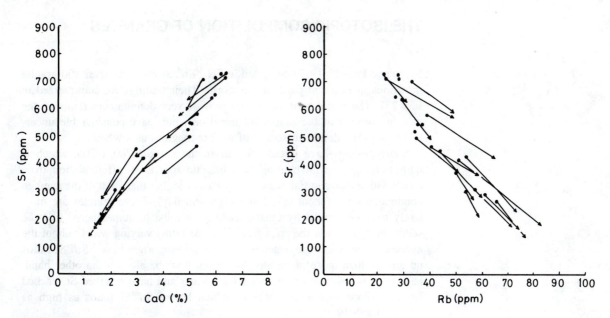

Figure 259. Plot of Sr against CaO and Sr against Rb for rocks from Crater Lake, Oregon (dots) and residual glasses (tips of arrows) (after Noble and Korringa 1974).

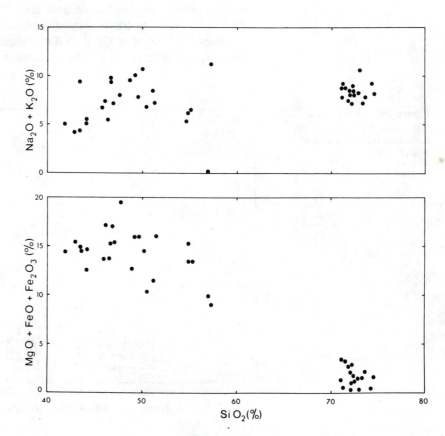

Figure 260. Compositional variation in the Variscan (late Carboniferous–early Permian) igneous rocks of south-west England (after Hall 1973a).

THE ISOTOPIC COMPOSITION OF GRANITES

Faure and Powell (1972) compiled initial $^{87}Sr/^{86}Sr$ ratios for over 100 granite intrusions in different parts of the world. Their findings are summarized in Fig. 261. The initial strontium ratios provide very definite constraints on the possible origins of the individual intrusions, but leave considerable uncertainty as to the dominant source of granite magmas as a whole.

Some granites have initial $^{87}Sr/^{86}Sr$ in the range 0.700–0.705, which is quite compatible with melting of a basaltic source or differentiation from basalt. Others have initial ratios as high as, or higher than, that of the average continental crust (about 0.720 at the present day). These granites are most likely to have formed by crustal melting. It must be remembered that the sedimentary rocks in the crust have $^{87}Sr/^{86}Sr$ ratios varying widely about the average value. Some greywackes of volcaniclastic origin have $^{87}Sr/^{86}Sr$ ratios no greater than those of basalts. Some sea-floor basalts, on the other hand, have been metasomatized by heated sea water and have received Sr leached from sea-floor sediments, some of which have $^{87}Sr/^{86}Sr$ ratios as high as 0.743 (Dasch 1969).

The relationship between initial $^{87}Sr/^{86}Sr$ ratio and chemical composition is particularly obvious for muscovite granites, some of which have very high initial ratios of 0.725 or more (Best *et al.* 1974; Whitney *et al.* 1976; Vidal *et al.* 1982). Melting of a peraluminous source material, i.e. pelitic sediment, is the most probable explanation of both their Al-rich composition and their high $^{87}Sr/^{86}Sr$ ratio. In contrast, trondhjemites and tonalites have uniformly low initial $^{87}Sr/^{86}Sr$ of 0.701 to 0.704 (Peterman 1979). These ratios lie within the mantle band of Fig. 261, and almost certainly indicate derivation

Figure 261. Plot of initial $^{87}Sr/^{86}Sr$ ratios of granites against their ages (after Faure and Powell 1972).

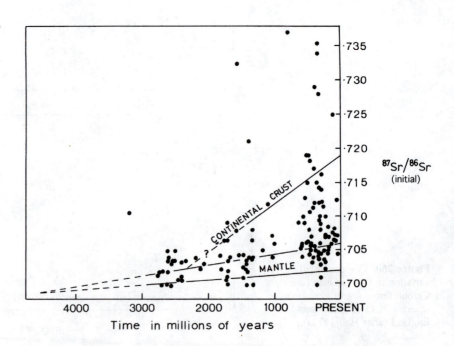

from a basaltic or mantle source, whether by partial melting or by differentiation.

Chappell and White (1974, 1992) proposed a two-fold division of granitic rocks into S- and I-types according to whether they were derived from predominantly sedimentary or igneous source materials. They believed that such a distinction could readily be made using Sr isotopes and other geochemical criteria. In particular, a sedimentary source would be characterized by high initial $^{87}Sr/^{86}Sr$ ratios and a peraluminous composition, and an igneous source by low initial $^{87}Sr/^{86}Sr$ and a relatively high Na_2O content. The classification of granites into S- and I-types has been widely adopted and elaborated.*

The oxygen isotopic composition of granites can also be used to infer the source of granitic magmas, but only in cases where the granite has not subsequently been affected by postmagmatic alteration. Mantle-derived rocks (e.g. basalt) normally have $\delta^{18}O$ in the range of +5 to +7, whereas sedimentary rocks have $\delta^{18}O$ from +5 to +30, so high values of $\delta^{18}O$ in a granite imply either a crustal origin or at least crustal contamination of the magma. However, $\delta^{18}O$ values are reduced by hydrothermal alteration (see Chapter 6), so oxygen isotopes cannot be used to infer the magma source if hydrothermal alteration is suspected. Unfortunately there are few granitic intrusions which are absolutely free of hydrothermal alteration; for example any granite which has a pink or red colour, or which contains even partly chloritized biotite or sericitized plagioclase has undergone some degree of hydrothermal alteration. Hydrothermal alteration is not confined to the obvious examples of granites which are cut by mineralized veins, but is often subtle and pervasive in its effects.

Unlike high $^{87}Sr/^{86}Sr$ ratios, which can originate in the mantle if there are local concentrations of Rb, high $\delta^{18}O$ values can only arise from near-surface processes such as are involved in the formation of sedimentary rocks. Thus the combination of Sr and O isotopic compositions is a better guide to possible contamination than $^{87}Sr/^{86}Sr$ alone. Figure 262 shows that many igneous complexes have a spread of Sr and O isotopic compositions suggestive of mixing between a mantle-derived component (like basalt) and a crustal component. The problem is to decide whether this spread is due to contamination, to magma mixing, or to melting of a mixed source.

Taylor (1980) has reviewed this problem, drawing attention particularly to the non-linear nature of the isotopic changes brought about by contamination. The exact compositional path taken by the contaminated rocks depends on the proportion of crystallizing cumulates to contaminating material, on the ratio of Sr contents in the magma and contaminant, and on the partitioning of Sr between the cumulates and the melt (the contents and partitioning of O are less crucial because all rocks contain similar amounts of oxygen). According to Taylor's calculations, a linear $^{87}Sr-^{18}O$ correlation is

---

* The S- and I-type classification of granites is not very soundly based. An ideal petrological classification should be based on observed characteristics, e.g. chemical or modal parameters, rather than on inferred origins. There is even less justification for muddying this classification with additional categories such as A-type (alkaline, anorogenic), C-type (charnockitic) or M-type (mantle-derived).

Figure 262. Variation of $\delta^{18}O$ with initial $^{87}Sr/^{86}Sr$ for various igneous rock suites (after Taylor 1980). Mantle-derived magmas (oceanic basalts) have $\delta^{18}O$ about +6 and $^{87}Sr/^{86}Sr$ about 0.703. Continental igneous suites extend from these values towards the higher $\delta^{18}O$ and $^{87}Sr/^{86}Sr$ compositions typical of sedimentary rocks.

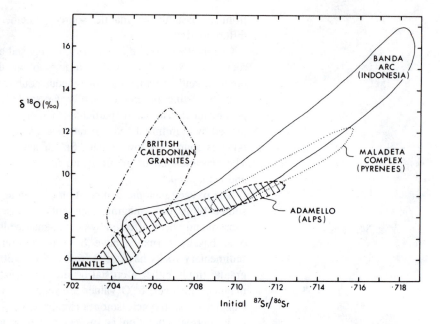

unlikely to be produced by simple contamination. Of the rock suites shown in Fig. 262, he considered the Adamello complex to be a possible example of contamination by virtue of the curved ^{87}Sr–^{18}O trend.

If crustal contamination is a widespread feature of granites, then it would be very difficult to use their isotopic compositions to deduce much about the original source of the magmas. In the opinion of Armstrong *et al.* (1977), the initial $^{87}Sr/^{86}Sr$ ratio is a better indicator of the nature of the enclosing crust than of the source of the magma itself, but Taylor (1980) argued that the Sr–O isotopic correlations were more likely to originate in the magmatic source region than by contamination.

Another reservation on the use of initial Sr isotope ratios to determine the origin of granites is that many published isochrons are based on insufficiently rigorous sample selection. In order to plot an isochron it is necessary to find rocks with a range of Rb/Sr ratios, and examination of the literature reveals that the published isochrons for some granite complexes are based on specimens ranging far beyond even the most liberal definition of a granite. There is a serious danger that the use of contaminated rocks in plotting isochrons may lead to error in measuring the $^{87}Sr/^{86}Sr$ ratios of the magmas (see Fig. 263).

In general, the isotopic composition of Pb in present-day rhyolites and granites is either more or less radiogenic than would be predicted by a single-stage model, and accordingly lies either to the right or to the left of the Geochron (present-day isochron) in Fig. 264. The enrichment or depletion of these magmas in radiogenic isotopes is an indication of either high U/Pb or low U/Pb in the magmatic source rocks, or in country rocks by which the magmas were contaminated. For example, the rhyolites of Yellowstone Park or the Tertiary granites of Skye are poor in radiogenic isotopes and lie to the left of the present-day isochron, whereas the Mesozoic

Figure 263. The possible effect of contamination on the initial $^{87}Sr/^{86}Sr$ ratio of a granite. A uniform body of magma with an assumed composition of $^{87}Sr/^{86}Sr = 0.713$ and $^{87}Rb/^{86}Sr = 1.4$ is contaminated by: (A) basalt with ratios $^{87}Sr/^{86}Sr = 0.7045$ and $^{87}Rb/^{86}Sr = 0.2$; and (B) shale with ratios $^{87}Sr/^{86}Sr = 0.723$ and $^{87}Rb/^{86}Sr = 2.8$. In both cases a range of contaminated rocks is produced which after time t will give rise to a false isochron corresponding to an apparent initial $^{87}Sr/^{86}Sr$ ratio of 0.703 (and an incorrect age).

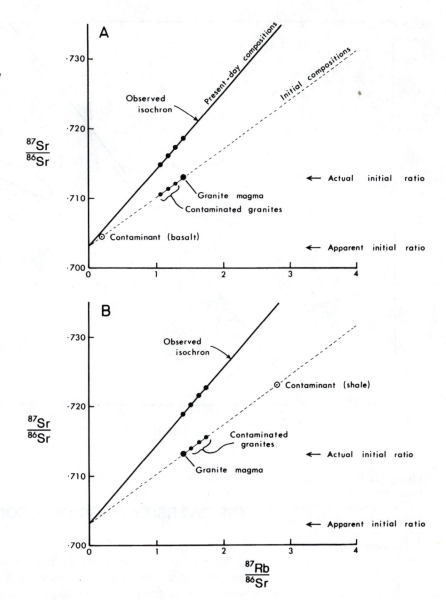

granites of California are rich in radiogenic isotopes and lie to the right of the present-day isochron. This does not help very much in determining whether the magmas were produced by crustal melting, because some parts of the crust have higher U/Pb ratios than the mantle and some have lower U/Pb ratios. In any case, according to Hogan and Sinha (1991) the Pb isotopic composition of a granitic melt may differ appreciably from that of its source rocks if the latter contained refractory accessory minerals rich in U and Th, i.e. zircon or monazite.

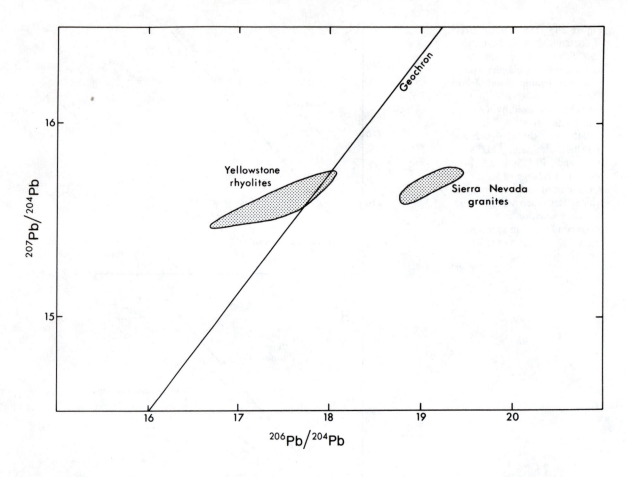

Figure 264. Lead isotopic compositions of granites from the Sierra Nevada batholith, California (Doe and Delevaux 1973) and rhyolites from Yellowstone Park (Doe *et al.* 1982).

THE DIVERSITY OF GRANITE COMPOSITIONS

Granitic rocks vary in many different ways, and many authors have interpreted their compositional variations as being primary (due to different source rocks or conditions of melting) or secondary (due to fractionation or contamination). There is no consensus of opinion on the relative importance of these processes in causing the diversity of granites as a whole. Some of the ways in which granites can vary are:

1. In colour index, from tonalites and granodiorites with a high proportion of mafic minerals to leucogranites with almost no mafic minerals.
2. In the relative proportion of the low-melting constituents (normative Q, Or and Ab) to one another.
3. In alumina-saturation.
4. In oxidation state.
5. In mineralogy. Two mineralogical features which are particularly inform- ative are the accessory mineral content and the nature of the alkali feldspar.

THE RANGE FROM FELSIC TO MAFIC

There is a continuous spectrum of granitic compositions from leucogranites to quartz–diorites. This not merely involves a variation in the proportion of dark-coloured minerals, but reflects the proportion of high- and low-melting temperature constituents. The low-melting constituents are quartz and alkali feldspars, and the high-melting constituents are biotite, hornblende and the calcic component of plagioclase.

Many petrologists have tried to interpret variation of this kind in particular granitic complexes, both between individual intrusive units and within each intrusive unit. If xenoliths of country rock are present, contamination may be suspected. If xenoliths are absent, then a crystallization differentiation model might be considered. Mixing is likely in cases where basaltic magma has been injected into a body of granite magma. More simply, the whole range of compositions may be primary magmas.

The presence of xenoliths can be very informative. Not all granites contain xenoliths, and only a few are rich in xenoliths, but they are often present and they can be very revealing. The traditional explanation of xenoliths is that they are fragments of country rock that have become surrounded by the magma, but in reality they are of at least three different types:

1. Pieces of country rock or roof rock that have fallen or been carried into the magma.
2. Blobs of a coexisting, more basic magma that have been dispersed through the granite.
3. Relicts of unmelted refractory material brought up from the magma source region.

In any particular pluton the xenoliths may all be of one type, but taking all granites together there are plentiful examples of all three types (Didier and Barbarin 1991). Type 1 are easily recognized in roof and wall zones of plutons where they can often be matched with specific country rock lithologies. However, they are often modified by reaction with the magma, and then their origin may not be so clear. Type 2 includes the 'microgranular mafic enclaves', which are relatively homogeneous and provide less evidence of their origin. However, the igneous origin of this type can sometimes be demonstrated by matching them with synplutonic dykes (Fig. 258). Type 3 are not easily distinguishable from modified country rock xenoliths of type 1, but the presence of inherited zircon is a good clue to their identity.

Primary variation

The melting experiments of Wyllie and his collaborators (e.g. Robertson and Wyllie 1971) have shown that during crustal anatexis magma compositions would progress from granite to granodiorite as temperatures rose beyond the solidus and increasing amounts of mafic material dissolved in the melt. Thus, granites and granodiorites occurring in association might well represent the products of different degrees of partial melting in the magmatic source region.

In addition to becoming more mafic as partial melting progressed, the melts would also become undersaturated with H_2O as the amount of melt continued to increase, and it is unlikely that temperatures in the crust would ever be sufficient for tonalite to be completely liquid under H_2O-deficient conditions. Tonalites have a greater temperature range of crystallization than leucogranites (Fig. 265), and higher liquidus temperatures. Thus to form in the crust, a granodiorite magma would sometimes (and a tonalite magma would usually) have to consist of a mixture of liquid and residual crystals rather than a pure liquid. Wyllie (1977) suggested that tonalites originating in the crust consist of a mush of granitic liquid carrying crystals of the complementary refractory residuum, and may even be enriched in residual crystals if some of the granitic liquid fraction escapes from the mush. Thus

Figure 265. Phase relations of (A) granite and (B) tonalite in the presence of excess water (after Stern *et al.* 1975).

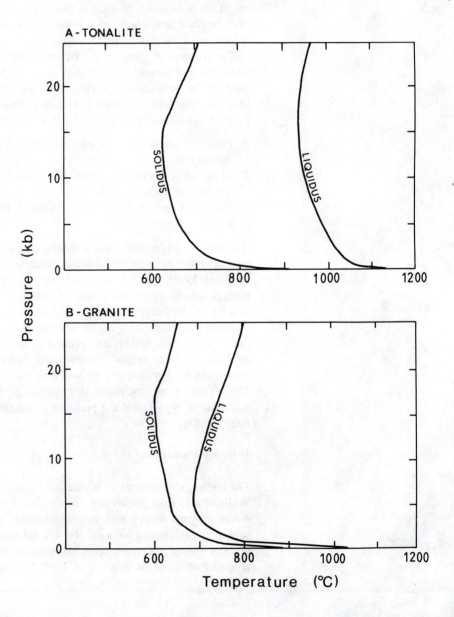

the proportions of granite and granodiorite in an intrusion might reflect the quantity of residual crystals carried by the magmas, or the efficiency with which the magmas divested themselves of their burden of residual crystals during their ascent.

The possibility that granite magmas may carry up a large amount of entrained residual crystals from the magmatic source region has become increasingly popular (Presnall and Bateman 1973; Wyllie 1977; White and Chappell 1977; Zeck 1992). Chappell *et al.* (1987) put the extreme view that most granites are a mixture of melt and residue (xenoliths and xenocrysts), that few granite magmas divest themselves completely from their restite component, and that more than 99% of xenoliths in granite are restitic source rock material. However, this view has been strongly disputed by Wall *et al.* (1987).

The Waldoboro granite in Maine has been suggested as a good example of a granite which was formed by partial melting of metasediments, which has not moved from its source, and which shows a transitional relationship with the adjacent migmatites. Barton and Sidle (1994) estimated that the granite was a mixture of 76% melt with 24% restite, while the associated gneissic granite could be a mixture of 55% melt with 45% restite. The gneissic granite contains many enclaves and rafts of the metamorphic country rocks, constituting up to 40% of the outcrops and in various stages of apparent digestion by the granite.

In an episode of partial melting in the crust it may be expected that the magma composition will vary during the melting episode. The initial and final batches of melt must have a near minimum-melting composition, while the magma product of the greatest extent of partial melting can contain a greater proportion of higher-melting constituents. In the Rosses granite complex in Donegal (Fig. 58), the sequence of magmas that were intruded is exactly consistent with the succession of magma compositions expected during the successive phases of a discrete episode of partial crustal melting, as shown in Fig. 266. Initially, dykes of the most felsic composition were intruded, then when melting reached its greatest extent a large body of comparatively mafic granite magma was intruded by repeated roof-foundering, followed by more felsic varieties of granite again in the waning stages of the magmatic episode.

Figure 266. Diagrammatic representation of the relationship between magma composition and order of emplacement in the Rosses granite complex, Ireland (Hall 1966b). The major events are: (1) the start of the magmatic episode; (2) the roof of the magma body cracks, and microgranite sheets are intruded; (3) the roof of the magma body founders, and the main granites are intruded; and (4) the end of the magmatic episode. A map of the complex is shown in Fig. 58.

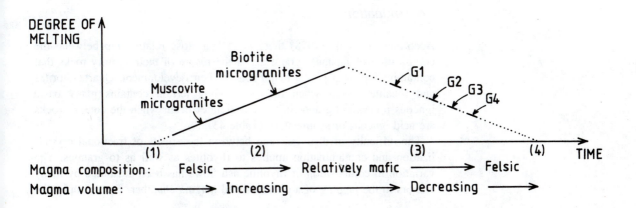

Fractionation

The relative role of fractionation and contamination is a matter of wide disagreement. There are two major problems about crystal fractionation: (1) granite magmas are too viscous, and (2) there are no cumulates. The occurrence of inherited zircon in some granites is also a problem, because a magma capable of carrying up restite crystals from the source region can hardly be undergoing crystal settling at the same time.

The rare occurrences of layering in granitic intrusions show that some gravitational differentiation is possible. On the other hand the high viscosity of granitic magmas would make crystal sorting or settling much slower and less effective than in basic magmas. Shaw (1965) believed that crystal settling rates would be high enough to be petrologically significant, but against this is the observation that granitic magmas can often support, and even carry up, large xenoliths of heavy country rocks. The most positive evidence in favour of fractionation comes from volcanic rocks. Nakada (1983) found that successive batches of dacite and rhyolite from Osuzuyama showed progressive enrichment in phenocrysts, suggesting the tapping of a magma chamber in which crystal settling had occurred. In contrast, Hildreth (1981) described zoning in many large bodies of rhyolitic magma that contained few or no phenocrysts when they were erupted, and strongly argued the case for liquid fractionation rather than crystal fractionation.

The Cathedral Peak granodiorite in Yosemite National Park is one of the few to show rhythmic layering (Reid *et al.* 1993). The layers, a few centimetres thick, consist of concentrations of magnetite, sphene, biotite, hornblende and zircon grading up into normal granite. The layering is confined to a few small areas in the granite. Cumulate textures and 'sedimentary' structures suggest that these dark layers are due to crystal settling. Other examples of layering in granites have been attributed to flow differentiation (Barrière 1981). The occasional examples of layering show that crystal sorting can take place locally during the crystallization of granite magma, but no one has been able to demonstrate any cryptic variation associated with visible layering in an acid intrusion. Furthermore the major and trace element variation between the dark and light layers in the Yosemite examples does not correspond to the large scale variation of their host granites.

Contamination

According to Ikeda (1978) there is such a close relationship between the composition of Japanese granites and the nature of their country rocks that contamination must play a major role in their development. Quartz–diorites and tonalites occur when the surrounding terrain contains many basic igneous rocks, and granodiorites and granites occur when the country rocks are acid igneous or sedimentary (Table 43).

The difficulty of distinguishing between the effects of fractional crystallization and contamination applies to rhyolites as well as to granites. The variation in composition of rhyolitic ash flows has been attributed by many authors to the tapping of a differentiated magma chamber. For example, the

Table 43. The relationship between the compositions of Japanese granitic rocks and the nature of their country rocks (after Ikeda 1978)

	Country rock composition			
Granitic lithology	*B*	*B>S*	*S>B*	*S*
Muscovite–biotite–granite	0	0	0	3
Biotite–granite	3	9	62	58
Granodiorite (biot > hbl)	0	17	12	19
Granodiorite (varied)	0	47	12	16
Granodiorite (hbl > biot)	0	15	9	2
Granodiorite and quartz–diorite	1	9	0	0
Quartz–diorite	96	4	4	2

B: comprises basalt, andesite, their plutonic and metamorphic equivalents, and ultramafic rocks.
S: comprises sedimentary rocks, acid igneous rocks, and their metamorphic equivalents.
The values given are percentages of the total area for each type of country rock.

Figure 267. Hypothetical section through the Long Valley caldera, California and its underlying magma chamber (after Bailey *et al.* 1976). The caldera was formed by a major eruption of rhyolitic pumice about 700,000 years ago, and minor eruptions of rhyolite and rhyodacite have continued intermittently up to the present day. Horizontal dashed lines show the silica variation in the compositionally zoned magma chamber; curved dotted lines show the extent of the remaining magma at various times up to the present.

distribution in space and time of different acid magma compositions in the Long Valley caldera of California led Bailey *et al.* (1976) to postulate tapping of a differentiating magma chamber along the lines indicated in Fig. 267. On the other hand, Johnson (1989) pointed out that many ash flows show isotopic variation indicative of contamination. Moreover, he inferred that because the amount of assimilation indicated by isotopic ratios often exceeds the abundance of phenocryts in the erupted magma, the processes of assimilation and fractional crystallization operate separately, and are not coupled as some authors have suggested.

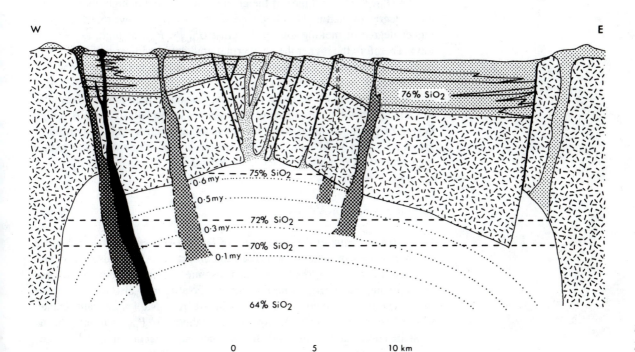

Zone melting

Zone melting may be of great importance in the evolution of granitic magmas. Melting of roof rocks, or more precisely the melting of stoped and foundered portions of the roof rocks, may modify the composition of a magma body as it advances upwards through the crust (Hall 1966b; Hamilton and Myers 1967; Barker *et al.* 1975). If this takes place on a large scale the magma will become enriched in low-melting constituents, partly by addition of melted material and partly by subtraction of the higher-melting constituents which are simultaneously crystallizing from the magma. It is even possible that only a small proportion of the magma that finally arrives at the final level of emplacement will have been present in the magma body that started on its ascent.

VARIATION IN THE RELATIVE PROPORTIONS OF FELSIC CONSTITUENTS

As Fig. 250 shows, the composition of a melt formed by partial melting or crystallization differentiation in the system Q–Or–Ab–H_2O will depend on P_{H_2O}. The higher the P_{H_2O}, the higher will be the normative albite content of the melt.

The P_{H_2O} during an episode of partial melting is largely determined by the geothermal gradient under which melting takes place (Fig. 268). If the geothermal gradient is low, the minimum melting temperature will only be attained in the lowest part of the crust and the degree of partial melting will be low ($P_{H_2O} \sim P_{total}$ = high). If the geothermal gradient is high, there will be a low degree of melting high in the crust ($P_{H_2O} \sim P_{total}$ = low) together with a higher degree of melting low in the crust ($P_{H_2O} < P_{total}$ = high, i.e. P_{H_2O} = low). Therefore the higher the geothermal gradient, the lower will be the P_{H_2O} of magma formation, and this will be reflected in high normative Q and low normative Ab in the resulting granites.

The composition of the rocks in the magma source region also contributes to variation in the composition of the magmas, especially if there is a high degree of partial melting. A geosynclinal terrain with a high proportion of greywackes can yield more Ab-rich magma compositions than one with a high proportion of normal shales and sandstones.

ALUMINA SATURATION

In *peralkaline granites* the cation ratio $(K + Na)/Al$ is greater than 1, and Na and K in excess of the amount needed to convert the Al to feldspar give rise to such minerals as riebeckite, arfvedsonite, aegirine, aenigmatite and astrophyllite; hedenbergite, hastingsite and fayalite may also occur in these rocks. The *peraluminous granites*, with an $Al/(K + Na + 2Ca)$ ratio greater than 1, normally contain muscovite, or cordierite if P_{H_2O} was low during crystallization; andalusite and topaz may also occur in peraluminous granites. The majority of granites are neither peraluminous nor peralkaline.

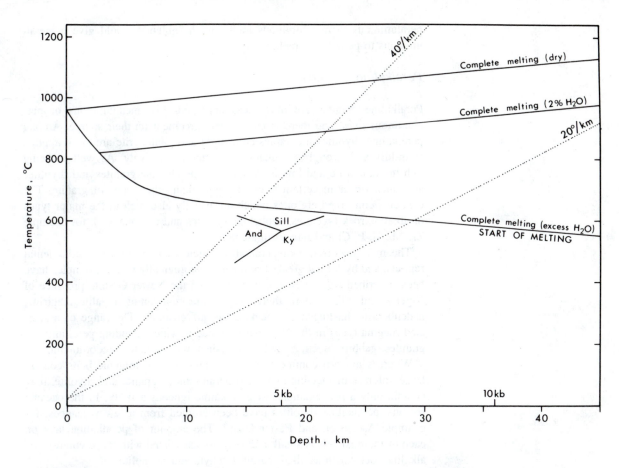

Figure 268. The variation in melting temperature of the minimum-melting fraction of granite in relation to the availability of water (after Hall 1972a). Two possible geothermal gradients are indicated. Melting could start at 600–700 °C in the presence of excess water, but in practice actual melting is deferred until some water becomes available, for example by the reaction muscovite + quartz → orthoclase + Al_2SiO_5 + H_2O. The position of the Al_2SiO_5 triple point is indicated approximately, as a guide to the temperatures and pressures that are reached in high-grade regional metamorphism.

Some biotite–granites are very slightly peraluminous, but many biotite–granites and virtually all granodiorites are *metaluminous*, i.e. (K + Na) < Al < (K + Na + 2Ca).

Peraluminous granites

Melting of metasediments is usually assumed to be responsible for the formation of peraluminous granites because argillaceous sediments are the main type of material with a very high aluminium to alkali ratio. Moreover, strongly peraluminous granitic liquids have been produced experimentally by the partial melting of pelitic metasediment (Patiño Douce and Johnston 1991). The peraluminous granites have high initial $^{87}Sr/^{86}Sr$ ratios, and their trace element compositions correspond to those expected from partial melting of a pelitic starting material (e.g. high B, Cs; low Ba, Sr). Peraluminous granites are also richer in ammonium than other types of granite, the ammonium ion being presumably derived from nitrogenous organic matter in sedimentary rocks (Hall 1987, 1988; Hall *et al.* 1995).

Whitney (1988) pointed out that if granite magma is generated in the crust by dehydration melting, then the Al-saturation of the melt is mainly governed by which of the hydrous silicates break down. Muscovite dehydration would yield a peraluminous melt, biotite would lead to a mildly peraluminous to metaluminous melt, and hornblende would give a metaluminous to peralkaline melt.

Peralkaline granites

Peralkaline granites are much less abundant than metaluminous or peraluminous ones, and there is no general agreement on their source. Among peralkaline rhyolites the names comendite and pantellerite are often applied to mildly and strongly peralkaline varieties respectively. As well as being rich in Na and Fe and low in Al and Mg, peralkaline granites and rhyolites are much richer in certain trace elements than other types of granite. The elements concerned are ones which are usually also high in the major types of alkaline rock (syenites, nepheline–syenites and carbonatites), for example Zr, Nb, U, F, Cl and rare-earth elements.

The main occurrence of peralkaline silicic rocks is in stable continental regions, and by far the greatest number of geologically young examples have been described from Africa, especially from the Newer Granite province of Nigeria and Niger. Here there are granites containing fayalite, aegirine, hedenbergite, hastingsite, riebeckite and arfvedsonite. The range of associated magma types in the Nigerian province is wide, including peraluminous granites, gabbros, syenites and anorthosites, with granites predominating.

Whereas modern continental margins and ancient orogenic belts contain large volumes of metaluminous and peraluminous granite and rhyolite, they contain only a few examples of peralkaline igneous activity. In the oceanic regions, peralkaline rhyolites have been erupted from a few volcanoes, for example Ascension and Easter Island. The amount of peralkaline lava on each of these islands is small, and they are associated with more voluminous alkaline rocks such as alkali basalt, trachyte and phonolite.

A crustal origin for the granites in the Nigerian Newer Granite province is indicated by high initial $^{87}Sr/^{86}Sr$ ratios (e.g. 0.721 in the Amo complex). In contrast, the pantellerites from Ethiopia described by Barbieri et al. (1975) have initial $^{87}Sr/^{86}Sr$ ratios not much greater than those of the accompanying basalts (0.706–0.707 compared with 0.705–0.706 respectively), and the intermediate rock types and trace element trends favour a differentiative relationship.

It is not known why some crustal melts should have a peralkaline character, but Bailey and Macdonald (1987) favour derivation from regionally metasomatized source rocks, possibly related to copious effusion of CO_2-rich fluids from the mantle.

OXIDATION STATE OF GRANITES

It is not easy to determine the oxygen fugacity under which granite magmas crystallize, but there is some evidence that granites show a greater range of oxidation state than other types of magma.

The oxygen fugacity does not correlate exactly with the Fe_2O_3/FeO ratio of the rock, and in any case there are analytical difficulties about measuring this ratio accurately. However, one can estimate the oxidation conditions from the accessory mineral assemblage or by measuring the Fe^{3+}/Fe^{2+} of biotite. In particular, Wones (1989) pointed to the significance of the equilibrium:

$$\text{hedenbergite} + \text{ilmenite} + \text{oxygen} \rightarrow \text{sphene} + \text{magnetite} + \text{quartz}$$

The assemblage on the right is the one which corresponds to relatively oxidizing conditions. Hedenbergite in this equation is a simple end-member ferromagnesian silicate; in practice it is representative of any pyroxene or amphibole.

Magnetite and primary sphene are therefore typical accessory minerals in granite formed under relatively oxidizing conditions and ilmenite is characteristic of more reducing conditions. Ishihara (1978) used these minerals to derive a two-fold classification of Japanese granites into an ilmenite-series and a magnetite-series. He considered the relatively reduced ilmenite-series rocks to have formed under more reducing conditions because their magmas incorporated carbon from originally organic-bearing metasediments. The two categories correspond in a very approximate way with the S- and I-type granites of Chappell and White.

MINERALOGICAL VARIATIONS

Unlike basalts and gabbros, acid igneous rocks have a wide variety of accessory mineral assemblages, some of which give useful information on the history of the magmas. Some accessory minerals (e.g. zircon, monazite) reflect the trace element content of the magmas, and hence possibly the nature of their source; some (e.g. andalusite, garnet, epidote) may be the product of contamination; some (e.g. garnet, topaz, hematite) are the products of late hydrothermal alteration; some are dependent on f_{O2}, or P_{H_2O} during crystallization (e.g. magnetite, ilmenite, andalusite).

Hypersolvus and subsolvus granites

These are two textural categories of granite which are distinguished by the character of their alkali feldspar.

The majority of granites are classified as *subsolvus granites*. They contain three felsic minerals (quartz, K-feldspar and plagioclase) in roughly similar proportions. A minority of granites (not necessarily different in chemical composition) are described as *hypersolvus granites*. They contain a similar amount of quartz to the subsolvus granites, but instead of two separate feldspars there is only one – it is a coarsely perthitic alkali feldspar in which both the K-rich and Na-rich components are abundant. This perthitic alkali feldspar crystallized from the magma as a single phase, i.e. $(Na,K)AlSi_3O_8$, and exsolved subsequently. There was no separate crystallization of plagioclase from the magma. Photomicrographs illustrating hypersolvus texture are given in Tuttle and Bowen's (1958) memoir.

The difference between the two categories is readily explained by phase relations in the system $Or–Ab–SiO_2–H_2O$. Minimum crystallization temperatures in this system at low P_{H_2O} (Fig. 250) are above the solvus in the system Or–Ab (Fig. 146), but at high P_{H_2O} they are below the solvus in the system Or–Ab. Therefore at low P_{H_2O} a single Na,K-feldspar crystallizes alongside quartz, but at high P_{H_2O} separate Na-feldspar (plagioclase) and K-feldspar crystallize together with quartz.

The main requirement for the development of hypersolvus texture is therefore crystallization at low P_{H_2O}, and this usually involves both low P_{total} and low magmatic water content. For this reason, hypersolvus texture is only found in granites intruded at a shallow level in the crust. The texture is not confined to granites; some syenites show hypersolvus crystallization.

GRANITES IN RELATION TO TECTONIC ENVIRONMENT

Some granitic magma is found in nearly all tectonic environments, but by far the greatest volume is produced in areas underlain by continental crust, and most granites are in orogenic regions. It appears that granite magmas can have more than one source, namely: (a) melting of continental crust; (b) melting of subducted oceanic crust or mantle; and (c) differentiation. Our problem is to decide on the relative importance of these sources, which may differ in their relative importance according to the tectonic environment.

OCEAN BASINS

The amount of rhyolite which occurs on islands which are not underlain by continental crust or subducted oceanic crust is very small, but such lavas do occur in one or two places, such as Iceland and Ascension Island. About 3% of all the volcanic rocks on Iceland are acidic, a smaller proportion are intermediate, and all the rest are basaltic, but in a few of the Icelandic central volcanoes rhyolite is more abundant (up to 20%) and intermediate compositions are less common.

There are two possibilities for the generation of acid magma in these oceanic islands: one is by fractionation of tholeiitic basalt, and the other is by the remelting of oceanic crust. Up to now, opinion has been evenly divided between those who favour fractionation (Furman *et al.* 1992) and those who favour remelting (Thy *et al.* 1990).

The Icelandic volcano, Hekla, has attracted attention by reason of its cyclic variation in magma composition. Although its lavas are predominantly basaltic andesite or icelandite (54 to 63% SiO_2), it has been found that each eruption starts with the eruption of acidic ash, followed rapidly by increasingly basic lava as the eruption continues (Fig. 269). Also, the amount of SiO_2 in the initially erupted magma is greatest when the volcano has been inactive for a long time. This suggests that the magma is undergoing fractionation in a sub-surface magma chamber in between eruptions, and that a build-up of volatiles associated with fractionation triggers off each new eruption (Baldridge *et al.* 1973).

Figure 269. Diagram showing the composition of the first magma erupted in each of seven historic eruptions of Hekla volcano, Iceland. Solid lines show the subsequent range in magma composition in each eruption, and dashed lines represent the increase in SiO_2 during quiescence as a linear function of time (after Thorarinsson and Sigvaldason 1973).

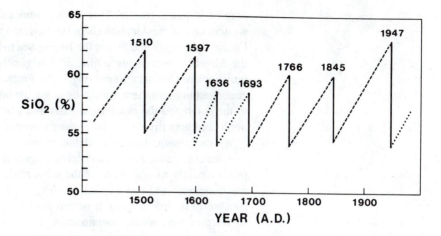

Strontium isotope evidence suggests that more than one source is involved. O'Nions and Grönvold (1973) found that in some Icelandic volcanoes the rhyolites and basalts have the same initial $^{87}Sr/^{86}Sr$ ratio (0.7033), but in others the rhyolites have very slightly more radiogenic strontium (0.7034 in rhyolite compared with 0.7033 in basalt). Those rhyolites with Sr ratios identical to the basalts may have been produced by differentiation and those with higher ratios may have been produced by partial melting of gabbro in the oceanic crust. Sigurdsson and Sparks (1981) examined the relationship between rhyolite, icelandite and basalt in the 1875 lavas of Askja. According to their calculations, it is possible to explain the icelandites by fractional crystallization of basalt but not the rhyolites, and there were no liquids intermediate between icelandite and rhyolite. They suggested that the formation of the rhyolites involved partial fusion of oceanic plagiogranite (trondhjemite), possibly derived from basaltic magma in an earlier magmatic episode. There are inclusions of partially fused trondhjemite in some of the rhyolites. Oxygen isotope evidence favours a crustal remelting origin for the Icelandic rhyolites because they have significantly lower $\delta^{18}O$ values than the associated basalts, which means that they must have come from a source which has undergone hydrothermal alteration, i.e. the oceanic crust (Sigmarsson *et al.* 1991).

Although ocean-floor volcanic rocks are overwhelmingly basaltic, small amounts of acidic material are sometimes present. At the Galapagos spreading centre in the Pacific Ocean there is a range of magma compositions from basalt through ferrobasalt and ferroandesite to rhyodacite, which appears to be a crystallization differentiation sequence (Byerly 1980).

ISLAND ARCS

In island arcs, rhyolites and granites are often present and are locally abundant, but there is an important difference between island arcs lying between two oceanic plates and those where an oceanic plate is subducted under continental crust. In the former, rhyolitic volcanism is very inconspicuous, but where continental crust is present the amount of acid magma erupted is much greater. This is very obvious in the Aleutian Islands. The western end of the Aleutian chain lies between the oceanic crust of the Pacific Ocean to the south and that of the Bering Sea to the north, and in this stretch of the Aleutians there is just a little acid magmatism, including the granodiorite intrusions of Unalaska Island (Fig. 270). Further east, the Aleutian arc and its continuation in southern Alaska lie along the boundary between oceanic crust to the south and the continental mass and continental shelf of Alaska to the north, and here rhyolite is much more common.

We can assume that in a subduction zone situated at a boundary between two oceanic plates, any primary felsic magma that is produced comes from the predominantly basaltic rocks of the subducted ocean crust. It is not likely that much oceanic sediment could be dragged down to the level of melting, because of its low density. It is also unlikely that the mantle adjacent to the subduction zone would contain enough low-melting constituents, especially potassium, to give rise to any acid magma. Initial $^{87}Sr/^{86}Sr$ ratios of island arc rhyolites and dacites are low, and are no greater than those of the associated basalts and andesites (Table 44).

Table 44. Initial $^{87}Sr/^{86}Sr$ ratios of rhyolites and dacites from three island arcs

	Izu Islands	Tonga	Aleutian Islands
Rhyolite	0.7033–0.7038	0.7036	—
Dacite	—	0.7042–0.7043	0.7030–0.7037
Andesite	0.7034–0.7050	0.7036–0.7042	0.7028–0.7038
Basalt	0.7034–0.7040	—	0.7031–0.7037
Reference	Pushkar (1968)	Ewart *et al.* (1973)	Singer *et al.* (1992)

Although experiments have shown that a small amount of acid magma could be produced directly by melting of subducted oceanic crustal material, such as quartz–eclogite of basaltic parentage, it seems more probable that the small amounts of rhyolite in island arcs are fractionation products of the associated tholeiites and andesites. Perfit *et al.* (1980) examined in detail the chemical variation in the Captains Bay pluton on Unalaska Island, which is zoned from a gabbroic rim to a granitic core. There appears to be a continuous gradation, and all the rocks have initial $^{87}Sr/^{86}Sr$ between 0.703 and 0.704, which is consistent with fractionation of a single parent magma with the composition of high-alumina basalt. The volcanic rocks of the Aleutians also show very little difference in Sr isotopic composition across the range from basalt to rhyodacite, and the rhyodacite compositions can be modelled by fractionation of the phenocryst phases from the associated basalts (Singer *et al.* 1992).

OROGENIC CONTINENTAL MARGINS

In orogenic continental margins, there is the possibility of melting either in a subduction zone, or in overlying continental crust, or in both. A much greater amount of rhyolitic or granitic magma is seen in this environment than in the oceanic island arcs, suggesting a more important role for melting in the continental crust.

Figure 270. The geology of Unalaska, Aleutian Islands (after Drewes *et al.* 1961).

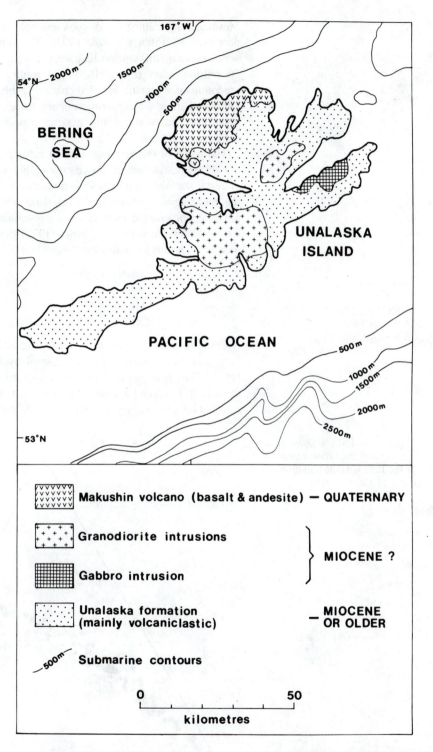

An important aspect of the magmatism of continental margins is its persistence in time. Granite formation has taken place in the Cordilleran regions of North and South America almost continuously throughout the

Mesozoic and Cainozoic. A good example is the Coastal Batholith of Peru, described by Pitcher (1978, 1993). The hundreds of intersecting plutons which make up this composite batholith were emplaced over a period of about 70 million years (from 102 to 34 million years ago). The batholith is predominantly granitic but also contains gabbros and diorites. The individual intrusions show a long-term tendency to become more acid in composition with time, but some mafic magma was available throughout the whole period.

The rocks of the Peruvian coastal batholith extend over a large compositional range, including hornblende–gabbros, tonalites, granodiorites and granites, but there is a distinct compositional hiatus between the gabbros and the more felsic rocks (Fig. 271). Like other Cordilleran batholiths, the predominant granitic rock types are tonalites and granodiorites rather than leucogranites. Cobbing and Pitcher (1972) gave the relative areas of different rock types in this batholith as:

Gabbro–diorite	15.9%
Tonalite	57.9%
Adamellite	25.6%
Granite	0.6%

Cobbing *et al.* (1981) attributed the formation of the acid magmas of the Peru batholith to crustal melting caused by influxes of basic magma from below. They believed that the continual introduction of heat caused the zone of melting to spread upwards into the crust, accounting for the increasingly acid nature of the magmas produced. On the other hand, Beckinsale *et al.*

Figure 271. FMA diagram for the rocks of the Lima segment of the Peru Coastal Batholith (after McCourt 1981).

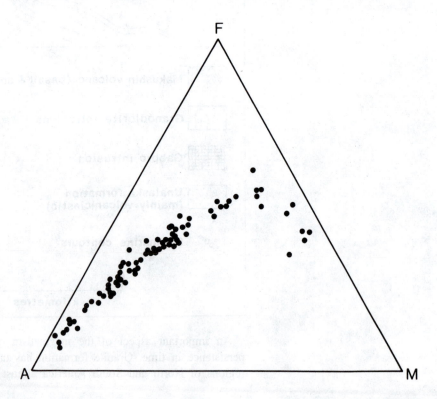

(1985) found the majority of rocks in this batholith to have initial $^{87}Sr/^{86}Sr$ ratios in the range 0.703–0.705, and favoured a sub-crustal source with differentiation giving rise to the more acid magmas.

Taking the Andean region as a whole, the granites have a range of initial $^{87}Sr/^{86}Sr$ from about 0.703 to 0.708 (Fig. 272), and there is a similar range in the granites of California and Japan. This range leads many authors to the conclusion that continental margin granites are of mixed origin, with contributions from both crustal and mantle sources.

The persistence of Cordilleran magmatism with time is difficult to reconcile with an origin by crustal melting alone, because although the crust is capable of yielding a considerable quantity of acid magma, it cannot go on providing melt indefinitely. Once a substantial episode of partial melting has taken place, the crust is left depleted in low-melting constituents and water, and its ability to provide further batches of magma is greatly diminished. The persistence of Cordilleran magmatism is much more easily reconciled with the repeated magma generation anticipated in a continuously fed subduction zone. Possibly the gradual depletion of the lower crust in fusible constituents is responsible for the fall in initial $^{87}Sr/^{86}Sr$ with time that is noticeable in the southern part of the Andean chain (Fig. 272).

Taking Sr and Nd isotopic evidence together, De Paolo (1981a) concluded that about 50% of the material in the Sierra Nevada and Peninsular Ranges batholiths must have come from the mantle and 50% from the crust, and suggested large-scale crustal contamination of mantle-derived magma as an

Figure 272. Plot of initial $^{87}Sr/$ ^{86}Sr ratio against time for the granitic rocks of the Andes (after Pankhurst *et al.* 1988).

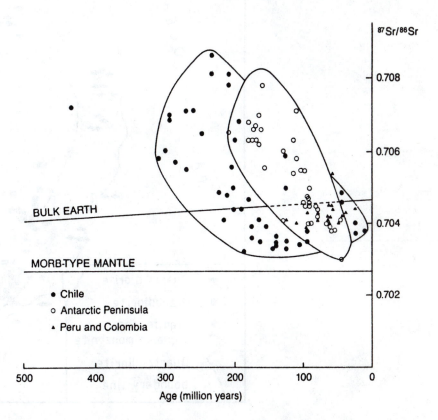

explanation. Alternatively, the large range of isotopic ratios may indicate that more than one type of parent magma is involved. It has been noticed that in the western United States and Canada the high $^{87}Sr/^{86}Sr$ ratios tend to occur in an inland belt and the low $^{87}Sr/^{86}Sr$ ratios occur nearer to the Pacific coast (Petö 1974; Miller and Bradfish 1980). The high-^{87}Sr inland granites are frequently muscovite-bearing and might be described as S-type in Chappell and White's classification. They have the character expected of crustal melts, whereas the low-^{87}Sr (I-type) granites are of more uncertain origin. Either melting of oceanic crust in a subduction zone, or melting of a continental margin greywacke succession might explain their isotopic composition. This two-fold division of the Cordilleran granites was anticipated by Moore (1959), who drew a line (the 'quartz–diorite boundary line') between a western area in which the predominant granitic rock type is quartz–diorite, and an eastern area in which granodiorite and quartz–monzonite are the predominant granitic rocks (Fig. 273).

Figure 273. The 'quartz–diorite boundary line' in the western United States (after Moore 1959). The line marks the eastern limit of quartz–diorite as the dominant granitic rock type.

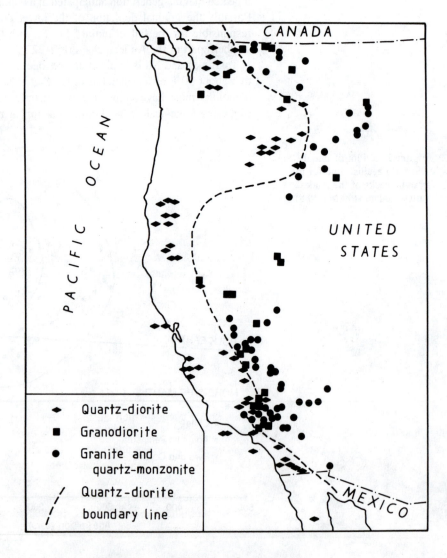

Sulphur isotope studies support the idea that there is more than one source of magma for the continental margin granites. Sasaki and Ishihara (1979) found a difference in $\delta^{34}S$ between the S-type ('ilmenite series') and I-type ('magnetite series') granites of Japan. The S-type granites have $\delta^{34}S$ in the range -11 to $+1‰$, suggesting the incorporation of light, biogenic sulphur from a sedimentary source. The I-type granites have $\delta^{34}S$ of $+1$ to $+9‰$ suggesting the incorporation of marine sulphate possibly derived from subducted sediments.

The Quaternary volcanic rocks of the Taupo zone in New Zealand have provided a testing ground for isotopic investigations into the origin of rhyolitic magmas. The volume of rhyolite in this province (16,000 cubic kilometres) and the preponderance of rhyolitic magma over andesite and basalt (98:2:0.01% respectively) provide a severe challenge to the hypotheses of a sub-crustal source or a differentiative origin. The strontium isotopes rule out a consanguineous relationship between the rhyolites and basalts but leave open the possibility of a close relation between the rhyolites and andesites. The range of $^{87}Sr/^{86}Sr$ values found by Ewart and Stipp (1968) is:

Basalt	0.7042–0.7044
Andesite	0.7046–0.7063
Rhyolite	0.7045–0.7062

Most investigators, including Ewart and Stipp, have favoured crustal anatexis as a source of the rhyolites, but the oxygen isotope study by Blattner and Reid (1982) has called this into question, because the $\delta^{18}O$ values of the rhyolitic rocks ($+7$ to $+9$) are lower than those of the sedimentary rocks which form the underlying crust ($+7$ to $+15$).

INTRA-CONTINENTAL OROGENIC BELTS

The intra-continental orogenic belts, such as the Alps or the Caledonides, correspond in plate tectonic terms to zones of continental collision. The early magmatism in these orogenic belts resembles that of the active continental margins, but the greatest intensity of granite formation occurs at the final culmination of the orogeny when continental suturing is complete and subduction is ceasing or has ceased.

The *Caledonian orogenic belt* of the British Isles has been mapped in great detail, and the history of magmatic activity during its development has been reconstructed from stratigraphic relationships and by isotopic dating. Following a long period of pre-orogenic sedimentation and volcanism (including rhyolitic volcanism of continental margin type) during the lower Palaeozoic, during which the pre-Caledonian Iapetus Ocean is believed to have been closed by subduction, the Caledonian orogeny culminated in the Silurian with intense folding and uplift. There were then intruded very many plutonic bodies of mainly granitic composition, accompanied by volcanic activity, in the late Silurian–early Devonian. The most intense granite-forming activity therefore coincides with or even follows the cessation of subduction, when the tectonic environment changed from continental margin to intra-continental.

The late-Caledonian magmatism is very varied with granites and grano-diorites predominating among the intrusive rocks, but accompanied by smaller amounts of diorite, gabbro and ultrabasic rocks, and even by minor amounts of alkaline rocks. Figure 100 in Chapter 3 shows the distribution of late-Caledonian igneous activity in a small part of the Caledonian belt in western Scotland. The more deeply eroded parts of the Caledonian orogenic belt show that regional metamorphism reached grades up to those corresponding to the beginning of melting, and migmatites are well developed in the areas of highest-grade regional metamorphism. Combined with the frequently high initial $^{87}Sr/^{86}Sr$ ratios of the Caledonian intrusions, these features suggest that there has been extensive crustal melting, of which the Caledonian granites are the product.

The Caledonian granites show a wide range of initial $^{87}Sr/^{86}Sr$, from 0.704 to 0.719. Furthermore there is some correlation between the Sr and O isotopic ratios of the granites and those of the country rocks into which the granites are intruded, as shown in Fig. 274. A similar relationship is shown

Figure 274. Plots of $\delta^{18}O$ against initial $^{87}Sr/^{86}Sr$ for the Caledonian granites and their country rocks in: (A) southern Scotland and northern England; (B) north-east Scotland (after Harmon 1983).

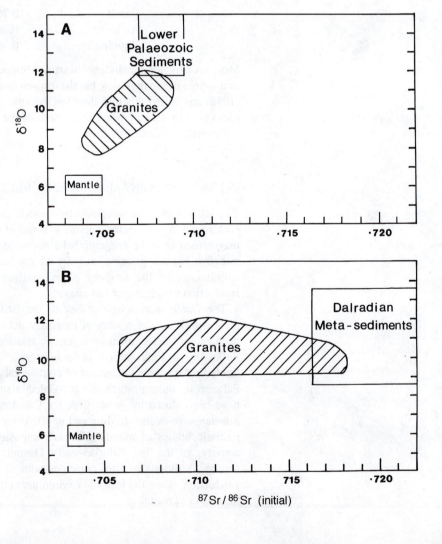

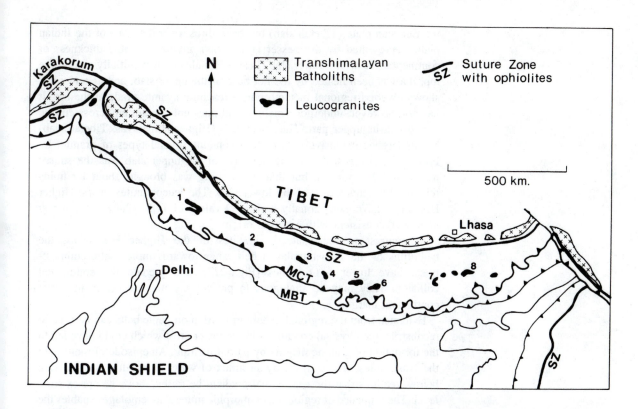

Figure 275. Distribution of granites in the Himalayan orogenic belt (after Blattner *et al.* 1983).

Figure 276. Hypothetical section through the Himalayas (after Le Fort 1975). MBT = Main Boundary Thrust, MCT = Main Central Thrust.

by the trace-element geochemistry. Regional stream-sediment geochemical surveys have been made of a large area of the Scottish Highlands, and the detritus derived from the granites shows the same pattern of high and low trace element concentrations as the detritus derived from the regional country rocks, implying that the magma source had a similar geochemical variation to the local crust (Johnstone *et al.* 1979).

The *Himalayan collision belt* between India and Eurasia contains late-Tertiary granites of several types (Fig. 275). A recent tectonic interpretation of the region is shown in Fig. 276. According to Le Fort (1975), the crust of

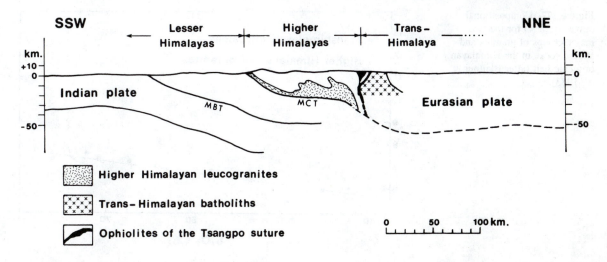

the Eurasian plate (Tibetan slab) has been thrust over the crust of the Indian plate (represented by the Lesser Himalayas), giving a double thickness of continental crust. The lower slab (Lesser Himalayas) was initially cold at the top, but has been heated by the hot base of the upper slab, and as a result it shows inverted regional metamorphism reaching kyanite grade at the top. It has not, however, undergone melting and does not contain any cross-cutting granites in its upper part. The upper slab (Higher Himalayas, Tibetan slab) has undergone extensive melting and contains several types of granite. Le Fort (1981) suggested that the hot base of the upper slab was the source region for the granites, but that the melting was brought about by fluids released by dehydration in the lower slab. The leucogranites of the Higher Himalayas have exceptionally high $\delta^{18}O$ values of 11.5–12.4‰ (Blattner *et al.* 1983), consistent with a crustal origin.

In contrast to the anatectic granites of the Higher Himalayas, the batholiths of the Transhimalayan belt range towards more mafic compositions, have lower $\delta^{18}O$ values (Fig. 277), and predate the continental collision. They may be analogous to present-day magmatism at an active continental margin.

Evidence from the regional metamorphism in orogenic belts can be used to reconstruct the physical conditions in the lower crust which could have led to the formation of granitic magma by partial melting. An episode of melting in the crust, unless triggered off by an influx of water or basaltic magma from below, requires that the geothermal gradient be raised above its pre-existing level. The sequence of regional metamorphic mineral assemblages enables the maximum geothermal gradient to be estimated. It is certain that in many orogenic belts geothermal gradients have risen to levels such that partial melting in the crust not merely could have happened, but must have happened.

Zwart (1967) pointed out that different metamorphic facies series predominate in the different Phanerozoic orogenic belts of Europe: a low pressure facies series in the Variscan (Hercynian), an intermediate facies series in the Caledonides, and a relatively high pressure facies series in the Alps. These metamorphic facies series correspond to high, medium and low geothermal

Figure 277. Compositional range and $\delta^{18}O$ for the two major groups of granites and related rocks in the Himalayan orogenic belt (after Blattner *et al.* 1983).

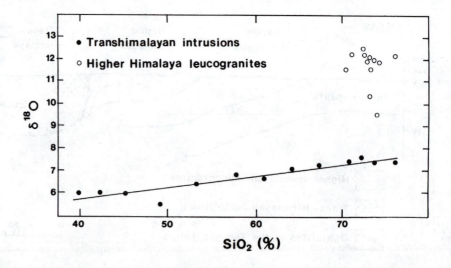

Table 45. The relationship between granites and regional metamorphism in different orogenic belts (after Zwart 1967; Hall 1971)

| | *Geothermal gradient* | | |
	High	*Medium*	*Low*
Metamorphic facies series (characteristic minerals)	Low pressure (andalusite)	Intermediate pressure (kyanite)	High pressure (glaucophane)
Granites	Very abundant	Fairly abundant	Rare
Granite composition	High Q Low Ab	Low Q High Ab	Very low Q Very high Ab
Example	Variscides	Caledonides	Alps

gradients respectively. He pointed out that granites are very abundant in the Variscan orogenic belt, less abundant in the Caledonides, and rare in the Alps. Thus in places where the highest temperatures have been reached in the lower crust, the greatest amounts of granite are found.

In the three orogenic belts studied by Zwart, there is not only a difference in the abundance of granites, but also in their compositions. The granites of the Variscan, Caledonian and Alpine orogenic belts have compositions corresponding to the minimum in the system Q–Or–Ab–H_2O at progressively higher P_{H_2O} values (Table 45). This is what would be expected if the magma had been produced at pressures and temperatures controlled by the relative geothermal gradients inferred from the metamorphic facies series. Similar relations exist on a smaller scale within both the Variscan and Caledonian orogenic belts (Hall 1972b, 1973b). The ranges of composition in these two belts are compared in Fig. 278.

Figure 278. Comparison of the mean and range of composition of Caledonian granites (circle and dotted line) and Variscan granites (triangle and dashed line), in relation to the system Q–Or–Ab–H_2O. Crosses indicate the isobaric minima at water pressures of 0.5 and 10 kilobars (after Hall 1972a).

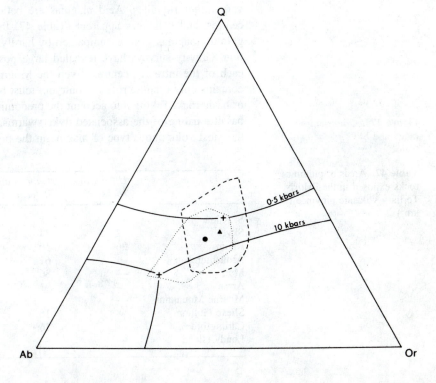

ANOROGENIC CONTINENTAL REGIONS

Unlike orogenic belts, where intermediate igneous rocks are commonplace, the anorogenic continental areas are characterized by a bimodal association of rhyolite and basalt, or granite and gabbro. A typical example is the Quaternary volcanism of Yellowstone Park. In such an association the lack of intermediate compositions is an obvious indication that the basic and acid magmas originated separately. This has been abundantly confirmed by their Sr and Pb isotopic compositions (Table 46).

Table 46. Initial $^{87}Sr/^{86}Sr$ ratios of the igneous rocks of Skye and Yellowstone Park

	Mean $^{87}Sr/^{86}Sr$	Range
Skye granites	0.712	0.709–0.716
Skye gabbros and basalts	0.706	0.703–0.708
Yellowstone Park rhyolites	0.712	0.708–0.727
Yellowstone Park basalts	0.706	0.703–0.708

The Skye data are from Moorbath and Bell (1965) and the Yellowstone Park data are from Doe *et al.* (1982).

The most thoroughly studied anorogenic province is the *British Tertiary volcanic province* (Fig. 99). The igneous activity in this region dates from the period of 54–66 million years ago, corresponding to the opening of the North Atlantic Ocean between Britain and Greenland, where igneous rocks of similar age occur. The volcanic rocks are predominantly basaltic with very minor rhyolites. Acid magmas are better represented in the plutonic centres, and it is very apparent (Table 47) that the abundant gabbroic and granitic magmas were accompanied by hardly any of intermediate composition. Gravity surveys have revealed large positive gravity anomalies under each of the intrusive centres. Even the Mourne Mountains complex, which contains only granitic rocks at outcrop, must be underlain by a large volume of basic rock. Taking into account the predominance of basaltic lavas and the basaltic nature of the associated dyke swarms, basalt must have been by far the most voluminous type of magma in the province.

Table 47. Areas of plutonic rocks exposed in the British Tertiary volcanic province (in km^2)

	Ultrabasic rocks	Gabbro and dolerite	Tonalite	Granite and granophyre
St Kilda	—	8	—	2
Skye	2	74	—	79
Rhum	26	10	—	17
Ardnamurchan	—	61	1	1
Mull	—	86	—	65
Arran	—	1	—	125
Mourne Mountains	—	—	—	143
Slieve Gullion	—	16	—	57
Carlingford	—	16	—	30
Lundy Island	—	—	—	5

The principal possibilities for generating the acid magmas in the British Tertiary province are either differentiation of a basic parent magma or partial melting of lower crustal rocks heated by the influx of the basic magma. The evidence for these alternatives has been reviewed by Thompson (1982). From gravity evidence, granite constitutes less than 8% of the volume of igneous rock in the Skye centre, and the heat content of the much larger volume of basaltic magma would be more than sufficient to cause the required amount of crustal melting. Crustal melting is supported by the initial $^{87}Sr/^{86}Sr$ ratios (0.703–0.708 for the basalts and gabbros; 0.709–0.716 for granites), and by the absence of intermediate magma types.

The evidence for a crustal melting origin is partly contradicted by the isotopic composition of the lead in the Skye granites (Fig. 279). The granites and basalts form a linear array which can be interpreted in either of two ways: as a mixing line or as a secondary isochron. If it is a mixing line, then the Pb in the rocks must be a mixture of Pb derived from the underlying crust (Lewisian) and from the mantle source region of the basalts. This would imply that less than half the Pb in the granite comes from older crust but that an appreciable amount of Pb in the basalts is of crustal origin, completely at variance with major element and Sr isotope evidence that the two types of magma are quite distinct. The alternative interpretation of the Pb isotope array as a secondary isochron implies that the Pb in both the granites and the basalts is derived from a source in which there is heterogeneity dating from about 2900 million years ago, which is roughly the age of the Lewisian basement rocks of the area.

Granitic rocks from the three intrusive centres of the Mull volcanic complex each have different initial $^{87}Sr/^{86}Sr$ ratios, ranging from 0.706 to 0.717, which suggested to Walsh *et al.* (1979) that both crustal melting and differentiation were involved in magma genesis, the two processes varying in importance from one centre to the next. The lowest $^{87}Sr/^{86}Sr$ ratios are in the

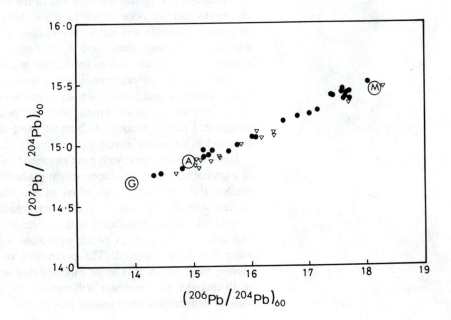

Figure 279. Lead isotopic composition of the Tertiary granites and basalts of the Isle of Skye, Scotland (after Dickin 1981). The isotopic compositions are shown for 60 million years ago, i.e. the time of emplacement. Triangles = granites, dots = basalts, G = mean composition of Lewisian granulite-facies gneiss, A = mean composition of Lewisian amphibolite-facies gneiss, M = estimated composition of the mantle beneath Skye.

Beinn Chaisgidle centre, where there is a complete range of magma compositions from gabbro through intermediate compositions to granite, and where there is the progression of trace element contents and rare-earth patterns that would be expected of a fractionation sequence (Walsh and Clarke 1982).

Taken as a whole, the anorogenic granites show a different range in composition from those of continental margins and orogenic belts. They are rarely peraluminous, are most commonly metaluminous, and are quite frequently peralkaline, as in the Nigerian Newer granite province. They more often have a hypersolvus texture than orogenic granites, and they sometimes contain amphibole, pyroxene or fayalite in addition to, or in place of, biotite. These features reflect low P_{H_2O} during crystallization and a high crystallization temperature.

PEGMATITES AND APLITES

Pegmatites and aplites are associated with most granite complexes. They have several different modes of occurrence:

(a) as sheet-like bodies emplaced at a late stage in the consolidation of a granitic intrusion, cutting either the granite or its country rocks;

(b) as upper border zones in roofed intrusions, as in Fig. 60;

(c) as small segregations completely enclosed by granite, as in Fig. 280; and

(d) as veins, lenses and irregular bodies in regional metamorphic terrains, as in Figs. 254 and 281.

Pegmatites and aplites are enriched in the minimum-melting constituents of granite and are poor in the higher-melting constituents such as ferromagnesian minerals and calcic plagioclase. They consist of quartz, alkali feldspars (including albite) and usually muscovite, and not uncommonly contain rarer minerals such as tourmaline and lepidolite.

These features of occurrence and composition identify pegmatites and aplites either as final liquid fractions from the crystallization of a large body of magma or as initial liquid fractions produced under conditions of progressive partial melting. At both of these stages granite magma can be expected to be comparatively rich in water.

Jahns and Burnham (1969) have described in detail how the water content of a crystallizing granite magma might gradually rise until the water content reached the miscibility limit, when an aqueous phase would separate and coexist with the magma. Pegmatite may readily be interpreted as granite magma which has crystallized in the presence of a separated aqueous fluid, and aplite as the product of an equivalent magma from which separated water has been removed. The pegmatites of migmatite areas show the reverse relationship. It is to be expected that in a zone of partial melting, a small quantity of free water will initially be available, resulting from the various dehydration reactions of progressive regional metamorphism. This

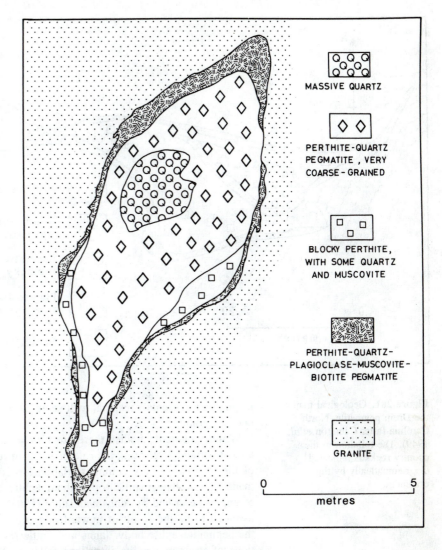

Figure 280. Geological map of a pod-like body of granite pegmatite, Robbins silica prospect, California (after Jahns 1953). The relative coarseness of the pegmatite units is indicated diagrammatically by the ornament.

MASSIVE QUARTZ

PERTHITE-QUARTZ PEGMATITE , VERY COARSE-GRAINED

BLOCKY PERTHITE, WITH SOME QUARTZ AND MUSCOVITE

PERTHITE-QUARTZ- PLAGIOCLASE-MUSCOVITE- BIOTITE PEGMATITE

GRANITE

0   5
metres

water may be more than enough to saturate and coexist with the first small fraction of melt to be formed. A higher degree·of partial melting would produce too much melt for the available water to saturate, and a separate aqueous phase could no longer be present. Pegmatites would therefore only be formed if the degree of partial melting remained relatively small.

Many individual occurrences of pegmatite are described in the reviews by Cameron *et al.* (1949) and Jahns (1955). The distinction between 'initial-melt' pegmatites associated with partial melting and 'final-melt' pegmatites associated with igneous intrusions is difficult to make in practice. Some of the pegmatites which occur in metamorphic surroundings are too large to represent initial melt fractions frozen in place, and are enclosed in schists or gneisses which are not migmatitic. They probably represent melts which have collected and migrated upwards for some distance from their original source. On the other hand, granitic intrusions are present in most areas of regional metamorphism, so there is always a possibility that a particular

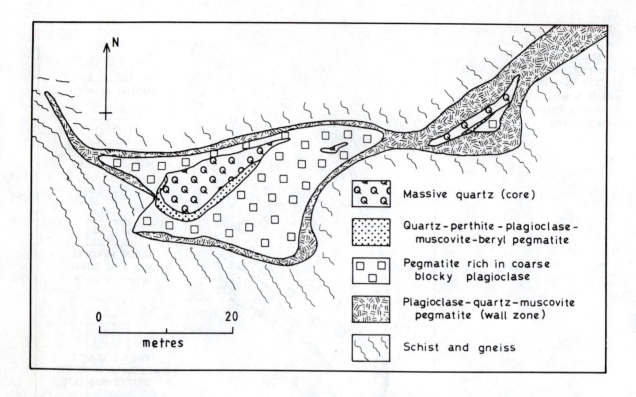

Massive quartz (core)

Quartz – perthite – plagioclase –
muscovite – beryl pegmatite

Pegmatite rich in coarse
blocky plagioclase

Plagioclase – quartz – muscovite
pegmatite (wall zone)

Schist and gneiss

Figure 281. Geological map of the Drum pegmatite, North Carolina (after Cameron *et al.* 1949). The foliation in the country rocks is indicated diagrammatically by the ornament.

pegmatite represents a residual melt segregation from a nearby or underlying body of granite.

Within pegmatite bodies there are pronounced compositional variations, almost certainly reflecting the role of water in redistributing the constituents of the magma. A very common relationship is for an inclined sheet to be made up of a pegmatitic top and an aplitic bottom, as in the excellent examples described by Jahns and Tuttle (1963). The pegmatitic tops may be assumed to be the result of upward migration of the aqueous phase within the pegmatite–aplite body. Jahns and Tuttle showed how the bulk compositions of pegmatite–aplite intrusions fall near the minimum of the system Q–Or–Ab–H$_2$O, but the pegmatite members are usually K-rich and the aplite members are Na-rich, a difference no doubt brought about by volatile transfer, i.e. redistribution of K and Na through the aqueous phase. Another common relationship in pegmatite bodies is for the pegmatite to pass into quartz-rich material similar to hydrothermal vein quartz. There is often a quartz core to enclosed pegmatite segregations in granite, and quartz vein-type material in the centre of symmetrically zoned pegmatite dykes. These quartz-rich rocks have presumably been deposited from concentrations of the aqueous phase which coexisted with the magma.

Many pegmatites are enriched in rare elements, such as B, Be, F, Li and Sn, resulting in the presence of such minerals as tourmaline, beryl, topaz, lepidolite and cassiterite. It is sometimes said that the rare elements have become concentrated in the residual magma fraction because they were unable to substitute isomorphously into rock-forming minerals at an earlier stage of crystallization. This is not correct. Most of them are well able to enter rock-forming minerals, for example lithium in micas or pyroxenes, and the normal behaviour of trace elements unable to enter the structure of a major rock-forming mineral is simply to crystallize as a small amount of an appropriate accessory mineral, for example Zr in zircon. The elements which are concentrated in pegmatites are in fact those which are strongly partitioned into the aqueous phase when an aqueous fluid coexists with a silicate magma.

The rarity of pegmatites of gabbroic composition is readily explained by the relatively low water content of basic magmas. There are small bodies of what have been called gabbro–pegmatites in many ophiolite complexes, but it is unlikely that they owe their coarse grain size to crystallization in the presence of water, particularly since their main ferromagnesian mineral is pyroxene and not amphibole. They perhaps owe their origin to slow cooling of a basaltic magma at a depth where the country-rock temperature was very high, and are not analogous in any way with granitic pegmatites. In contrast, nepheline–syenite intrusions are often accompanied by pegmatites of similar composition, which probably originated in the same way as granite pegmatites. They are often rich in rare and exotic minerals, but a different suite from those found in granite pegmatites and reflecting the different trace element concentrations of nepheline–syenite magmas.

CHAPTER 10

Andesites

OCCURRENCE AND ASSOCIATIONS

Andesites occur in association with basalts, dacites and rhyolites in the volcanoes of modern island arcs and continental margins, and in orogenic belts of all ages. In island arcs which are not underlain by continental crust they are associated with abundant basalts and scarce dacites and rhyolites. In volcanic regions which are underlain by continental crust they are associated with less abundant basalts and with voluminous dacites and rhyolites.

Table 48 gives some estimates of the relative abundance of acid, intermediate and basic rock types in different provinces. These are only a very approximate indication of the importance of the various magma types. Measurements of volcanic complexes tend to underestimate the importance of acid pyroclastic materials, whereas plutonic complexes are under-representative of the basic magmas which pass through to the surface without crystallizing in bulk at depth. Despite these limitations, the figures do indicate the great abundance of intermediate magma in the orogenic environment compared with other tectonic situations.

Table 48. The proportions of acid, intermediate and basic igneous rocks in some orogenic provinces

	Japan	Oregon Cascades	Taupo, N.Z.	Lesser Antilles	Central Peru	Southern California
Rock types	volcanic	volcanic	volcanic	volcanic	plutonic	plutonic
Measurement	volume	volume	volume	area	area	area
Basic rocks (%)	12	85	<0.1	17	16	7
Intermediate rocks (%)	73	13	<2.5	42	58	63
Acid rocks (%)	15	2	>97.4	41	26	30
Reference	(1)	(2)	(3)	(4)	(5)	(6)

Sources of data: (1) Sugimura (1968); (2) McBirney and White (1982); (3) Cole (1981); (4) Brown *et al.* (1977); (5) Cobbing and Pitcher (1972); (6) Larsen (1948).

Relatively few rocks described as andesites occur outside the island arc and continental margin environment. Some intermediate rocks have been given the name andesite in islands such as Iceland and Hawaii and in parts of the ocean floor, but they are dissimilar to the bulk of orogenic andesites, rarely containing hornblende and being comparatively low in Al_2O_3. Some, associated with alkali basalts and trachytes, would nowadays be classified as trachybasalt, hawaiite or mugearite. Others, associated with tholeiite and rhyolite, would now be termed icelandite. Only a few are petrographically indistinguishable from orogenic andesites. Although non-orogenic andesites have some slight geochemical differences from orogenic ones, the most notable characteristic of anorogenic andesites is simply their rarity.

In island arc and continental margin volcanic provinces, the basalts which are associated with andesites show a systematic compositional variation. Those on the oceanic side of an arc or continental margin are tholeiitic; further away from the ocean they are of high-alumina type; furthest from the ocean they are alkali basalts, as shown in Fig. 246. Jakes and White (1972) believe that the associated andesites and dacites vary in a similar way, and according to them the volcanic rocks of island arcs can be grouped into three associations:

1. The tholeiitic association: tholeiitic basalt–'tholeiitic andesite' (icelandite)–'tholeiitic dacite'
2. The calc-alkaline association: 'calc-alkaline basalt' (high alumina basalt)–basaltic andesite–andesite–dacite
3. The shoshonitic association: 'shoshonitic basalt' (absarokite)–'shoshonitic andesite' (banakite)–'shoshonitic dacite' (latite)

These three associations are said to follow one another laterally and stratigraphically in recent island arcs, but this may be an oversimplification. According to Aramaki and Ui (1982), the calc-alkaline series is present right across Japan, being accompanied by a tholeiitic association close to the Pacific Ocean and by an alkalic association away from the Pacific Ocean.

The most important chemical differences between the rocks of the three series are in their alkali contents and iron/magnesium ratios (Table 49). The

Table 49. Representative compositions of island arc volcanic rocks (examples from Jakes and White 1972)

	SiO_2	Al_2O_3	Na_2O	K_2O	$\Sigma Fe/Mg*$
Tholeiitic association					
Basalt	51.57	15.91	2.41	0.44	1.45
Andesite	57.40	15.60	4.20	0.43	2.51
Dacite	79.20	11.10	3.40	1.58	3.94
Calc-alkaline association					
High-Al basalt	50.59	16.29	2.89	1.07	0.98
Andesite	59.64	17.38	4.40	2.04	1.33
Dacite	66.80	18.24	4.97	1.92	1.51
Shoshonitic association					
Shoshonite	53.74	15.84	2.38	2.57	1.27
Latite	59.27	15.90	2.67	2.68	0.99
Abyssal tholeiite (for comparison)					
	49.34	17.04	2.73	0.16	1.23

*$\Sigma Fe/Mg$ denotes the ratio $(FeO + Fe_2O_3)/MgO$.

tholeiitic, calc-alkaline and shoshonitic associations are characterized by successively higher alkali and especially K_2O contents and lower FeO/MgO ratios. An alternative classification of island-arc volcanic series based solely on their K_2O contents is shown in Fig. 282.

The term 'calc-alkaline' is frequently used in describing orogenic rock series and requires some explanation. The term was originally proposed by Peacock (1931) to describe the relative variation of $(Na_2O + K_2O)$ and CaO in a magma series. If a series of rocks is plotted on a Harker diagram, as in Fig. 283, the value of $SiO_2\%$ at which the $(Na_2O + K_2O)$ and CaO trends cross is called the alkali-lime index. In Peacock's classification, a rock series is called 'calc-alkalic' if the alkali-lime index lies between 56 and 61. Many of the orogenic rock series to which the name calc-alkaline has been applied do not meet this criterion, the series plotted in Fig. 283 being an example, and it is difficult to see any justification for its continued use. In fact the whole classification of orogenic lavas into series is questionable, since it groups together basalts, andesites and rhyolites which may not be consanguineous, while playing down the relationships between different types of basalt or different types of andesite.

It appears that there is a relationship between the K_2O content of modern orogenic andesites (and the basalts and rhyolites associated with them) and the depth of focus of earthquakes in the underlying Benioff zone. In each volcanic arc there is a range of lava types becoming richer in K_2O from basalt to rhyolite, but for any particular SiO_2 content there is said to be a positive correlation between K_2O ('K') and the depth of earthquake hypocentres ('h'). Figure 284 shows the K–h relationship for andesitic volcanoes as a whole.

Although taken for granted by many workers, the K–h relationship is not as clear-cut as it might seem. In many island arcs and continental margins all

Figure 282. Classification of island arc volcanic rock series according to their K_2O content (after Taylor *et al.* 1981).

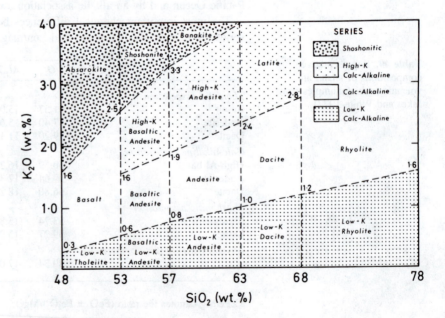

Figure 283. The variation of CaO and Na₂O + K₂O against SiO₂ for the Quaternary volcanic rocks of the central Oregon Cascades, based on the average compositions of basalts, basaltic andesites, andesite, dacites and rhyolites compiled by McBirney and White (1982).

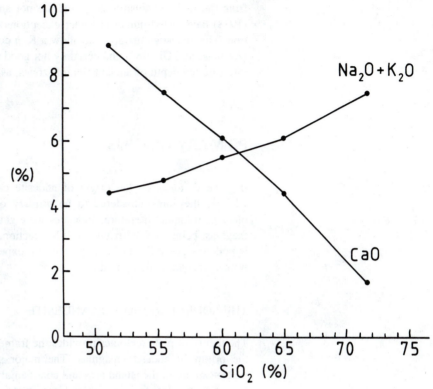

the active volcanism is along a line parallel to the oceanic trench and there is a no spread of 'h' values that would enable a K–h relationship to be properly plotted. In Central America all the active andesitic volcanoes lie along a line situated 150 kilometres above the Benioff zone and there is no K–h correlation (Carr *et al.* 1979); a few very alkaline volcanoes lie far inland

Figure 284. The relationship between K (% K_2O at 57.5% SiO_2) and h (depth to the top of the inclined seismic zone in kilometres) for 64 volcanoes in 14 different volcanic arcs. Circles are continental margin volcanoes and dots are intra-oceanic arcs (after Dickinson 1975).

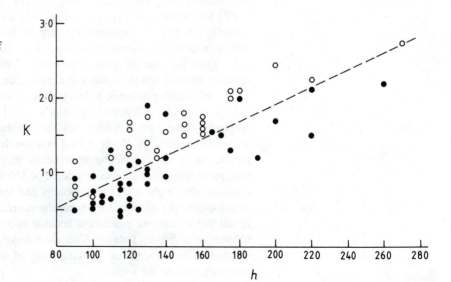

from the inclined seismic zone and are not andesitic. Arculus and Johnson (1978) have listed numerous other exceptions to the supposed K–*h* correlation. The andesites of Japan do show a K–*h* correlation, but it is a poor one (Aramaki and Ui 1982) and certainly not good enough to serve as a basis for inferring the depth of ancient Benioff zones, as some authors have attempted to do.

PRIMARY ORIGINS

In general, ideas on the origin of andesite can be classified according to whether they are considered to be primary or secondary magmas. In the older petrological literature, andesites were generally regarded as secondary magmas, being derived from basalt by fractionation or contamination. These hypotheses are still widely supported, but experimentation has shown that a primary origin is also possible.

THE SOURCE REGION OF ANDESITE

The theory of plate tectonics provides the framework within which to discuss the origin of andesite magmas. The major sites of present-day andesite volcanism are in the island arcs and continental margins, which are underlain by subduction zones in which oceanic crust is descending into the mantle. There are many indications of a relationship between the occurrence of andesites and the existence of the underlying seismic zone. The correlation between the K_2O and trace element contents and the depth of the seismic zone is one expression.

Barazangi and Isacks (1976) have drawn attention to differences in the volcanism of regions overlying seismic zones of different inclination. In segments of the Andes underlain by a very shallow-dipping seismic zone (~10°) recent volcanism is absent (Fig. 285), and elsewhere the nature of the volcanic products (for example the ratio of basalts, andesites and rhyolites) may be related to the angle of subduction (Lopez-Escobar *et al.* 1977). Also, the higher the rate of plate convergence, the greater the proportion of andesite erupted and the smaller the proportion of basalt (Sugisaki 1976).

The relationship between andesitic volcanism and seismic activity at depth has been vividly illustrated by Blot's (1981) study of earthquakes and eruptions in the New Hebrides and New Zealand. He showed, for instance, that each eruption of White Island volcano is preceded by an earthquake originating in the subduction zone below the volcano but occurring some time previously. Earthquakes at a depth of 150 kilometres are followed by an eruption after eight and a half months and those at 250 kilometres by an eruption after 14 months. Of course the source of the earthquake is not the actual site of magma generation because rocks in a melting region do not behave in a brittle fashion. The earthquakes must be related either to subduction movements or the fracturing of the rocks through which the magma passes to the surface.

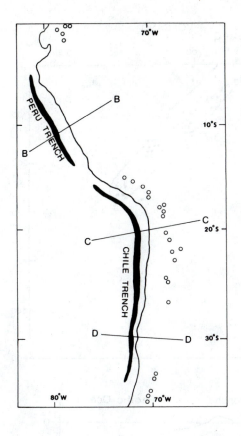

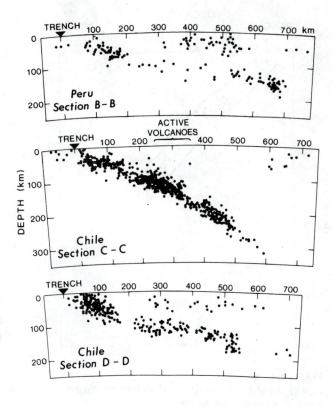

Figure 285. Cross-sections through the inclined seismic zone which underlies the western edge of South America. The map on the left shows the position of the offshore trench and the location of active volcanoes (as circles). The composite sections at 10 °S and 30 °S (B–B and D–D), where there are no active volcanoes, show less seismic activity than the section at 20 °S (C–C), where there are many active volcanoes. After Barazangi and Isacks (1976).

More detailed study of the Benioff zone underneath the Andes suggests that the positions of volcanoes may be related to aseismic gaps in the Benioff zone (Fig. 286), which can be interpreted as parts of the zone where partial melting is taking place. In both the Andes and the Tonga–Kermadec island arc this aseismic gap is located at a depth of about 150 kilometres beneath the active volcanoes (Hanuš and Vaněk 1978, 1979).

In Japan, volcanicity starts abruptly 270 kilometres to the west of the trench axis where the centre of the Benioff zone is at a depth of 140 kilometres; 81% of the volume of magma erupted is within 50 kilometres of the volcanic front and 99% within 150 kilometres of the volcanic front (Fig. 287). Below the volcanic region and above the Benioff zone is a wedge-shaped section of the asthenosphere characterized by absence of seismic activity, low Q value and high heat flow, perhaps indicative of partially molten mantle (Aramaki and Ui 1982).

The observation that andesites occur above a subduction zone does not necessarily imply that the magmas actually originate in the subduction zone, although this is the most obvious possibility. At a depth of 100 kilometres or more, a subduction zone is overlain by a wedge of mantle (and in some cases also by continental crust), where magma generation could also occur.

The main characteristics of the subducted oceanic crust are: (a) that it is mainly of basaltic (i.e. oceanic tholeiite) composition; (b) that it is at least partly metamorphosed and hydrated; (c) that because of the rate of

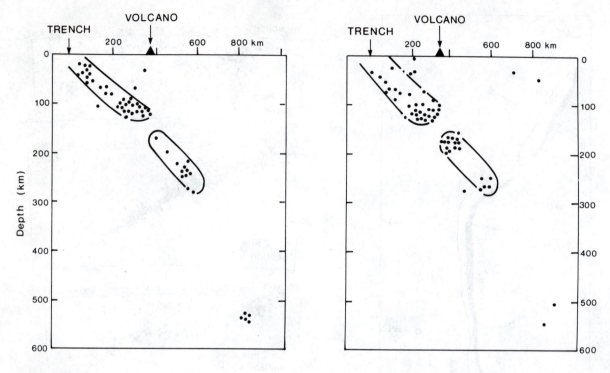

Figure 286. Two sections through the inclined seismic zones of western South America which show a seismic hiatus immediately underneath actual volcanoes (after Hanuš and Vaněk 1978). Dots show the foci of earthquakes stronger than ISC magnitude 4.

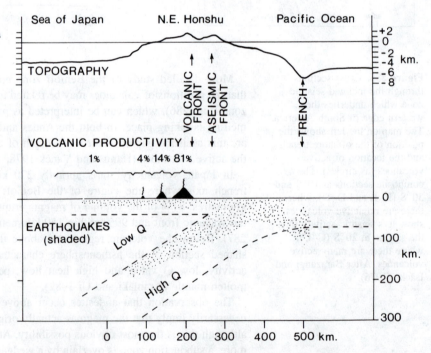

Figure 287. Cross-section through Japan showing the position of the oceanic trench, the distribution of seismic activity, and the location and relative intensity of volcanic activity (after Aramaki and Ui 1982).

subduction it is colder than the mantle above and below it; (d) that it is underlain by mantle depleted of low melting constituents which has travelled with it from the spreading axis at which the oceanic crust was generated, and it may also be immediately underlain by ultrabasic cumulates devoid of low-melting constituents; and (e) that it is initially overlain by thin deposits of oceanic sediment and by the occasional oceanic island containing alkali basalts. The main possibilities for magma generation at a subduction zone are therefore partial melting of the subducted oceanic crust itself (with or without accompanying sediment), and partial melting of the overlying mantle under the influence of water expelled from the oceanic crust as it is brought to higher temperatures at depth.

The involvement of oceanic sediment is a complicating factor, because it would be relatively rich in potassium and other incompatible elements, and could make an important difference to the composition of the magmas produced. Because of their low density and lack of rigidity oceanic sediments are probably not subducted, but are scraped off the surface of the oceanic crust at the top of the subduction zone. However, isotopic and trace element evidence has occasionally been adduced for a small contribution of oceanic sediment to the magma source, and Ryabchikov *et al.* (1982) have shown experimentally that andesitic liquid can be produced by partially melting a mixture of basalt and argillaceous sediment. Nevertheless, Stern and Ito (1983) asserted that sea-floor sediments and alkali basalt seamounts are not accumulating in the Marianas trench and must therefore be undergoing subduction, even though there is very little chemical or isotopic

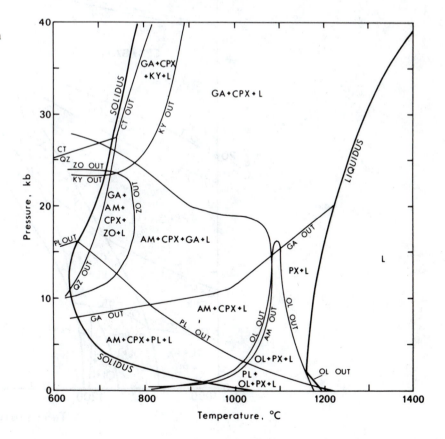

Figure 288. Generalized phase relations of tholeiitic basalt with 5% H_2O (after Green 1982). Abbreviations: AM = amphibole, CPX = clinopyroxene, CT = coesite, GA = garnet, KY = kyanite, L = liquid, OL = olivine, PL = plagioclase, PX = pyroxene, QZ = quartz, ZO = zoisite.

evidence for a sedimentary contribution to the magmas of this particular island arc.

Recent experimentation on andesite development in or adjacent to subducted oceanic crust has been concerned with the following possibilities: (a) partial melting of amphibolite-facies basaltic material at shallow depth, or amphibole-controlled fractionation of basaltic magma; (b) partial melting of eclogite-facies basaltic material at greater depth, or eclogite-controlled fractionation of basaltic magma; or (c) partial melting of mantle peridotite in the presence of water.

Figures 288 and 289 show the generalized phase relations of basaltic rocks under hydrous and anhydrous conditions. Because of mineralogical and field evidence for a high water content of andesite magmas it is assumed that partial melting or fractionation would have taken place in the presence of water. At the same time it is very unlikely that melting took place in the presence of excess water, because of the inadequate water content of the potential source materials.

Figure 289. Generalized phase relations of anhydrous tholeiitic basalt (after Green 1982). Abbreviations are the same as in Fig. 288.

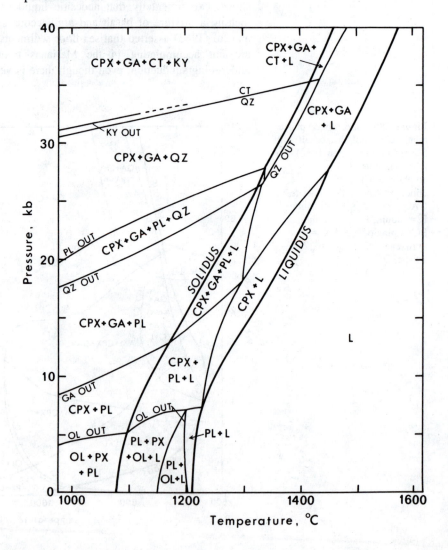

PARTIAL MELTING OF AMPHIBOLITE

Experimental studies have shown that in a basic rock melted under hydrous conditions, amphibole would coexist with melt up to a pressure of about 25 kilobars and plagioclase would coexist with melt up to about 15 kilobars (Fig. 288). There is some doubt about whether the temperature in a subducting slab would be high enough for melting to occur in the 0–25 kilobar range in which basaltic rocks would have an amphibole-bearing mineralogy. A subducted plate is colder than the surrounding mantle, and according to most thermal models it would not reach the temperature required for melting to begin before it reached a depth where the pressure was over 25 kilobars, although this depends on what assumptions are made about the rate of subduction and the heating mechanisms in a subduction zone. The temperatures needed for melting in the amphibolite facies are most likely to be reached if the subducting slab is relatively young and therefore hot at the time of subduction.

The rate of subduction and the angle of dip of the subduction zone also have an influence on melting, by determining the depth at which the subducted material reaches its melting temperature. Slow subduction, or a gently dipping subduction zone, would permit the subducted material to start melting at a depth where its hydrated basaltic components still had an amphibolite mineralogy, but faster subduction or a steeper dip would bring them into the eclogite facies and dehydrate them before melting could begin.

Delany and Helgeson (1978) calculated the temperatures and pressures of dehydration reactions in a subducting slab of hydrated oceanic crust, and concluded that dehydration would mostly take place in the pressure range 20–30 kilobars at between 300 and 700 °C. These temperatures are below the solidus and, if geologically applicable, would virtually eliminate the melting of subducted amphibolite-facies basaltic rocks as a source of andesite magma. In contrast, Allen and Boettcher (1978) have argued the case for melting of amphibolite. They point out that the depth of the inclined seismic zone beneath Andean volcanoes is 90–150 kilometres, but that the earthquakes originate in the cold interior of the subducted slab and not on the top edge where the subducted oceanic crust is. They relate the formation of andesite to melting due to the breakdown of amphibole, which would occur at a depth of about 75 kilometres. Melting in a subduction zone is most likely to occur when relatively young (and therefore relatively hot) oceanic lithosphere is subducted, a possible example being at Mount St Helens (Defant and Drummond 1993).

Holloway and Burnham (1972) carried out hydrous melting experiments on a tholeiitic basalt at pressures of 5–8 kilobars, and obtained liquids of broadly andesitic composition at an appropriate degree of melting. However, these pressures are too low to be applicable to melting in a subduction zone. Sen and Dunn (1994) carried out dehydration melting experiments on amphibolite at 15–20 kilobars, and found that although liquids intermediate in SiO_2 content could be obtained they were too low in CaO and MgO to be described as andesitic.

PARTIAL MELTING OF ECLOGITE

The Benioff zone actually lies at a depth of about 150 kilometres below those volcanoes from which the greatest volumes of andesite magma are erupted, i.e. in the region of 40–50 kilobars total pressure. Above about 25 kilobars amphibole is no longer stable, and subducted oceanic crust at this depth would have an eclogitic mineralogy and would have lost most of its water content. It may not be completely anhydrous because of the presence of serpentinite, which can retain some H_2O up to pressures as high as 130 kilobars (Ringwood 1975).

Partial melting experiments, for example by Green and Ringwood (1968), suggest that under anhydrous conditions an andesitic liquid can be obtained by the partial melting of quartz–eclogite (coesite–eclogite above 35 kilobars). As Fig. 290 shows, the primary liquidus phases of anhydrous andesite magma at high pressures are garnet and clinopyroxene, i.e. andesite magma is in equilibrium with the mineralogical constituents of eclogite. Unfortunately, all the evidence suggests that actual andesite magmas are not anhydrous, for example their sometimes explosive mode of eruption and the frequent presence of hornblende.

Whereas the primary liquidus phases of anhydrous andesite magma at 25–40 kilobars are garnet and clinopyroxene, under hydrous conditions garnet alone is present at the liquidus. This implies that a hydrous andesite magma could not be in equilibrium with a garnet–pyroxene (i.e. eclogitic) source material.

Considering both the low temperatures in the upper part of a subducted lithosphere slab and its mainly dehydrated nature at depth, most authors do not favour an origin for andesite magmas in the subducted oceanic crust but prefer a source in the overlying mantle.

HYDROUS MELTING OF PERIDOTITE

Melting of peridotite might take place in the mantle overlying a subduction zone, under the influence of water expelled from hydrated basaltic rocks and serpentinites as they undergo dehydration with increasing temperature and pressure. Under dry conditions, partial melting of peridotite in the mantle could be expected to yield a basaltic magma, but Kushiro and Yoder (1969) predicted on the basis of phase relations in the synthetic system MgO–SiO_2–H_2O that in the presence of water a more silica-rich melt might be produced. Attempts by various workers to confirm this experimentally have given ambiguous results. Mysen and Boettcher (1975) were able to produce intermediate liquids by the hydrous melting of natural peridotites at pressures up to 25 kilobars and over a wide temperature range above the solidus, but the liquids were not strictly andesitic, being very rich in normative plagioclase and poor in Fe and Mg. A closer approach to andesitic melt compositions was achieved when a small amount of CO_2 was present in addition to H_2O. However, when the molar fraction of CO_2 equalled or exceeded that of H_2O the melts became alkaline, and were not at all similar to andesite.

Figure 290. The liquidus of andesite at various concentrations of H_2O, showing the phases crystallizing at or close to the liquidus (after Stern *et al.* 1975).

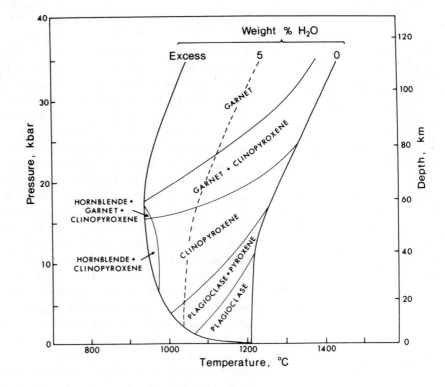

Figure 290 shows which minerals are present immediately below the liquidus of andesite at various pressures and degrees of H_2O-undersaturation. The predominant liquidus minerals are plagioclase at crustal pressures, clinopyroxene at uppermost mantle pressures, and garnet at deeper mantle pressures. These experimental observations suggest the unlikelihood of andesite magma being produced by the melting of mantle peridotite, because if such a melt were produced by the partial melting of olivine-rich material leaving an olivine-rich residuum it would almost certainly be ready to crystallize olivine at its liquidus.

There is one particular group of high-magnesian andesites which have been proposed as primary mantle melts (Kuroda *et al.* 1978). These are the boninites, which are glassy lavas with high Mg/Fe ratios and with orthopyroxene phenocrysts. Tatsumi (1981) described a high-magnesian andesite from south-west Japan which is in equilibrium with olivine and orthopyroxene when melted in the presence of water at 11–15 kilobars, and which could therefore be a primary melting product of peridotite. It has a high Mg/Fe ratio and contains olivine phenocrysts of composition Fo_{87-91}. Tatsumi and Ishizaka (1982) recorded another andesite from the same area which contains olivine phenocrysts more magnesian than those in the accompanying basalt, from which it could not therefore have been derived by crystal fractionation.

FORMATION OF ANDESITE MAGMA BY MELTING IN THE CRUST

Because of the association of andesites with subduction zones, it is usually assumed that the magma must originate in these zones or in the adjacent mantle, but there have been suggestions that andesite magma could be formed by melting of basic rocks within the continental crust. This could not account for more than a limited proportion of andesites because of the absence of continental crust beneath many island arcs where andesite is abundant, but it is worth considering as a minor source of andesite. Burnham (1979) proposed the injection of basalt magma from beneath the crust as a possible heat source for this process, and Gust and Arculus (1986) described andesites from Arizona which may have originated in this way.

The partial melting of basalt under hydrous conditions at crustal pressures has been shown experimentally to produce liquids of intermediate composition (Holloway and Burnham 1972), but the temperatures required are higher than those that would normally exist in the lower crust. As in the case of metasediments, the temperature required for the melting of amphibolites is crucially dependent on the availability of water. The H_2O contents of amphibolites and pelitic schists are similar, but whereas both types of rock start to melt at about the same temperature, a much higher temperature is needed to produce a completely liquid andesite than a completely liquid granite. Table 50 indicates the much wider melting intervals of andesite and basalt compared with granite. To produce an andesite liquid with 2% water at a pressure of 10 kilobars (i.e. near the base of the crust in an orogenic area) would require a temperature of about 1020 °C.

Table 50. Approximate solidus and liquidus temperatures of granite, andesite and basalt at a pressure of 10 kilobars

	Solidus (°C)	Liquidus (°C)
Granite (water-saturated)	620	710
Andesite (water-saturated)	630	940
Basalt (water-saturated)	640	1060
Andesite (dry)	1070	1220
Basalt (dry)	1140	1250

Of course the bulk composition of the whole continental crust is not very different from that of andesite (Table 51) and very extensive melting in the crust would give an intermediate magma. This is held by some authors to be

Table 51. The average composition of the continental crust compared with the average composition of andesite

	Continental crust	Andesite
SiO_2	61.5	57.9
TiO_2	0.7	0.9
Al_2O_3	15.1	17.0
Fe as Fe_2O_3	6.3	7.8
MgO	3.7	3.3
CaO	5.5	6.8
Na_2O	3.2	3.5
K_2O	2.4	1.6
Reference	Wedepohl (1994)	Le Maitre (1976)

the origin of tonalites and quartz–diorites, but it is doubtful whether melting could ever be so extensive as to cause complete fusion. Quartz–diorite magmas are probably crystal–liquid mushes rather than pure liquids. Very extensive crustal melting would also involve a large sedimentary contribution to the andesite magma and this is not consistent with the strontium isotope composition of andesites.

SECONDARY ORIGINS

FRACTIONATION

It is generally supposed that if andesite magma is formed by differentiation, its parent magma must be basaltic. The phase diagrams for basalt (Figs 288 and 289) show what possibilities exist for crystal fractionation in a basaltic magma. At high pressures, fractionation would involve garnet and/or pyroxene ('eclogite fractionation'). At low pressures, fractionation would involve pyroxene, plagioclase and olivine in the absence of water, or these minerals plus amphibole if water was present.

It has been observed that where basaltic magma is actually seen to undergo fractional crystallization, as in the examples of the Skaergaard intrusion or Alae lava lake (Chapter 7), the liquid line of descent does not lead to andesite; instead it leads towards ferrobasalt. The reason for this is that during low-pressure anhydrous fractionation the mafic minerals which crystallize are Mg-rich olivine and pyroxene, and their separation causes an increase in the Fe/Mg ratio of the residual liquid.

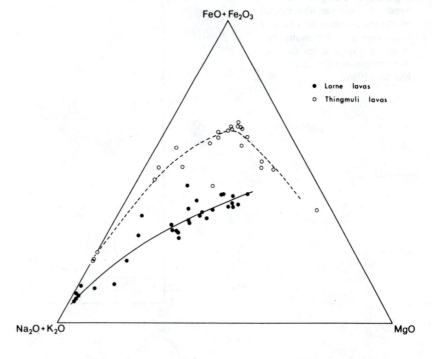

Figure 291. FMA diagram for the orogenic lava series of Lorne, Scotland (a 'calc-alkaline' series), compared with the non-orogenic lavas of Thingmuli, Iceland (a 'tholeiitic' series). The data are from Groome and Hall (1974) and Carmichael (1964).

It is only in the rare non-orogenic examples (e.g. Thingmuli – see Fig. 291) that the trend from basalt towards intermediate or acid magma involves a degree of Fe-enrichment. The normal orogenic andesites are not significantly enriched in Fe compared with the associated basalts.

The only way in which fractionational crystallization can lead to an increase in SiO_2 without an increase in Fe/Mg (i.e. towards an andesitic composition) is if the ferromagnesian minerals which crystallize are iron-rich. Osborn (1959, 1969, 1979) proposed that under oxidizing conditions the crystallization of magnetite might remove iron from the magma and prevent iron-enrichment. He suggested that in a hydrous magma the dissociation of H_2O and the loss of H_2 by diffusion would be a possible cause of increased f_{O_2}. Some authors have criticized Osborn's hypothesis on the grounds that magnetite is not a liquidus phase during crystallization of basic magmas or that it is not a common phenocryst mineral, but neither of these objections is correct.

Figure 292 shows two possible liquid lines of descent from basalt. Curve **A** shows initial Fe-enrichment due to the crystallization of olivine or pyroxene, followed by a sharp change in direction when magnetite starts to crystallize. Curve **B** shows the effect of magnetite crystallizing alongside olivine or pyroxene from the start. The two contrasting trends of differentiation have been called the Fenner trend (Fe-enrichment) and the Bowen trend (Si-enrichment), and the fractionation products would correspond to the tholeiitic and calc-alkaline magma series respectively.

A significant water content is required for fractionation of basalt to lead to andesite. Reference to Figs 288–290 will show that the separation of crystals

Figure 292. Two possible trends of liquid variation caused by the fractionation of a basaltic liquid (see text). For examples of actual magma series showing these trends, see Grove and Kinzler (1986).

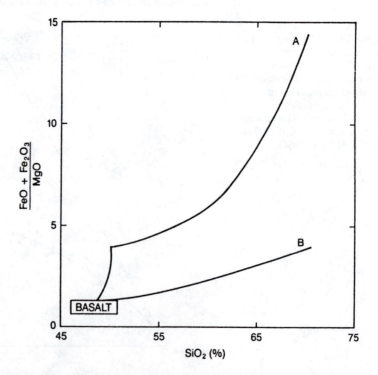

from a basaltic magma would only be appreciable at temperatures in the order of 1100°C, whereas anhydrous andesite is only completely liquid above about 1220°C at atmospheric pressure. Furthermore, the only phase which crystallizes from andesite for about 70° below the liquidus at atmospheric pressure is plagioclase, which is not what you would expect if its parent magma had been fractionating pyroxene or olivine as well. In contrast, the production of andesitic liquids by the fractionation of H_2O-saturated basalt has been reproduced experimentally (Sisson and Grove 1993).

Differentiation of basaltic magma under hydrous conditions would involve the separation of amphibole in addition to olivine and pyroxene, and hornblende phenocrysts are often present in andesites. Some hornblende-rich meladiorites, such as the appinites described later in this chapter, have been identified as having formed by hornblende accumulation from andesite or basalt magma (Groome and Hall 1974). Arculus and Wills (1980) have shown by calculation that the andesites of the Lesser Antilles island arc could be related to the basalts by fractionation of amphibole-bearing cumulate assemblages similar to those found in ejected plutonic blocks, and the very identification of these ejecta as cumulates shows that some fractionation has taken place. Many other andesites contain hornblende-rich cumulate xenoliths which can be interpreted as complementary differentiates from a hydrous basalt parent magma (Conrad and Kay 1984; Yagi and Takeshita 1987).

The implication that the basaltic magma in island arcs is water-rich offers a possible solution to two problems regarding fractionation: namely why andesite is so restricted to the orogenic environment when its presumed parent magma (basalt) is so common elsewhere, and why andesites are so different from the actual fractionation products of basalt observed in Hawaiian lava lakes or intrusions such as Skaergaard.

Although fractionation of hornblende or magnetite might account better for the compositional trends among orogenic magmas, it is plagioclase which is by far the dominant phenocryst phase in andesitic lavas, and Gill (1981) has reiterated the case for andesite being derived from basalt magma by low pressure fractionation of the common phenocryst phases, i.e. plagioclase + orthopyroxene (and/or olivine) + augite + magnetite. At the very least, the abundance of plagioclase phenocrysts in andesites suggests that most of them have undergone some low-pressure crystallization and perhaps fractionation before eruption. As an example, the lavas of the 1982 eruption of Galunggung (Java) were observed to change gradually during the course of the eruption from andesite to basalt, and their compositional variation can be modelled by the fractionation of plagioclase, olivine, clinopyroxene and magnetite; Harmon and Gerbe (1992) suggested that the fractionation took place in an upper crustal magma chamber which was progressively drained by the eruption. Many andesites have similar Sr isotopic compositions to the associated basalts (Table 52) and could be related to them by fractionation, but others have higher ${}^{87}Sr/{}^{86}Sr$ ratios and could not be related by fractionation alone.

Table 52. Initial $^{87}Sr/^{86}Sr$ ratios of andesites and associated basalts

	Andesite	Basalt	Reference
Inter-oceanic island arcs			
Mariana arc	0.7035	0.7035	Woodhead *et al.* 1987
South Sandwich arc	0.7040	0.7039	Gledhill and Baker 1973
Aleutians	0.7031	0.7031	McCulloch and Perfit 1981
Tonga	0.7038	0.7044	Ewart and Hawkesworth 1987
Kermadec Is.	0.7035	0.7037	Ewart and Hawkesworth 1987
New Britain	0.7033	0.7033	De Paolo and Johnson 1979
Lau Ridge, Fiji	0.7031	0.7030	Cole *et al.* 1990
Continental margins			
South-east Alaska	0.7044	0.7031	Myers and Marsh 1981
Mount St Helens	0.7035	0.7031	Smith and Leeman 1993
S. Cascades	0.7037	0.7033	Mertzman 1979
Taos, New Mexico	0.7046	0.7044	McMillan and Dungan 1988
Mexico	0.7039	0.7037	Moorbath *et al.* 1978
Central Andes	0.7073	0.7062	Harmon *et al.* 1984
Southern Andes	0.7040	0.7040	Harmon *et al.* 1984
Taupo zone, N.Z.	0.7055	0.7042	Ewart and Stipp 1968
South-west Japan	0.7051	0.7044	Ishizaka and Carlson 1983
Hokkaido	0.7038	0.7033	Katsui *et al.* 1978
Central Chile	0.7040	0.7040	Gerlach *et al.* 1988

CONTAMINATION

Contamination of basaltic magma by more acid crustal material was at one time much favoured as an origin for andesitic magma. There are two main difficulties about this explanation: (a) the widespread occurrence of andesite in island arcs not underlain by continental crust (e.g. the Solomon Islands, Tonga, New Hebrides, South Sandwich Islands); and (b) the low initial $^{87}Sr/^{86}Sr$ ratios of most andesites, which are often not much higher than those of associated basalts (Table 52). Like any other type of magma, andesites may show some contamination by sedimentary material, but this process cannot be responsible for the formation of the magma type itself.

Despite this, there are small differences in composition between the andesites of inter-oceanic island arcs and those of continental margins which imply that some continental crust has been assimilated in the latter environment. The continental andesites are richer in Na, K and Al, and poorer in Ca and Fe than the oceanic ones (Maaløe and Petersen 1981). Many andesitic volcanoes show specific evidence of contamination. For example, Paricutin (Mexico) has erupted andesites containing xenoliths of partially fused basement rocks and shows chemical and isotopic evidence of magma contamination, although the magma is thought to have already been andesitic before contamination (McBirney *et al.* 1987).

MIXING OF MAGMAS

Mixing of magmas has attractions for those who seek to explain the origin of all igneous rock types in terms of only two parent magmas, basaltic and granitic. This hypothesis provides a means of accounting for the disequilibrium mineralogies of many andesites, particularly if mixing involved magmas which had each already begun to crystallize before they were mixed. The minerals of andesites often show non-equilibrium features, such as a wide range of plagioclase and pyroxene compositions in the same rock (commonly with reversed or oscillatory zoning) and corrosion or resorption of crystals. It is possible that some of these features could be explained by variations in P_{H_2O} during the course of eruptive activity, rather than by mixing.

Anderson (1976) found positive evidence for magma mixing in the tiny glassy inclusions in some geologically young volcanic rocks. His analyses for a porphyritic andesite from the Cascade province are shown in Table 53. Although the rock is andesitic and the residual glass in the groundmass is andesitic, the glass inclusions in the phenocrysts range almost from basaltic to dacitic compositions. Myers and Marsh (1981) and Nixon (1988) have presented chemical and isotopic evidence for the production of andesites by the mixing of basalt magma with acid liquids produced by crustal melting or fractionation.

Table 53. Compositions of glasses from porphyritic andesite cinders of Medicine Lake Highland, California (from Anderson 1976)

	Groundmass	*Residual glass*	*Inclusion in olivine*	*Inclusion in plagioclase*
SiO_2	58.3	64.6	54.4	70.5
Al_2O_3	16.8	13.6	18.5	14.8
Fe as FeO	5.3	7.2	6.6	2.3
MgO	3.3	1.9	4.9	1.0
CaO	6.5	4.3	6.8	2.3
Na_2O	8.5	2.9	4.3	4.9
K_2O	1.9	2.8	1.3	4.2

It may not be possible to make a simple choice between fractionation, contamination and mixing as mechanisms for producing andesite, since it is likely that the processes often operate together. For example, Grove *et al.* (1988) described evidence that in Medicine Lake volcano (California) the processes of fractionation of basalt, melting of granitic crust, and the mixing of fractionation and contamination products all contributed to generating the andesite that was erupted.

The major objection to magma mixing on a large scale lies in the volume relationships of andesite and the presumed basaltic and rhyolitic parent magmas. The amount of rhyolitic lava erupted in some of the andesitic island arcs is small, implying that it has been nearly all used up in the mixing process, whereas the evidence of acid–basic magma relationships where they are seen together is that mixing of acid and basic magmas takes place only with great difficulty and to a limited extent. For island arc andesites, it is easier to postulate a primary andesite magma than to think of a source for the rhyolitic constituent required by the mixing hypothesis.

GEOCHEMISTRY OF ANDESITES

ISOTOPIC COMPOSITION

Most orogenic andesites have initial $^{87}Sr/^{86}Sr$ ratios that are low (Table 52) and generally differ little from those of associated basalts. Where there is a difference, it is usually in a continental margin environment rather than an island arc, and this suggests contamination as a possible explanation. For example the Quaternary andesites of the Taupo volcanic zone in New Zealand have ratios of 0.7046–0.7063 compared with 0.7042–0.7044 for the basalts (Ewart and Stipp 1968), and a similar difference was found by Ishizaka and Carlson (1983) in south-west Japan.

Slightly raised $^{87}Sr/^{86}Sr$ ratios in andesites compared with the average basalt, or in one andesite compared with another, are not necessarily indicative of secondary contamination, because the magmas could come from the melting of subducted oceanic crust containing variable amounts of radiogenic strontium as a result of sea-floor metasomatism. The $^{87}Sr/^{86}Sr$ ratio of sea water has varied over the range 0.707–0.709 during Phanerozoic time, and modern sea-floor basalts have been found with isotopic ratios raised by submarine metasomatism (Satake and Matsuda 1979; Staudigel and Hart 1983).

The combination of Sr and Nd isotopic ratios is particularly informative. Figure 293 shows the compositions of some Quaternary andesites from Indonesia. Those from Java lie on the main mantle trend and are no different isotopically from basalts of the same area, but the andesites of the Banda arc diverge towards ^{87}Sr-rich compositions. Two possible reasons for the extra ^{87}Sr are indicated. Enrichment in $^{87}Sr/^{86}Sr$ alone without a change in $^{143}Nd/^{144}Nd$ could be due to melting of a basaltic source material to which ^{87}Sr had previously been added by sea-floor alteration. High $^{87}Sr/^{86}Sr$ combined with low $^{143}Nd/^{144}Nd$ could be due to a sedimentary contribution to the magma.

The andesites of the Banda arc are not only high in $^{87}Sr/^{86}Sr$ (up to 0.7095), but also in $\delta^{18}O$ (up to 8.2‰). Magaritz *et al.* (1978) attributed this to the presence of some crustal material in the source region, pointing out that neither mantle heterogeneity nor disequilibrium melting can account for high $\delta^{18}O$ values. Crustal contamination is the most probable explanation for high $^{87}Sr/^{86}Sr$ in a number of andesite provinces. Nohda and Wasserburg (1981) estimated that the difference in Sr and Nd isotopic compositions between the basalts of the Izu arc (not underlain by continental crust) and the andesites of Japan (underlain by continental crust) could be accounted for by 10–20% contamination of the latter by continental material. Harmon *et al.* (1984) compared the particularly high $^{87}Sr/^{86}Sr$ ratios of recent andesites from northern Chile (0.7056–0.7143) with those of Ecuador (~0.7042), and thought the greater crustal thickness of northern Chile (70 kilometres compared with less than 40 kilometres in Ecuador) suggestive of greater contamination.

In inter-oceanic island arcs, direct contamination by continental crust is not a possibility, but there may be a contribution to the magma from oceanic

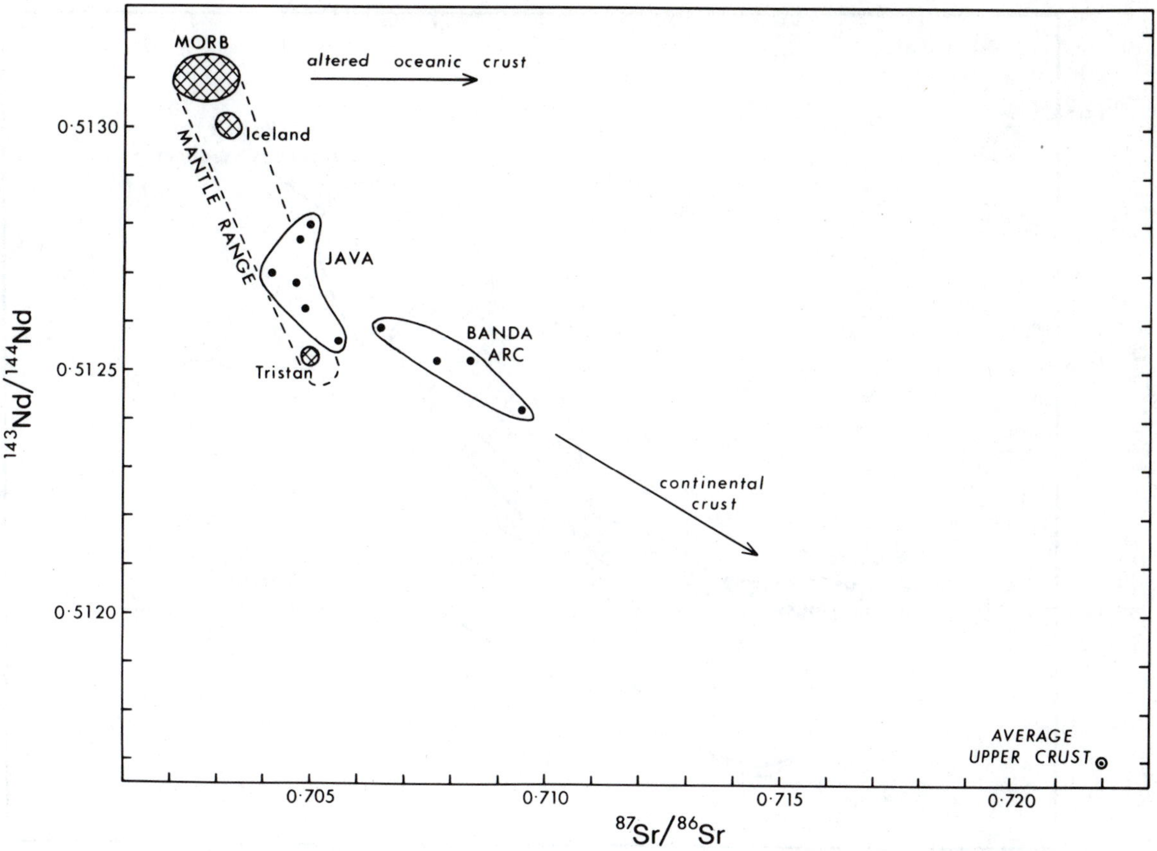

Figure 293. Sr and Nd isotopic compositions of Recent Indonesian andesites (after Whitford *et al.* 1981).

sediment subducted into the source region. Lead isotopes have been used by Oversby and Ewart (1972) and Meijer (1976) to estimate that not more than 1% of the magmas in the Tonga, Kermadec and Marianas arcs can have been derived from oceanic sediment. In contrast, lavas from the Aleutian and Cascade provinces have Pb isotopic compositions intermediate between those of ocean-floor basalts and continentally derived oceanic sediments (Fig. 294). Similar results have been obtained from various other island arc andesites, but the Pb content of oceanic sediments is generally much higher than that of basalts, so only a small sedimentary contribution is required to influence the isotopic ratios of the magmas (Barreiro 1983).

An additional technique which can be used to identify the sedimentary contribution to island arc magmas is to measure the relatively short-lived isotope ^{10}Be (half-life = 1.5 million years). This is produced in the atmosphere by cosmic radiation and is present in young marine sediments. Traces of ^{10}Be have been found in Recent lavas from the Aleutians and Central America, but it is absent from some other orogenic lavas and from all lavas not underlain by subduction zones (Morris and Tera 1989). It is said to have entered the magmas by way of subducted oceanic sediment, although ocean-floor basalt metasomatized by sea water seems an alternative possibility.

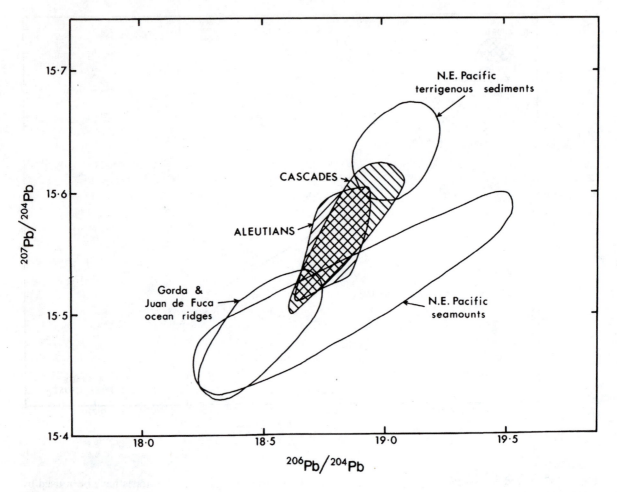

Figure 294. Pb isotopic compositions of andesites and related lavas from the Aleutian Islands and the Cascade province, compared with oceanic basalts and sea-floor sediments (after Kay *et al.* 1978).

TRACE ELEMENT CONTENTS

Trace element studies have been extensively used to test models of andesite petrogenesis, but so far with rather disappointing results. Two groups of elements are of special interest: the rare earths, and the first-series transition elements. The value of the rare earths lies in the fact that the minerals most likely to be involved in magma generation differ in their ability to accommodate the individual rare-earth elements. For example, garnet is best able to accommodate the heavy rare earths and plagioclase the light rare earths, while under reducing conditions europium has a particular preference for feldspars. The importance of the transition elements, especially Ni and Cr, is that although they tend to behave in a similar way to iron and magnesium during partial melting, they are much more abundant in peridotite than in basalt, so that their concentrations should provide some indication of the parent material in the magmatic source region.

So far, the results of trace element studies have been very ambiguous. Andesites do not have as much Cr and Ni as would be present in a direct partial melting product of a peridotite source material. This does not necessarily exclude a peridotite source because the Cr and Ni contents could be lowered by secondary differentiation from a primary magma which did have a peridotite source. Mysen and Popp (1980) suggested that this difficulty could be also overcome if the primary magma contained about 1% of sulphur. Separation of an immiscible sulphide phase during ascent of the magma could then bring down the transition element contents to the observed levels because of the strong partitioning of these elements into the sulphide phase.

The rare-earth distribution patterns should reveal whether garnet was involved in the development of the magma, either as a residual phase during partial melting or as a fractionating phase during secondary evolution, but attempts to use this feature as a petrogenetic criterion have been inconclusive, partly because of uncertainties about the rare-earth partition coefficients and partly because of the influence of metasomatizing fluids on the distribution of REE in the source region (Mysen 1981). Abnormally low concentrations of the heavy rare earth elements (e.g. Yb, Lu) and the absence of a Eu anomaly were taken by Kay *et al.* (1993) as evidence that some andesite magmas were separated from a garnet-rich, feldspar-free residue, i.e. they originated by the partial melting of eclogite.

Boron is an interesting trace element because it is very strongly enriched in seawater and marine sediments compared with most rock types of the crust and mantle. It can therefore be used as a tracer to indicate the presence of subducted sediment or altered oceanic crust in the source region of magmas. Boron concentrations are approximately 1–2 ppm in mid-ocean ridge and oceanic island basalts, but frequently over 10 ppm in volcanic arc basalts and andesites (Ryan and Langmuir 1993). Nitrogen is another element which may reflect the contribution of subducted sediments in the magmatic source region (Kita *et al.* 1993).

DISCUSSION OF ANDESITE PETROGENESIS

The experiments on melting of basalt and eclogite have revealed a variety of ways in which andesite could be produced as a primary magma, but the consensus of opinion is that orogenic andesites are normally produced by fractionation of hydrous basalt magma. Isotopic and trace element studies have been applied to the problem, but the isotopic evidence is not very conclusive, because whether andesite magma is primary or secondary it can be expected to have low initial $^{87}Sr/^{86}Sr$ ratios, similar to basalt. The main value of isotopic studies has been to indicate the limited role of oceanic sediment in the source region, and of crustal contamination subsequently.

Many authors have resorted to complex models of magmatic development involving more than one source material or secondary modification of a range of primary magmas. One such example is that of Ringwood (1974), who presented a two-stage model of magma formation resulting from subduction of oceanic crust. At depths of 80–100 kilometres, dehydration of amphibolite releases water, which causes partial melting in the overlying mantle peridotite giving hydrous tholeiitic magma and its differentiates, i.e. the tholeiitic island arc association of Jakes and White. At a depth of 100–150 kilometres, hydrous partial melting of quartz–eclogite gives rhyodacite–rhyolite magmas, which react with overlying mantle peridotite converting it to garnet–pyroxenite. Diapiric rise of the garnet–pyroxenite results in its partial melting and differentiation of the melts to give the various members of the 'calc-alkaline' magma series.

There are many explanations for the relationship between the K_2O content of andesites and the depth of the subduction zone. The explanations vary according to whether the magmas are thought to originate in the subduction zone itself or in the overlying mantle. In subducted oceanic crust potassium would most likely be held as sanidine, whereas in peridotite mantle it is more likely to be present as phlogopite or K-richterite. The amount of potassium in the magmas is governed by the breakdown of these minerals during melting. The potassium content of subducted oceanic crust is well above that of normal basalt because of the addition of K_2O by sea-floor alteration (Bloch and Bischoff 1979). Nielson and Stoiber (1973) showed that there is a different, although always positive, correlation between K_2O content and the depth of the seismic zone in different regions (Fig. 295). There are various reasons why this might be so, for example differing K_2O content of the ocean-floor rocks subducted in different Benioff zones.

According to Mysen and Boettcher (1975) there is no relationship between the K_2O content of liquids produced by partial melting of phlogopitic peridotite and their pressure of formation, but Kushiro (1980b) believed that the observed variation in K_2O with seismic depth could be related to variation in degree of melting in mantle peridotite as a function of pressure.

Figure 295. The K–*h* relationship for andesites (SiO_2 ~60%) from several different island arcs and continental margins (after Nielson and Stoiber 1973).

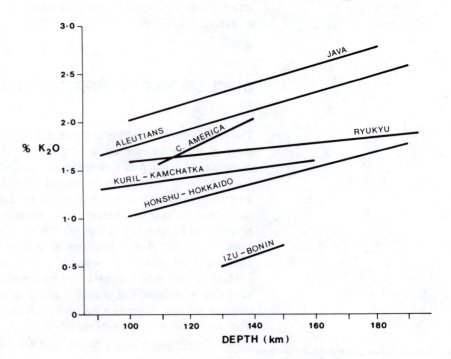

Ninkovich and Hays (1972) and Best (1975) suggested that hydrous melts produced in the mantle above a subduction zone (or hydrous fluids expelled from subducted oceanic crust) scavenge mantle-incompatible elements such as potassium on their way up through the mantle; the deeper their origin, the greater the thickness of mantle from which the K_2O is scavenged. Meen (1987) considered that K-rich magmas were due to fractionation of orthopyroxene rather than olivine from dry basaltic magma, which he thought more likely to occur at the high pressures at the base of thick crust.

For the relatively small number of andesitic rocks (icelandites) that are not associated with subduction zones the petrogenetic possibilities are far fewer, and most workers believe that they are produced by the fractionation of dry basaltic magma. This explanation is consistent with the results of both experimental petrology and trace element modelling (Macdonald *et al.* 1990; Thy and Lofgren 1994).

DIORITES

Whereas many diorites are simply the intrusive equivalents of andesite lava, this is not true of them all. Some diorites are very heterogeneous, and it is common for bodies of diorite to be associated with granite or gabbro in such a way as to suggest that they are derived from granitic or basaltic magma by contamination, fractionation or mixing.

Diorites differ from andesites in having a higher H_2O+ content, averaging 1.15% for diorite compared with 0.83% for andesite. This is manifested by the greater abundance of hornblende, which is the characteristic ferromagnesian mineral of diorites. The lower content of H_2O+ in andesites is no doubt due to the loss of water on extrusion. The H_2O+ content of diorites is greater than those of either granites or gabbros, and suggests that before extrusion andesite is the most water-rich of the major magma types.

DIORITES AS CONTAMINATED ROCKS

Granitic intrusions commonly show a gradation from an inner leucocratic facies to an outer facies of tonalitic composition, with xenoliths and petrographic relationships indicating that the variation is due to contamination of an originally granitic magma. It is not possible for a contaminated granite magma to incorporate a large amount of calcic plagioclase and ferromagnesian minerals into the actual melt because of their relatively high melting temperatures, and we may assume that the tonalitic composition of contaminated facies has been attained by dispersal of crystalline material in the magma rather than by complete assimilation into the melt. These contaminated tonalites could obviously not have any direct extrusive equivalent, but they do sometimes approach an andesitic composition. Walker and Davidson (1935) described a gradual progression from adamellite through quartz–monzonite into quartz-free monzonite and diorite as a result of contamination by limestone and schist in the Dorback granite in Scotland. In most cases, the contamination of granite does not lead as far as a true diorite, but contamination of granite to form tonalite or quartz–diorite is common.

A more basic, i.e. gabbroic, magma may also be converted into diorite by contamination. In the Peninsular Ranges batholith of California, Walawender *et al.* (1979) have recognized a progression from gabbronorite into quartz–diorite attributable to assimilation of granitic country rocks by the basic magma.

De Albuquerque (1979) noted that high-K (biotite-bearing) diorites of Nova Scotia were not identical in composition to any type of extrusive andesite and suggested that they were the products of mixing of granitic and tholeiitic magmas before emplacement. Magma-mixing has also been invoked to explain the layered diorites of the Tigalik intrusion in Labrador (Wiebe and Wild 1983).

DIORITIC CUMULATES

Just as the fractionation of basaltic magma leads to the formation of accumulative rocks, well seen in the layered basic complexes, so we can expect that fractionation of andesitic magma may lead to the formation of dioritic cumulates. Anhydrous andesite magma is more viscous than basalt, and crystal fractionation correspondingly more difficult, but in practice

Figure 296. Dioritic and appinitic intrusions in the Appin area, Scotland (after Bowes and Wright 1967).

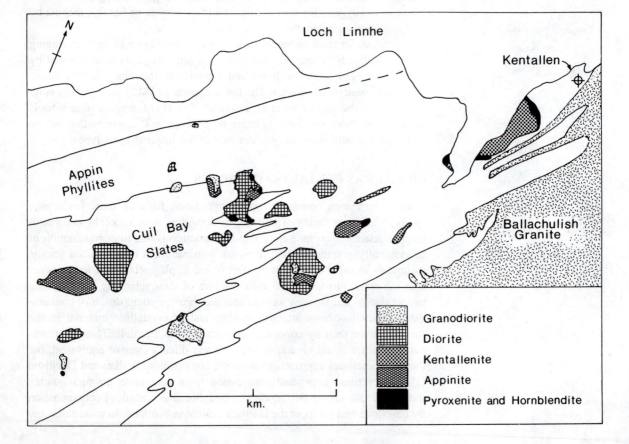

natural andesite magmas have high water contents and the reduction of viscosity by water should enable fractionation to take place readily under plutonic conditions.

Two types of plutonic rock association are suspected of being the products of accumulation from hydrous andesite or hydrous basalt magmas. These are the appinite association, named originally after the Appin area in Scotland (Bailey and Maufe 1960; Hall 1967), and the zoned ultramafic intrusions described as being of 'Alaskan-type' (Taylor 1967; Irvine 1974).

The appinitic intrusions form small stocks, often associated with a large body of granite, as shown by the type area (Fig. 296). The rock types which are present range from diorite to dunite, including dunite, peridotite, pyroxenite, gabbro, hornblendite, meladiorite and diorite. Hydrous minerals such as hornblende and biotite are more abundant than in other basic and

Figure 297. Zonation in the dioritic and ultramafic rocks of (A) the Mulnamin More intrusion, Ireland; (B) the Kilrean intrusion, Ireland; and (C) the Hall Cove area of the Duke Island complex, Alaska. A and B are after French (1966), and C is after Irvine (1974).

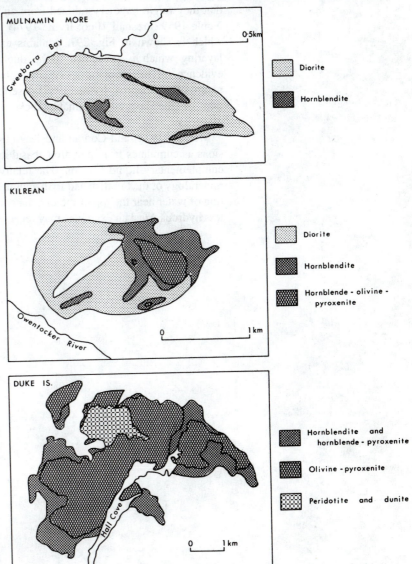

ultrabasic rocks, and hornblendite is a characteristic member of the association. The name appinite refers to the hornblende-rich meladiorites found in the type area. Appinitic intrusions are often zoned in a roughly concentric way, with ultramafic rocks forming a core, surrounded by hornblende-rich diorites (Fig. 297). The appinitic rocks which have been studied in most detail are those of the Appin area itself (Bowes and Wright 1967) and the suite situated around the Ardara granite in Ireland (French 1966; Hall 1967). The intrusions in Appin are contemporaneous with late-Caledonian andesite volcanism in the surrounding region (Groome and Hall 1974).

The zoned ultramafic intrusions of Alaskan type are also hornblende-bearing although not as hornblende-rich as the appinites, and they contain a higher proportion of ultrabasic rocks, which has led some workers to believe that they might have crystallized from an ultramafic magma (Ruckmick and Noble 1959; Presnall 1966; Irvine 1974). The Duke Island complex, part of which is shown in Fig. 297, contains extremely well developed rhythmic layering which is presumed to be due to crystal settling. There is also evidence for metasomatic recrystallization, which may reflect the hydrous nature of the magma. Irvine (1987) described the post-cumulus recrystallization of the layered ultramafic rocks as 'the magmatic equivalent of diagenesis'.

Murray (1972) and Conrad and Kay (1984) interpreted the Alaskan intrusions as cumulates from a hydrous basaltic magma, with andesite lava as the complementary liquid fraction. The relatively hydrous (i.e. hornblende-rich) mineralogy of the Scottish and Irish appinites might be attributed to concentration of water near the top of the magma conduits before eruption, whereas the less hydrous Alaskan examples may represent crystallization at deeper levels.

CHAPTER 11

Alkaline igneous rocks

THE NEPHELINITE–CARBONATITE ASSOCIATION

Nephelinite is the commonest of the ultra-alkaline mafic lavas. Erosion of nephelinite volcanoes in East Africa and elsewhere has revealed the corresponding intrusive rocks of the ijolite suite, often differentiated to give feldspathoid-rich rocks and pyroxenites. Carbonatite, where it occurs, is nearly always associated with ijolitic rocks, so that Le Bas (1977) writes of a carbonatite–nephelinite association, in which carbonatites, nephelinites, ijolites, pyroxenites and fenites are the principal members, and of which fenitization is a characteristic feature.

TYPICAL ROCK TYPES

Nephelinites are lavas composed of nepheline and clinopyroxene. Olivine, leucite or melilite may also be present. Plutonic rocks composed of nepheline and clinopyroxene show a much wider range of composition than their volcanic counterparts, from rocks composed largely of nepheline to rocks composed largely of clinopyroxene. Names commonly used for these rocks include urtite (>70% nepheline), ijolite (nepheline > clinopyroxene) and melteigite (clinopyroxene > nepheline). Collectively these rock types are called the ijolite series, and chemically the average nephelinite is most similar to melteigite.

The peridotites, pyroxenites and anorthosites of layered gabbroic complexes have their counterparts in the peridotites, pyroxenites and urtites which are associated with ijolites. The peridotites are mainly composed of olivine but often contain large amounts of magnetite and phlogopite. The pyroxenites are typically composed of clinopyroxene (diopside or titanaugite), often with abundant magnetite, apatite and perovskite. Crystal settling is more effective in nephelinitic than in basaltic magmas because of their lower viscosity, and fractionation may involve not only settling but also flotation, as in the nepheline-rich rocks of the Usaki complex in Kenya (Le Bas 1977).

Melilite-bearing rocks are associated with many nephelinites and ijolites. Melilitite is the most silica-undersaturated of all lava types (other than carbonatites). Melilitites have extremely low SiO_2 contents (30–40%), high CaO (~15–20%), and Na_2O in a similar range to basalts. They contain melilite, clinopyroxene and usually olivine as essential constituents, and perovskite, apatite and calcite are characteristic accessory minerals. Melilitites grade into nephelinites, and in East Africa there are also varieties containing leucite or kalsilite. Alnoites are hypabyssal melilite-bearing rocks with alkali contents higher then melilitites and containing biotite as a potassic phase. There are all gradations between alnoite and kimberlite. Uncompahgrite and melilitolite are the plutonic equivalents of melilitite.

The carbonatites are essentially carbonate rocks, and may be roughly divided into those composed mainly of calcite (sövites) and those composed mainly of dolomite or ankerite. Most carbonatites contain some silicate minerals, and magnetite or hematite are also common. Many unusual minerals occur in carbonatites, including monticellite, melilite, apatite and pyrochlore, of which the last two may be abundant enough to form economic mineral deposits. Some of the carbonatites in the Kola peninsula in Russia contain cumulate deposits of apatite, olivine and magnetite.

FENITIZATION

Around most carbonatites, and around some alkaline ultrabasic intrusions, the country rocks are not only thermally metamorphosed but are metasomatized. This contact metasomatism is called fenitization, after its occurrence in the Fen complex in Norway. Fenitization is essentially alkali metasomatism. Its principal effects are the replacement of quartz by alkali feldspars, and of feldspars by nepheline, and the development of sodic ferromagnesian minerals such as aegirine and arfvedsonite.

A great many carbonatites are intruded into Pre-Cambrian shield areas where quartzo-feldspathic gneisses are a common type of country rock. The metasomatized derivatives of these gneisses are often similar in appearance to syenites or in extreme cases foyaites, so that in badly exposed areas it is not always possible to discriminate between felsic intrusive rocks and fenitized country rocks. There are great variations in the intensity of fenitization around different carbonatites and ijolites, and some differences between the fenitization associated with the two magma types (Morogan 1994).

OCCURRENCE OF NEPHELINITE AND CARBONATITE

Nephelinite is widely distributed in both oceanic and continental areas. In the Atlantic Ocean, nephelinite lavas are present on Trinidade and in the Cape Verde Islands, in both cases associated with alkali basalts and phonolites. In the Pacific Ocean, the best known occurrence is in the Hawaiian Islands, the bulk of which are made up of basalts. The greatest continental development of nephelinite volcanism is in East Africa, where the hundreds of Tertiary to

Recent volcanoes include some that have erupted carbonatite. Basalts, trachytes and phonolites also occur throughout this region, but in eastern Uganda nephelinite predominates over all other lava types. Another extensive continental development is the late Tertiary to Quaternary province of Central Europe, stretching across France, Germany and the Czech Republic, which contains a similar range of rock types to East Africa. The Tertiary–Recent volcanic province of south-eastern Australia contains a complete range of mafic magma types from quartz–tholeiite, through olivine–tholeiite and basanite, to nephelinite, olivine–nephelinite and olivine–melilitite.

The most completely described nephelinite volcano is *Kisingiri* in western Kenya (Le Bas 1977). This is a partially dissected volcanic cone originally about 3000 metres high. It consists of interstratified lavas, agglomerates and tuffs in the following volume proportions:

nephelinite and melanephelinite lava	36%
melilitite lava	2%
mugearite lava	5%
nephelinite agglomerate and tuff	51%
melanite–nephelinite agglomerate	6%

More felsic rocks, i.e. phonolites, do not occur in the lava pile but are present as dykes and plugs. Beneath the centre of the dissected cone, erosion has revealed an intrusive complex consisting of ijolite, uncompahgrite and pyroxenite, with carbonatite at the centre intruded as ring dykes, cone sheets and plugs.

Only 300 kilometres to the north of Kisingiri are the rather more eroded remains of *Napak* volcano in Uganda (Fig. 298), which shows a very similar arrangement of rock types. The volcanic rocks at Napak are predominantly pyroclastic. Lavas make up only about 3% of the extrusive succession, and they are nephelinites and melanephelinites, including both olivine- and melilite-bearing varieties. The central intrusive complex, which is about 2 kilometres in diameter, consists of an outer ring of ijolite and related rocks (pyroxenite, melteigite, urtite) and a central plug of carbonatite. The latter is a sovite containing some siderite and magnetite. The basement gneisses which underlie the volcano are exposed around the central complex and can be seen to be shattered and fenitized for about 2 kilometres around the intrusive complex.

The volcano *Oldoinyo Lengai* in Tanzania is one of the few Recent volcanoes to have erupted carbonatite lava. The bulk of the volcano is composed of ijolitic tuffs and agglomerates in which there are blocks of nephelinite, phonolite, urtite, ijolite, melteigite, jacupirangite, biotite–pyroxenite

Figure 298. Section across Napak volcano, eastern Uganda (after King 1965).

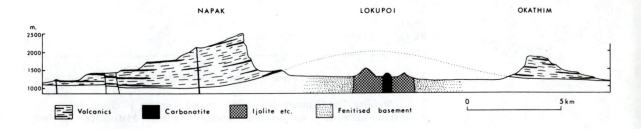

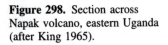

and fenite. There are a few silicate lava flows, of nephelinite, melanephelinite and phonolite. The extrusion of carbonatite lavas has been observed in several eruptions in recent times (Dawson *et al.* 1994). These lavas behave in just the same way as silicate lavas; there are very mobile pahoehoe flows and more viscous aa flows, as well as carbonatite ashes.

One of the most interesting features of the Oldoinyo Lengai lavas is that they are natrocarbonatites, consisting of sodium and calcium carbonates (Table 54), unlike intrusive carbonatites in which sodium is only a minor constituent. This observation suggests the possibility that intrusive carbonatites could have crystallized from a magma which originally had a high sodium content but lost it to the wall rocks in the process of fenitization, or in the case of extrusive rocks they could have lost sodium carbonate by leaching. The relative solubility of carbonatites renders them vulnerable to weathering and hydrothermal alteration of all kinds (Anderson 1984). There are natrocarbonatite inclusions in the apatites of some otherwise Na-poor carbonatite rocks (Le Bas 1981), but petrographic evidence suggests that most carbonatites really did crystallize as calcite, dolomite or ankerite, and that natrocarbonatites are rare (Keller 1989; Cooper and Reid 1991). Natrocarbonatite magmas may be uncommon, but they are probably the only carbonate melts capable of reaching the surface without dissociation; pure $CaCO_3$ dissociates at atmospheric pressure at 894 °C and $CaMg(CO_3)_2$ at a lower temperature.

Table 54. Carbonatite analyses

	1	2	3
SiO_2	trace	3.36	1.87
TiO_2	0.10	0.30	0.00
Al_2O_3	0.08	1.69	0.20
Fe_2O_3	—	6.13	2.80
FeO	0.26	2.99	10.40
MnO	0.04	0.31	5.00
CaO	12.74	44.35	19.26
BaO	0.95	0.10	0.10
SrO	1.24	—	2.48
MgO	0.49	3.10	14.47
Na_2O	29.53	0.04	—
K_2O	7.58	0.50	—
P_2O_5	0.83	3.26	0.80
H_2O	8.59	0.30	0.00
CO_2	31.75	32.80	38.36
F	2.69	0.28	—
Cl	3.86	0.02	—
SO_3	2.00	0.06	0.12
S	—	0.42	—
Nb_2O_5	—	0.80	—
$RE_2O_3 + ThO_2$	—	—	1.18

1. Carbonatite lava, pahoehoe flow erupted October 1960, Oldoinyo Lengai, Tanzania (Dawson 1966).
2. Sövite, Söve, Fen Complex, Norway (Barth and Ramberg 1966).
3. Ankeritic carbonatite, Kangankunde carbonatite complex, Malawi (Garson 1966).

Figure 299. The geology of Chilwa Island, Malawi (after Garson 1966).

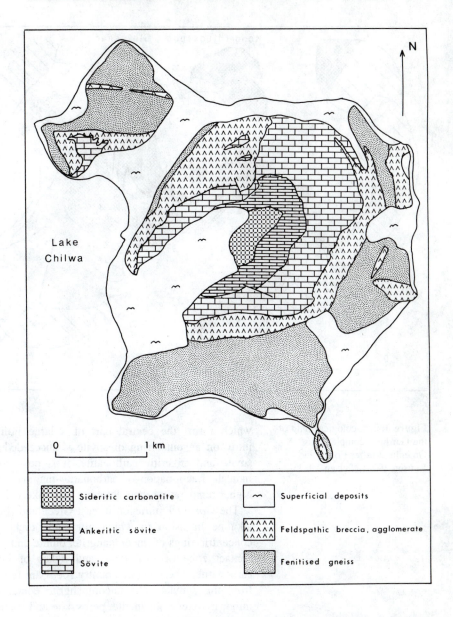

Lake Chilwa

N

0   1 km

▦ Sideritic carbonatite	～ Superficial deposits
▤ Ankeritic sövite	ʌʌʌʌ Feldspathic breccia, agglomerate
▥ Sövite	▨ Fenitised gneiss

Several carbonatites show internal differentiation. In the *Chilwa* complex, Malawi (Fig. 299), an outer ring of sovite is followed by inner rings of ankeritic sovite and siderite–carbonatite. The outer ring of sovite consists of ring dykes separated by screens of feldspathic breccia. The sovite is cut by dykes and sheets from the later ankeritic sovite. The central part of the carbonatite is composed of manganiferous sideritic carbonatite. There is no major silicate component in this complex, but there are small plugs, dykes and cone-sheets of nepheline–syenite, ijolite, phonolite, nephelinite and alnoite. The intrusion as a whole is surrounded by a wide zone of brecciation and fenitization. The sequence of carbonatite types from early outer sovite to later iron- and magnesium-rich varieties is shown by other composite carbonatite intrusions, for example the Spitskop intrusion in South Africa,

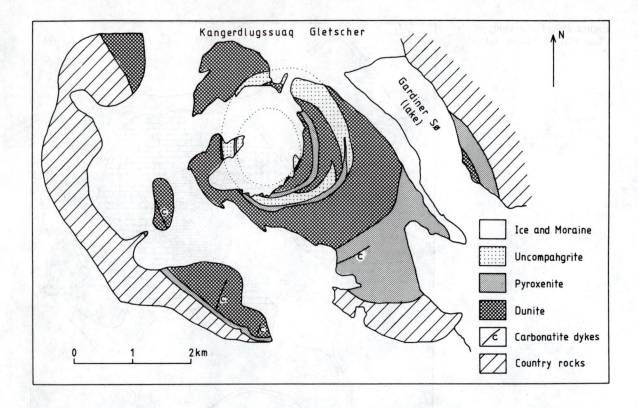

Figure 300. Geological map of the Gardiner complex, east Greenland (after Frisch and Keusen 1977; Nielsen 1981).

which forms the central part of a large ijolite–foyaite complex. In this intrusion an outer ring of sovite is succeeded by inner rings of dolomitic sovite and ankeritic carbonatite. It is possible that this relationship may indicate fractionation of carbonatite magma, with calcite crystallizing at a higher temperature than dolomite and ankerite.

The *Gardiner* intrusion in east Greenland (Fig. 300) is an example of an alkaline intrusion in which ultramafic rocks predominate. It consists of concentric rings of uncompahgrite, dunite and pyroxenite. There is a narrow contact zone of melteigite, and dykes of ijolite, nepheline–syenite and carbonatite. The gneissic country rocks are fenitized for only about 1 metre from the contact. The uncompahgrite contains about 75% melitite, with minor pyroxene, magnetite, perovskite and apatite, and is strongly layered in places.

A very wide range of ultramafic rocks is found in the *Phalaborwa* (Palabora) complex in South Africa (Fig. 301), which contains a carbonatite core and peripheral syenite bodies. The greater part of the intrusion consists of pyroxenite. This is a diopside-rock containing a large but variable amount of mica and apatite. In places the mica occurs as schlieren; elsewhere it is so abundant as to constitute a mica-rock (glimmerite). Near the outer contacts the pyroxenite is feldspathic. In the northern part of the intrusion the core consists of mica-peridotite, which is very coarse-grained ('pegmatoid'). Most of the mica has altered to vermiculite and the olivine to serpentine. Carbonatite occurs in the centre of the intrusion, and its average modal composition is 60% carbonate, 30% magnetite, 10% apatite and 3%

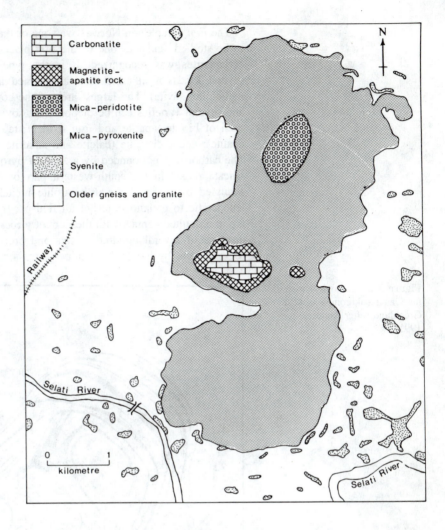

Figure 301. Geological map of the Phalaborwa complex, Transvaal (after Hanekom *et al.* 1965). A more detailed map is given by Eriksson (1989).

chalcopyrite. It is surrounded by a ring of magnetite-apatite rock ('phoscorite') with the average composition 35% magnetite, 25% apatite, 22% serpentine and mica, and 18% carbonate. This in turn is surrounded by a ring of very coarse-grained micaceous pyroxenite. There are many gradational and inter-banded contacts between these rock types but sharp contacts can be found between carbonatite and phoscorite, and between phoscorite and pyroxenite. Veins of carbonatite extend out into all the other rock types of the complex.

The country rocks of the Palabora complex are gneisses and granites, which are fenitized in proximity to the intrusion. They contain many small bodies of aegirine- and arfvedsonite-bearing syenite, some of which may be fenites but some of which are definitely intrusive. The rocks of the complex are of great economic value. The carbonatite contains about 1% of copper and is economically workable for this metal; reserves are estimated at more than 300 million tons of ore. The phoscorite is mined as a source of phosphate, along with some apatite-rich pyroxenite. Vermiculite, baddeleyite (ZrO_2) and magnetite are also worked in large quantities.

The *Fen* complex in Norway was one of the first ultra-alkaline complexes to be studied, and was one of the first places where the igneous nature of carbonatites was recognized. It is the type-locality for fenitization, and several kinds of alkaline rock are named after places in the area (e.g. melteigite, sövite). The largest area of the complex consists of carbonatite, and several varieties can be mapped out: sövite (calcite with a few per cent each of biotite, apatite and iron oxides); rauhaugite (ankerite with a little apatite); and rödbergite (calcite and ankerite plus about 30% of hematite). The carbonatite is bounded by a zone of pyroxenites and mixed carbonate–silicate rocks. In the south-western part of the complex there is an area occupied by rocks of the urtite–ijolite–melteigite series, i.e. ranging from leucocratic to melanocratic but all feldspar-free. Where the intrusive rocks are in unfaulted contact with their country rocks, the latter are fenitized for a distance of several hundred metres and brecciated in places. The country rocks are cut by hundreds of minor intrusions. These are of several kinds:

Figure 302. Geological map of the Qaqarssuk complex, west Greenland (after Knudsen 1991).

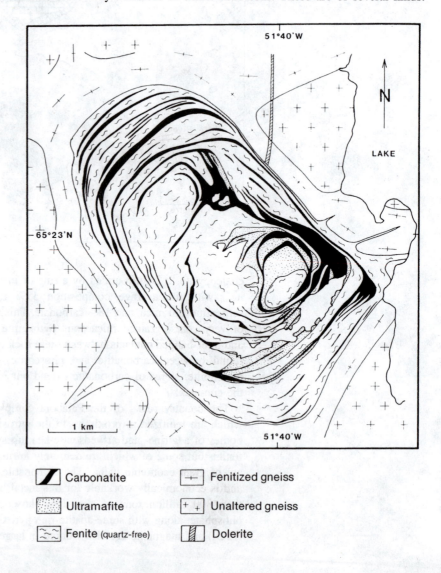

damkjernite (biotite–pyroxenite) plugs or dykes, breccias with a damkjernitic matrix, breccias with a granitic matrix, and tinguaite dykes. Some of the breccias have only a very small amount of igneous matrix, suggesting that they formed by gas-drilling.

One of the most thoroughly studied carbonatite intrusions is the *Qaqarssuk* complex in west Greenland (Fig. 302). Unlike many other carbonatites this complex is accompanied by very few intrusive silicate rocks. It consists of a large number of concentric ring dykes occupying about 15% of the total area of the complex, the intervening basement gneiss being heavily fenitized. Some of the fenites are very rich in phlogopite. The complex contains large amounts of apatite-carbonatite and some Nb and REE mineralization.

A general feature of mafic alkaline magmatism is that its greatest development is in continental rift zones, which are also regions of repeated crustal uplift or doming. The most spectacular example is in the great Rift Valleys of East Africa, but other well-known examples are in the Rhine and Oslo grabens in Europe and at the intersection of the St Lawrence and Ottawa grabens in North America. About half of all the carbonatite occurrences that are known are in Precambrian shield areas of Africa. Many of these are in the East African rift system, but they are also numerous in Angola, Namibia and Transvaal. Carbonatites occur in the Canadian and Baltic shields, but there are only a few occurrences outside shield areas.

ORIGIN OF NEPHELINITE MAGMA

There is no doubt that nephelinite and carbonatite magmas originate in the mantle. They occur in both oceanic and continental areas, and have consistently low initial $^{87}Sr/^{86}Sr$ ratios. Nephelinites are often associated with basalts, but in some provinces (for example western Kenya) nephelinite occurs without accompanying basalt, and in others the basaltic and nephelinitic magmas appear to be independent of one another, or occur together but have slightly different isotopic compositions (Francis and Ludden 1990).

Experiments on the melting of possible mantle materials, reviewed in Chapter 8, have indicated that highly undersaturated melts are most likely to be formed under any or all of the following conditions: (a) a low degree of partial melting; (b) high pressure; (c) high P_{CO_2}. It is now generally assumed that nephelinite can be a primary magma type under one or more of these conditions. For example, Frey *et al.* (1978) calculated that the range of major and trace element compositions in the alkali basalts and nephelinites of south-east Australia could be accounted for by different degrees of partial melting of a 'pyrolite' mantle source, with the nephelinites and melilitites being products of the lowest degrees of melting (~4%). If nephelinite is the initial product of partial melting of a mantle which on further melting would yield basalt, then one might expect to see nephelinites as the first magmas erupted during an episode of predominantly basaltic volcanism. Such a relationship is observed in the Karroo province of southern Africa (Cox *et al.* 1965).

Clague and Frey (1982) made a very detailed geochemical and isotopic study of the Honolulu Volcanic Series, which are volumetrically minor lavas erupted on the flanks of the much larger Koolau tholeiitic shield volcano. The Honolulu lavas range from alkali basalt through basanite and nephelinite to nepheline–melilitite. Their geochemistry can best be explained by a low degree of partial melting of a garnet–lherzolite source (only 2% melting for the nepheline–melilitite), but isotopic differences suggest that this source was not the same as that of the Koolau tholeiites ($^{87}Sr/^{86}Sr = 0.7033$ for the nephelinites, 0.7037 for the tholeiites).

Le Bas (1977, 1987) believes that there is an important distinction between olivine–nephelinites (which are associated with basalts, as in Hawaii), and olivine–poor nephelinites (which are associated with carbonatites, as in East Africa). This difference may perhaps reflect the availability of CO_2 in the magmatic source region. Certainly the presence of CO_2 is necessary for partial melting of peridotite to yield such highly undersaturated melts as carbonated nephelinite or melilitite (Brey and Green 1975; Mysen and Boettcher 1975; Eggler and Mysen 1976). The uniform Ca/Mg ratio of melilitites from all parts of the world is considered by Brey (1978) to be a possible reflection of the presence of dolomite in the source material.

A low degree of melting and high CO_2 may not be sufficient to explain the unusual abundance of very alkaline mafic lavas in the East African Rift Valley. Ultramafic xenoliths from these lavas are heavily veined by amphibole and phlogopite (Dawson and Smith 1988), and this suggests that there has been unusually extensive metasomatism in the upper mantle underlying the rift.

The high sodium content of nephelinite magmas (averaging ~5% Na_2O) requires the presence of a sodium-bearing phase in their source rocks. In addition to sodic clinopyroxenes, amphiboles are a possibility, and pargasite and kaersutite have both been suggested. Their magnesian end-member compositions are:

pargasite: $NaCa_2Mg_4AlSi_6Al_2O_{22}(OH)_2$
kaersutite: $NaCa_2Mg_4TiSi_6Al_2O_{22}(OH)_2$

Both minerals are hydrated and are capable of carrying many of the trace elements which are concentrated in alkaline rocks. Pargasites are so close to olivine–nephelinites in composition that their chemical analyses could not easily be distinguished. These observations led Varne (1968) to suggest that the nephelinites in East Africa were produced by the melting of hydrated volatile-rich portions of the mantle containing pargasitic hornblendite, and that basalts were produced from more normal anhydrous mantle. Olafsson and Eggler (1983) subsequently produced nephelinitic liquids experimentally by the partial melting of amphibole–peridotite and amphibole–carbonate–peridotite at high pressures. Francis (1991) drew attention to the significance of interstitial veins and patches of glass which occur in the ultramafic xenoliths of alkaline volcanoes: glass in contact with amphibole–garnet pyroxenite is highly undersaturated in silica, in contrast to the glass in lherzolite xenoliths. This again suggests that nephelinite could be produced by partial melting in amphibole-bearing parts of the mantle, and that the range of compositions from nephelinite to alkali basalt represents mixing with less alkaline melts from surrounding lherzolite.

ORIGIN OF CARBONATITE MAGMA

There are three known ways in which a carbonatite magma might originate: (a) as the initial product of partial melting of mantle containing a small proportion of carbonate minerals; (b) as the final product of differentiation of a silicate melt containing dissolved carbonate; or (c) by liquid immiscibility from an initially homogeneous carbonate-bearing silicate melt.

Primary magma

Carbonatite might well be a primary magma, produced by partial melting of carbonate-bearing mantle under appropriate conditions. Increasing pressure raises the temperature at which carbonate minerals can exist. The temperature at which dolomite decarbonates by the reaction En + Dol $\rightarrow$ Di + Fo + CO_2 intersects the solidus in the system CaO–MgO–SiO_2–CO_2 at a pressure of about 25 kilobars (Fig. 303). Above this pressure, dolomite can be present in peridotite up to the melting temperature, and partial melts in this system would be carbonate-rich with a low silicate content (Wyllie and Huang 1976; Eggler 1978; Falloon and Green 1990; White and Wyllie 1992). The carbonatite liquids would only be the initial products of partial melting because with increasing temperature they would become more silicate-rich as more of the silicate constituents of the source rock became melted. Wendlandt and Mysen (1980) actually observed the formation of a carbonatitic melt as the first product of melting of a natural garnet–lherzolite in the presence of excess CO_2 at a pressure of 30 kilobars. The same material melted in the presence of excess CO_2 at 15 kilobars gave an initial melt which was tholeiitic.

Huang *et al.* (1980) suggested that the reason why some parts of the mantle are carbonate-bearing could be the deep subduction of limestones or carbonated oceanic basalts. This is consistent with the belief of many isotope workers that deeply subducted material is present in the enriched source of alkali basalts (see Chapter 8).

A major difficulty about accepting carbonatite as a primary magma is that carbonate melts in equilibrium with mantle peridotite would be expected to have high Mg contents, as has been confirmed experimentally by Dalton and Wood (1993). Some Mg-rich carbonatites do occur, for example the ankeritic carbonatite in Table 54, but most carbonatites are calcite-rich. Dalton and Wood suggested that Mg-rich carbonatite melts could become more Ca-rich by metasomatic reaction with upper mantle harzburgite, and more Na-rich by reaction with upper mantle lherzolite, as they rose towards the crust.

If carbonatites are primary magmas, then their extremely low viscosity must be a significant factor in their behaviour. McKenzie (1985) suggested that carbonate melts could be extracted from their source rocks even if the melt fraction was as low as 0.1%, in contrast to silicate melts which would remain within their source rocks at this degree of melting.

Final melt fraction

Fractional crystallization of a carbonated silicate melt is a theoretical

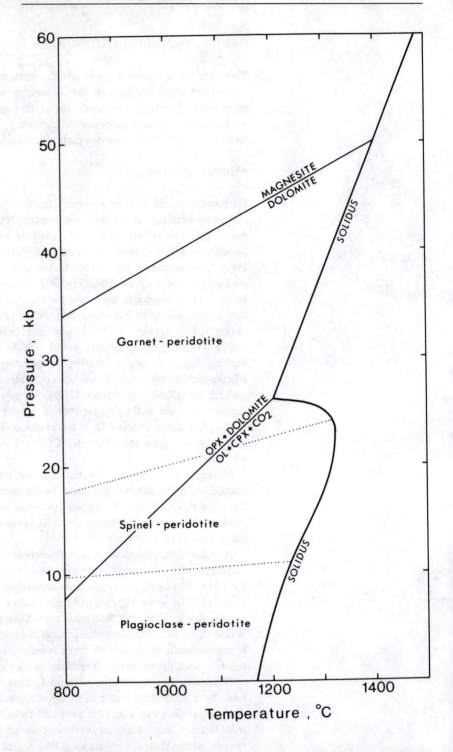

Figure 303. Schematic phase relations of peridotite with a small amount (~0.1%) of CO_2 (after Wyllie 1978).

possibility, but has not found much favour because although many carbonatites and nephelinites occur together, rocks of intermediate carbonate–silicate composition are rare. On the basis of their experimental studies, Wyllie and Huang (1976) showed that at low pressures the crystallization of a basic magma with dissolved CO_2 would yield silicates and free CO_2 rather than a carbonate-rich liquid, but at higher pressures differentiation to give carbonatite liquid would be possible. The separation of carbonatite melt from an associated silicate, whether by fractional crystallization or liquid immiscibility, may well be very rapid because of the very low density of carbonate liquids (2.2–2.3 g/cm^3).

Liquid immiscibility

Liquid immiscibility between silicate and carbonate melts has been found in a number of synthetic systems, and in melting experiments on mixtures of silicate and carbonate rocks. Freestone and Hamilton (1980) showed the immiscibility of carbonate melts with both nephelinite and phonolite, and measured the extent of immiscibility over a wide range of temperature and pressure. They found that the carbonate liquids coexisting with nephelinite or phonolite were richer in Na than the silicate liquids (Fig. 304), in contrast to natural plutonic carbonatites which are Na-poor. If plutonic carbonatites do separate from their associated ijolites by liquid immiscibility, then large quantities of alkalis must subsequently be lost from the magma, possibly by the fenitization of country rocks. Figure 305 shows the extent of liquid immiscibility in compositions intermediate between albite, anorthite, Na_2CO_3 and $CaCO_3$ at 5 kilobars; the extent of immiscibility is somewhat less at lower pressures. It can be seen that Na-rich silicate melts coexist with Na-rich carbonate melts, and Ca-rich silicate melts with Ca-rich carbonate melts. It is easy to see how a natrocarbonatite magma might exsolve from a

Figure 304. The distribution of Na_2O and CaO between immiscible carbonate and silicate liquids determined experimentally at 2.9 kilobars and 1100 °C by Freestone and Hamilton (1980).

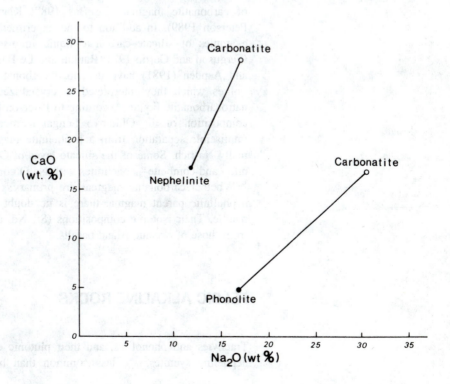

Figure 305. The extent of the 2-liquid field in the system albite–anorthite–Na_2CO_3–$CaCO_3$ at 5 kilobars and 1250 °C, determined by Kjarsgaard and Hamilton (1989). Tielines connect the starting compositions (crosses) with the compositions of the coexisting liquids (dots).

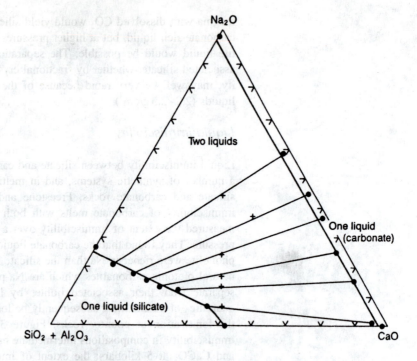

carbonated nephelinite, but a Na-poor carbonatite melt would seemingly require a Na-poor parent magma, perhaps carbonated melilitite.

Immiscible separation of carbonatitic and nephelinitic liquids from a common parent is now the most widely favoured explanation for the origin of carbonatite magmas (Le Bas 1987; Kjarsgaard and Hamilton 1989; Peterson 1989). In addition to the experimental evidence, various natural examples of silicate–carbonate liquid immiscibility have been described (Ferguson and Currie 1971; Rankin and Le Bas 1974). In particular, Le Bas and Aspden (1981) have described carbonate inclusions in East African ijolites which they interpreted as crystallized droplets of an immiscible natrocarbonatite liquid. According to Peterson (1989), the unusually Na-rich composition of the Oldoinyo Lengai natrocarbonatite magma is due to immiscible separation from a nephelinite magma which was itself abnormally Na-rich. Some of the silicate lavas of Oldoinyo Lengai are wollastonite– and combeite–nephelinites with Na_2O contents up to 15%.

Whether carbonatite magmas are primary, or derived from a carbonated nephelinite parent magma, there is no doubt that their source lies in the mantle. Their isotopic compositions (Sr, Nd, and Pb) are indistinguishable from those of oceanic island basalts.

FELSIC ALKALINE ROCKS

Trachytes and phonolites, and their plutonic equivalents, the syenites and nepheline syenites, are less common than basalts or rhyolites, and are generally thought of as derivative rather than parental magmas, i.e. secondary rather than primary.

Phonolitic and trachytic rocks occur in all tectonic environments, but they are most abundant in non-orogenic continental areas such as the Monteregian province of Quebec or the rift valleys of East Africa. Where intrusive, they have been emplaced by mechanisms similar to those of granites, for example as ring complexes or as stoped stocks.

In oceanic areas, phonolites and trachytes are minor associates of alkali basalts. In continental areas, they are associated either with nephelinites, ijolites and carbonatites, as in many of the East African examples, or with peralkaline granites, as in New England, the Oslo region of Norway and the Gardar province of Greenland. In addition to the obviously magmatic syenites and nepheline syenites it may be borne in mind that rocks of similar mineralogy may be formed by fenitization, and not all rocks mapped under these names are necessarily igneous.

PERALKALINE NEPHELINE SYENITES

An important distinction can be made between those phonolites and nepheline syenites in which the cation ratio (Na+K)/Al is greater than 1 (agpaitic) and those in which the ratio is less than 1 (miaskitic). Most phonolites and nepheline syenites are miaskitic, but a minority are agpaitic and this is reflected by acmite in the norm and by the actual presence of sodic pyroxenes and amphiboles. Table 55 gives the modal compositions of some agpaitic and miaskitic examples. The Vishnevogorsk rocks are miaskitic, the Khibina rocks slightly agpaitic, and the Lovozero rocks strongly agpaitic.

Table 55. Average modal compositions of three Russian nepheline–syenite intrusions (after Vlasov *et al.* 1966)

	Vishnevogorsk (miaskitic)	*Khibina*	*Lovozero (agpaitic)*
Feldspars	70	56	39
Nepheline	21	34	22
Biotite	4	—	—
Aegirine + arfvedsonite	—	8	21
Zeolites	1.2	—	5.5
Eudialyte	—	0.2	4
Sodalite	—	—	4.5
Calcite	1.2	—	—

Agpaitic nepheline syenites may contain aegirine, fluor-arfvedsonite, polylithionite, sodalite and analcime as essential minerals in addition to alkali feldspars and nepheline. They are rich in a number of trace elements, particularly Be, Li, Nb, rare earths, Th, U, Zr, Cl and F, giving rise to unusual accessory minerals such as eudialyte, villiaumite, rinkite and steenstrupine. Miaskitic nepheline syenites are less distinctive in their mineralogy and chemistry, containing such minerals as hornblende, biotite, magnetite and zircon, although rarer accessories such as melanite, aeschynite and pyrochlore may also occur.

Figure 306. Simplified geological map of the Ilímaussaq intrusion, west Greenland (after Larsen and Sørensen 1987).

Agpaitic nepheline syenites often show very well-developed rhythmic layering, presumably because the high volatile content of their magmas lowers the viscosity and permits easy crystal fractionation. The *Ilímaussaq* intrusion (Fig. 306) is a composite intrusion in which these features are extremely well displayed. Figure 307 shows a generalized section through

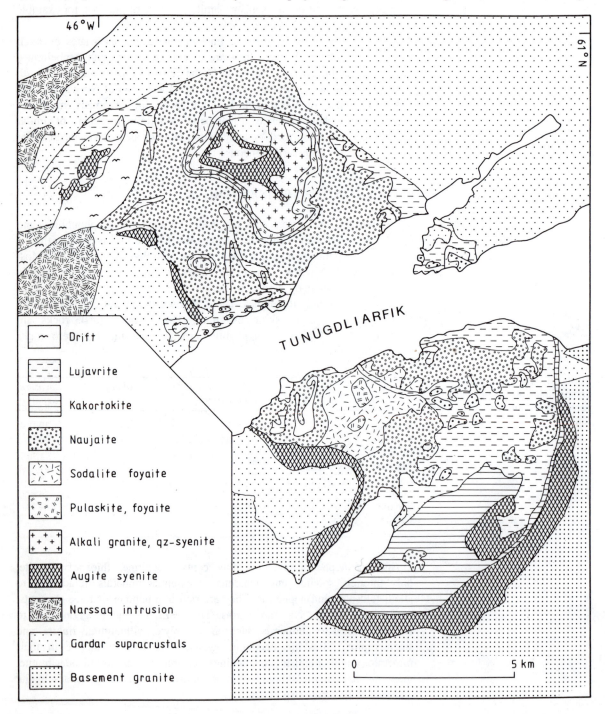

Drift

Lujavrite

Kakortokite

Naujaite

Sodalite foyaite

Pulaskite, foyaite

Alkali granite, qz-syenite

Augite syenite

Narssaq intrusion

Gardar supracrustals

Basement granite

TUNUGDLIARFIK

0   5 km

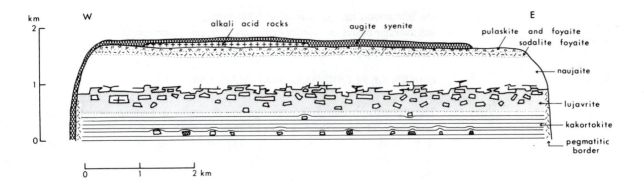

Figure 307. Diagrammatic cross-section through the southern part of the Ilímaussaq intrusion, Greenland (after Andersen *et al.* 1981). The horizontal and vertical scales are equal.

this intrusion, which is well exposed in the dissected fjord coastline of south-west Greenland. The complex was intruded in three stages. The first phase consists of a shell of augite–syenite adjoining the walls and roof of the intrusion. This rock contains over 70% alkali feldspar together with fayalite, hedenbergite-rich clinopyroxene and biotite. The second phase consists of two sheets of peralkaline granite and quartz–syenite intruded into the augite–syenite of the roof zone. The third phase makes up the bulk of the intrusion and consists of agpaitic nepheline syenite. The nepheline syenite is highly fractionated, most of the rocks being obviously accumulative.

The nepheline syenites appear to have crystallized from the roof downwards, in the sequence pulaskite–foyaite–sodalite foyaite–naujaite. The kakortokites which occur in the lowest part of the intrusion may represent a later intrusion of magma; they are strongly layered cumulates with an alternation of black (arfvedsonite-rich), red (eudialyte-rich) and white (feldspar-rich) layers. They pass up into lujavrite (mafic nepheline syenite) which contains many xenoliths of naujaite and some of augite–syenite.

The essential mineralogy of the Ilímaussaq rocks is summarized in Table 56. Many rare minerals are also present, reflecting the high level of certain trace elements, especially Zr, Nb and the rare earth elements. These include some of the elements thought of as 'mantle-incompatible elements' which are normally enriched in alkalic as opposed to tholeiitic basalts (for example Rb, Zr, Nb), but not others (for example Ba, Sr). The average Zr content of

Table 56. Average modal compositions of the nepheline syenites of the Ilímaussaq intrusion, Greenland (after Ferguson 1970)

	Sodalite foyaite	Naujaite	Kakortokite	Green lujavrite	Black lujavrite
Alkali feldspars	32.8	25.7	37.7	39.5	31.1
Nepheline	10.4	9.9	22.0	15.0	20.2
Aegirine	9.4	6.6	8.0	32.2	5.4
Arfvedsonite	9.2	5.6	13.8	3.3	36.8
Eudialyte	3.2	4.9	10.9	10.0	6.5
Sodalite	18.7	40.2	0.7	—	—
Analcime	7.4	3.6	4.6	—	—
Natrolite	8.8	1.7	2.0	—	—

Eudialyte is $(Na,Ca,Fe)_6ZrSi_6O_{18}(Cl,OH)$.

the Ilímaussaq intrusion is over 5000 ppm, which is very much higher than the average basalt or granite (109 and 246 ppm respectively).

ORIGIN OF FELSIC ALKALINE MAGMAS

The essential minerals of this group of rocks are alkali feldspars and nepheline. Their melting behaviour may be represented by phase relations in the system $NaAlSiO_4$–$KAlSiO_4$–SiO_2–H_2O. Figure 308 shows the form of the liquidus surface in this system at various water pressures. There are two temperature minima, one on the silica–feldspar boundary at a composition corresponding to that of rhyolite or granite, and the other on the feldspar–nepheline boundary at a composition corresponding to phonolite or nepheline syenite. The two minima are separated by a ridge along the albite–orthoclase join.

The normative compositions of most nepheline syenites and phonolites are clustered around the nepheline–alkali feldspar minimum (Fig. 309), leading us to infer that these liquids must be the products of crystal–liquid equilibrium, i.e. either crystallization differentiation or partial melting. The normative compositions of alkali syenites and trachytes lie on the saddle between the oversaturated and undersaturated minima, and the generation of these liquids requires either incomplete fractionation, a higher degree of partial melting, or some process other than fractionation or partial melting.

Because of their comparatively small abundance, and the absence of suitable starting materials to give rise to such liquids by partial melting, most petrologists believe that phonolites and trachytes are not primary magmas but are the products of crystal fractionation. Because alkaline felsic lavas are nearly always associated with alkaline mafic lavas, it is generally assumed that they are residual differentiated fractions from alkali basalt or nephelinite. Initial $^{87}Sr/^{86}Sr$ ratios of trachytes, phonolites and syenites are given in Table 57, together with the ratios for associated basalts. Although there are some felsic alkaline rocks with high initial ratios, most are in the range characteristic of mantle-derived magmas.

Table 57. Initial $^{87}Sr/^{86}Sr$ ratios of phonolites and trachytes and their plutonic equivalents, and associated basalts

	Phonolite	*Trachyte*	*Basalt*	*Reference*
St Helena	0.704	0.703	0.703	Grant *et al.* 1976
Kerguelen	0.706	0.706	0.705	Weis *et al.* 1993
Dunedin	0.703	0.704	0.703	Price and Compston 1973
Monchique	0.704	0.705	0.703	Rock 1976
Mt Kenya	0.704	0.704	0.704	Rock 1976
Kangerdlugssuaq	0.704	0.704–0.709	—	Pankhurst *et al.* 1976
Klokken	—	0.703	0.703	Blaxland and Parsons 1975
Ilímaussaq	0.710	0.703	—	Blaxland *et al.* 1976
Trans–Pecos	0.704–0.712	0.703–0.709	—	Barker *et al.* 1977
Red Hill	0.703–0.707	0.703	—	Foland and Friedman 1977
Tristan da Cunha	0.705	—	0.705	Le Roex *et al.* 1990
Marangudzi	0.707–0.710	0.707–0.719	—	Foland *et al.* 1993
Lundy Island	—	0.705	0.705	Thorpe and Tindle 1992

Figure 308. The system silica–nepheline–kalsilite–water at P_{H_2O} = 1 kilobar and 5 kilobars (after Tuttle and Bowen 1958; Hamilton and Mackenzie 1965; Morse 1969b).

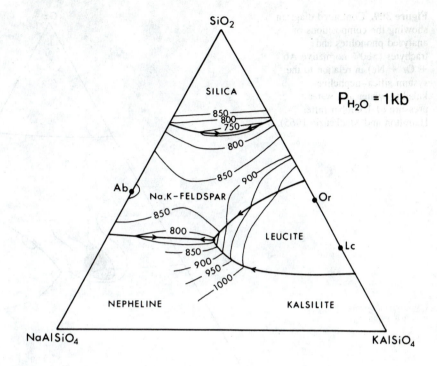

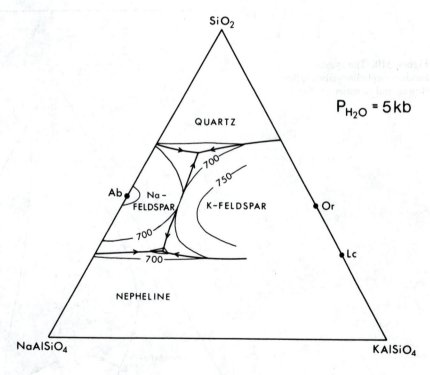

Figure 309. Contoured diagram
showing the compositions of
analysed phonolites and
trachytes (>80% normative Ab
+ Or + Ne) in relation to the
system silica–nepheline–
kalsilite–water at a water
pressure of 1 kilobar (after
Hamilton and Mackenzie 1965).

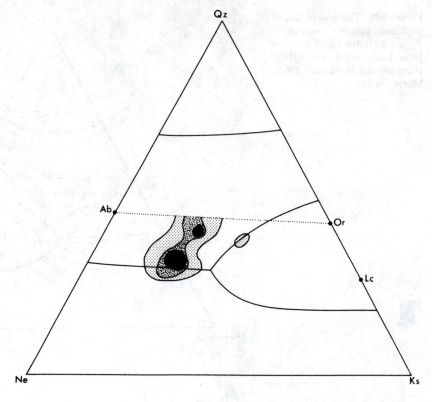

Figure 310. The system
fayalite–nepheline–silica (after
Bowen and Schairer 1938).

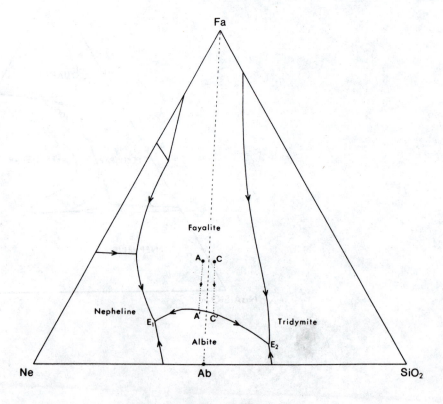

The system fayalite–nepheline–silica (Fig. 310) illustrates in a simplified way the relationships between the principal mafic and felsic magmas. There are two eutectic points in this system (E_1 and E_2), and fractional crystallization of any liquid in the system leads eventually to a residual melt with one of these compositions. Eutectic point E_1 crystallizes as a mixture rich in feldspar, with smaller amounts of nepheline and mafic mineral (olivine), and is analogous to a phonolitic rock. Eutectic point E_2 crystallizes as a mixture rich in feldspar, with smaller amounts of silica and mafic mineral, and is analogous to a rhyolite. More mafic compositions, analogous to basalts, lie within the primary olivine field.

Compare the effects of fractional crystallization of liquids A and C in Fig. 310. The separation of fayalite drives the composition of initial melt A towards the fayalite–albite boundary at A', while melt C goes towards the boundary at C'. Once the boundary has been reached, the residual melt from A heads towards E_1 by the removal of fayalite and albite, and the residual melt from C heads in the opposite direction towards E_2, also by the separation of fayalite and albite.

This system shows that melts of rather similar composition can fractionate towards quite different residua, depending on which side of the line Fa–Ab they originally lie. This line is in effect a barrier, which cannot (at low pressures and in the absence of other constituents) be crossed by a liquid during fractional crystallization. So a slightly alkaline mafic melt fractionates to a 'phonolitic' residuum and a slightly oversaturated mafic melt fractionates to a 'rhyolitic' residuum, and both melts pass through a 'trachytic' composition (around A' and C' in Fig. 310). More strongly alkaline mafic melts could fractionate towards the 'phonolitic' minimum without passing so close to the feldspar-rich 'trachytic' composition.

To be more realistic, fractionation would have to be represented by a system with more components, to allow for other constituents such as calcium and magnesium. Phase relationships analogous to those of Fig. 310 are shown by the systems diopside–nepheline–silica, forsterite–nepheline–silica and anorthite–nepheline–silica, in each of which fractionation leads towards one or other of two residual compositions, one rich in nepheline + albite and the other rich in silica + albite (Schairer 1957; Schairer and Yoder 1961).

Figures 311 and 312 show possible fractionation trends from basalt to trachyte and phonolite. The central trend shown in these diagrams runs from a basalt which is neither oversaturated nor undersaturated with silica (no normative Q or Ne) to a trachyte which is likewise neither oversaturated nor undersaturated. Intermediate compositions on this trend would correspond to rocks called hawaiite, mugearite and benmoreite.

The naming of the rocks intermediate between basalt and trachyte has given rise to a number of problems. Hawaiite and mugearite are relatively straightforward, because as their analyses show (Table 58) they do not contain normative Q or Ne. Cox *et al.* (1979) made the reasonable suggestion that trachybasalt should be a broad term for the relatively basic rocks lying between basalt and trachyte. Unfortunately, as the average composition shows (Table 58), the rocks to which the name trachybasalt has most often been given are Ne-normative. Carmichael *et al.* (1974) described

Figure 311. Possible fractionation trends from basaltic magma to felsic derivatives by analogy with Fig. 310.

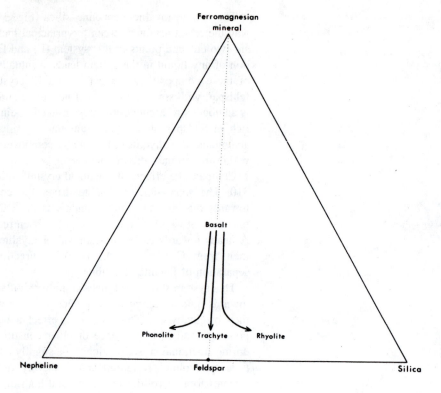

Figure 312. Possible fractionation trends from basaltic magmas to felsic derivatives. Alkali basalts are presumed to fractionate towards phonolite, and tholeiites towards rhyolite. Compositions on the direct line from basalt to trachyte are labelled H (hawaiite), M (mugearite) and B (benmoreite).

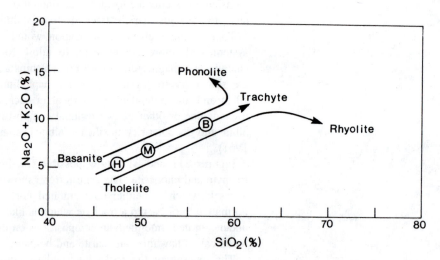

the terms trachybasalt and trachyandesite as roughly synonymous, but the rocks to which the term trachyandesite has most often been applied are Q-normative. Thus most 'trachybasalts' would lie on the alkali basalt–phonolite trend, whereas most 'trachyandesites' would lie on the tholeiite–rhyolite trend.

Table 58. Chemical and normative compositions of trachybasalt and similar rock types (averages from Le Maitre 1976)

	Hawaiite	Mugearite	Trachybasalt	Trachyandesite
SiO_2	47.5	50.5	49.2	58.2
Al_2O_3	15.7	16.7	16.6	16.7
Fe_2O_3	4.9	4.9	3.7	3.3
FeO	7.4	5.9	6.2	3.2
MgO	5.6	3.2	5.2	2.6
CaO	7.9	6.1	7.9	5.0
Na_2O	4.0	4.7	4.0	4.4
K_2O	1.5	2.5	2.6	3.2
Q	—	—	—	7.8
Or	9.0	14.6	15.1	19.0
Ab	33.6	40.1	29.4	36.8
An	20.6	17.1	20.1	16.6
Ne	—	—	2.2	—
Di	10.9	6.2	11.8	4.0
Hy	0.1	2.2	—	6.1
Ol	9.3	4.1	8.3	—

In the light of the experimental evidence, one can predict the existence of different fractionation sequences depending on the silica saturation of the initial parent magma, and the corresponding magma series are readily identifiable in modern volcanoes. In order of increasing alkalinity we can distinguish:

1. Tholeiite → trachyte → rhyolite. This association is observed in the volcanic rocks of Aden, Ascension Island, and Bouvet Island. The rhyolitic end-member is normally peralkaline.
2. Alkali basalt → trachyte → phonolite. This is the commonest association in oceanic island volcanoes. Examples are seen in Mauritius, St Helena, and Tenerife.
3. Nephelinite → phonolite. In the most alkaline volcanoes trachyte is absent and nephelinite appears to have differentiated directly to phonolite. This association is seen in the island of Trinidade, in several of the East African volcanoes, and in the Balcones province of Texas.

In the larger volcanic provinces more than one of these trends may be represented. In other alkaline volcanoes, trachyte is the most fractionated lava type that is observed and neither phonolite nor rhyolite is present.

Direct evidence of differentiation can be seen in a number of alkaline intrusions. In the Gardar province of south Greenland there is a suite of 'Giant Dykes', up to 1 kilometre in width, in which layered alkali–gabbroic cumulates grade through syenogabbro into syenite (Upton and Thomas 1980). Some of the giant dykes in this region contain ultramafic (picritic) cumulates and others contain foyaitic differentiates, although not at the same localities. In several of the intrusions of the Monteregian province (Fig. 313), rhythmically layered alkali–gabbros are associated with alkaline felsic rocks.

The Brome intrusion contains a range of compositions from alkali–gabbro to syenite and foyaite (Valiquette and Archambault 1970). Other Monteregian intrusions (Bruno, Yamaska) contain nepheline-bearing or nepheline-normative ultramafic rocks as well as alkaline felsic rocks. The Mount Johnson intrusion shows rhythmically layered alkali–gabbro (essexite) grading into syenite (pulaskite).

A differentiative relationship between basalt, trachyte and phonolite has not been universally accepted. Chayes (1963) objected to such a relationship on the grounds that lavas intermediate between basalt and trachyte are very rare in oceanic volcanoes. This apparent lack of intermediate compositions was discussed long ago by Daly and has come to be known as the 'Daly gap'. In a differentiation series, liquids of all the intermediate compositions should be produced at some stage. Several suggestions have been made as to why the intermediate compositions are apparently uncommon, such as inadequate sampling of the relatively inconspicuous trachybasalts and trachyandesites, or the difficulty of erupting intermediate magmas with higher viscosity than basalt but lower volatile content than trachyte. More careful sampling has shown that the intermediate lava compositions are not

Figure 313. The Monteregian igneous province, Quebec (after Gold 1967). The shaded area is the city of Montreal.

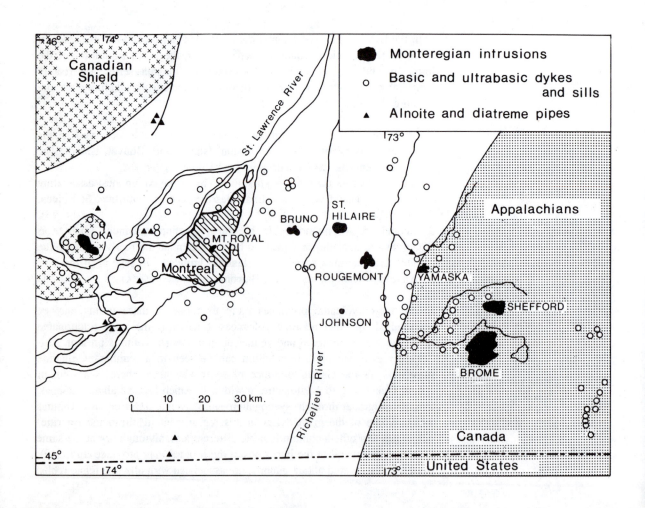

as rare as was once thought. For example Baker (1968) carefully measured the relative amounts of different rock types on St Helena and estimated the relative volumes of subaerial lavas as basalts 70–80%, trachybasalts 15–25%, trachyandesites 4%, and trachytes + phonolites 1%. Nevertheless, Chayes (1977) maintained that the Daly gap is real, based on a very large collection of chemical analyses of samples from the Canary Islands and elsewhere. To reconcile the shortage of intermediate rocks with fractionation, Wyllie (1963) pointed out that the shape of the liquidus surface may critically influence the amounts of different liquids produced during fractionation. On a gently sloping liquidus surface, a large change in liquid composition can result from a small drop in temperature, and during fractional crystallization the intermediate liquid compositions might be produced in only small amounts. According to the calculations of Wilshire (1967) on the Prospect alkali–dolerite intrusion, the residual melt composition changed rapidly at the stage when about 95% of the magma had consolidated. A rapid change of this sort from basaltic to trachytic composition would leave almost no opportunity for an intermediate melt to separate and be erupted.

In continental alkali-basalt–trachyte–phonolite provinces the proportion of felsic to mafic magma that is erupted is greater than in oceanic islands. This is seen in the relative abundance of syenites and felsic lavas in the Gardar province of Greenland, the Quaternary lavas of the Laacher See district in Germany, or the Eastern Rift Valley of Kenya. Gill (1973) suggested that the thickness of low-density continental crust acts as a deterrent to the rise of high-density basaltic magma while permitting uprise of its lighter, more felsic fractionation products.

The large size of many felsic alkaline intrusions is surprising considering the comparative rarity of alkaline igneous rocks as a whole. The Kangerdlugssuaq intrusion (Fig. 316 below) has an outcrop area ($700\,km^2$) larger than most granite intrusions, while trachytic and phonolitic eruptions are often comparable in scale to rhyolitic ones. The trachytic Neapolitan Yellow Tuff, which was erupted 12,000 years ago and underlies the city of Naples, has a volume equivalent to $46\,km^3$ of magma (Scarpati *et al.* 1993). Assuming that such magmas were produced by fractionation, the implication is that their parent magmas undergo prolonged fractionation in very large magma chambers before eruption. This may be related to their occurrence in extensional tectonic environments, which provide the optimum conditions for the formation of large magma chambers.

Assuming that trachytes and phonolites are produced by fractionation, several authors have shown quantitatively that fractionation of particular combinations of minerals can account for the differences in major and trace element content between alkali basalts and their associated trachytes. Maaløe *et al.* (1986) examined the alkaline lava series of Jan Mayen Island, which contains a continuous range of compositions from ankaramite through trachybasalt to trachyte. The compositional variation could be modelled by the fractionation of crystal assemblages which varied during fractionation, but were initially of olivine and clinopyroxene, with plagioclase and magnetite playing a greater part as the liquids became more felsic. Experimentally determined phase relations of the most mafic ankaramite showed that olivine was only stable in the presence of liquid at pressures below 19.5 kbar and plagioclase only below 11 kbar, from which it was deduced that fractionation occurred at low pressures, i.e. at shallow depths below the volcano.

Widom *et al.* (1992) showed by major element modelling that a trachytic pumice from the Azores could have differentiated from the associated alkali basalt by approximately 70% fractionation of augite and plagioclase, with smaller amounts of magnetite, kaersutite, olivine and apatite. The trachyte is itself highly variable, suggesting further fractionation of sanidine, with minor biotite, clinopyroxene and accessory minerals. These authors also used U-series disequilibrium to investigate the time-scale of fractionation. For example, they calculated that the whole process, from the formation of the parental alkali basalt magma by melting in the mantle to the eruption of the evolved trachyte, took 90,000 years. They suggested that trachytic magma chambers evolved and erupted several times during the last 15,000 years, each one becoming gradually more zoned as a result of fractionation, the magma finally being erupted when it became saturated with volatiles.

CROSSING THE SYENITE DIVIDE

The correspondence between phonolites and rhyolites and the two minima of the system kalsilite–nepheline–silica (Fig. 308) has been pointed out. It is not surprising that phonolite and rhyolite compositions are concentrated around these two positions. However, the abundance of trachytic or syenitic compositions is more surprising, since they lie across a thermal divide falling away to liquidus minima on either side.

Many igneous provinces display a range of magma compositions which includes trachytes but also extends towards undersaturated (phonolitic) and oversaturated (rhyolitic) compositions. Figure 314 shows the association of rock types in the Red Hill intrusion in New Hampshire, one of the complexes of the White Mountain plutonic series. Individual components of the complex include syenite, nepheline syenite and granite, and their compositions extend right across the syenite thermal divide (Fig. 315).

The simplest explanation for such a range is that the felsic magmas differentiated independently from mafic parents which themselves varied from oversaturated to undersaturated. This explanation was given by Barker (1977) for the syenites of the Trans-Pecos province in Texas, which include both nepheline-bearing and quartz-bearing types. A different explanation was given by Chapman (1968) for the gabbro–syenite–granite association in the White Mountain magma series. He envisaged the syenite as having fractionated from the gabbroic magma, but the granite as being produced by melting of country rocks above the advancing gabbro. Foland and Friedman (1977) confirmed by Sr and O isotopic study that the syenite, nepheline syenite and granite of the Red Hill complex are not comagmatic. The syenites lie on a single isochron corresponding to an initial $^{87}Sr/^{86}Sr$ ratio of 0.7033, but the granites and nepheline syenites have higher initial ratios in the range of 0.7033 to 0.707 which could be attributed to crustal contamination.

Figure 314. The principal components of the Red Hill intrusive complex, New Hampshire (after Size 1972).

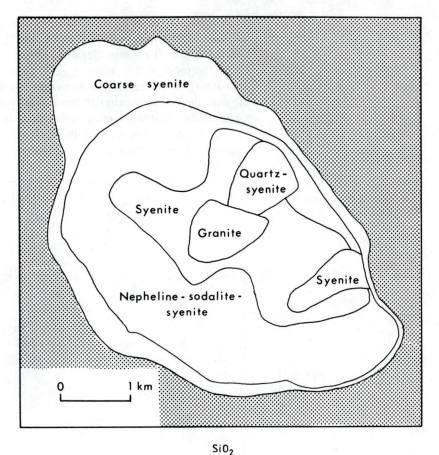

Figure 315. Normative compositions of the rocks of the Red Hill igneous complex in relation to the system silica–nepheline–kalsilite–water at P_{H_2O} = 1 kilobar (after Size 1972).

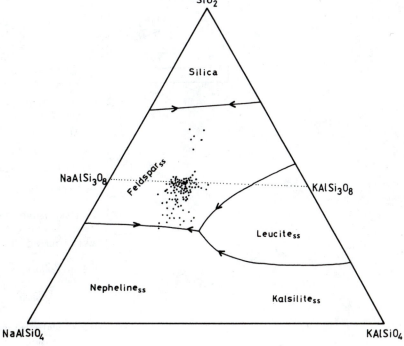

A more serious problem is posed by the apparently direct differentiative relationship between nepheline- and quartz-bearing syenites in some individual intrusions. The Kangerdlugssuaq intrusion in Greenland is shown in Fig. 316. The rocks of this intrusion grade from a lower and outer nordmarkite (quartz–syenite) through pulaskite (syenite with no quartz or nepheline) into foyaite (nepheline syenite) in the central upper part of the intrusion. The syenites have a strong igneous lamination, dipping everywhere inwards, and the distribution of rock types thus resembles a pile of saucers, becoming smaller and more undersaturated upwards. The saucer-shaped disposition of

Figure 316. The Kangerdlugssuaq intrusion, east Greenland (after Wager 1965).

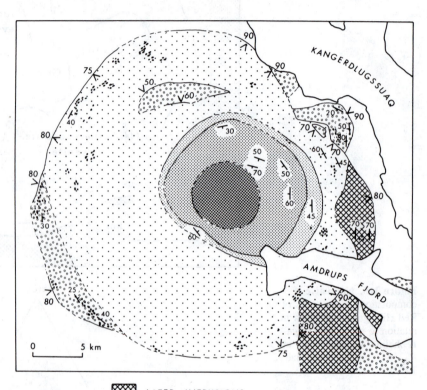

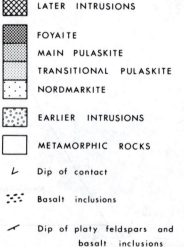

LATER   INTRUSIONS

FOYAITE

MAIN   PULASKITE

TRANSITIONAL   PULASKITE

NORDMARKITE

EARLIER   INTRUSIONS

METAMORPHIC   ROCKS

∟ Dip of contact

Basalt   inclusions

Dip of platy feldspars and of
basalt   inclusions

the various petrographic types is brought out by zones of basalt xenoliths which dip inwards at angles of between 30 and 60 degrees, as well as by the platy parallelism of the feldspar crystals. The order of solidification of the various syenites is shown by nepheline-bearing veins, presumably related to the foyaite, which cut the nordmarkite; there are no quartz-bearing veins cutting the foyaite. The spatial relationships suggest that the nordmarkite makes up nearly 90% of the volume of the intrusion, and the foyaite not much more than 1%. The parent magma was therefore presumed by Wager (1965) to be of quartz–syenite composition and the foyaite was regarded as a differentiate of the nordmarkite magma.

The compositions of the Kangerdlugssuaq rocks are shown in Fig. 317 in relation to the system $KAlSiO_4$–$NaAlSiO_4$–SiO_2. The means by which a slightly oversaturated felsic magma might become undersaturated during fractional crystallization, or vice versa, are not clear. Kempe and Deer (1976) discussed the possible depression of the thermal barrier by a change in the water pressure. It can be seen from Fig. 318 that the height of the barrier is relatively small compared with the effect that a change in water pressure could have. Contamination offers additional possibilities. Brooks and Gill (1982) suggested that the oversaturated rocks of the Kangerdlugssuaq intrusion were formed by contamination and not differentiation, because the cryptic variation in the mineralogy of the intrusion is too irregular for a crystal fractionation sequence; they regarded the central foyaite as the original and least contaminated magma.

In the Marangudzi intrusive complex in Zimbabwe, the major units are

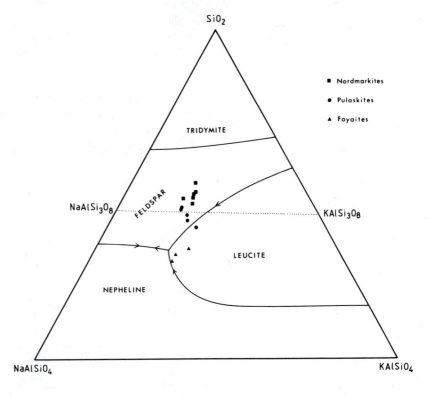

Figure 317. Normative compositions of the rocks of the Kangerdlugssuaq intrusion, Greenland (after Wager 1965).

Figure 318. Section through the liquidus surface of the system $KAlSiO_4$–$NaAlSiO_4$–SiO_2–H_2O, drawn approximately through the phonolite and rhyolite minima (after Schairer 1957; Morse 1969b).

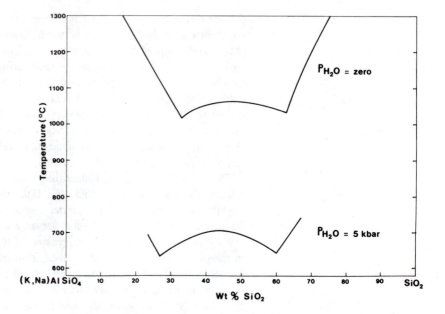

gabbro, quartz–syenites and nepheline syenites. The nepheline syenites have relatively low initial $^{87}Sr/^{86}Sr$ ratios (~0.708) whereas the quartz–syenites have initial ratios ranging from 0.708 up to 0.719, implying crustal contamination of the latter (Foland *et al.* 1993).

It is noticeable that whereas continental igneous complexes often show an association of granite with nepheline syenite, oceanic volcanoes do not show an analogous association between phonolite and rhyolite. A very interesting group of volcanoes is the 'Cameroon line', a string of Tertiary–Recent volcanoes which extends across the continental margin of West Africa (Fitton 1987). All the volcanoes in this line contain alkali basalts, but those built on oceanic crust also erupt trachyte and phonolite but no rhyolite, whereas those standing on continental crust also have rhyolite. These relationships suggest that the presence of rhyolite in the continental section must be due to crustal contamination.

CHAPTER 12

Kimberlites and ultrapotassic igneous rocks

KIMBERLITES

Kimberlite is a serpentinized and carbonated mica–peridotite of porphyritic appearance, which frequently contains xenoliths of ultrabasic rock-types characterized by a high-pressure mineralogy. The commonest phenocrysts (or xenocrysts) are olivine, garnet (pyrope, chrome-pyrope or pyrope–almandine), pyroxene (enstatite or chrome-diopside), amphibole, mica (phlogopite, biotite, or the related layer silicate vermiculite), and ilmenite. The groundmass is fine-grained and may contain many different minerals, including serpentine, pyroxene, amphibole, mica, melilite, monticellite, zeolites, calcite, magnetite, chrome-spinel, apatite and perovskite.

Kimberlites occur in pipe-like bodies, in dykes, and more rarely in sills. The intrusions are small; pipes are up to about a kilometre in diameter and usually much less, and dykes are up to about 10 metres wide. The pipes are diatremes, i.e. they have been drilled by a mixture of gas and solid particles in a fluidized condition.

Three types of kimberlitic rock can be recognized:

1. Massive kimberlite, intruded as dykes or forming part of the filling of pipes. This type of kimberlite has flow features and fine-grained margins and contains few xenoliths.
2. Intrusive kimberlite breccia, occurring mainly in pipes, and consisting of abundant fragments (kimberlite and/or xenoliths) in a kimberlite matrix.
3. Kimberlite tuffs and tuff–breccias, occurring only in pipes, and consisting largely of fragments (kimberlite and/or xenoliths) cemented by secondary minerals.

The xenoliths are an important feature of kimberlites, and not only because of their abundance. They comprise a sample of rocks from all levels of the crust penetrated by the kimberlite diatremes (both above and below the level of emplacement), together with ultramafic rocks which are apparently of mantle origin. Many of the large crystals which give kimberlites their porphyritic appearance are xenocrysts rather than phenocrysts, so kimberlite may be regarded as a hybrid rock whose variability largely results from the uneven mixture of magmatic and xenolithic or xenocrystic constituents.

It is not universally agreed that kimberlites were emplaced as magmas, although most authors believe that they were. Kimberlites rarely show contact metamorphism of their country rocks, but the intrusions are rather small and large aureoles would not necessarily be expected. In Illinois and Pennsylvania, where kimberlites cut coals and bituminous shales, devolatilization of the organic materials indicates intrusion temperatures of several hundred degrees centigrade (Watson 1967). Other kimberlites show what are thought to be chilled margins.

The relationship between kimberlite pipes and dykes appears to be along the lines indicated in Fig. 319, with breccia-filled diatremes rising from dykes of massive kimberlite. A possible interpretation of this relationship is that the emplacement of gas-rich magma changed from passive to explosive as the magma approached the surface and decreasing pressure caused the exsolution of magmatic gases.

Many kimberlites, even massive ones and those occurring in dykes, have a microbreccia texture. For example the Moses Rock dyke in Utah is filled by a breccia or microbreccia in which only about 12% of the constituents are kimberlitic, and only about 1% could be mapped as kimberlite (McGetchin

Figure 319. Diagrammatic representation of the relationship between kimberlite dykes and diatremes.

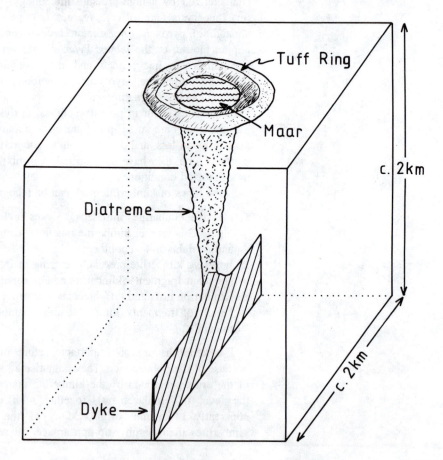

and Silver 1970). The fine-grained contact zone (1–2 centimetres thick) consists of serpentinized material. McGetchin and Silver believed that the Moses Rock kimberlite was emplaced not as a melt but as a fluidized system consisting of solid particles suspended in a low density fluid, gaseous near the surface. In Siberia and in Tanzania the phlogopite in some kimberlites gives a K/Ar age much greater than the stratigraphic age of the country rocks (Davidson 1964, 1967; Edwards and Howkins 1966), suggesting either that the kimberlite was intruded in a solid condition or that at least the phlogopite is xenocrystic.

Kimberlites are chemically very variable. Analyses show ranges of SiO_2 20–50%, MgO 14–40%, CaO 2–14% and K_2O 0–5%. Some average compositions are given in Table 59. Dawson (1980) reviewed the analytical data and concluded that some of this variability is due to the presence in analysed samples of comminuted xenolithic material, to variations in megacryst/matrix ratio, and in the abundance of volatile constituents. Even so, there is no reason to doubt that some of the variability is primary. In particular, South African kimberlites have long been divided into two categories: Group I (formerly called basaltic kimberlites) and Group II ('micaceous kimberlites'), the latter being particularly rich in phlogopite.

In the Maimecha–Kotui area in northern Siberia there occur thick flows of meimechite, a picritic lava consisting of serpentinized olivine phenocrysts in an amygdaloidal glassy groundmass. Some samples contain phlogopite, and this lava is considered to be a possible extrusive equivalent of the kimberlites which occur in the same area (Egorov 1970). A comparison of the glassy chilled margin of a meimechite dyke with its porphyritic centre, represented by the analyses in Table 59, suggests that the meimechites were formed by accumulation of olivine crystals in a magma whose composition approximates to that of some potassic melilitites. Meimechites or related extrusive

Table 59. The compositions of kimberlite and related rock types

	Siberian kimberlite	Lesotho kimberlite	S. African kimberlite	Alnoite	Melilitite	Meimechite (margin)	Meimechite (centre)
SiO_2	27.6	33.2	36.4	30.0	36.7	38.6	38.2
TiO_2	1.6	2.0	1.0	3.2	3.5	4.8	3.3
Al_2O_3	3.2	4.4	5.1	8.3	7.8	7.9	5.1
$FeO + Fe_2O_3$	8.2	10.2	7.7	12.9	13.2	14.8	13.6
MgO	24.3	22.8	17.4	16.7	9.3	12.3	24.7
CaO	14.1	9.4	11.2	15.0	18.2	12.4	7.6
Na_2O	0.2	0.2	0.4	0.4	2.3	0.6	0.4
K_2O	0.8	0.8	1.5	1.5	1.8	2.6	1.4
P_2O_5	0.6	0.6	0.6	1.5	1.1	0.8	0.4
H_2O+	7.9	8.0	—	6.4	3.0	—	—
CO_2	10.8	4.6	—	2.8	—	0.2	0.4

Sources of data: Gurney and Ebrahim (1973) for means of 623 Russian, 25 Lesotho and 80 South African kimberlite analyses; Von Eckermann (1961) for mean of two alnoites from Alno, Sweden; Le Bas (1977) for mean of six melilitites from western Kenya; and Vasil'yev and Zolotukhin (1970) for analyses of the glassy chilled margin and the centre of a meimechite dyke from the Maymecha River, Siberia.

rocks have also been found in Quebec (Dimroth 1970), Mali (Hawthorne 1975) and Tanzania (Reid *et al.* 1975).

Kimberlites only occur in continental areas, and mostly in Pre-Cambrian shields, although the kimberlites themselves are often much younger than the Pre-Cambrian. Kimberlites are more abundant in southern Africa (Fig. 320) than in other parts of the world, and there are also important kimberlite provinces in West Africa and Siberia, but some kimberlites are known from all the continents. Despite the rarity of kimberlite, there are several areas (in Siberia, South and West Africa, Greenland and Tanzania) where kimberlite emplacement has recurred at widely separated times during geological history.

In most areas where they occur, kimberlites are accompanied by more commonplace alkaline basic magmas. Normal basaltic rocks may even occur in the same intrusive bodies as kimberlite. For example, the kimberlite pipe of the Premier diamond mine in Transvaal contains gabbro, and the Diamantkop pipe in Namibia contains dolerite as well as kimberlite. In the Czech Republic, kimberlites occur in a Tertiary volcanic province which includes alkali basalts, nephelinites, melilitites and leucitites. Elsewhere they are associated with the plutonic equivalents of these lava types, and with minettes, monchiquites and alnoites. In Arkansas, Siberia, Sweden, and various parts of southern Africa they occur alongside carbonatites, or contain carbonatite veins and segregations. The Arvida dykes in Quebec show every gradation from kimberlite to carbonatite (Gittins *et al.* 1975), and composite

Figure 320. Occurrences of kimberlite and related rock types in southern Africa (after Dawson 1980).

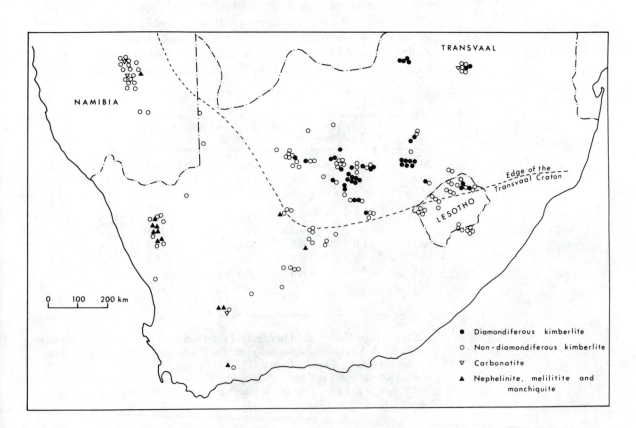

sills of kimberlite and carbonatite occur at Benfontein in South Africa (Dawson and Hawthorne 1973).

Various types of isotopic study have been carried out on kimberlites. Their $^{87}Sr/^{86}Sr$ initial ratios are very variable (from 0.703 to 0.718), as are their Nd and Pb isotopic ratios, showing that their source is very heterogeneous. Their high K_2O content shows of course that they have come from a source which is richer in K and Rb than that of ordinary basalts, so their Sr might be expected to have a greater radiogenic component than other mantle-derived rocks. The oxygen isotopic compositions of kimberlites also vary: the highest $\delta^{18}O$ values are found in kimberlites of fragmental texture, which suggests an interaction with meteoric water during emplacement (Kobelski *et al.* 1979).

XENOLITHS IN KIMBERLITE

The ultramafic xenoliths in kimberlite are of great interest because their mineralogy and composition provide many clues to their depth of origin and to the nature of the upper mantle. The commonest ultramafic xenoliths are of dunite, harzburgite, spinel–lherzolite, garnet–lherzolite and eclogite, together with their metasomatized derivatives and monomineralic nodules of olivine, pyroxene and garnet. Garnet–lherzolite is probably the most abundant type of xenolith in the South African kimberlites but in some other regions eclogite is more common.

The compositions of the coexisting minerals in these xenoliths can be used to estimate their temperatures and pressures of equilibration. The pyroxenes are especially valuable for this purpose. The solvus between diopside (clinopyroxene) and enstatite (orthopyroxene) becomes narrower as temperature increases, and so the amount of Ca in coexisting clinopyroxene and orthopyroxene can be used as a geothermometer. The position of the solvus does not depend on temperature alone, but also on the pressure, as can be seen from Fig. 321. The presence of Al_2O_3 and FeO also has an effect on the solvus, although the pyroxenes in peridotites contain only small amounts of these constituents.

Measurement of pressure is mainly based on the Al_2O_3 content of orthopyroxene. It was originally observed that in the presence of garnet, which is an aluminous phase, the Al_2O_3 content of enstatite varies with pressure, and it was assumed that the same would be true of enstatite coexisting with other aluminous phases such as plagioclase or spinel. It is now known that the Al_2O_3 content of orthopyroxenes is dependent on temperature as well as pressure, and that the relationship between Al_2O_3 content and pressure depends on the nature of the aluminous phase which is present, as indicated in Fig. 322. The orthopyroxene geobarometer is therefore applicable to garnet–lherzolites, but not to spinel–lherzolites because the pyroxenes in the spinel–lherzolite assemblage do not show a sufficient pressure-dependence to estimate their conditions of formation. The extent of solid solution of Al_2O_3 in clinopyroxene is also dependent on pressure and temperature (Gasparik 1984), and may be useful for either geothermometry or geobarometry, depending on the aluminous phase with

Figure 321. The diopide–enstatite solvus at 20 and 30 kilobars (after Mori and Green 1976). Additional data are given by Carlson and Lindsley (1988).

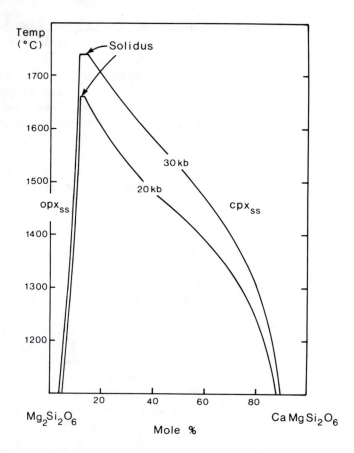

which it coexists. Additional geothermometers and geobarometers applicable to ultramafic xenoliths are described by Nickel and Green (1985), Carswell and Gibb (1987a), and Brey and Köhler (1990).

Other mineralogical features of kimberlites which are indicative of their depth of origin are the presence of diamond and the absence of the high-pressure polymorphs of olivine. Diamond is the stable form of carbon at pressures above about 45 kilobars in the temperature range 1000–1200 °C (Kennedy and Kennedy 1976), and is present in the kimberlites of some regions but not others. Olivine starts to transform to either the β- or γ-polymorph, depending on its composition (Ringwood 1975), at a pressure of about 125 kilobars (assuming a temperature of about 1500 °C), and since the high-pressure forms of olivine have not been found in xenoliths this may be taken as a maximum pressure of formation.

Boyd (1973) was the first to use the CaO and Al_2O_3 content of pyroxenes to measure the temperature and pressure of equilibration of lherzolite and harzburgite nodules in kimberlite, and his results are shown in Fig. 323. The range of pressures indicates that the xenoliths have been picked up by the kimberlite at different levels during its ascent, and the maximum pressures are an indication of the depths below which the kimberlite must have originated. The xenoliths define a trend of temperatures and pressures which can be interpreted as representing the geothermal gradient at the time the

Figure 322. Calculated isopleths of Al content of orthopyroxenes in plagioclase–, spinel– and garnet–peridotite mineral assemblages in the system CaO–MgO–Al$_2$O$_3$–SiO$_2$ (after Obata 1976). The isopleths are expressed as mole % of MgAl$_2$SiO$_6$ end-member. A more recent version of this diagram is given by Gasparik (1984) and data for the FeO-bearing system by Harley (1984).

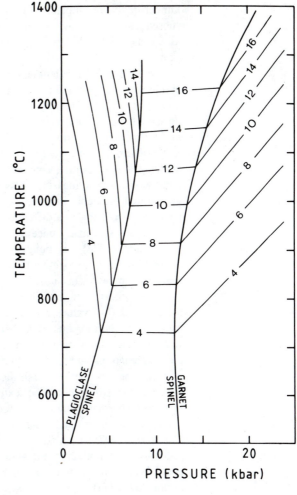

Figure 323. Estimated temperatures and depths of equilibration of lherzolite nodules from the kimberlites of Lesotho (after Boyd 1973). Solid and open circles represent sheared- and granular-textured nodules respectively. The transition from graphite to diamond is after Kennedy and Kennedy (1976). The accuracy of these estimates has been reviewed by Finnerty and Boyd (1984).

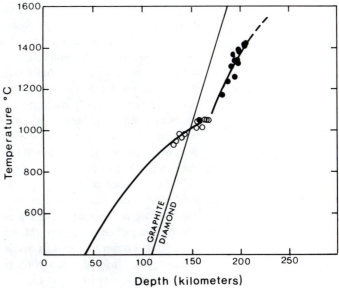

kimberlites were emplaced (during the Cretaceous). A similar gradient was subsequently found by Danchin and Boyd (1976) in the much older Premier kimberlite pipe in South Africa, which has been dated at about 1115 million years.

The geothermal gradient shown in Fig. 323 has a sharp kink, implying an abrupt steepening of the geothermal gradient. There is a relationship between this kink and the textures of the xenoliths. The low-temperature, low-pressure xenoliths are coarse-grained and granular, and are severely depleted in low-melting constituents such as Na, Ca, Al, Fe and Ti. The high-temperature, high-pressure xenoliths are sheared or mylonitized and are less depleted in low-melting constituents. Similar results have been obtained by other workers. The measurements by Macgregor (1975) show three different geothermal gradients in three different areas, with a kink in the calculated geotherm occurring at a different depth in each area.

There are still many uncertainties about the geotherms calculated in this way. Harte (1978) has pointed out that the inflected geotherms are only obtained by plotting together the calculated P–T conditions from a number of kimberlite pipes in each area; individual pipes show only a small range of P–T conditions with no obvious inflection. Carswell and Gibb (1987b) recalculated P–T values for the Lesotho kimberlites using revised data for the various thermometers and barometers, and obtained a palaeogeotherm with no inflection. However, Finnerty and Boyd (1984, 1992) have defended their calculations and asserted that the inflections are genuine.

Assuming that they are real, the cause of the inflections in the geotherms is not known for certain. The textural difference suggests that the inflection in the geothermal gradient may correspond to the boundary between rigid mantle (lithosphere) and plastically deformed mantle (asthenosphere). Boyd and Nixon (1975) pointed out that a sharply inflected geothermal gradient probably does not represent a steady-state situation, and suggested that it could be due to heating caused by deformation of the sheared lherzolites during the course of lithospheric plate movement. Movement of the lithosphere over the asthenosphere might also explain the differences in bulk composition between the granular and sheared xenoliths, which could represent different sectors of the mantle which have been brought laterally into superposition. Several alternative hypotheses have been proposed, including the arrival of a plume or diapir into the kimberlite source region from a greater depth. The lower temperature limb of an inflected geotherm may represent a steady-state geotherm in a region traversed by the rising kimberlite, while the higher temperature limb may reflect abnormal influx of heat into the kimberlite source region. According to Mercier (1979) the deformation of the sheared xenoliths is not an original feature of these rocks, but is a product of the kimberlite eruption mechanism.

The occurrence of diamond is a particularly interesting feature of certain kimberlites. The diamonds occur in both eclogite and peridotite xenoliths, and are not necessarily related genetically to the kimberlite host magma which has picked them up (Meyer 1985). In fact, diamonds from the Cretaceous kimberlite pipes in South Africa have been dated at 3200–3300 million years from the Sm–Nd isotopic ratios of their garnet inclusions (Richardson *et al.* 1984). It can be seen from Fig. 323 that the Lesotho

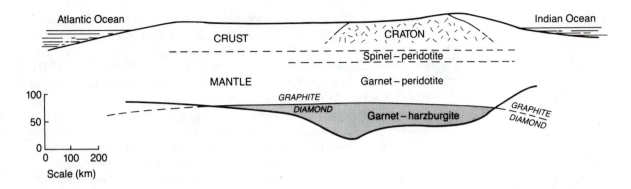

Figure 324. Diagrammatic section through the lithosphere beneath southern Africa (after Boyd 1987).

xenoliths have equilibrated at pressures in the diamond stability field, and like other kimberlites in the Transvaal craton these intrusions are diamond-bearing. In parts of southern Africa lying outside the Transvaal craton the geothermal gradients revealed by xenoliths are higher and do not intersect the diamond stability field, so it is not surprising to find that these kimberlites are not diamond-bearing (Boyd and Nixon 1979). Figure 324 shows a diagrammatic section through the lithosphere under southern Africa representing Boyd's interpretation of his xenolith measurements.

The xenoliths in kimberlite include representatives of all the rocks encountered by the rising magma from its source up to the level of emplacement, including some picked up from the uppermost mantle and lower crust. The whole xenolith assemblage of any particular kimberlite may be used to construct possible mantle sections beneath the areas where kimberlites occur. General features of such models are the transition from spinel–peridotite to garnet–peridotite with increasing depth, and a zone of dunite or harzburgite depleted in low-melting constituents overlying relatively 'fertile' lherzolite. The nature and source of eclogite xenoliths is more problematical; they are not generally regarded as normal mantle material, and since many of them are entirely free of potassium they are unlikely to represent partial melts. They may possibly be either igneous cumulates or remnants of subducted oceanic crust (Schulze and Helmstaedt 1988).

ORIGIN OF KIMBERLITE

Their mode of emplacement suggests that kimberlites ascended very rapidly, assisted by a flow of gas. Rapid ascent has not permitted the same degree of fractionation as may have taken place in other magmas derived from the mantle. Unassimilated xenoliths of deep origin are present in quantity, and are not just cumulates or melting residues but include fragments of fertile mantle which still contain low-melting (basaltic) constituents (Mysen and Kushiro 1977). Kimberlite may therefore be regarded as a primary magma type.

The most distinctive features of kimberlite compared with other mantle-derived magmas are the high contents of potassium and of volatiles, both CO_2 and H_2O. The high potassium content is presumed to be due to the presence of a potassium-bearing mineral in the mantle of the magmatic source region. Phlogopite is the most likely candidate, and is also a potential source of some of the trace elements in kimberlite. It is stable at pressures of up to 50 kilobars at 1200 °C (Wendlandt and Eggler 1980); at very high pressures it breaks down to a potassium-bearing garnet (Kushiro *et al.* 1967). The carbonate in kimberlites is a primary constituent and not of supergene origin. It occurs in the freshest specimens, has $\delta^{13}O$ values similar to the diamonds which are undoubtedly of deep origin, and it sometimes forms typically magmatic acicular or dendritic 'quench' crystals (Kobelski *et al.* 1979). The presence of either carbonates or a carbonate liquid in the rocks of the source region is indicated by the carbonate inclusions in pyrope reported by McGetchin *et al.* (1973); inclusions of clinohumite, mica and amphibole also indicate the presence of hydrous phases in the source region.

A possible origin for kimberlite magma is therefore that it formed by the high-pressure partial melting of a garnet–peridotite source rock containing small amounts of phlogopite and a carbonate. It is very likely that these minor constituents could have been introduced into the peridotite by metasomatism, involving upward-migrating fluids or possibly low-volume melts containing CO_2 and H_2O as well as potassium and associated trace elements. Many xenoliths in kimberlite do in fact show veins, zones and patches rich in phlogopite, amphibole, ilmenite or sulphides which must be the result of metasomatism, and in some cases this dates from long before the eruptive event (Gurney and Harte 1980; Boettcher and O'Neil 1980; Menzies and Hawkesworth 1987; Nielson *et al.* 1993). Such metasomatized mantle is a potential source not only of kimberlite but of ultrapotassic basic magmas such as minettes and leucite–basalts, and could explain the great compositional variability of all these rock types.

It is difficult to decide whether the presence of phlogopite is a local peculiarity of the source regions of kimberlite, or is more widespread. The results of melting experiments on kimberlites, and their high concentration of mantle-incompatible elements, point to formation by a low degree of partial melting, but if this were the only factor resulting in melts having a kimberlite composition there would be small amounts of kimberlite magma found in most basaltic provinces. The repeated occurrence of kimberlite in certain geographical regions, but not others, suggests that the mantle in those regions is particularly rich in the appropriate constituents. South Africa not only has many kimberlite occurrences, but kimberlite or related ultrabasic rocks have been intruded repeatedly during the course of geological time, from the Pre-Cambrian to the Cretaceous. This repetition also suggests that the kimberlite magmas originate in the upper (lithospheric) part of the mantle, because the sub-continental lithosphere is coupled to the continental crust during plate movements, whereas the underlying asthenosphere is not. The sub-continental lithosphere is very thick beneath cratonic areas, and like the crust of such areas may be very old.

It is not universally agreed that kimberlites do originate in the lithospheric part of the mantle, although they have certainly picked up many xenoliths from the lithosphere. Taylor *et al.* (1994) suggested that some kimberlite magmas originate well below the lithosphere, and they also suggested that the K-bearing phase in the source region could be K-richterite (amphibole) rather than phlogopite (mica).

Moore *et al.* (1991) found that some diamonds from a South African kimberlite contain syngenetic inclusions of majorite garnet. These garnets have a substantial proportion of the constituents of pyroxene in solid solution, and pressures corresponding to depths of 200 to 480 km are required to permit such extensive substitution of Si for the normal trivalent cations in the garnet structure. However, garnet becomes unstable at very high pressures (equivalent to a depth of 650 km). We can therefore assume that the garnet inclusions crystallized in the seismic transition zone of the mantle between 400 and 650 km. It is not yet clear whether the diamond xenocrysts containing majorite garnets were brought up from this deep source to a higher level before being picked up by the kimberlite magma, as suggested by Haggerty and Sautter (1990), or whether the kimberlite itself could have had a very deep source, as suggested by Ringwood *et al.* (1992).

Figure 325 shows how the sub-solidus mineralogy of a mantle peridotite might vary with pressure in the presence of a small amount of CO_2 and H_2O. At a pressure of about 20–30 kilobars, depending on the temperature, there is a major change in the phase relations affecting the carbonate content of mantle rocks. Above this pressure carbonate minerals and liquids are stable,

Figure 325. Experimentally determined phase relations of a peridotite in the presence of approximately 0.3% H_2O and 0.7% CO_2 (after Olafsson and Eggler 1983). Additional data for this system are given by Wallace and Green (1988).

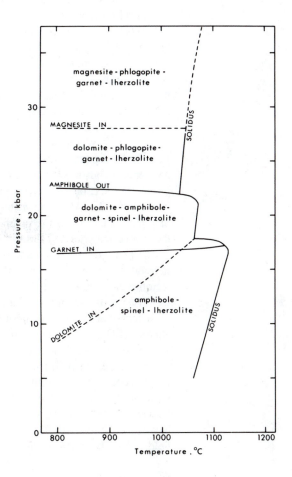

but below this pressure they dissociate to give free CO_2 or a vapour or liquid containing dissolved CO_2. The reaction is:

$$2Mg_2SiO_4 + CaMgSi_2O_6 + 2CO_2 \rightleftharpoons CaMg(CO_3)_2 + 2Mg_2Si_2O_6$$
$$\text{(Fo)} \qquad \text{(Cpx)} \qquad\qquad \text{(Dol)} \qquad \text{(Opx)}$$

At a higher pressure, dolomite gives way to magnesite, by the reaction:

$$CaMg(CO_3)_2 + Mg_2Si_2O_6 \rightleftharpoons 2MgCO_3 + CaMgSi_2O_6$$
$$\text{(Dol)} \qquad \text{(Opx)} \qquad \text{(Mag)} \qquad \text{(Cpx)}$$

Up to about 25 kilobars the most probable H_2O-bearing phase is amphibole, but above that pressure amphibole is not stable, although H_2O can still be held by phlogopite if there is enough potassium present.

Because of these variations in the sub-solidus mineralogy, the solidus itself is drastically affected by the presence of CO_2 and H_2O. Figures 227 and 228 show the separate influences of H_2O and CO_2 on the melting of peridotite. Their combined effects depend on their relative concentrations and total amount, and the solidus shown in Fig. 325 could vary considerably in practice.

During partial melting, the small amounts of phlogopite and carbonate likely to be present in the source region would enter the magma close to the solidus and be concentrated in the first liquids produced. Holloway and Eggler (1976) showed experimentally that the initial melt fraction of a phlogopite–dolomite–peridotite was rich in the constituents of phlogopite and dolomite, and if the temperature were raised only a small amount above the solidus it would rapidly become either potassium-rich or carbonate-rich depending on the initial phlogopite/dolomite ratio of the source material. At higher pressures, the stability field of phlogopite above the solidus shrinks, so that the constituents of phlogopite enter the melt more rapidly, leading to a greater likelihood that partial melts would be kimberlitic rather than carbonatitic (Wendlandt and Eggler 1980). Experiments on an average Lesotho kimberlite composition (Eggler and Wendlandt 1979) showed that at 55 kilobars (but not at 30 kilobars), a kimberlitic liquid coexists with those phases (garnet, olivine, orthopyroxene, clinopyroxene) that occur as megacrysts in the Lesotho kimberlites, consistent with kimberlite being a primary magma. At this pressure, magnesite would be the solid carbonate phase in the source material and the solidus temperature would be about $1200\,°C$.

The phase relations of carbonate-bearing peridotites are shown in Fig. 326 in relation to the range of geothermal gradients encountered in different tectonic environments. It is only at ocean ridges that peridotite can melt in the absence of CO_2 and H_2O, and the product is a tholeiitic basalt. In regions of successively lower geothermal gradient, melting is only attained if CO_2 and/or H_2O are present, and the melting product varies accordingly. Increasing pressure or increasing CO_2/H_2O ratio leads to undersaturated melts, the degree of silica-undersaturation increasing in the sequence alkali basalt–nephelinite–melilitite. Thus the lower the geothermal gradient, the more undersaturated the melt products are likely to be. The boundary between dolomite and magnesite is crucial because the liquids produced will have much higher Mg/Ca ratios if the solidus is crossed in the magnesite field. This boundary may correspond to a change from melilitite to

Figure 326. Simplified phase relations of peridotite with or without a small amount of CO_2 and H_2O in relation to a range of possible geothermal gradients, based on the same data as Figs 235 and 325.

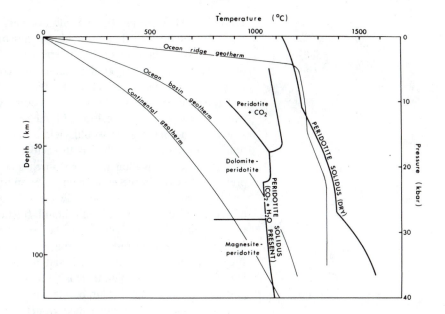

kimberlite magma in the region of lowest geothermal gradient (Brey *et al*. 1983).

During the ascent of kimberlite magma towards the surface, it crosses the carbonate-decomposition curve at about 60 kilometres depth (20 kilobars), and its dissolved carbonate is converted to dissolved CO_2 and partly exsolved. Exsolution of CO_2 (as a dense supercritical fluid) continues as the magma rises, and contributes to the explosive way in which kimberlites are erupted.

McGetchin *et al*. (1973) presented a model for the kimberlite diatremes in Utah which although partly superseded by more recent experimental data has some interesting features. The kimberlite and carbonatite in these intrusions were considered to have been emplaced as a single supercritical carbonate-rich fluid in which solid particles of comminuted mantle rock were suspended. Ascent was attributed to the buoyancy of the low-density fluid, the hydraulic head driving the fluid up the crack towards the surface being dependent on the rate of propagation of the crack above the ascending fluid. Expansion of the fluid as it reached levels of lower pressure would contribute to its acceleration. McGetchin envisaged that the kimberlite reached the surface as a mixture of cold gas and hot particles travelling at a speed of over 350 metres/second (about 800 miles per hour), having taken only 2 hours to travel from a depth of 200 kilometres. Spera (1984) believed that the propellant effect of exsolved volatiles is not as great as some previous workers suggested, and he calculated that the ascent rate of a typical kimberlite might increase from 28 metres/second at a depth of 30 kilometres to 59 metres/second at a depth of 3 kilometres, accelerating rapidly only during the last few kilometres of the upward journey. Very rapid emplacement of kimberlites is certainly indicated by the preservation of the metastable form of carbon (i.e. diamond), by annealing textures in peridotite xenoliths (Mercier 1979), and by the very large size of some of the ultramafic nodules carried up by kimberlites (up to 400 kilograms).

Harris and Middlemost (1970) interpreted kimberlite magmas as the products of zone melting in the mantle. The constituents in which kimberlite is most conspicuously enriched (K, Ba, Rb, Sr, H_2O, CO_2) are just those which would be enriched by such a process. However, zone melting fails to explain the isotopic composition of kimberlites, since zone melting of a large volume of normal mantle would yield a magma with the isotopic composition of normal mantle, whereas the actual isotopic composition of kimberlites indicates derivation from a Rb-rich source. Alibert *et al.* (1983) also preferred to explain the enrichment in K and other minor elements by zone melting, rather than by a low degree of partial melting, because of the difficulty of extracting a small liquid fraction from the partially melted source. They also considered that the zone melting hypothesis removes the need for metasomatism of the source material. They made the additional suggestion that the series alkali basalt–melilitite–kimberlite represents melting of source rocks with an increasing ratio of garnet to clinopyroxene. Their postulated source rock for kimberlite would be almost free of clinopyroxene (i.e. garnet–harzburgite rather than garnet–lherzolite), and this idea helps to explain the low Na content of kimberlites since clinopyroxene is the main host of Na in the mantle. On the other hand, Edgar and Charbonneau (1993) found no orthopyroxene in high-pressure melting experiments on a natural kimberlite and inferred that it could not have been present in the source, i.e. the source could not have been lherzolite or harzburgite; they preferred a garnetite or olivine–garnetite source. Anderson (1989) also preferred a garnet-rich (eclogite) source, although most workers still assume the source to be some type of peridotite.

Although the kimberlites undoubtedly originate deep in the mantle, many of them have high $^{87}Sr/^{86}Sr$ ratios (Fig. 327). This is not unexpected, since the source region of any K-bearing magma can also be expected to have a high Rb content, including high ^{87}Rb and hence high ^{87}Sr. There is a noticeable difference in isotopic composition between the Group I kimberlites (mica-poor) and Group II kimberlites (mica-rich). The former have isotopic compositions similar to many oceanic island basalts, whereas the latter resemble olivine–lamproites.

Neodymium isotope studies have thrown more light on the kimberlite source region and the relationship between kimberlites and their xenoliths. Basu and Tatsumoto (1979) carried out high precision measurements on kimberlite matrices from seven pipes of varying ages in three continents. The initial $^{143}Nd/^{144}Nd$ ratios were calculated by making a correction for the independently determined emplacement age (by K/Ar, Rb/Sr or U/Pb). They also measured the initial ratios and Sm and Nd contents of the two main types of garnet–lherzolite xenolith from one of the pipes, i.e. granular and sheared. They found that both types of xenolith had different Sm and Nd contents and $^{143}Nd/^{144}Nd$ ratios from their host kimberlite, suggesting that both types were accidental mantle xenoliths and not derived from the same source region as the kimberlite, which they presumed to lie much deeper.

The $^{143}Nd/^{144}Nd$ ratios of the kimberlite matrices gave an even more interesting result. They showed a linear correlation with age (Fig. 328). Basu

Figure 327. Initial $^{87}Sr/^{86}Sr$ and $^{143}Nd/^{144}Nd$ ratios for kimberlites and ultrapotassic lavas, from data compiled by Nelson *et al.* (1986), Menzies and Kyle (1990), and Conticelli and Peccerillo (1992).

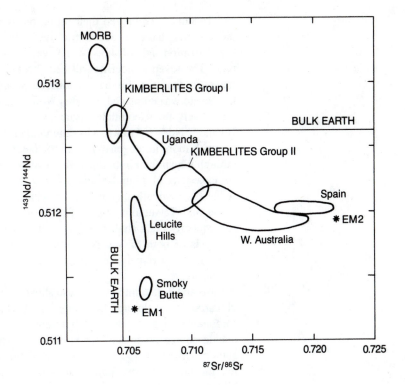

Figure 328. Initial $^{143}Nd/^{144}Nd$ ratios of kimberlites plotted against their age of emplacement. The straight line represents the independently determined $^{143}Nd/^{144}Nd$ evolution of the Juvinas basaltic achondrite. After Basu and Tatsumoto (1979).

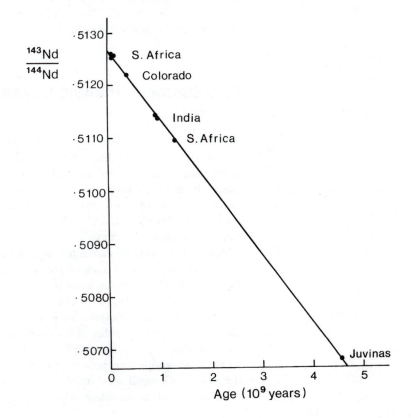

and Tatsumoto compared their initial ratios with the corresponding ratios of the Juvinas basaltic achondrite meteorite. This meteorite has given a crystallization age of 4.56×10^9 years and has a near-chondritic Sm/Nd ratio. The seven kimberlites all had $^{143}Nd/^{144}Nd$ ratios that were the same as the ratio that the Juvinas achondrite would have had at the time each kimberlite was emplaced. In other words the kimberlites are all derived from a relatively undifferentiated primeval mantle reservoir with chondritic Sm/Nd and $^{143}Nd/^{144}Nd$ ratios. Thus the kimberlite source in the deep mantle may be much more homogeneous than the heterogeneous lherzolitic upper mantle through which the kimberlites were eventually emplaced. Taken together, the compositions of granular and sheared lherzolite xenoliths and kimberlites suggest a vertically zoned mantle which is increasingly fertile with depth (Basu and Tatsumoto 1980). The uniformity of the kimberlite source has subsequently been challenged by McCulloch *et al.* (1983), who asserted on the basis of both Sr and Nd isotopic composition that the source of these magmas is extremely 'enriched' (i.e. high Rb/Sr, Nd/Sm) and probably composite.

Basu and Tatsumoto (1980) also showed that the Nd isotopic compositions of carbonatites do not lie along the Juvinas line of Fig. 328, and that carbonatite magmas do not therefore come from the homogeneous deep mantle source postulated for kimberlites. This supports the belief of some authors (Mitchell 1979; Boctor and Boyd 1981) that there is no genetic relationship between kimberlites and carbonatites, even though many kimberlite magmas are carbonate-bearing and the two rock types tend to occur in the same regions. The high K/Na ratio of kimberlites also contrasts strongly with the low K/Na ratio of some carbonatites.

POTASSIUM-RICH BASIC LAVAS

Most basalts have low potassium contents (average $K_2O = 1.1\%$) and their K_2O/Na_2O ratio is less than 1. Basaltic rocks with high K_2O and K_2O/Na_2O greater than 1 are rare but distinctive, and have often been given special names. In a few igneous provinces there are potassium-rich basic and ultrabasic lavas of even more extreme composition which cannot be described as 'basaltic' at all. Table 60 gives an idea of the variability of some of the named varieties.

Most of these varietal names are regarded as obsolete, but because of the extreme variability of the rocks it has proved difficult to arrive at a simpler classification. The examples listed in Table 60 are divided into three groups. The rocks of Group I are low in Na_2O, CaO and Al_2O_3 compared with basalt and are deficient in plagioclase. They are not necessarily undersaturated with silica; for example wolgidite is undersaturated, jumillite is close to saturation, and orendite is oversaturated. Some of them contain leucite or sanidine but the most conspicuous potassium mineral is mica (phlogopite or biotite). The rocks of Group II are thoroughly ultrabasic, and have high CaO, which often shows itself as melilite in addition to pyroxene. They all contain either

Table 60. Compositions of some potassium-rich basic lavas, compared with ordinary basalt

		SiO_2	Al_2O_3	MgO	CaO	Na_2O	K_2O	
I	Wolgidite	44.9	6.6	11.4	4.7	0.6	7.7	West Kimberley
	Orendite	52.6	9.8	6.3	2.8	0.4	10.8	West Kimberley
	Madupite	43.1	8.5	11.0	12.1	0.8	7.6	Wyoming
	Wyomingite	52.4	10.6	6.8	4.6	1.3	11.0	Wyoming
	Orendite	54.1	9.7	7.7	3.7	1.2	11.5	Wyoming
	Jumillite	47.7	7.7	16.3	7.1	1.7	5.0	South-east Spain
	Cancalite	55.4	9.3	12.0	3.7	1.5	8.8	South-east Spain
	Fortunite	56.7	10.8	10.2	2.7	2.2	6.9	South-east Spain
II	Katungite	35.4	6.5	14.1	16.8	1.3	4.1	East Africa
	Mafurite	39.1	8.2	17.7	10.4	0.2	7.0	East Africa
	Ugandite	43.8	7.3	15.4	11.1	2.5	3.3	East Africa
	K-ankaratrite	38.6	9.3	10.8	15.7	2.3	2.5	East Africa
	Venanzite	40.5	10.4	12.6	16.2	1.1	7.4	Italy
	Coppaelite	41.4	7.6	11.2	16.0	0.6	5.3	Italy
	Olivine–lamproite	42.8	3.3	24.9	4.3	0.5	4.2	Western Australia
III	Absarokite	51.3	13.2	10.2	8.1	2.1	3.9	Wyoming
	Ottajanite	49.4	18.6	5.6	9.1	2.6	5.9	Italy
	Shoshonite	54.7	18.2	3.7	6.6	3.5	3.8	Wyoming
	Vicoite	48.8	18.5	3.6	8.6	2.4	8.4	Italy
	Leucite–basanite	47.7	17.6	4.8	9.4	2.8	7.6	Italy
Average basalt		49.2	15.7	6.7	9.5	2.9	1.1	—
Average nephelinite		40.6	14.3	6.4	11.9	4.8	3.5	—

Data are from compilations by Jaques *et al.* (1984), Johannsen (1938), Nicholls and Carmichael (1969), Sahama (1974) and Le Maitre (1976).

leucite or kalsilite, and depending on their mineralogy could also be classed as leucitites, kalsilitites or melilitites. The rocks of Group III are the leucite–basalts, with plagioclase and clinopyroxene as major constituents, accompanied by leucite and sometimes sanidine. Foley (1992) proposed that the rocks of Group I should be called lamproites, those of Group II should be called kamafugites, and those of Group III should be called plagioleucitites. Although Table 60 only lists mafic rocks with $K_2O>Na_2O$, many ordinary nephelinites and melilitites also have high K_2O contents and may contain leucite or kalsilite, although their K_2O/Na_2O ratios are less than 1.

In considering the mineralogy of these rocks it is important to realize that leucite, being a very low-density mineral, is not stable at high pressures, so that a plutonic rock which contains phlogopite may be chemically equivalent to a volcanic rock which contains leucite and olivine. Furthermore, the dissociation of carbonates at atmospheric pressure means that a plutonic rock that contains a Ca- or Mg-carbonate mineral might be equivalent to a comparatively silica-undersaturated volcanic rock. Thus a magma which crystallized as a quartz-bearing intrusion at depth might crystallize as a leucite-bearing lava at the surface.

There are a small number of igneous provinces in the world in which ultrapotassic lavas are relatively abundant. They include Italy, Wyoming, and the Western Rift Valley of East Africa. Most of the occurrences are in a

Figure 329. Plot of K_2O against SiO_2 for the lavas of Roccamonfina volcano, Italy (after Appleton 1972). The dots represent leucitite, leucite–tephrite and leucite–phonolite; the circles represent olivine–basalt, trachybasalt and latite; the crosses represent trachyte and trachytic ash. The rocks shown by open circles have initial $^{87}Sr/^{86}Sr$ of 0.7064–0.7075; those shown by dots have ratios of 0.7084–0.7100 (Hawkesworth and Vollmer 1979).

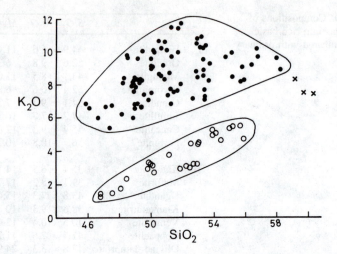

non-orogenic continental environment, for example in East Africa or the West Kimberley region of Western Australia. A few are in orogenic continental margins, for example in Indonesia (north Java, Sumbawa) and in south-east Spain, and several ancient occurrences are late-orogenic, for example in south-west England. On the whole the rocks of Group III tend to occur near convergent plate margins, and the rocks of Groups I and II in extensional tectonic settings, but there are many exceptions to this rule.

An exceptionally wide array of potassium-rich compositions is found in Roccamonfina volcano in the Campanian province of Italy. The variation diagram (Fig. 329) shows that there are two series of rocks: a high-K series from leucitite through leucite–tephrite to leucite–phonolite, and a low-K series (but nevertheless still strongly potassic) from olivine basalt through latite to trachyte.

Among intrusive rocks, the commonest K-rich mafic rock type is minette, the mica-bearing lamprophyre. Lamprophyre is a general name for mafic dykes with a strongly porphyritic texture. Table 61 shows how the

Table 61. Average compositions of lamprophyres compared with other types of basic rock

	SiO_2	Al_2O_3	MgO	CaO	Na_2O	K_2O
Minette	51.2	13.9	6.9	6.6	2.1	5.5
Vogesite	51.1	14.4	6.8	7.0	3.0	3.8
Kersantite	51.8	14.8	6.3	6.2	3.0	3.7
Spessartite	52.4	15.4	6.3	7.4	3.3	2.5
Camptonite	44.7	14.4	7.0	9.4	3.0	1.9
Monchiquite	40.7	13.2	9.2	11.0	3.1	2.2
Trachybasalt	49.2	16.6	5.2	7.9	4.0	2.6
Basanite	44.3	14.7	8.5	10.2	3.6	2.0
Basalt	49.2	15.7	6.7	9.5	2.9	1.1

The average compositions of the six types of lamprophyre are from Métais and Chayes (1963), and the basalt, trachybasalt and basanite data are from Le Maitre (1976).

compositions of minettes and other lamprophyres are related to those of basalts. Some of the lamprophyres (spessartite and camptonite) are not much different from alkali basalt or trachybasalt, but minettes are much more potassium-rich. In south-west England and in the Navajo province of Arizona minette dykes are associated with potassic trachybasalt lavas. In Arizona, minette is also very closely associated with kimberlite and the two rock types occur together in diatremes. Some of the Arizona minettes contain garnet–lherzolite nodules similar to those of alkali basalts and kimberlites.

The olivine–lamproites of the Kimberley region in Western Australia provide a link between the ultrapotassic lavas and kimberlites. They are phlogopite-bearing and olivine-rich, although they have more SiO_2 than kimberlites (see Tables 59 and 60) and they contain a glassy matrix. They carry ultramafic xenoliths, and some of them are diamond-bearing (Jaques *et al*. 1990). They occur in volcanic pipes and infilled craters and are associated with leucite–lamproites of less mafic composition.

The essential characteristic of all these magmas is their richness in K_2O compared with other mafic magmas, but the high K_2O is also accompanied by high levels of many of the other 'incompatible' elements, e.g. Rb, Cs, Ba, Th, U, Zr, Ti and the light rare earth elements. As a result they sometimes contain rare accessory minerals such as wadeite, $K_2ZrSi_3O_9$ or priderite, $(K,Ba)(Ti,Fe)_8O_{16}$.

ORIGIN OF K-RICH MAFIC MAGMAS

The hypotheses advanced for the origin of the ultrapotassic magmas are more varied than for any other type of igneous rock. Among the suggestions that have been put forward, they have been regarded as: (a) primary magmas formed by the partial melting of a mantle material abnormally rich in potassium; (b) products of fractionation of either basalt, picrite or kimberlite magma; (c) products of zone-refining of basaltic magma; or (d) products of contamination of basalt, nephelinite or carbonatite magma by either granitic or sedimentary material.

From the presence of ultramafic xenoliths with high-pressure mineralogy and from their association with kimberlites we can assume that the ultrapotassic magmas originate in the mantle. They have very radiogenic isotopic compositions (Fig. 327) showing that they cannot have been derived from ordinary basalts by any such process as fractionation or zone refining, and the complete absence of the K-rich magma types from many basaltic provinces suggests that unusual local circumstances are required for their formation.

Compared to other major igneous rock types, the ultrapotassic lavas are an exceptionally variable group of rocks. Although rocks intermediate between K-rich mafic lavas and ordinary basalts can be found, there is no regular trend from K-poor to K-rich compositions. The compositions of the ultrapotassic lavas are very scattered with respect to Si, Al, Ca, Mg and Fe. It does not appear that any systematically occurring process leads from normal basaltic magmas to ones rich in potassium. Some K-rich rocks are high in Mg and others not; some are Ca-rich and others not.

Any explanation for the formation of the ultrapotassic magmas must account for two features: (1) the high abundance of K_2O and volatiles; and (2) the low abundance of Na_2O and Al_2O_3 in rocks of Groups I and II (Table 60). The first feature is frequently attributed to metasomatism of the source material by fluids rich in potassium. The second feature is most easily explained by a previous depletion of the source material by a melting event; mineralogically it may correspond to a deficiency in clinopyroxene, which is the main host mineral for Na in mantle rocks. Thus a possible protolith for the ultrapotassic magmas would be a metasomatized harzburgite (Mitchell *et al.* 1987).

In a typical interpretation of their composition, Van Kooten (1980) suggested that the ultrapotassic basanites of the central Sierra Nevada in California were derived from a very low degree of partial melting (1.0–2.5%) of phlogopite-bearing upper mantle. Experiments by Barton and Hamilton (1982) on the high-pressure phase relations of orendite suggested that the breakdown temperature of phlogopite would be close to the anhydrous solidus of peridotite, so the initial melt produced by a low degree of partial melting of phlogopite-bearing peridotite would be strongly enriched in the constituents of phlogopite.

The initial $^{87}Sr/^{86}Sr$ ratios of K-rich basic rocks vary greatly from province to province, and also within each igneous province, for example 0.703–0.721 in Italy, 0.704–0.711 in the Western Rift Valley, and 0.710–0.719 in West Kimberley (Nelson *et al.* 1986). The neodymium and lead isotopic compositions are also highly variable. This variation immediately eliminates all hypotheses requiring a single source of material, i.e. differentiation from a common parent magma. It also contradicts the idea of extensive zone refining, which would lead to a melt enriched in K_2O compared with the average mantle but with isotopic ratios approximating to the average mantle that had been zone-melted.

The high $^{87}Sr/^{86}Sr$ ratios do not necessarily indicate crustal contamination of the magmas or even necessarily a high $^{87}Sr/^{86}Sr$ ratio in the mantle source rocks. If the mantle source contained phlogopite, which has a high Rb/Sr ratio, and if the decay of ^{87}Rb had been enriching the phlogopite in ^{87}Sr for a long period without isotopic exchange with the surrounding phases, then partial melting might result in a magma with a high K_2O content and $^{87}Sr/^{86}Sr$ ratio so long as the phlogopite was an early melting phase.

Carlson and Irving (1994) measured the isotopic compositions of pyroxenite, peridotite and glimmerite (mica-rich) xenoliths in Eocene minettes from the Wyoming craton, and found them to have extremely radiogenic compositions. Some of these xenoliths plot on isochrons indicating episodes of melt infiltration or removal during the Pre-Cambrian. They interpreted this in terms of Pre-Cambrian metasomatism of the lithospheric mantle beneath the Wyoming craton. This metasomatized material is inferred to be the source of the Eocene minette magmas.

In addition to the evidence from radiogenic isotopes, some of the ultrapotassic rocks have high $\delta^{18}O$ values, suggesting the influence of a sedimentary contribution to the magma compositions. For example, Ferrara *et al.* (1986) recorded high $^{87}Sr/^{86}Sr$ ratios (0.710–0.717) and $\delta^{18}O$ from +6 to +14 in lavas of the Roman province, which they interpreted in terms of derivation from metasomatized mantle, combined with subsequent crustal contamination. However, Conticelli and Peccerillo (1992) argued against a significant role for crustal contamination on the grounds that it cannot account for the major element compositions of the most mafic rocks and because improbable amounts of contamination would be required. Like most workers in this field, they attribute the high K_2O of the ultrapotassic lavas to the presence of subducted sedimentary material in the mantle source region, or to metasomatism of the mantle source region by fluids derived from subducted sedimentary material.

The consensus of opinion is that the main chemical and isotopic characteristics of the ultrapotassic lavas are primary, i.e. they reflect the nature of the mantle source material. Some secondary variation, i.e. fractionation or contamination, may be superimposed upon this. There is evidence of small-scale contamination in a few well-known volcanoes which erupt K-rich mafic magmas, for example the vesuvianite-bearing limestone xenoliths of Vesuvius or the partially fused granite xenoliths in the lavas of Nyiragongo, but it is unlikely that this contamination had a major influence on magma development. For example in Vesuvius the limestone source of the skarn ejecta has a $^{87}Sr/^{86}Sr$ ratio (0.7077) no higher than that of many of the lavas (0.7069–0.7079) so that it cannot be responsible for their high ratios (Cortini and Hermes 1981).

Figure 330 shows the Sr isotopic compositions of dated historical lavas of Vesuvius. There is a very clear correlation of $^{87}Sr/^{86}Sr$ with age, but along two separate trends. This suggests that the Vesuvian eruptions of the last 200 years have been fed from two different magma systems or reservoirs. Cortini and Hermes suggested that each of these trends is due to magma mixing. The end-members of the trends do not differ in major element chemistry, which again rules out bulk crustal contamination as the cause of the variation. It is

Figure 330. Variation of $^{87}Sr/$ ^{86}Sr with age of the historically dated lavas of Vesuvius (after Cortini and Hermes 1981).

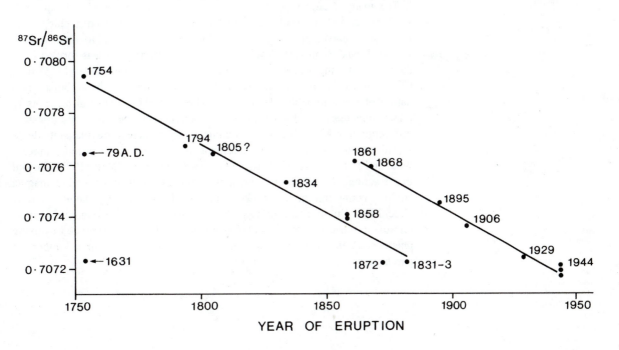

difficult to avoid the conclusion that despite some secondary fractionation and contamination the primary magmas of the K-rich lavas are derived from a source in the mantle which is very heterogeneous and contains a high-K component.

Some occurrences of K-rich basic lavas occur above subduction zones, for example the leucite–basalts of Muriah and Batu Tara in Indonesia, or of Stromboli and Vulcano in Italy. Ninkovich and Hays (1972) regarded these potassium-rich lavas as the culmination of the trend of K-enrichment with earthquake depth that is shown by volcanic arc basalts and andesites. In these areas the leucite-bearing lavas occur in places where the subduction zone is relatively deep. On the other hand, major occurrences of K-rich basic magmas such as those of East Africa are not related to any known subduction zone, past or present, nor are the major occurrences of kimberlites in southern Africa. Several authors who favour the involvement of subducted oceanic crust have drawn attention to the observation that some subduction zones flatten out beneath continental areas, and in such regions the continental lithosphere may be extensively underplated by former oceanic crust. Modern subduction zones of this type are those underlying south-west Japan and central Peru (Sacks 1984). The actual cause of K-enrichment in the lavas overlying the deepest part of subduction zones is variously identified as the scavenging of potassium from the overlying mantle wedge by water or hydrous magma (Ninkovich and Hays 1972), the extreme fractionation of ordinary basalt (Meen 1990), or the introduction of K_2O from subducted sediment (Nelson 1992). Although the anorogenic K-rich mafic lavas show a considerable chemical and isotopic overlap with the examples related to recent subduction, Nelson believed that their lead isotopes differed in such a way as to suggest long-term (>1000 m.y.) storage of subducted sediment in the mantle prior to magma generation.

The many chemical features shared by ultrapotassic lavas, minettes and kimberlites, and the probability that all of them come from a phlogopitic ultramafic source, pose the question of their relationship to one another. Bachinski and Scott (1979) suggested that minette magma was formed from the same starting materials as kimberlite under comparatively H_2O-rich, CO_2-poor conditions, but this seems unlikely in view of the carbonate-rich nature of some minettes (Hall 1982). There have been several attempts to relate minettes and kimberlites by different schemes of crystallization differentiation. Kimberlite is an obvious choice of parent magma as its deep-seated xenoliths and rapid emplacement indicate it to be a primary magma type, and minettes are often associated with kimberlite. Scott (1979) showed that fractionation of olivine and titanomagnetite from a kimberlite magma could account for the composition of the Holsteinsborg minettes in west Greenland. Upton and Thomas (1973) also described phlogopite-bearing ultramafic intrusions from south Greenland in which the presence of olivine phenocrysts indicated the possibility of differentiation from a kimberlitic parent magma.

CHAPTER 13

Peridotites

Ultramafic rocks can originate in three ways: as magmas, as cumulates, and as melting residues.

Very few ultramafic rocks have crystallized directly from ultramafic magmas. Lavas comparable to peridotite are not being erupted at the present day, and a very high temperature would be needed to maintain a liquid of this composition. Forsterite (Mg_2SiO_4) melts at 1890 °C and enstatite ($MgSiO_3$) melts (incongruently) at 1557 °C, and most peridotites contain a high proportion of these constituents. The clearest examples of ultramafic magmas are the komatiite lavas, and there may also be a few intrusive examples such as the Nordre Bumandsfjord intrusion (see Chapter 8). In addition, some of the very large layered basic intrusions such as the Bushveld complex may also have formed from a magma more mafic than basalt, although not as mafic as peridotite. The intrusion temperature of such a magma would be high enough to cause extensive crustal melting, with consequent contamination of the magma, and there is plenty of evidence for this in the Bushveld complex.

Ultramafic cumulates are a familiar feature of many basic intrusions. They occur as layers in layered basic complexes, e.g. Stillwater, Bushveld; as ultramafic stocks of the 'appinitic' or 'Alaskan' type, e.g. Appin, Duke Island; and as components of basic alkaline complexes, e.g. Fen, Palabora. These cumulate rocks are obviously related to identifiable magmas, i.e. basaltic, andesitic, nephelinitic, or komatiitic.

Consideration of the petrogenesis of basalts suggests that many basaltic magmas would leave behind a melting residue in the source region, and that this would be peridotitic in composition. Possible examples of residual peridotite are found among the ultramafic inclusions in alkali basalts.

The three categories of peridotite described above may all be easy to recognize in favourable circumstances, but in practice there are many large peridotite massifs whose origin is difficult to determine because they have been emplaced tectonically into their present position or because they are serpentinized. This is especially true of the peridotites of orogenic belts, which are the subject of this chapter.

ALPINE-TYPE PERIDOTITES

Large bodies of peridotite and other ultramafic rocks are scattered throughout orogenic belts such as the Alps and the Appalachians (Figs 331 and 338), and have been called 'Alpine-type' or 'orogenic' peridotites to distinguish them from the peridotites of layered basic intrusions. Geological mapping shows that they were mostly emplaced into their present position tectonically rather than by magmatic intrusion, and their interpretation is greatly complicated by the faulting, internal deformation and serpentinization that they have undergone.

Among the suggestions that have been made as to the origin of Alpine ultramafic rocks are that they are segments of oceanic crust, or of oceanic crust and mantle, or of sub-continental mantle. Some may be genuine intrusive magmatic rocks, some may be cumulates, some may be melting residues, and some may be unmodified mantle material. They certainly do not all have the same origin.

An important category of Alpine-type peridotites are those that form part of an 'ophiolite' assemblage. This is an association of serpentinized peridotite with gabbros, amphibolites, basic lavas and basic dykes which closely resembles the presumed composition of oceanic crust. There are easily recognizable examples of ophiolitic complexes in all the orogenic belts, for example the Alps or the Appalachians, but these orogenic belts also contain many 'Alpine-type' peridotites which are definitely not part of an ophiolite assemblage, as well as ultramafic rocks occurring in differented gabbroic, appinitic or Alaskan-type intrusions.

Figure 331. The distribution of peridotites (shown in black) in the Western Alps. AA = Alpe Arami, F = Finero.

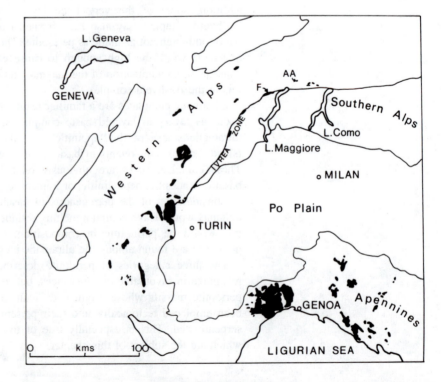

ULTRAMAFIC ROCK TYPES

There are many types of ultramafic rock, and the predominant varieties found in Alpine-type massifs differ from those in stratiform intrusions. Figure 332 shows the nomenclature of peridotites and pyroxenites. The ultramafic layers in large stratiform intrusions such as Stillwater or the Great Dyke are most commonly orthopyroxenite, harzburgite and dunite, and have cumulus textures. The Alpine-type massifs also contain these rocks, but in addition there are clinopyroxene-bearing varieties such as lherzolite, and there are many rocks with non-cumulate textures. The Alpine-type peridotites are more often serpentinized than the stratiform ones, although this is only a rough generalization and exceptions are common.

Lherzolites commonly contain an Al-bearing mineral besides olivine and pyroxenes, and experimental studies have shown that in lherzolite compositions the nature of the Al-bearing phase which is present depends mainly on the pressure, as shown in Fig. 333. For successively increasing pressures the lherzolite mineral assemblages are:

Low pressure – olivine + orthopyroxene + clinopyroxene + plagioclase
Medium pressure – olivine + orthopyroxene + clinopyroxene + spinel
High pressure – olivine + orthopyroxene + clinopyroxene + garnet

This mineralogical variation provides valuable clues as to the depth of origin of the lherzolite varieties. In particular, garnet lherzolites must have formed within the mantle.

Most peridotites have undergone some degree of serpentinization, and in many cases the rock is almost completely serpentinized. Oxygen and hydrogen isotope studies show that serpentinization is a low-temperature process, but that different serpentine minerals form at different temperatures (220–460 °C for antigorite and 85–185 °C for lizardite and chrysotile). It is probable that antigorite is formed as a metamorphic mineral under greenschist facies conditions, whereas lizardite and chrysotile are produced by hydrothermal activity at low temperatures (Wenner and Taylor 1974).

Figure 332. The nomenclature of peridotites and pyroxenites according to the IUGS classification (Streckeisen 1976).

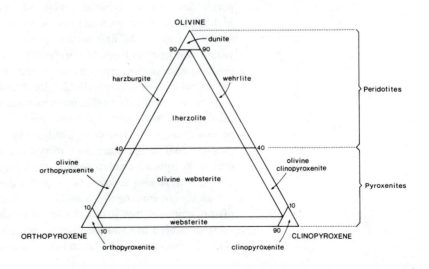

Figure 333. Lherzolite facies. The stability fields of plagioclase–, spinel– and garnet–lherzolite with respect to temperature and pressure. The anhydrous solidus is after Takahashi and Kushiro (1983). The spinel–garnet facies boundary is after O'Hara *et al.* (1971), and the plagioclase–spinel boundary is after Saxena and Eriksson (1983). In the field labelled 'spinel–lherzolite' there may not be any actual spinel present if the Al_2O_3 content of the rock is small and the temperature is close to the solidus; this is because the pyroxenes are able to accommodate all the Al themselves and a separate Al-bearing phase does not then develop.

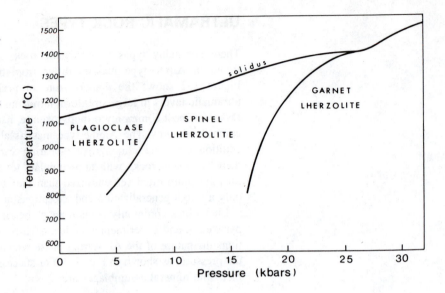

EMPLACEMENT OF PERIDOTITES

The absence of peridotite lavas (with the exception of the rare Pre-Cambrian komatiites) has always cast doubt on the existence of peridotite magma. If peridotites are not intruded as magmas then some other mechanism of emplacement is required. Field studies are not as decisive as one might expect, because the evidence for a high temperature of intrusion is often ambiguous and the contact relationships are frequently obscured by serpentinization and faulting.

One category of Alpine peridotites which has been proposed is that of 'high-temperature peridotite diapirs'. It has been suggested that these peridotites have been intruded vertically as diapiric bodies, and that their high temperature of intrusion can be recognized from contact metamorphic aureoles. One of the first intrusions to be interpreted in this way was the Lizard peridotite in England (Green 1964), and of the examples described in this chapter a similar explanation has been put forward for the Ronda and Mt Albert peridotites (Loomis 1972; Macgregor and Basu 1976). In the Lizard and Mt Albert intrusions this interpretation was supported by evidence from petrography and from pyroxene geobarometry, which indicated that a peridotite mineral assemblage originally in equilibrium at a very high pressure and temperature had recrystallized to a lower *P–T* assemblage during the upward emplacement of the peridotite. Geophysical and structural evidence were also adduced to support a diapiric form.

Subsequent investigations of these peridotites have undermined so much of this evidence that it now looks rather doubtful whether there is such a thing as a high-temperature diapiric peridotite. Deep drilling and structural studies support a sheet-like rather than diapiric shape for the Lizard and

Ronda peridotites (Styles and Kirby 1980; Lundeen 1978), the pressure estimates have been called into question by revision of the pyroxene geobarometer (Wilshire and Jackson 1975; Obata 1976), and the high-temperature aureoles are also open to alternative interpretation (Styles and Kirby 1980). There is also no reason why a solid peridotite diapir should diapirically intrude country rocks that are lighter than itself unless it was either partially melted or serpentinized. Some of the examples quoted either lack evidence of large-scale partial melting (Lizard), or lack extensive serpentinization (Ronda, Mt Albert), or if serpentinization is present it is post-emplacement (Lizard).

There is, of course, no objection to diapiric intrusion of a peridotite after it has been serpentinized, because the density of serpentinite ($2.5-2.6 \, \text{g/cm}^3$) is much less than that of peridotite and lower than that of most metamorphic or basic igneous rocks. Coleman (1980) has described a possible diapiric serpentinite body from the Californian Coast Ranges.

The great majority of structural studies on Alpine-type peridotites now indicate that these rocks were tectonically transported into place, and not intruded like other plutonic igneous rocks. Where the country rocks show contact metamorphism, this is generally held to indicate that the peridotite was intruded as a hot body, albeit solid. This conclusion leads to two possible interpretations: (a) that the peridotite originated at a much greater depth than its country rocks; or (b) that the magmatic events leading to the crystallization of the peridotite did not long pre-date its eventual emplacement.

Quite apart from the likelihood that many Alpine-type peridotites are emplaced tectonically, subsequent serpentinization may cause further movement. Serpentinite is structurally a very incompetent rock, and in any terrain in which it occurs it will be the most easily deformed material. Serpentinites have been aptly described as 'the ball-bearings of an orogeny'. This adds to the difficulty of field interpretation because movement along the contacts is an invariable feature of serpentinized ultramafic rocks.

OPHIOLITES

Alpine-type peridotites commonly form part of an assemblage which has come to be known as the 'ophiolite suite'. This term has never been clearly defined, but it usually includes most of the following components: ultramafic rocks (peridotites and pyroxenites, often serpentinized), gabbros, amphibolites, greenschists, basic dykes, and basic pillow lavas. In a typical ophiolite these would be arranged in a sequence, starting with the ultramafic rocks at the base, followed by gabbros and ultramafic cumulate rocks, followed by a sheeted mafic dyke complex, followed by the volcanic rocks at the top. The overlying sedimentary rocks may include deep-sea types, such as bedded cherts, shales and umbers (Fe–Mn-rich chemical sediments). Some examples are described below.

REPRESENTATIVE OPHIOLITES

Troodos, Cyprus

The basic–ultrabasic massif of the Troodos mountains (Fig. 334) is made up of three units: the Troodos plutonic complex, the sheeted intrusive complex, and the Troodos pillow lava series.

The Troodos plutonic complex forms the core of the massif, and consists of gabbros and ultrabasic rocks. These are arranged in a roughly concentric fashion with the ultrabasic members in the centre. According to Gass (1980) there has been substantial post-emplacement uplift in the centre of the massif, so that the outer units of the complex represent successively higher levels of the ophiolite sequence. In the core of the plutonic complex is a tectonized harzburgite. This is strongly foliated and contains large lenticular masses of dunite and smaller bodies of gabbro and lherzolite. The harzburgite has been interpreted as the residue of partial melting of lherzolite, from which basaltic melts have been extracted. The tectonized harzburgite is followed by a series of ultramafic cumulates, which contain cyclic successions of dunite, wehrlite, pyroxenite and troctolite with cumulate textures. The ultramafic cumulates pass upwards into layered gabbro. The gabbro contains some veins and small irregular bodies of granophyre and quartz–diorite.

The sheeted intrusive complex consists of an immense number of nearly vertical tholeiite dykes, from 0.1 to 5 metres wide. They are intruded into basic rock of a similar composition to the dykes, and are so numerous that they form more than 90% of the outcrop. The host rock is structureless lava in the lower part of the complex and pillow lava in the upper part, but in places the dykes occupy 100% of the ground and there is no other country rock. The dykes strike north–south and the swarm extends for at least 100 kilometres from west to east. Most of the dykes have suffered hydrothermal alteration or low grade metamorphism but their igneous textures are preserved.

Figure 334. Geological map of the Troodos massif, Cyprus (after Gass 1980).

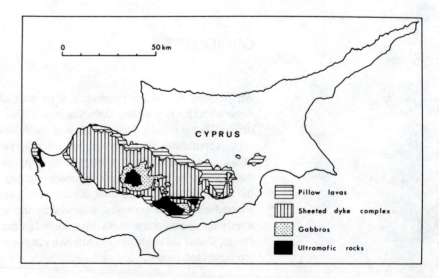

CYPRUS

▭	Pillow lavas
⦀	Sheeted dyke complex
⦂	Gabbros
▮	Ultramafic rocks

0   50 km

The Troodos pillow lava series consists of a pile of basaltic pillow lavas in the order of 2–3 kilometres thick, accompanied by minor dykes and flow breccias. They can be divided into two units. The lower pillow lavas are similar in composition to the underlying tholeiite dykes. The upper pillow lavas are olivine basalts and picrites and represent a separate and later phase of volcanic activity. The pillow lavas have all been affected by low grade hydrothermal metamorphism, and oxygen isotope evidence suggests that this was brought about by introduced sea water. There are also sulphide ore bodies and deposits of Fe–Mn hydroxide sediment (umber) which have been attributed to the action of a sea-floor hydrothermal convection system.

The Oman ophiolite

The Oman ophiolite is one of the largest in the world as well as being one of the best exposed. It extends for 500 kilometres through the mountains of northern Oman. It contains a complete assemblage of ophiolitic rock types, including both tectonized and cumulate peridotites, gabbros, a dolerite dyke swarm and volcanic rocks. The ophiolite does not outcrop continuously but as separate sheets, in some of which the sequence is internally repeated by thrust faulting. The ophiolite as a whole has been thrust over a sedimentary succession, and it is immediately underlain by a thick mélange consisting mainly of fragmented Mesozoic sedimentary rocks.

The peridotites compose 60% of the igneous outcrop and have been extensively serpentinized. They are mainly harzburgites, with minor lherzolites, dunites and chromitites, and towards the base they have been mylonitized. There is an upward transition from peridotites into gabbros through a zone of interlayered cumulate rocks (peridotite, dunite, anorthosite, gabbro, norite, troctolite). Layered gabbros are followed upwards by massive gabbro. In the overlying sheeted dyke swarm, there is no country rock between the dykes, which range from 5 centimetres to several metres in width. The dolerites are partly fresh and partly affected by greenschist-facies metamorphism. The volcanic rocks make up only 3% of the complex. At the contact between the dyke swarm and the lavas there are screens of lava between the dykes, and some dykes pass up into the lavas, to which they act as feeders. The lavas are mainly pillow lavas, but they also contain massive flows and sills. They have all been affected by zeolite- or greenschist-facies alteration. A diagrammatic section of the ophiolite showing its possible correlation with oceanic crust is given in Fig. 335.

Appalachian ultramafic belt

Over a thousand small bodies of ultramafic rock are scattered along the length of the Appalachian fold belt for a distance of nearly 3000 kilometres from Alabama to Newfoundland. Some of these form part of an ophiolitic assemblage and others do not.

The Bay of Islands complex in Newfoundland is clearly ophiolitic (Fig. 336). From the top downwards it contains the succession: sediments – pillow basalts – sheeted dykes and breccias – gabbro (mostly layered) – ultramafic cumulates (mainly dunite) – ultramafic tectonites (mainly harzburgite), above

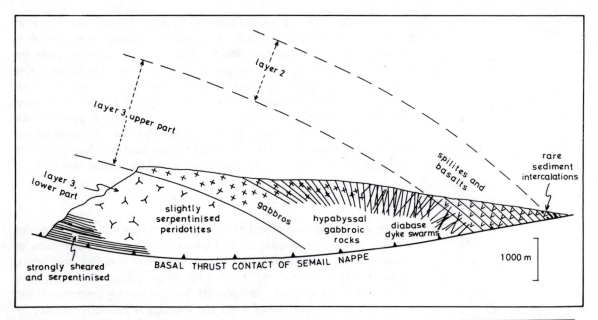

Figure 335. Diagrammatic section of the Semail nappe, Oman and its suggested correlation with oceanic crust (after Glennie *et al*. 1974). For detailed maps and sections of the Oman ophiolite see Lippard *et al*. (1986) and Boudier and Nicolas (1988).

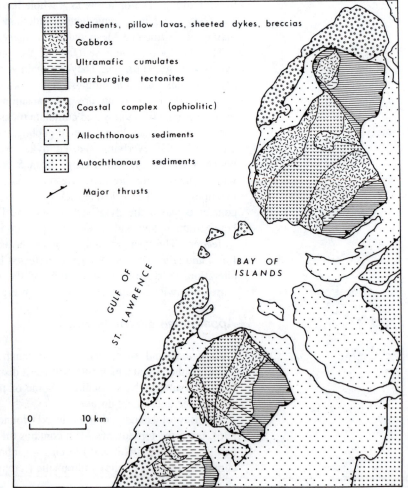

Figure 336. Geology of the Bay of Islands area, Newfoundland (after Church and Riccio 1977; Karson *et al*. 1983).

a basal thrust. The country rocks have suffered contact metamorphism to greenschist, amphibolite and pyroxene–granulite facies within 200 metres of the lower contact. Apart from the main masses of ultrabasic rock in this area, there are numerous small bodies, some of which are intrusive and have basaltic chilled margins. These may perhaps have been intruded as crystal mushes from mobilized ultramafic cumulates (Karson *et al.* 1983).

Further south in the Appalachians the ophiolitic nature of the ultramafic rocks is much less certain. Figure 337 shows the distribution of ultramafic rocks in the State of Vermont. They include both concordant and discordant bodies, mainly tabular or lensoid in shape, of varying sizes up to several kilometres but mostly very small, often under 100 metres across. The country rocks are folded and metamorphosed Palaeozoic sediments. The ultramafic rocks are peridotites, dunites and pyroxenites, and are mostly serpentinized. Along with the ultramafic rocks there occur greenstones and amphibolites of basaltic and andesitic composition. The simplest interpretation of these isolated ultramafic bodies is that they are ultramafic cumulates in the feeder intrusions of higher-level basaltic volcanoes, but they have also been interpreted as fragments of a tectonically disrupted ophiolite (Stanley *et al.* 1984).

Figure 337. The distribution of ultramafic rocks (shown in black) in the State of Vermont (after Chidester 1968).

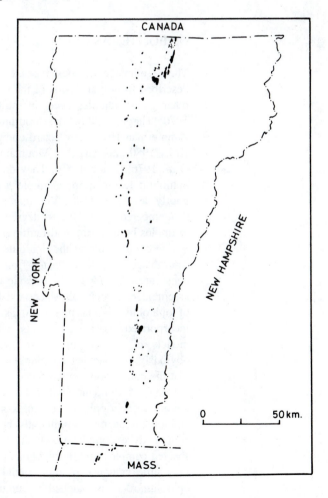

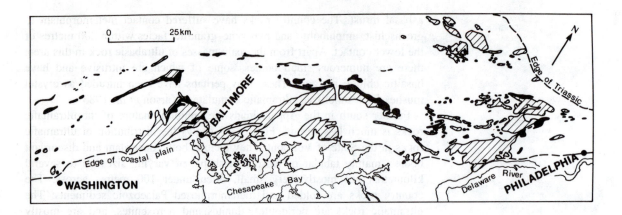

Figure 338. The distribution of ultramafic rocks (black) and associated basic rocks (shaded) in the area between Washington and Philadelphia (after Larrabee 1966).

Basic and ultrabasic rocks are abundant in the region between Washington and Philadelphia, shown in Fig. 338, where they are severely deformed and metamorphosed. Some of these are ophiolitic in character but others are not. Shaw and Wasserburg (1984) were able to recognize different types on the basis of their Nd isotopic composition.

OPHIOLITES AS OCEANIC CRUST

The assemblage of rock types characteristic of ophiolites has now been described from many parts of the world. Apart from those mentioned above, other good examples occur in the Californian Coast Ranges (Bailey *et al.* 1970; Harper 1980), the Vourinos complex in Greece (Moores 1969; Zimmerman 1972), the Lizard complex in Cornwall (Flett 1946; Styles and Kirby 1980), and the Dun Mountain ophiolite belt in New Zealand (Coombs *et al.* 1976; Sinton 1980). They do not always show the complete range of ophiolitic features; for example a sheeted dyke swarm is absent or only poorly developed in the Vourinos and Lizard examples. Most ophiolites are dismembered to a greater degree than the Troodos and Bay of Islands examples by thrusting and faulting.

The components of the ophiolite suite bear a striking resemblance to the assemblage of rocks collected from sea-floor dredging and now believed to constitute the bulk of the oceanic crust. They are also very similar to the assemblage of rocks on Macquarie Island (Chapter 1). This has led students of ophiolites to interpret them as segments of oceanic crust tectonically incorporated into the continents, and marine geologists in turn to construct models of the oceanic crust based on continental ophiolites. This is obviously a somewhat circular line of argument, although it is quite likely that the conclusions are correct.

Deep-sea sediments, pillow lavas and dyke swarms are all features which are common in ophiolite complexes, but they are not confined to ophiolites. Ophiolites can most confidently be equated with oceanic crust when they contain dyke complexes which are 100% dykes, such as in the Troodos sheeted complex and in the Oman ophiolite, but some other ophiolites show very little evidence of this crucial feature. The cumulate nature of some ophiolitic rocks is matched by cumulates which have been dredged from the

Figure 339. Hypothetical section through an oceanic spreading axis showing the relationship between melted, unmelted and residual peridotites in the suboceanic upper mantle (after Leblanc *et al.* 1980).

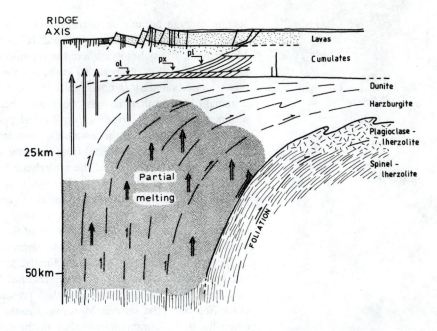

ocean floor (Tiezzi and Scott 1980), and the cumulates may well have formed in magma chambers at the oceanic spreading axes. Figure 339 shows how the tectonized harzburgites which lie below the ultramafic cumulates in most ophiolites may be interpreted as partial melting residues of an originally lherzolite upper mantle.

Figure 340 shows a typical ophiolitic sequence and its interpretation as a

Figure 340. Schematic section of a typical ophiolite sequence.

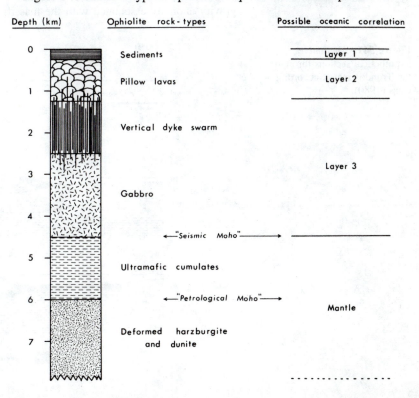

section through oceanic crust and mantle. The nature of the oceanic crust–mantle boundary is not straightforward. It has customarily been taken as the Mohorovičić discontinuity, where the seismic velocity V_p rises to its normal upper mantle value of about 8.1 km/s, but the petrological identification of this seismic boundary is not simple. If it is a boundary between mafic and ultramafic rocks, then it must be drawn between the gabbros and their ultramafic cumulates. From a petrological point of view there is a much greater affinity between the ultramafic cumulates and the overlying gabbro, both of which have crystallized from the same magma, than there is between the ultramafic cumulates and the underlying harzburgite tectonite, which is presumed to be residual mantle. For this reason, some petrologists refer to the boundary between ultramafic cumulates and ultramafic tectonites as the 'petrological Moho', even though it may not be seismically obvious. Geophysicists are not entirely happy about this; as far as they are concerned the Moho is a seismic discontinuity by definition.

There are further complications. Firstly, some ultramafic tectonites may not be mantle residues, but cumulates that have been deformed during emplacement of the ophiolites into their present positions in the continental crust. Secondly, the seismic Moho in oceanic areas may not correspond to either the mafic–ultramafic boundary or the cumulate–tectonite boundary of Fig. 340, but may lie at a boundary between ultramafic rocks that are serpentinized and unserpentinized (Clague and Straley 1977).

Most of the workers who have identified ophiolite complexes with oceanic crust have appealed to some sort of thrusting mechanism to emplace them into their present positions. Figure 341 shows two interpretations of the major structure of the Troodos massif. There is an extremely large positive gravity anomaly associated with the massif, which indicates that it must be

Figure 341. Alternative hypothetical sections through the Troodos massif according to Gass (1980).

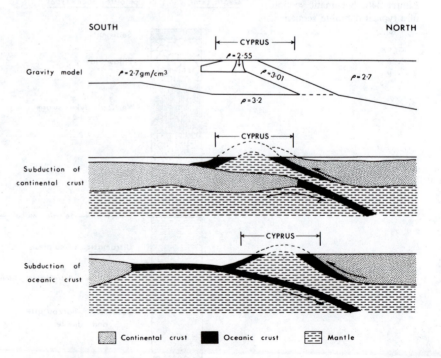

SOUTH   NORTH

Gravity model

underlain by a large mass of very dense rock at shallow depth. A thickness of over 10 kilometres and density of over 3.3 g/cm³ (i.e. ultrabasic composition) has been inferred, which obviously corresponds to the buried extension of the Troodos plutonic complex. The regional isostatic equilibrium requires that this ultrabasic mass be separated from the main body of the mantle by a layer of relatively lighter rocks beneath it, and the cross-sections in Fig. 336 show possible ways in which this could have been brought about.

Evidence for thrusting is present in most of the major ophiolite complexes, and in several (e.g. the Oman ophiolite, the Vourinos complex, the California Coast Ranges) the individual outcrops can be mapped as thrust slices. Other emplacement mechanisms have also been proposed. Lockwood (1972) suggested that serpentinite, which is of low density and easily deformed, could be intruded upwards in a solid condition along major fault zones, lubricated by interstitial water. As these serpentinite 'protrusions' reach a shallow level in the crust they could be propelled explosively towards the sea floor by the release of steam. Once on the sea floor the serpentinite debris is then capable of being transported and deposited as a sedimentary material. Lockwood described several examples of undoubted sedimentary serpentinites, and suggested that the small serpentinite bodies of Vermont might be of this type. There are several examples of serpentinite diapirs in the Coast Ranges of California (Oakeshott 1968; Coleman 1980), and Dickinson (1966) has described one body of extrusive serpentinite breccia from this region.

Nearly all regions that show regional metamorphism of high pressure type (glaucophane–schist facies) also contain ophiolites. Examples are the Franciscan terrain in California and parts of Japan and the Western Alps. These metamorphic facies require an unusually low geothermal gradient (combination of high pressure with relatively low temperature), i.e. the physical conditions characteristic of a subduction zone, where cold near-surface material is moving downwards. In such zones, convergence of oceanic and continental crust is continually taking place, but it remains to be explained why some of the oceanic crust should be thrust on to the continent (obducted) and why it is not all subducted.

In addition to the deformation associated with transport and incorporation into the continental crust, Alpine-type ultramafic bodies may also show earlier deformation of possible ocean-floor origin. Some of the mylonite zones which occur in ophiolites may have originated in oceanic fracture zones, where the oceanic crust is cut by transform faults (Smewing 1980). It is also believed that some serpentinites are intruded up through these oceanic fracture zones to reach the sea floor, and analogous serpentinites have been identified in ophiolite complexes (De Wit *et al.* 1977).

The normal thickness of mature oceanic plates is 60–100 kilometres whereas the largest exposed ophiolite complexes are only about 10 kilometres thick, so either only a small part of the oceanic plate has been obducted or the obducted oceanic plate was abnormally thin. Christensen and Salisbury (1975) suggested that the obducted oceanic material might be derived from the region of a spreading ridge, where the oceanic plate is thin and mechanically weak. This suggestion goes a long way to explaining the presence of contact metamorphic aureoles underneath some ophiolitic peridotites (Jamieson 1980; Karamata 1980). Temperatures would still be high in the oceanic plate if it were emplaced on to the continent soon after formation at a ridge crest.

It may be possible not only to identify ophiolites as oceanic crust but to say from whereabouts in the oceans a particular ophiolite assemblage has come. This might perhaps be done by identifying the associated sediments as having come from an abyssal plain or a continental margin, or by recognizing the basic volcanic rocks as coming from an ocean ridge, an island arc or a back-arc part of the ocean.

Isotopic studies on the origins of ophiolitic rocks are greatly complicated by the alteration that most of them have undergone. Oxygen isotope studies have shown that sea water has been involved in the hydrothermal metamorphism of the Troodos and Oman ophiolites (Spooner et al. 1974; Gregory and Taylor 1981), and this type of alteration together with later serpentinization also affects Sr isotope ratios. The $^{87}Sr/^{86}Sr$ ratios of Alpine-type peridotites as a whole show a surprisingly wide range, averaging 0.711 but ranging from 0.702 to 0.729. The high ratios are almost certainly due to serpentinization. The Sr contents of Alpine-type ultramafic rocks are so low (usually less than 10 ppm) that the isotopic ratios are very easily modified by small amounts of introduced strontium. An isotopic study of the Oman ophiolite by McCulloch et al. (1980) revealed that the isotopic composition of Nd is much less sensitive to alteration, a water/rock ratio of 10^5 being needed to alter the initial $^{143}Nd/^{144}Nd$ ratio. The Nd isotopes in the Oman ophiolite show the mafic and ultramafic rocks to have had an identical mantle source, similar to that of mid-ocean ridge basalts but different from that of oceanic islands or continental flood basalts. Shaw and Wasserburg (1984) used Nd isotopes to show that some of the mafic and ultramafic bodies in the Appalachian orogenic belt had the depleted mantle signature of oceanic crust and others did not.

NON-OPHIOLITIC ALPINE PERIDOTITES

There are many Alpine peridotites which do not form part of an ophiolitic assemblage. It would not be surprising if some of the ophiolite components were occasionally missing in view of the tectonic mode of emplacement of ophiolites, but there are some Alpine peridotites which are associated with none of the other ophiolitic rock types. These peridotites almost certainly do not come from the oceanic crust. Some examples are described below.

LHERZ

The peridotite of Etang de Lers (Lherz) in the Pyrenees is a small body (about 1 square kilometre), one of several which occur within a belt of Cretaceous marble (Fig. 342). Much of the peridotite is brecciated, especially

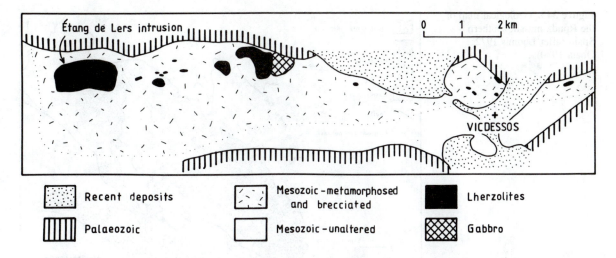

Recent deposits

Palaeozoic

Mesozoic – metamorphosed and brecciated

Mesozoic – unaltered

Lherzolites

Gabbro

Figure 342. Geological map of the ultramafic bodies near Lherz in the Pyrenees (after Avé Lallemant 1967).

around the edge, as is the adjoining marble. During the brecciation, which accompanied the emplacement of the peridotite, fine-grained peridotitic material with a few fragments of carbonate was injected as dykes and net veins into larger masses of marble and peridotite.

The Lherz peridotite is predominantly made of foliated spinel–lherzolite, and contains abundant layers of spinel–pyroxenite 2–4 centimetres thick. Some of these layers are zoned, with olivine-rich centres and pyroxene-rich margins. In places the layers are tightly folded. There are also five sill-like bodies of a garnet–pyroxenite which cut across the spinel–pyroxenite layering. The peridotite near the contact of these bodies is dunite or harzburgite, and not lherzolite. Hornblendite dykes and veins also cut across all the other rock types. Bodinier *et al.* (1988) interpreted the pyroxenite layers as crystal segregates from basaltic magmas which had traversed the peridotite while it was at a high temperature and pressure.

There is no evidence that the peridotite has been emplaced tectonically. There are no mappable faults in the country rocks or shear zones or slickensides in the lherzolites. The breccias appear to be explosion breccias rather than fault breccias. This suggested to Avé Lallement (1967) that the intrusion of solid peridotite was due to the presence of gas, but the presence of sapphirine and kornerupine in the aureole suggests that emplacement took place at a very high temperature and low pressure, so the brecciation is almost certainly due to decarbonation of the carbonate country rocks (Minnigh *et al.* 1980).

RONDA

The Ronda peridotite in the Betic orogenic belt of southern Spain is one of the largest peridotites known (Fig. 343). It consists mainly of harzburgite or lherzolite with a small amount of dunite, and is only slightly serpentinized. It has a wide metamorphic aureole, showing that it was emplaced at a high enough temperature to cause severe contact metamorphism.

The peridotite varies internally. It consists of olivine + orthopyroxene

Figure 343. Geological map of the Ronda massif, southern Spain (after Loomis 1972; Obata 1980).

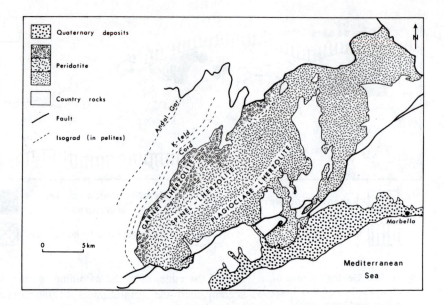

(enstatite) + clinopyroxene (diopside) + an aluminous phase (garnet, spinel or plagioclase). In the north-western part of the massif, pyropic garnet occurs as an aluminous phase, and in the south-eastern part calcic plagioclase occurs, while spinel is present throughout. Thus different mineralogical facies of the peridotite can be mapped out, as shown in Fig. 343.

A notable feature of the Ronda peridotite is the presence of mafic layers. These vary from garnet pyroxenites to olivine gabbros, depending on the facies of the enclosing peridotite, and many of them have major element compositions similar to basalt or picrite. The mafic layers make up about 5% of the massif but are more abundant (up to 30%) in the western part of the massif and less abundant (<1%) in the east. Suen and Frey (1987) studied the mafic layers and decided that they were not partial melt segregations in the peridotite, but cumulates deposited by magmas migrating through the peridotite.

Obata (1980) proposed that the mineralogical variation in the peridotite reflects the influence of decreasing pressure during the ascent of the peridotite from the upper mantle. Some of the pyroxenite layers in the Ronda peridotite contain graphite in the form of octahedral aggregates, indicating that it is pseudomorphing diamond (Davies *et al.* 1993). The amount of graphite in the most graphitic of these rocks implies an original diamond content of up to 15%! The former presence of diamond implies that the peridotite was emplaced into the crust in a solid condition from a depth of more than 150 km, but its method of emplacement is still uncertain. Van der Wal and Vissers (1993) interpreted it as having originally formed part of the mantle wedge overlying a subduction zone.

BENI-BOUSERA

The Beni-Bousera massif in Morocco (Kornprobst *et al.* 1990) occupies a similar tectonic position to the Ronda peridotite on the other side of the

Mediterranean. It is an elongated dome of ultrabasic rocks outcropping over about 50 square kilometres, overlain by granulite-facies metamorphic rocks. The peridotite appears to have been emplaced at a very high temperature, and the adjoining country rocks include kinzigites, which are interpreted as partial melting residues. The ultrabasic rocks, which are schistose and isoclinally folded, are layered peridotites, with pyroxenite bands making up about 3% of the total volume. The predominant type of peridotite is a spinel–lherzolite, made up of olivine and orthopyroxene with minor clinopyroxene and spinel.

The pyroxenites contain varying proportions of orthopyroxene, clinopyroxene, garnet and spinel, and have the composition of picritic basalts. Graphitized diamonds have been found in some of the pyroxenite layers, showing that they crystallized at pressures above about 45 kilobars (Pearson *et al.* 1993). The pyroxenites were originally thought to represent liquids produced by partial fusion of the peridotite, but this is now considered unlikely. Their O, C and S isotopic compositions all suggest that the magmas from which the pyroxenites crystallized were derived from the melting at depth of hydrothermally altered subducted oceanic lithosphere, possibly including some sediment.

MOUNT ALBERT

The Mt Albert ultramafic massif (Fig. 344) is in the Gaspé peninsula, Quebec. It has a semicircular outcrop with a radius of about 5 kilometres, truncated in the south by a large fault. It is notable for its well-developed contact metamorphic aureole.

Figure 344. Geological map of the Mount Albert intrusion, Quebec and its metamorphic aureole (after Macgregor and Basu 1979).

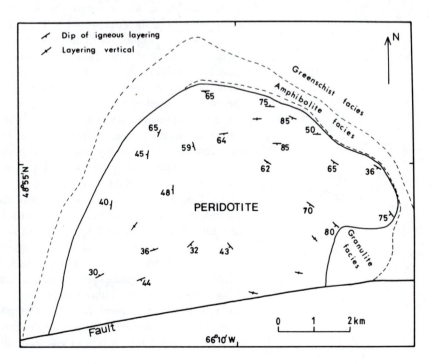

The massif itself consists almost entirely of harzburgite and dunite, which occur as alternating layers from a few centimetres to several metres thick. The attitude of the layering is not related to the margins of the intrusion but appears to form a fold structure. The ultramafic rocks vary texturally from a relatively undeformed core to more deformed margins. Serpentinization is restricted to the margins.

The metasedimentary and metavolcanic country rocks are conformable in foliation and schistosity with the contacts of the intrusion. Regionally they are of greenschist facies, but as the contact is approached they show progressive metamorphism to epidote–amphibolite, almandine–amphibolite, and eventually granulite facies rocks.

On the basis of its structural features, its mineralogy, and geophysical evidence of a cylindrical form, Macgregor and Basu (1976) interpreted the intrusion as a diapir of mantle material emplaced in a solid condition. On the other hand, the structural position of the Mt Albert massif in the Appalachian belt is analogous to that of the serpentinites in southern Quebec which Laurent (1975) regards as ophiolitic.

FINERO

The Western Alps contain many examples of Alpine-type peridotites (Fig. 331), the majority showing a typically ophiolitic association with gabbros and spilites. A very different aspect is shown by the Finero peridotite near the northern end of Lake Maggiore, which is not associated with any volcanic rocks and is in a much higher-grade metamorphic environment.

The Finero peridotite is situated on the south side of the Insubric line, the major fault zone which forms the southern limit of Tertiary regional metamorphism in the Alpine-fold belt. To the south of this major fault are high-grade metamorphic rocks and basic and ultrabasic rocks making up the 'Ivrea-Verbano zone' (Fig. 345). A northward traverse from Lake Maggiore towards the Insubric line reveals: (1) metasedimentary gneisses of increasing grade, intruded by some granites; (2) kinzigites and stronalites – high grade gneisses composed of quartz and feldspar + garnet (in stronalite), biotite (in kinzigite), sillimanite and graphite – which are possibly melting residua; (3) a zone in which ultrametamorphic gneisses are intimately mixed with basic rocks composed of plagioclase and clinopyroxene + orthopyroxene + olivine + garnet + hornblende; and (4) several bodies of peridotite, of which the Finero peridotite is the largest.

The Finero peridotite is considerably altered, particularly around its margins, but in its least altered central part it is a phlogopite-bearing spinel–peridotite. This is separated by a band of hornblende-bearing metagabbros from an outer zone of hornblende–lherzolite (Fig. 346). The various components of the Finero complex are folded into an antiform (Steck and Tièche 1976). The phlogopite–peridotite consists of olivine (60–90%), orthopyroxene (5–20%), clinopyroxene (0–5%), amphibole (0–5%), phlogopite (10–15%), and spinel (0–5%). The phlogopite and hornblende are irregularly distributed but are occasionally concentrated in thin bands. There are also monomineralic layers of enstatite, diopside, amphibole and spinel.

Figure 345. Geological map of the northern part of the Ivrea zone, Italy (after Schmid 1968; Hunziker and Zingg 1982).

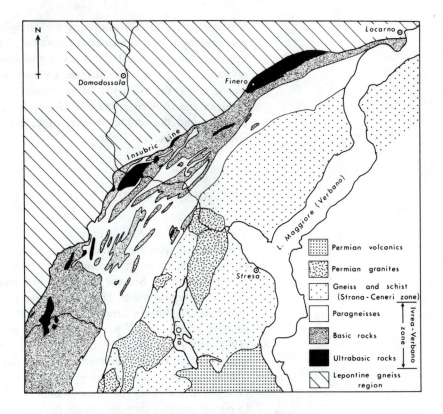

Figure 346. Geological map of the Finero peridotite, Italy (after Vogt 1962; Lensch 1968).

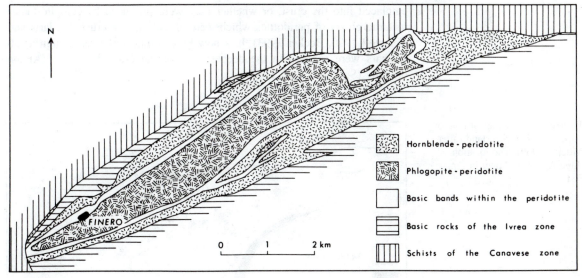

According to Cawthorn (1975) the phlogopite–peridotite has a cumulate texture, with euhedral cumulus olivine set in a matrix of diopside, enstatite, phlogopite, spinel, magnetite and amphibole, but Exley *et al.* (1982) interpreted the phlogopite–peridotite as a highly deformed rock in which the presence of phlogopite was due to metasomatism in the upper mantle.

Voshage *et al.* (1987) attributed the phlogopite to metasomatism of the peridotite by K-rich fluids derived from the local crust. Cawthorn interpreted the peridotites and metabasic rocks as consanguineous products of fractional crystallization of a basic magma, and not as either a fragment of upper mantle nor as the lower part of an ophiolite assemblage, whereas according to Siena and Coltorti (1989) the phlogopite–peridotite is a slice of mantle tectonite and the amphibole–peridotite and metagabbros form a layered igneous complex.

ALPE ARAMI, SWITZERLAND

The peridotite of Alpe Arami (Fig. 347) is a small body, tectonically interstratified with high-grade gneisses, and is one of a number of such bodies which occur in the same area. It is a garnet–lherzolite, and is of particular interest because pyroxene geobarometry suggests a very high pressure of equilibration. The garnet is pyrope-rich and is partially altered to spinel and amphibole. Around the margin of the peridotite there is a band of eclogite which locally passes into hornblendite or clinopyroxenite. Ernst (1981) calculated that the minerals in the peridotite last equilibrated at a pressure of 40 kilobars, corresponding to a depth well down in the mantle.

Many such bodies of garnet–peridotite are found in orogenic belts, and the presence of garnet suggests equilibration at very high pressures. It is difficult to decide whether they are mantle peridotites which have been tectonically emplaced into the crust, or whether they were produced by eclogite-facies metamorphism of peridotites which were already in the crust (Medaris and Carswell 1990, Becker 1993). It is now known that pressures as high as 30 kilobars were reached in the deepest part of the crust during the Alpine orogeny (Schreyer *et al.* 1987).

Figure 347. Geological map of Alpe Arami, Switzerland (after Möckel 1969). The location of Alpe Arami is shown in Fig. 331.

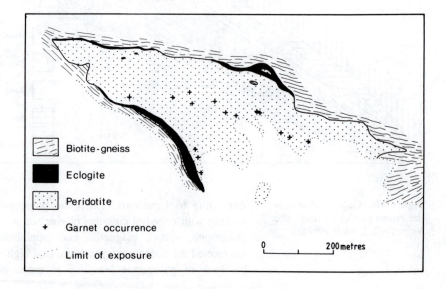

ORIGIN OF NON-OPHIOLITIC ALPINE PERIDOTITES

There are differences between the non-ophiolitic Alpine-type peridotites which suggest that they may not all have the same origin, but it is possible to make some generalizations. They are all highly deformed and rarely contain cumulate textures. They mostly have pyroxene compositions indicating initial crystallization at a very high pressure followed by re-equilibration at lower pressures. Some of them are garnet-bearing and some contain diamond, both of which indicate a high-pressure origin and suggest that they are bodies of upper mantle material.

The examples which have been described here are composed predominantly of lherzolite (with the exception of the Mt Albert intrusion), in contrast to the harzburgites and dunites which predominate among ophiolitic peridotites. This difference is significant, because lherzolites are much richer in fusible constituents than harzburgites and dunites. As can be seen from Fig. 348, it would be possible to extract a basaltic melt from lherzolite and leave a harzburgitic or dunitic residuum. Thus the lherzolites might be considered as 'fertile' or 'undepleted' mantle material, whereas the harzburgites and dunites are more likely to be melting residues or cumulates.

Nicolas and Jackson (1972) distinguished two major types of Alpine peridotite in the Mediterranean region. Those of the Alps, Pyrenees and the Betic Cordillera are mainly lherzolites with metamorphic textures, associated with granulites (e.g. Finero, Lanzo, Lherz, Ronda, Beni-Bousera). Those in the Balkans and eastern Mediterranean are mainly metamorphic harzburgites and cumulus peridotites associated with gabbros and other constituents of the ophiolite suite. They considered the first type to represent sub-continental upper mantle and the second type to be sub-oceanic. A similar distinction

Figure 348. The difference in relative Al, Ca and Mg content of the main types of peridotite (after Nicolas and Jackson 1972).

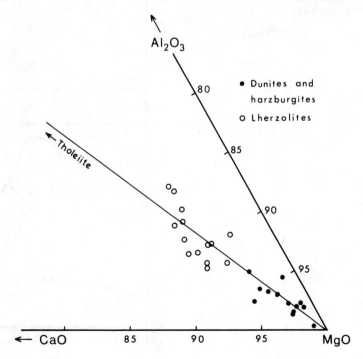

was made by Chidester and Cady (1972) in the Appalachians, between peridotites formed beneath oceanic crust (e.g. Newfoundland) and those formed beneath continental crust (e.g. those in southern Quebec and New England). More recently, some workers have used the term 'Alpine' in a more restricted sense so as to exclude ophiolite assemblages, distinguishing between Alpine and ophiolitic peridotites (Menzies *et al.* 1977) or between orogenic and ophiolitic peridotites (Menzies 1984).

The position of the Finero peridotite is particularly interesting. Together with the adjoining rocks of the Ivrea zone it meets all the requirements for a section through lower continental crust and mantle (Voshage *et al.* 1990; Zingg *et al.* 1990). The sequence of increasing metamorphic grade approaching the peridotite from the south side simulates a downward section through the lower crust.

Gravity and seismic studies have shown the Ivrea zone to be underlain by a thick mass of very dense rocks, which geophysicists interpret as a slab of upthrust lower crust and perhaps some attached mantle (Giese 1968; Berckhemer 1969). Figure 349 shows a section across the zone. If the Finero

Figure 349. A cross-section through the Ivrea zone consistent with gravity and seismic evidence (after Berckhemer 1969). The shaded area represents an upward extension of high-density material from the mantle. The thick lines in the lower diagram indicate where the top of the high-density material has been identified seismically.

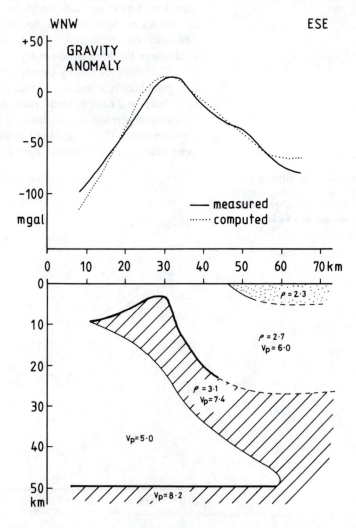

peridotite is considered to be a segment of upthrust mantle, then its south-eastern margins represent an exposure of the continental Mohorovičić discontinuity. The Balmuccia peridotite, situated further south in the Ivrea zone, has also been interpreted as a piece of sub-continental mantle uplifted into the crust (Shervais and Mukasa 1991).

Quantitative estimates of the temperature and pressure of equilibration of peridotite mineral assemblages can be made from the compositions of the pyroxenes present, as described in Chapter 12. The temperatures and pressures of equilibration of some Alpine-type peridotites estimated in this way are shown in Fig. 350. The garnet–peridotites show the highest pressures of formation. For the Alpe Arami garnet-peridotite the estimates range from 29–33 kilobars (Carswell and Gibb 1980) to as high as 46 kilobars (Ernst 1978, 1981). There are great discrepancies between different methods of calculation, and some of the earlier *P–T* data obtained by this method are highly suspect, because it was not fully realized that the pressure-dependence of Al_2O_3 in orthopyroxenes depends on which Al-bearing phase the orthopyroxene coexists with, i.e. plagioclase, spinel or garnet (Wilshire and Jackson 1975). In addition, most of the peridotites have re-equilibrated at lower pressures during their emplacement. No pressure estimates can be made for spinel–lherzolites, other than that they must lie within the pressure range of spinel stability. Plagioclase–peridotites show the lowest equilibration pressures, as might be expected.

The pressures estimated for the Alpe Arami peridotites correspond to depths of 90–140 kilometres, i.e. well below the base of the crust. From their *P–T* position in Fig. 350, close to a continental shield geotherm, it seems that they equilibrated in sub-continental mantle before being brought up into the crust (Ernst 1978). In contrast, Carswell and Gibb (1980) think that they equilibrated in the garnet–lherzolite facies after their emplacement into the surrounding gneisses, implying that this portion of the crust has been down to a depth of 100 kilometres at some time in its history and that these rocks are simply the products of very high pressure metamorphism.

Figure 350. Some estimates of the temperatures and pressures of equilibration of Alpine peridotites, based on pyroxene compositions. The lherzolite facies and solidus are outlined as in Fig. 333. The oceanic ridge axis geotherm and continental geotherm are as in Fig. 235.

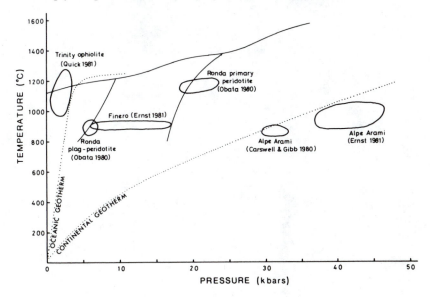

The positions of the Ronda and Finero peridotites in Fig. 350 correspond to conditions of equilibration equivalent to steep oceanic rather than continental geotherms, but the pressures calculated by pyroxene geobarometry only record the conditions of uplift. The original presence of diamond in the Ronda peridotite shows that the rocks actually formed at a pressure of more than 45 kilobars. The combination of high temperature and low pressure estimated for the Trinity ophiolite can only be consistent with a very high ocean-ridge geotherm and is what would be expected for a peridotite formed at an oceanic spreading axis.

CHAPTER 14

Anorthosites

Anorthosite, like peridotite, is a rock type which has no extrusive equivalent at the present day. Moreover the great majority of anorthosites which have been described are of Pre-Cambrian age, which suggests that the conditions required for anorthosite formation may no longer exist or may be very rare. Three main occurrences of anorthosite can be distinguished: (1) as cumulate layers in layered basic intrusions, for example Bushveld or Stillwater; (2) as large, dominantly anorthositic massifs within Pre-Cambrian metamorphic terrains, as in the Adirondacks and south-west Norway (the Proterozoic massif-type anorthosites); and (3) as thin layers in very ancient and highly deformed metamorphic complexes (the Archaean megacrystic anorthosites). The first type of anorthosite is easily accounted for by differentiation of a basic magma *in situ*, but the other two are much more problematical. In terms of abundance, the second type is by far the most important.

The distinction between the three types of anorthosite occurrence is by no means clear-cut. Gabbroic anorthosite and gabbro are often present in the Proterozoic massif-type anorthosites, and in many Pre-Cambrian complexes which have been deformed and metamorphosed the relationship between anorthosite and associated basic rocks can be difficult to determine. The most important distinguishing feature of the Proterozoic massif-type anorthosites is that they show a much more subordinate presence of gabbroic rocks than the layered basic type, and their plagioclase is much more sodic (see Table 62).

ANORTHOSITE IN LAYERED BASIC INTRUSIONS

Many basic intrusions contain some layering, with light and dark minerals being concentrated into alternating bands. However, in contrast to the Skaergaard intrusion (Chapter 7) in which most of the rock could be described as gabbro, there are some layered basic intrusions that contain very thick bands of ultrabasic rocks and anorthosite. The best known of these are the Bushveld intrusion in South Africa and the Stillwater intrusion in

Table 62. A comparison of some well-known anorthosite occurrences (mainly after Ashwal 1993)

	Area (km²)	Age (m.y.)	Plagioclase (mol% An)
Layered basic intrusions			
Stillwater, Montana	>4,400	2701	74–78
Bushveld, South Africa	66,000	2060	47–82
Dufek, Antarctica	>50,000	172	52–85
Sept Iles, Quebec	5,000	544	63–76
Abontorok, Niger	5	400	58–62
			Mean = 68
Proterozoic massif-type			
Marcy, Adirondacks	3,000	1288	43–55
Morin, Quebec	2,500	1160	42–54
Lac St-Jean, Quebec	17,000	1156	45–62
Labrieville, Quebec	250	—	34–39
Michikamau, Labrador	2,000	1450	44–70
Harp Lake, Labrador	10,000	1460	46–59
Nain, Labrador	7,000	1305	37–64
S. Rogaland, Norway	580	1500	40–55
			Mean = 49
Archaean megacrystic type			
Bad Vermilion Lake, Ontario	100	2747	75–81
Fiskenaesset, Greenland	500	2869	75–98
Sittampundi, India	>30	—	80–100
Shawmere, Ontario	560	>2765	65–90
			Mean = 83

Montana. There are only a few of these intrusions, and although they are large, massive anorthosite forms only a small proportion of each (6–7% in Bushveld, 18% in Stillwater).

Stillwater, Montana

The Stillwater intrusion (Fig. 351) contains the most spectacular develop-

Figure 351. Geological map of the Stillwater complex, Montana (after Todd *et al.* 1982).

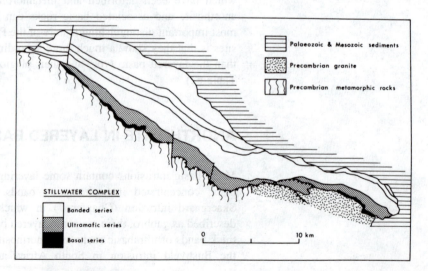

ment of anorthosite in a layered complex. Within its Banded Series it contains two major horizons of almost pure anorthosite, one 400 metres thick and the other 550 metres thick, as well as many plagioclase-rich troctolites, norites and gabbros. In the whole 2500-metre thickness of the middle Banded Series, including its gabbroic and anorthositic layers together, the average rock contains 84% of feldspar, but the whole assemblage of rocks in the intrusion, and particularly its lower margin of tholeiitic dolerite, suggests the parent magma or magmas to have been basaltic.

The intrusion extends for 45 kilometres, but is partly covered by later sedimentary rocks so that the form of the intrusion cannot be fully determined. The exposed layered rocks are nearly 6000 metres thick, although because of faulting a complete section is not seen in any one place. Three main divisions can be distinguished: (1) the Basal Series, whose contact with the underlying hornfelses can be seen; (2) the Ultramafic Series; and (3) the Banded Series, whose top is concealed by the overlying sediments. It is the Banded Series that contains the thick anorthosite layers.

The Basal Series (Page 1979) is mostly between 60 and 240 metres thick. It comprises two subdivisions: a lower noritic member, locally contaminated by sediments, and an upper bronzite cumulate. The Ultramafic Series follows the Basal Series conformably, averages 1070 metres in thickness (Page 1977), and consists of cyclic repetitions of olivine, olivine–orthopyroxene, and orthopyroxene cumulates. The corresponding rock types are dunite, harzburgite and bronzitite (where consolidation of the cumulates was by adcumulus growth), or slightly more feldspathic rocks (where consolidation of the cumulates was by crystallization of intercumulus liquid). The Banded Series has a maximum thickness of over 4400 metres (McCallum *et al.* 1980); its base is marked by the appearance of cumulus plagioclase. It consists of alternating layers of the following types of cumulate (in decreasing order of abundance): (1) plagioclase–orthopyroxene–clinopyroxene; (2) plagioclase; (3) plagioclase–orthopyroxene; (4) plagioclase–olivine.

In terms of lithology the predominant rock types are norite and gabbronorite in the lower Banded Series, anorthosite in the middle Banded Series, and gabbronorite in the upper Banded Series. The two major anorthositic units in the middle Banded Series are about 400 metres and 550 metres thick. In these units plagioclase is the only cumulus mineral. The stratigraphy of the various cumulate layers can be traced for long distances, although there are variations in thickness. The J–M platinum-bearing reef which occurs in the lower part of the Banded Series can be followed for 40 kilometres even though it is only 1–3 metres thick, which is an indication of the huge size of the magma body.

The rhythmic layering is so well developed, and the rock types so varied, that the cryptic variation can only be measured as changes in the composition of the cumulus minerals. There is a trend of iron-enrichment of the ferromagnesian minerals as in the Skaergaard intrusion, but it cannot be seen whether it leads eventually to ferrogabbros or more siliceous rocks because the upper part of the intrusion is hidden. Cumulus plagioclase occurs throughout the Banded Series and varies from An_{82} at the base to An_{62} at the top, but in this range there are several minor reversals of the trend (Fig. 352). Also in the Banded Series cumulus olivine disappears and reappears several

Figure 352. The lithological divisions of the Stillwater complex, and the variation in composition of its cumulus plagioclase (after Keith *et al.* 1982; McCallum *et al.* 1980).

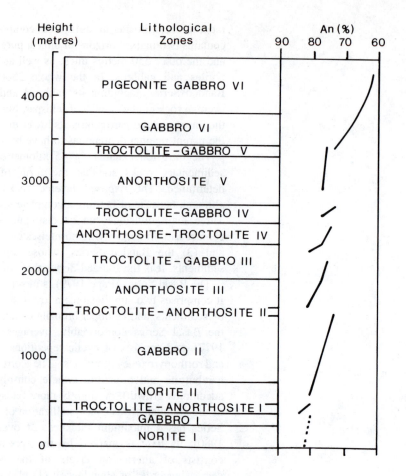

times which would not be expected in simple progressive fractionation sequence.

It is generally agreed that the Stillwater intrusion did not crystallize as a single body of magma. According to one interpretation (McCallum *et al.* 1980) there was repeated injection of fresh magma at various stages during crystallization, but according to another (Keith *et al.* 1982; Todd *et al.* 1982) there were two distinct parent magmas involved. One is characterized by the crystallization sequence olivine–bronzite–plagioclase–augite and is held responsible for the formation of peridotite, bronzitite, norite and two-pyroxene gabbro. The other is characterized by the crystallization sequence plagioclase–olivine–augite–bronzite and is held responsible for the formation of anorthosite, troctolite, olivine gabbro and two-pyroxene gabbro.

Aïr Massif, Niger

In the Aïr massif there are several very unusual occurrences of anorthosite as a component of sub-volcanic ring complexes of Paleozoic age (Bowden *et al.* 1987; Brown *et al.* 1989). The Aïr volcanoes mainly erupted peralkaline rhyolite (as ash flows), together with minor basalt and trachyte, but no magma resembling anorthosite. The underlying intrusive complexes contain granite, syenite, gabbro and anorthosite. In the Ofoud intrusion the basic rocks range from anorthosite to leucogabbro and ordinary gabbro. The anorthosite is layered in places, and among the layers are some made up of titanomagnetite + olivine with a cumulate texture. An interesting feature of the gabbros in these intrusions is that many of them are ferrogabbros with high Fe/Mg ratios, suggesting that they have undergone severe fractionation under reducing conditions.

PROTEROZOIC MASSIF-TYPE ANORTHOSITES

The Proterozoic massif-type anorthosites form extremely large intrusive complexes, some of them covering thousands of square kilometres, but despite their size they are almost entirely restricted to the middle Proterozoic (1.0–1.6 billion years). The anorthosite in these massifs is often associated with leuconorite, leucogabbro and leucotroctolite, but more mafic rocks such as norite, gabbro and troctolite are present in only small amounts. A suite of intermediate to acid plutonic rocks often accompanies the anorthosites, and the relationship of these rocks to the anorthosite has been the subject of much debate.

OCCURRENCE

There is no single anorthosite massif which can be considered typical; they vary in the amount and nature of the associated basic and acid igneous rocks and in size and shape. The larger anorthosite massifs resemble granitic batholiths in that when they are mapped in detail they are found to be composite, i.e. made up of individually emplaced smaller plutons. Their mode of intrusion cannot usually be determined because of the effects of later deformation, and because their external contacts are often obscured by the associated bodies of acid and intermediate material. There is no general agreement as to the tectonic environment in which the Proterozoic anorthosites formed, although a majority of anorthosite specialists believe it was anorogenic.

Adirondack Mountains, New York

The Adirondack anorthosites have been studied in more detail than any others, mainly because of their accessibility and because of the efforts over many years of Buddington and his collaborators, although they are not an ideal example because they have suffered granulite-facies metamorphism. The Adirondacks lie at the southern end of the Grenville tectonic province (Fig. 353), in which anorthosite intrusions are unusually abundant compared with all other parts of the world. The largest of the Adirondack anorthosite bodies is the Marcy massif which is shown in Fig. 354.

The rocks of the Adirondack massif are grouped into two series. The

Figure 353. The distribution of anorthosite massifs in the eastern part of the Canadian shield. A = Adirondacks, HL = Harp Lake, L = Labrieville, MI = Michikamau, MO = Morin, N = Nain, SI = Sept Iles, SJ = Lac St-Jean.

anorthosite series includes anorthosite, gabbroic or noritic anorthosite, gabbro, norite, oxide-rich gabbro or norite, pyroxene–ilmenite–magnetite rock, and magnetite and ilmenite ores. The feldspar-rich rocks predominate in this series, and all of the rocks have been metamorphosed. The *mangerite–charnockite series* consists of metamorphic gneisses with the bulk compositions of pyroxene–syenite, quartz–pyroxene–syenite, and quartz–syenite. These rocks were formerly described as the 'quartz–syenite' series.

The form of the Marcy massif is thought from structural and geophysical evidence to be sheet- or slab-like, and the negative gravity anomaly shows that it is not underlain by gabbro (Simmons 1964). Tabular offshoots of anorthosite and gabbroic anorthosite occur in the metasedimentary country rocks above the roof of the main intrusion. Anorthosite generally constitutes the core, and gabbroic anorthosite the outer border facies of the Marcy massif, and there is some evidence for a gabbroic roof zone. Figure 355

Figure 354. Geological map of the Marcy massif, Adirondack Mountains, New York (after Ashwal 1982).

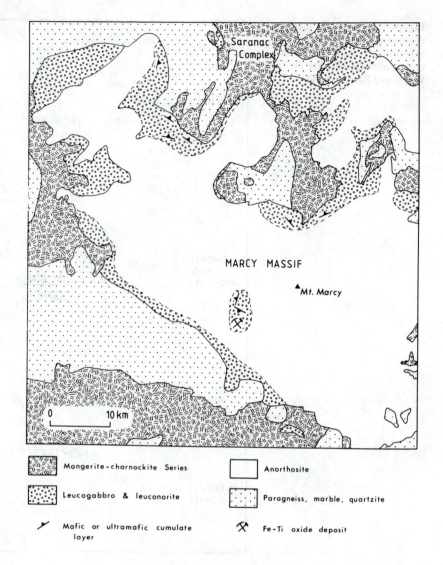

Legend:
- Mangerite-charnockite Series
- Leucogabbro & leuconorite
- Mafic or ultramafic cumulate layer
- Anorthosite
- Paragneiss, marble, quartzite
- Fe-Ti oxide deposit

shows some relationships which were interpreted by De Waard and Romey (1969) as evidence for the coexistence of a plagioclase cumulate and a noritic magma during the crystallization of the complex.

The most difficult problem in the Adirondack massif is to determine whether the anorthosite series and the mangerite–charnockite series are of comagmatic origin. Buddington (1972) favoured an independent origin for the latter ('quartz–syenite') series and pointed out that the latter rocks have a cross-cutting intrusive relationship to the anorthosites. He attributed the anorthosite series to flow differentiation of a gabbroic anorthosite magma with much plagioclase in suspension. Figure 356 shows the chemical variation in the two series of rocks. Trace element studies on the whole support an independent origin for the anorthositic and 'syenitic' rocks (Seifert *et al.* 1977).

Figure 355. Schematic diagram illustrating relationships between anorthosite and norite in the Adirondacks (after De Waard and Romey 1969).

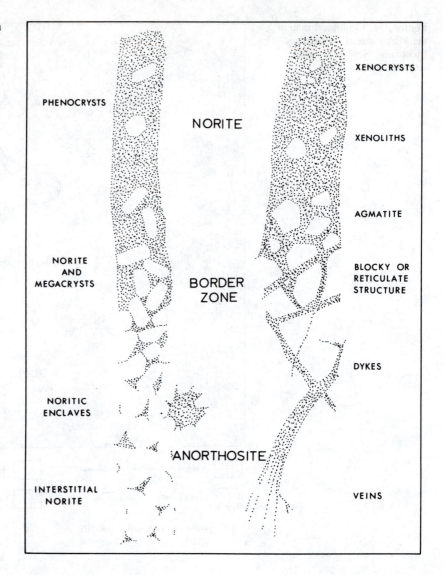

Harp Lake, Labrador

The Harp Lake complex (Fig. 357) is one of the largest of the great anorthosite massifs in the Canadian Shield, and is unmetamorphosed and undeformed. The major part of the massif consists of anorthosite, leuco-troctolite, leuconorite and minor leucogabbro. Tholeiitic gabbro occurs as a thin margin to the complex and as several smaller bodies within the anorthosite. Large masses of adamellite and granite adjoin the anorthosite and are thought to have been intruded at about the same time, although they are not necessarily comagmatic.

The anorthosite is not a single intrusion but a complex of about 20 separate intrusive units. The anorthositic rocks show adcumulate texture, implying that they formed from a magma that was sufficiently liquid for crystals to accumulate and one that was precipitating only plagioclase. The mineral compositions of the anorthosite are consistent with crystallization

Figure 356. Chemical variation in the anorthosite and quartz-mangerite series of the Marcy massif (after Buddington 1972). Circles = anorthosite series; dots = quartz-mangerite series.

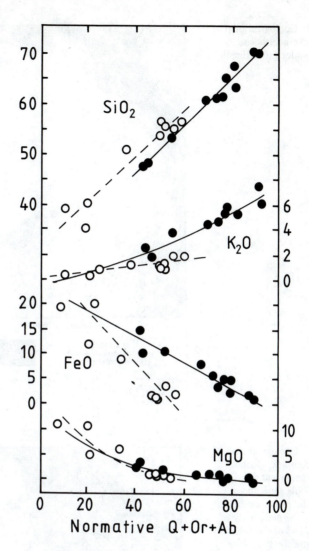

from a gabbroic magma, and Emslie (1980) believed that the parent magma was a high-Al gabbro formed by prior fractionation of pyroxenes and possibly olivine at depth from a more normal gabbro. The adamellites contain clinopyroxene and fayalitic olivine and have evidently formed from a magma very undersaturated in water.

Morin, Quebec

The Morin complex lies in the Grenville province about 50 kilometres north of Montreal. The range of rock types which are present includes anorthosite, leuconorite, troctolite, ferrogabbro, jotunite, mangerite and farsundite. The complex contains many economically important ilmenite–hematite and titaniferous magnetite deposits. According to Martignole and Schrijver (1970), the complex consists of three tectonic units – a dome, a diapir and a nappe. Each of these units is composed of a core of anorthosite and leuconorite surrounded by an irregular mantle of jotunite and mangerite (Figs

Figure 357. Geological map of the Harp Lake anorthosite massif, Labrador (after Emslie 1980).

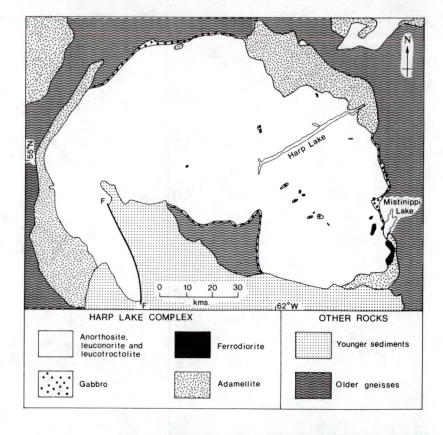

HARP LAKE COMPLEX | OTHER ROCKS

Anorthosite, leuconorite and leucotroctolite

Ferrodiorite

Younger sediments

Gabbro

Adamellite

Older gneisses

Figure 358. Geological map of the Morin complex, Quebec (after Martignole and Schrijver 1970). Symbols show the attitude of the planar cumulate structures in the main body of anorthosite, and of the foliation in the deformed eastern lobe.

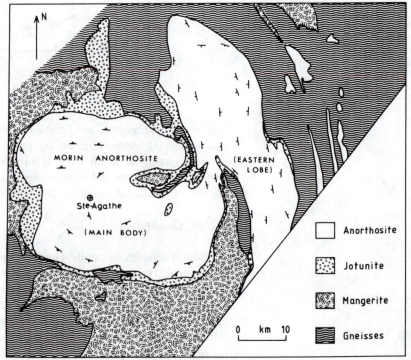

Anorthosite

Jotunite

Mangerite

Gneisses

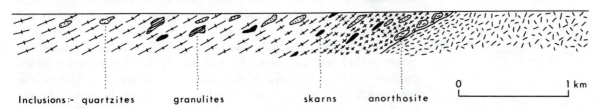

Inclusions:- quartzites   granulites   skarns   anorthosite

Figure 359. Schematic section across the boundary of the Morin massif (after Martignole and Schrijver 1970).

358, 359). The troctolite occurs as a small isolated layered body. Except in the core of the dome, the rocks have been severely deformed. They are granulated and foliated, and in the eastern part of the area the anorthosite and mangerite of the nappe unit are preserved only in the cores of synforms. Martignole and Schrijver suggested that the anorthosite and leuconorite were intruded in a largely crystalline condition, but were separated from the country rocks by the magma which later crystallized to give the jotunite and mangerite. The latter rocks contain xenoliths of both the country rocks and the anorthosite and the leuconorite.

Subsequently, Martignole (1974) investigated the complex geochemically and found a compositional gap between anorthosites and farsundites. He suggested an independent, possibly anatectic, origin for the farsundites. Isotopic evidence also shows that the mangerites crystallized somewhat later than the anorthosites (Barton and Doig 1977). Kearey (1978) carried out a gravity survey of the area and showed that the anorthosite is not underlain by mafic material at depth.

Labrieville, Quebec

The Labrieville anorthosite (Fig. 360) is a dome-shaped body approximately 10 kilometres across. It is composed of three divisions: (a) a domical core

Figure 360. Geological sketch map of the Labrieville massif, Quebec (after Anderson 1966).

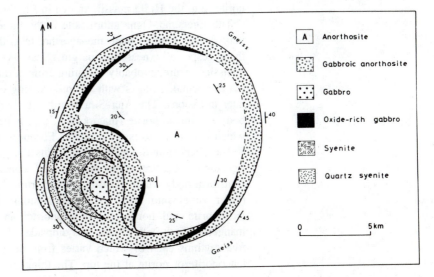

composed mainly of strongly foliated anorthosite; (b) a border facies a few hundred metres thick mostly composed of weakly foliated gabbroic anorthosite; and (c) the funnel-shaped Sault-aux-Cochons complex made up of massive gabbroic anorthosites, anorthosite, oxide-rich gabbro and syenite.

The lithological sequence anorthosite–gabbroic anorthosite–oxide-rich gabbro–syenite was considered by Anderson (1966) to represent a differentiation sequence from an initial magma which was very rich in the constituents of feldspar, highly oxidized, and poor in water. In the anorthosite there are several bodies of hemo-ilmenite, some of which are rich in apatite, which may possibly have separated from the silicate magma by liquid immiscibility.

The massif is surrounded by an aureole of plagioclase-rich hypersthene– and orthoclase–gneisses. Further away, the country rocks are biotite–hornblende–microcline granitic gneisses. Within the aureole several mineralogical changes can be recognized; for example ferrian ilmenite away from the intrusion is replaced by hemo-ilmenite close to the intrusion.

Oxygen isotope studies on coexisting ilmenite and plagioclase in the anorthosite suggest a very high crystallization temperature ($\sim 1100\,°C$). If crystallization took place in the present structural setting then large amounts of the country rock would have been melted. Anderson suggested that the plagioclase–hypersthene gneiss around the intrusion may be a refractory residuum resulting from this process.

South Rogaland, Norway

The South Rogaland complex occupies an area of 1500 square kilometres in south-west Norway. It is a composite mass, consisting of many separate intrusions of anorthosite and related rock types. A geological map is given in Fig. 361. The South Rogaland complex consists of three large anorthosite massifs (Egersund–Ogna, Håland–Helleren, and Åna–Sira), a layered leuconoritic intrusion (Bjerkreim–Sokndal), and several smaller bodies of leuconorite (e.g. the Hidra massif, shown in Fig. 362).

The Egersund–Ogna anorthosite is the largest of the anorthosites. It is a very coarse grained andesine–anorthosite with a small amount of hypersthene, especially near the margin. Near its border the anorthosite has a gneissic texture, probably resulting from a diapiric mode of emplacement, but the actual contacts with the pre-existing country rocks are obscured by later intrusions. The Åna–Sira anorthosite is notable for containing a large body of ilmeno-norite (plagioclase 36%, hypersthene 15%, ilmenite 39%), which is mined and is the principal European source of titanium oxide.

The Bjerkreim–Sokndal lopolith cuts the gneissic country rocks and the Egersund–Ogna intrusion sharply, and contains many xenoliths. It is a stratiform body which is differentiated from the base (outer zone) to the top (inner zone) into three main subdivisions: (a) hypersthenic anorthosite, leuconorite and norite; (b) monzonorite; and (c) mangerite and quartz–mangerite (both with olivine and diopside). The first subdivision consists of five rhythms each of which varies from anorthosite at the base to banded leuconorite or norite at the top. The thicknesses of the three lower rhythms

Figure 361. Geological map of the South Rogaland igneous complex, Norway (after Michot and Michot 1969). Anorthosite massifs: I Egersund–Ogna, V Haaland–Helleren, VI Aana–Sira. Leuconorite bodies: VII Hidra, VIII Garsaknatt. The Bjerkreim–Sokndal lopolith: IIIa–c.

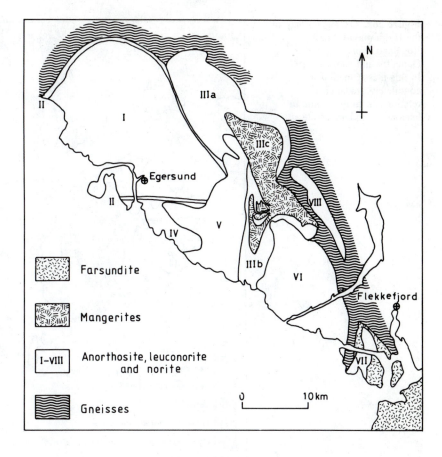

decrease from the axis to the flanks of the lopolith so as to suggest that the synclinal form of the lopolith was developing during the period that crystallization was taking place. Igneous lamination and fine-scale rhythmic layering are well developed in this intrusion, and slump and trough structures indicate deposition of crystals from a strongly convecting magma.

The S. Rogaland country rocks are quartzites, cordierite– and sillimanite–gneisses, granitic gneisses, pyroxene granulites and charnockitic gneisses, and there is much evidence of anatexis, with many conformable bodies of granitic composition. All the rocks of the region, both intrusions and country rocks, are characterized by a granulite-facies mineralogy: hypersthene is the main ferromagnesian mineral, biotite is rare, and muscovite is absent. This mineralogy is a reflection of the very high temperature and anhydrous conditions of crystallization. The anorthositic rocks have not been metamorphosed, although there is some internal deformation associated with their emplacement.

In the Egersund area the complex contains anorthositic, noritic and granitic rocks in the proportions of about 70:25:5, and the absence of a positive gravity anomaly shows that there is no concealed gabbroic body that could be regarded as complementary to the anorthosite (Smithson and Ramberg 1979).

Figure 362. Geological map of the Hidra massif, South Rogaland, Norway (after Demaiffe and Hertogen 1981). In this massif there is a completely gradational relationship from jotunite to leuconorite to anorthosite.

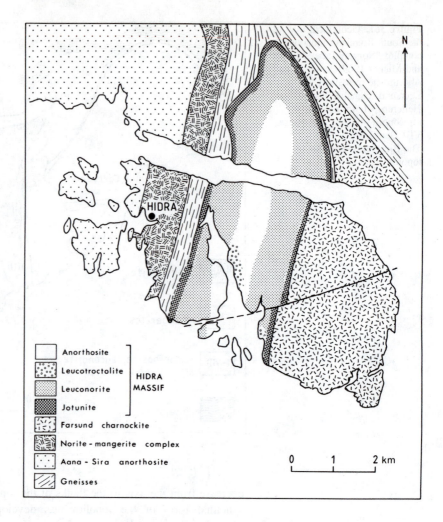

Anorthosite ⎤
Leucotroctolite ⎥ HIDRA
Leuconorite ⎦ MASSIF
Jotunite
Farsund charnockite
Norite - mangerite complex
Aana - Sira anorthosite
Gneisses

0   1   2 km

Nain, Labrador

Anorthosites and related rocks occupy several thousand square kilometres around Nain, in northern Labrador (Fig. 353). The rock types of this region have been divided by Wheeler (1969) and De Waard (1976) into the following groups, in decreasing order of abundance: (1) anorthositic rocks – anorthosite, leuconorite and leucogabbro; (2) 'adamellitic' rocks – mostly plagioclase-rich intermediate rocks containing K-feldspar, but ranging from syenodiorite to granite; (3) troctolitic rocks – made of plagioclase, olivine and pyroxene with abundant olivine; and (4) jotunitic rocks – fine grained, mesocratic plagioclase–pyroxene rocks with K-feldspar. Contact metamorphism of the surrounding paragneisses has produced mineral assemblages of the pyroxene–hornfels facies. Berg (1977) adduced mineralogical evidence that the conditions of contact metamorphism varied up to 915 °C at 3.7 to 6.6 kilobars, corresponding to depths of 13–24 kilometres. According to Wiebe (1980) the gneissic country rocks have locally undergone partial melting for a distance of 10–20 metres from anorthosite contacts giving rise to anatectic granite veins. Some of the anorthosite bodies have anorthositic chilled margins and were therefore intruded as anorthosite magma, but most do not and were probably intruded as crystal mushes.

The relationship between the major rock types is not yet fully known. Some contacts between anorthosite and the 'adamellitic' rocks are sharp, with the latter rock types being later, but other contacts are transitional through jotunite. The adamellites are enriched in iron relative to magnesium, and despite their leucocratic nature some of them contain olivine (fayalite-rich) and clinopyroxene (hedenbergite-rich). The largest body of basic rocks in the area is the Kiglapait intrusion, which is a classic example of a layered basic intrusion with extreme crystal fractionation (see Chapter 7).

An important aspect of the Nain anorthosites is that they were not emplaced as part of any orogenic episode. The regional metamorphism of the country rocks long pre-dates the period of anorthosite intrusion.

Michikamau, Labrador

The Michikamau intrusion in Labrador occupies an area of about 200 square kilometres (Emslie 1965, 1969, 1970) and is shown in Fig. 363. It is unmetamorphosed and undeformed, and has given K/Ar ages of about 1400 million

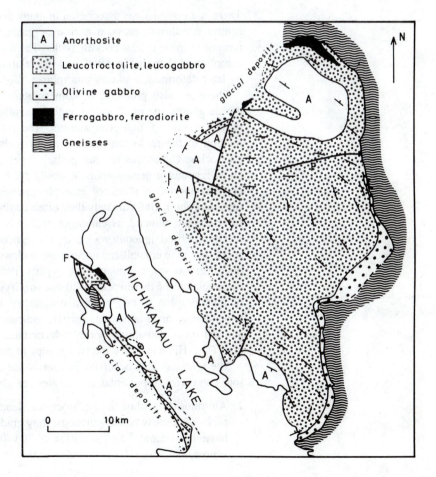

Figure 363. Geological map of the Michikamau intrusion, Labrador (after Emslie 1969).

years. The country rocks are quartzo-feldspathic gneisses, with mineral assemblages of the pyroxene–hornfels facies adjacent to the intrusion.

The principal components of the intrusion, in order of their crystallization, are a border zone of olivine-gabbro which is chilled at the contact; layered leucotroctolite, making up the bulk of the intrusion; anorthosite, which grades upwards into leucogabbro and gabbro; and finally quartz–ferrogabbro or ferrodiorite. The rocks of the intrusion show many examples of rhythmic layering, gravity stratification, and planar orientation of plagioclase. Textures are clearly indicative of crystal accumulation.

In many ways the Michikamau intrusion is transitional between a layered basic intrusion and the more anorthosite-rich massifs described above. The resemblance between this and other layered basic intrusions, and the gabbroic margin, are indicative of crystallization from an initially basaltic magma, but the abundance of plagioclase-rich rocks and the concentric arrangement of a gabbroic margin around a leucocratic core are features of many anorthosite massifs.

IGNEOUS ROCKS ASSOCIATED WITH ANORTHOSITE

There is a conspicuous association in many anorthosite complexes between a central anorthosite mass, a noritic or gabbroic border, and a surrounding zone of syenitic or charnockitic rocks. The distribution of rock types is more complicated than this in detail, and the relationships are very often obscured by later deformation and metamorphism.

There are also problems of nomenclature. For example, the gabbroic or noritic anorthosite of American authors is often identical to the leucogabbro or leuconorite of European authors. The most serious difficulty arises in relation to the rocks that contain alkali feldspar and quartz in addition to plagioclase ('charnockite', mangerite', 'syenite', etc.). If these were clearly either igneous or metamorphic, it would not be difficult to give them igneous or metamorphic names, for example granite (igneous) or granitic gneiss (metamorphic). Unfortunately they often combine igneous compositions with metamorphic mineral assemblages and it is not certain whether they are metamorphosed igneous rocks, or non-igneous rocks, or perhaps igneous rocks that have crystallized under a particular set of conditions (i.e. high total pressure, low water pressure). De Waard (1969) attempted to rationalize the nomenclature of these rocks. Without specifying whether they were igneous or metamorphic in origin, he distinguished between: (a) rocks containing hypersthene, or fayalite plus quartz, indicating crystallization at low P_{H_2O}; and (b) rocks containing hornblende or biotite, indicating crystallization at high P_{H_2O}. His names for the two groups of rocks are shown in Fig. 364.

The igneous rocks occurring in anorthosite complexes can be grouped into four categories (representative analyses are given in Table 63):

1. Anorthosite, gabbro, norite, troctolite, leuconorite, leucotroctolite. These rocks are clearly consanguineous; they grade into one another and they all have low initial $^{87}Sr/^{86}Sr$ ratios (0.703–0.706) indicative of a mantle source.

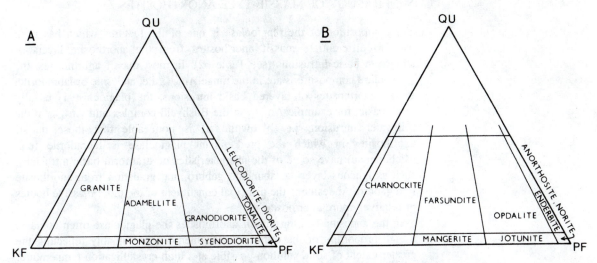

Figure 364. The nomenclature of felsic rocks associated with anorthosite massifs (after De Waard 1969). **A**: nomenclature of rocks containing hornblende or biotite; **B**: nomenclature of rocks containing hypersthene (or fayalite plus quartz). Abbreviations: KF = K-feldspar, PF = plagioclase, QU = quartz.

2. Jotunite (Figs 358, 362), ferrogabbro, ferrodiorite (Figs 357, 363), monzonorite, oxide-rich gabbro (Fig. 360). These are only minor constituents in any particular complex, but they have very distinctive compositions, being very rich in Fe, Ti and P, and with very high FeO/MgO ratios.

3. Mangerite (Figs 354, 358, 361), syenite (Fig. 360). These rocks are rich in alkali feldspar and not always clearly separable from the next category which resemble granite and contain quartz. However, they also have very distinctive compositions, high alkalis being combined with low Si and high Fe.

4. Charnockite (Fig. 354), farsundite (Fig. 361), quartz–mangerite, granite, adamellite (Fig. 357). These rocks are rich in alkali feldspar and quartz, and have broadly granitic compositions, although they often contain orthopyroxene rather than biotite.

Table 63. Representative analyses of rock types occurring in Proterozoic anorthosite massifs, compared with the average basalt

	1	*2*	*3*	*4*	*5*	*6*
SiO_2	49.20	53.27	53.83	48.35	56.00	68.64
TiO_2	1.84	0.37	0.27	3.17	1.51	1.05
Al_2O_3	15.74	23.46	25.67	14.36	15.51	13.14
$FeO + Fe_2O_3$	10.92	4.61	2.75	15.54	11.46	4.63
MgO	6.73	4.25	1.16	2.82	1.47	0.70
CaO	9.47	10.02	9.27	8.67	3.53	2.82
Na_2O	2.91	3.70	5.12	3.03	4.72	3.14
K_2O	1.10	0.62	0.97	1.09	4.82	5.40
P_2O_5	0.35	0.03	0.01	1.53	0.60	0.27

1. Average basalt (Le Maitre 1976).
2. Leucotroctolite, Marcy, Adirondacks (De Waard 1970).
3. Anorthosite, Marcy, Adirondacks (De Waard 1970).
4. Jotunite, Morin, Quebec (Owens *et al.* 1993).
5. Mangerite, Bjerkreim-Sokndal, Norway (Duchesne *et al.* 1987).
6. Quartz–mangerite, Morin, Quebec (Philpotts 1966).

MINERALOGY OF MASSIF-TYPE ANORTHOSITES

The composition of the plagioclase is one of the features which has been used to differentiate massif anorthosites from the anorthosite layers of stratiform basic intrusions (see Table 62). In most massif anorthosites, the plagioclase compositions are in the range An_{35-65}, i.e. andesine or labradorite. In the anorthosites of layered basic intrusions, the plagioclase is usually more basic, for example An_{47-82} in the Bushveld complex and An_{74-78} in the Stillwater intrusion, i.e. bytownite. It is noticeable that those massif anorthosites in which the predominant plagioclase is labradorite (e.g. Michikamau) have some of the characteristics of stratiform basic intrusions, such as igneous layering, abundant gabbro, and gradation from anorthosite into gabbro. Andesine is the principal constituent of the more massive bodies of relatively pure anorthosite.

In the andesine-bearing massif anorthosites the plagioclase often contains a high proportion of the orthoclase end-member, presumably reflecting the greater extent of solid solution possible at a high crystallization temperature: $An_{45}Or_6$ at Allard Lake, Quebec, $An_{37}Or_8$ at Labrieville, Quebec, and $An_{23}Or_{25}$ at San Gabriel, California.

The pyroxenes of massif-type anorthosites also have a distinctive composition. These rocks frequently contain very large megacrysts of aluminous orthopyroxene (up to 9% Al_2O_3) showing exsolution of plagioclase and sometimes of spinel or garnet. It has been suggested that these might indicate crystallization at a very high pressure, followed by adjustment to lower pressures, but in the light of present knowledge (Fig. 322) it is safe only to say that these megacrysts probably crystallized at an extremely high temperature.

Titaniferous magnetite is the principal iron oxide mineral of labradorite–anorthosites, but in andesine–anorthosites the main ore mineral is hemoilmenite, i.e. ilmenite with 20–35% hematite in solid solution. This is a much higher content of ferric iron than is found in the ilmenites of basalts, and indicates a high oxygen fugacity during crystallization (Anderson and Morin 1969). A more detailed study of oxidation conditions during crystallization of some of the South Rogaland anorthosites by Duchesne (1972) showed more reducing conditions, which varied during crystallization and showed discontinuities possibly indicative of periodic magma replenishment.

ORIGIN OF PROTEROZOIC MASSIF-TYPE ANORTHOSITES

Proterozoic massif-type anorthosites have initial $^{87}Sr/^{86}Sr$ ratios of between 0.703 and 0.706 (Heath and Fairbairn 1969), which are within the range that might be expected for mantle-derived magmas. Emslie (1978) compared the available Sr isotope data for anorthosites and associated granitic rocks, and showed that the initial $^{87}Sr/^{86}Sr$ ratios for the latter are appreciably higher than those of the anorthosites, ranging from 0.704 to 0.712.

Anorthosite as a melting residuum

Partial melting in the crust is capable of leading to a residuum of anorthositic composition if the starting material includes the right constituents. Winkler and Von Platen (1960) obtained a plagioclase-rich residuum from the partial melting of a calcareous clay, and De Vore (1975) calculated that removal of a granite-like melt from a variety of greywackes would leave residues with the compositions of anorthosite, norite, syenite and mangerite. Frith and Currie (1976) also showed by experiment that partial melting of tonalite could leave an anorthositic residuum, and invoked this process to explain the enormous extent of anorthosite in the Grenville province. The low $^{87}Sr/^{86}Sr$ ratios of most anorthosites are incompatible with a calcareous clay source, but many greywackes and tonalites have low enough initial $^{87}Sr/^{86}Sr$. Given the abundance of greywackes and tonalites, anorthosite must occasionally be produced by crustal anatexis, as suggested by Berg (1969) for the Bitterroot Range in Montana.

The anorthosites in the Bitterroot Range occur as a group of about 20 small bodies ranging up to 2.5 square kilometres in a predominantly gneissic terrain. Their average composition is 95% labradorite, 3% biotite and 2% quartz. The anorthosites contain numerous small areas of quartz-diorite (24% quartz, 66% andesine) which grade into the anorthosite. In both the anorthosites and the surrounding country rocks there are bodies of quartz–monzonite gneiss which have sharp contacts against the anorthosite. The country rocks are quartz–feldspar–biotite–sillimanite–gneisses in the amphibolite grade of regional metamorphism. Unlike many other anorthosite massifs, there are no basic rocks in these intrusions and orthopyroxene is not present as a mafic constituent. Berg tentatively ascribed the quartz–monzonite gneiss and the anorthosite to partial melting of the more albite-rich portions of the sillimanite gneiss, the quartz–monzonite representing the melt and the anorthosite representing the residuum.

The intrusive relationships of anorthosites are not necessarily incompatible with a residual origin because although a plagioclase-rich residuum would be denser than the complementary melt it would be lighter than most other granulite-facies materials in the lower crust. Despite these considerations, most anorthosite specialists regard anorthosites as magmatic rather than residual rocks.

Existence of anorthosite magma

There has been considerable doubt as to the existence of anorthosite magma. There are no anorthositic lavas, and as Bowen pointed out many years ago the anhydrous melting temperatures of plagioclase feldspars are much higher than the observed temperatures of known magmas such as basalt or rhyolite. A dry anorthosite magma would have to have a very high intrusion temperature, around 1300 °C depending on the exact composition (see Fig. 122), and a wet anorthosite magma is not consistent with the anhydrous mineralogy of anorthosites.

The field relationships of anorthosites do not always provide clear evidence for magmatic emplacement. Some have been deformed since crystallization, while others may have been intruded diapirically in a solid condition. Woussen *et al.* (1981) listed the following evidence that the Lac St-Jean massif in Quebec, which is one of the largest anorthosite complexes (>20,000 square kilometres), was emplaced diapirically: (a) there is a subvertical foliation in both the anorthosites and the surrounding gneisses; (b) sub-solidus reaction between plagioclase and olivine shows that the original crystallization took place at a depth (~25–30 kilometres) greater than that at which the anorthosite was eventually emplaced; and (c) dykes of diorite, amphibolite and leucotroctolite cutting the anorthosite have a mineralogy indicating crystallization at successively decreasing *P–T* conditions from granulite to amphibolite facies. The Labrieville and Morin intrusions (Figs 358 and 360) also have many of the characteristics of diapiric intrusions.

Anorthosite dykes are the most direct evidence of anorthositic magma, although they are not common. Wiebe (1990) described dykes from the Nain complex which are truly anorthositic (80–95% plagioclase). Other evidence cited by various authors favouring magmatic anorthosite intrusion includes angular and disorientated xenoliths, chilled margins, and cumulate textures.

Despite this evidence for the magmatic origin of anorthosites, there is a possibility that many anorthosite bodies were formed as crystal accumulates from magmas that were not themselves anorthositic. Even for those anorthosite massifs for which a magmatic origin is well established, there is no unanimity about the composition of the initially intruded magma. Whereas the Nain anorthosite dykes and chilled margins do indicate an anorthositic magma, the chilled margins of some of the anorthosite bodies in South Rogaland are jotunitic (Duchesne and Demaiffe 1978).

A primary anorthosite magma?

One of the strongest reasons for taking seriously the possibility of a primary anorthosite magma is the enormous quantity of anorthosite which occurs in those areas where anorthosite massifs are found. Several of the anorthosites in eastern Canada have outcrop areas of over 10,000 square kilometres. Basalt, andesite and granite are the only other magmatic rocks that occur in such large volumes. If anorthosite is a primary magma type it is necessary not only to find a source for it, but also to explain why it has apparently become scarce since the Pre-Cambrian.

Anorthositic liquids can be produced by the partial melting of peridotite or basalt under suitable conditions. Whereas anhydrous peridotite would give rise to basaltic liquid on partial melting, experiments on the hydrous partial melting of peridotite show that the melts produced are not basaltic but are much richer in the constituents of plagioclase. For example Mysen and Boettcher (1975) partially melted garnet–lherzolite in the presence of excess H_2O at 15 kilobars pressure and produced melts with up to 84% of normative plagioclase. Re-melting of basaltic rocks at a high pressure is also a way of producing a magma rich in the constituents of plagioclase. As a simple analogy, consider the system diopside–anorthite (Fig. 365). In this system the minimum-melting composition (i.e. the eutectic composition) becomes richer in the constituents of anorthite as the pressure is increased. A similar relationship is apparent in the system plagioclase–diopside–enstatite (Fig. 366). In the absence of water the high-pressure liquidus temperatures are very high, but at high water pressures the liquidus temperatures are greatly

Figure 365. The system diopside–anorthite at various pressures with or without the presence of water (after Yoder 1969).

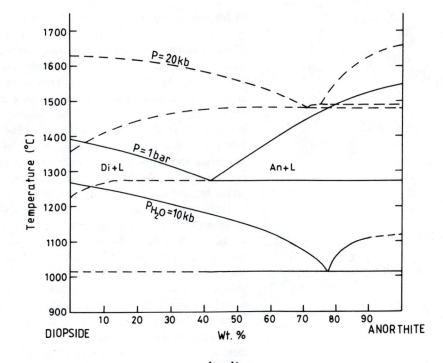

Figure 366. Liquidus relationships in the system plagioclase (Ab₄₀An₆₀)–diopside–enstatite at 1 bar (light lines) and 15 kilobars (heavy lines). After Emslie (1971).

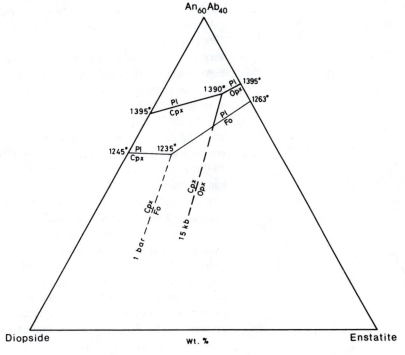

reduced while the minimum-melting composition remains plagioclase-rich. Despite these considerations, it is very unlikely that the formation of anorthosite magma involves the presence of H_2O because anorthosites themselves and their associated rocks have a largely anhydrous mineralogy.

A fractionation origin?

There is strong evidence for a basaltic parent magma for some anorthosites. Border facies of several anorthosite intrusions are gabbroic (Michikamau, Harp Lake). The $^{87}Sr/^{86}Sr$ and $^{18}O/^{16}O$ ratios of anorthosites are in the basaltic range. On the other hand, gravity as well as field studies of anorthosite masses have failed to reveal the presence of the expected ultramafic complementary differentiates (Simmons 1964). The scarcity of anorthosites younger than 1000 million years contrasts with the abundance of basaltic intrusions and lavas, and the mineralogy of some anorthosites (presence of hemo-ilmenite) conflicts with the relatively unoxidizing conditions under which basaltic magmas usually crystallize.

There are two quite contrasting possibilities for the production of anorthosite by differentiation of basalt magma. The first is by concentration of the constituents of plagioclase in the liquid by fractionation under high pressure, i.e. the reverse of the partial melting hypothesis considered above. The enlargement of the pyroxene fields in Figs 365 and 366 at high pressures, with or without the pressure of water, means that the residual melt would become richer in the constituents of plagioclase as pyroxene was removed. The second possibility is the separation of plagioclase crystals as an anorthositic cumulate. Many workers believe that anorthosites crystallize from a magma rich in suspended plagioclase crystals. There is a spectacular occurrence of what are believed to be anorthosite flotation cumulates in the gabbroic dykes of the Gardar province in Greenland (Bridgwater and Harry 1968). These gabbros contain a great number of anorthosite xenoliths and very large plagioclase megacrysts, ranging in abundance from zero up to 80% of the whole rock. They are thought to have crystallized at depth from the same basaltic magmas that gave rise to their host rocks, and may have accumulated in the roof zone of an underlying magma chamber.

Very interesting evidence on the possible conditions of fractionation came from the discovery in the Adirondack anorthosite of very large orthopyroxene megacrysts (up to 40 cm) with exsolved garnet (Jaffe and Schumacher 1985). Before exsolution these megacrysts must have been Al-bearing, and it can be inferred that they must therefore have crystallized at a high pressure (20–25 kbar according to Jaffe and Schumacher). This implies that they were carried up from a magma chamber in which fractionation was occurring at a depth of 70–80 km. Other orthopyroxene megacrysts in anorthosite contain exsolved plagioclase, also an indication of original crystallization at high pressure. On the basis of this evidence, Wiebe (1986) proposed that the magma of the Proterozoic anorthosites formed by the fractionation of basaltic magma ponded at the base of the crust. Separation of olivine and pyroxene caused the magma to decrease in density despite a simultaneous increase in its Fe/Mg ratio, and enabled it to rise into the crust as an anorthositic magma.

The role of pressure may be very important for another reason. The density of basalt magma increases with pressure more rapidly than that of plagioclase (Fig. 184), so that plagioclase crystals which remained suspended in the magma at low pressures would readily float at the high pressures prevailing in the lower crust. This may help to explain why some layered basic complexes contain a more substantial development of anorthositic cumulates than others, but if it is to account for the formation of massif-type anorthosites there need to be complementary ultramafic cumulates, for which very little evidence has been found.

Relationship between anorthosites and associated rock types

Each of the three rock suites associated with anorthosites (jotunites, mangerites, charnockites) can be shown to be magmatic in at least some of their occurrences. Subsequent metamorphism or deformation has often obscured this, but there are localities where good evidence can be found, such as intrusive relationships, cumulate textures or layering. Particularly good examples of igneous layering in mangerites have been described by Griffin *et al.* (1974).

There is a very close spatial relationship between the jotunites and the anorthosites (gradational in the case of the Hidra massif – Fig. 362) which suggests that their magmas are consanguineous. The very high Fe/Mg ratios of jotunites suggest that these magmas are residual liquids from advanced crystal fractionation. By analogy with the ferrogabbros of the Skaergaard intrusion, one may suspect that the parent magma was basaltic and that fractionation took place under reducing conditions (the Fenner trend). Wiebe (1984) considered that the monzonorites (jotunites) of the Bjerkreim–Sokndal lopolith were a residual liquid product of fractionation of the leuconorites, whereas Demaiffe and Hertogen (1981) considered that jotunite was parental to anorthosite in the Hidra massif. Subsequently the same group of workers (Duchesne *et al.* 1989) concluded that the Rogaland jotunites were not co-magmatic with anorthosite at all.

The very high total Fe, Ti and P content of jotunites (see Table 63) is of interest because many anorthosites contain bodies of Fe–Ti ore, some of which are of major economic importance, and in the Roseland anorthosite in Virginia there are several bodies of ilmenite–apatite ('nelsonite'), magnetite–apatite and rutile–apatite rock. There is a strong possibility that liquid immiscibility has played some part in generating rocks of this composition (see Chapter 7).

Mangerites are relatively abundant compared with jotunites, but are not so closely associated spatially with anorthosite. Mangerites share the high Fe/Mg ratios of jotunite. Some workers consider them to have formed separately from the anorthosites and jotunites, although Owens *et al.* (1993) believe that the jotunites associated with the Quebec anorthosites are transitional members of a co-magmatic sequence from anorthosite to mangerite. The distribution of rare earth elements in the Hidra massif (Fig. 367) suggests that the anorthosites and mangerites ('charnockites') are complementary products of fractional crystallization, for which Demaiffe and Hertogen (1981) suggested a jotunite (monzodiorite) parent magma.

It is rather unlikely that the acid rocks (charnockites, quartz–mangerites) are comagmatic with the anorthosites. They have higher initial $^{87}Sr/^{86}Sr$ ratios than the associated anorthosites (Emslie 1978), and the scatter and lack of continuity in the intermediate rock compositions shown in Fig. 368 suggests a separate origin. It is possible that anorthosites were intruded at a

Figure 367. Chondrite–normalized rare-earth patterns of rocks from the Hidra anorthosite massif (after Demaiffe and Hertogen 1981).

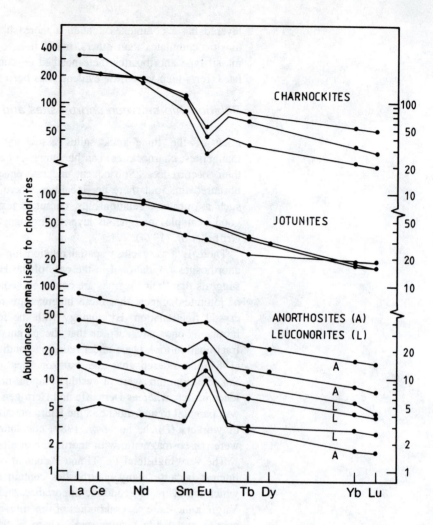

Figure 368. (below) Modal compositions of felsic rocks from the Marcy, Morin and Nain massifs (after De Waard 1969). KF = K-feldspar, PF = plagioclase, QU = quartz.

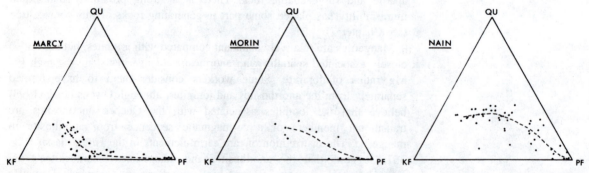

sufficient temperature to partially melt their surrounding rocks, and Wiebe (1984) considered that the quartz–mangerites were partial melting products of the country rocks. Alternatively the charnockite magmas could have been produced deeper in the crust as a result of heating by the parent magmas of the anorthosites (Emslie 1980).

ARCHAEAN MEGACRYSTIC ANORTHOSITES

Ashwal (1993) drew attention to features of Archaean anorthosites that suggested they should be classified separately from the Proterozoic massif-type anorthosites. They have some features in common with the anorthosites of layered basic intrusions and of Proterozoic massifs but also some differences.

The most distinctive feature of Archaean anorthosites is the presence of large euhedral phenocrysts or megacrysts, typically 1–5 cm across but sometimes much larger, set in a gabbroic matrix. This texture is not shared by other types of anorthosite, and they also differ from Proterozoic massif-type anorthosites in containing a much more calcic plagioclase.

Most Archean anorthosites have been severely deformed and metamorphosed, so their field relationships and rock associations are difficult to determine. Most of them appear to have been emplaced as sills or at least sheet-like bodies, and are associated with gabbros and mafic volcanic rocks. Some contain ultramafic zones or show layering. One of the best-known examples is that of Bad Vermilion Lake, Ontario (Ashwal *et al.* 1983), which is relatively well preserved and is in a lower grade of metamorphism (greenschist–amphibolite facies) than any of the other examples that have been studied.

Because of the high degree of metamorphism that the Archaean anorthosites have undergone, it is difficult to estimate the composition of their primary plagioclase, but Ashwal (1993) estimated it to be around An_{85} in most cases. This is comparable to the plagioclase in some layered basic intrusions, but much more calcic than the plagioclase in Proterozoic massif-type anorthosites. Layers of Fe–Ti oxides and occasionally of chromite are another feature linking some of the Archaean anorthosites to layered basic intrusions.

All the evidence suggests that the Archaean anorthosites are derived from a mafic magma, but the mechanism by which they were differentiated and in particular the conditions giving rise to their megacrystic texture are still unknown. Phinney *et al.* (1988) reviewed their composition, and suggested that the anorthosite magma was produced by high-pressure fractionation of tholeiitic or picritic magma, perhaps in a magma chamber at the crust–mantle boundary. However, the same explanation has been advanced by Wiebe (1986) for the Proterozoic massif-type anorthosites, which have quite different plagioclase compositions and textures.

LUNAR ANALOGIES

The discovery that the Moon has an anorthositic crust opened up new possibilities for the origin of terrestrial anorthosites. The surface of the lunar highlands (terrae) has been found to consist predominantly of anorthositic rocks, and the surface of the flat areas (maria) of basalt. Isotopic and chemical studies show that the mare basalts are the products of volcanism which occurred between 4200 and 3900 million years ago, whereas the anorthosites date back to 4600 million years, i.e. close to the original time of formation of the Moon and Earth. It is believed that early in the Moon's history, possibly while accretion was still going on, the outermost several hundred kilometres of the Moon underwent melting, probably as a result of the heat generated by planetesimal bombardment. Formation of the anorthositic crust is assumed to result from gravitational crystallization differentiation during the cooling of this outermost layer, while the deeper complementary ultramafic layer is the most probable source region of the later mare basalts.

The original nature of the Earth's crust has been obscured by the many geological processes which have been in operation during the last 4000 million years, but there is a strong possibility that the Earth's early history was similar to that of the Moon, in which case the Earth may also have had an anorthositic crust at one time.

Many authors have discussed the possibility that existing anorthosites may be derived from, or be relicts of, an early terrestrial anorthosite crust, but it seems unlikely that the large Proterozoic massif-type anorthosites described here are in any way analogous to the anorthosite of the lunar highlands. They differ in two important respects. Firstly, the massif anorthosites contain relatively sodic plagioclase, whereas the plagioclase of the lunar anorthosites is very calcic (An_{94-97}). Secondly, some of the massif anorthosites show mineralogical evidence of having crystallized under very oxidizing conditions, whereas the lunar anorthosites crystallized under exceptionally reducing conditions. The Archaean megacrystic anorthosites are closer to the lunar anorthosites in composition (and age), but it seems unlikely that any of the terrestrial anorthosites now visible are directly analogous to those on the Moon.

APPENDIX

A key to some common abbreviations

This Appendix gives brief notes on the meaning of some of the abbreviations and acronyms in common use in igneous petrology. Where possible a reference is given to a further explanation in the preceding chapters.

A.S.I.
: The aluminium saturation index. Equals the cation ratio $Al/(Na+K+2Ca)$ and the molecular ratio $Al_2O_3/(Na_2O+K_2O+CaO)$.

BABI
: Basaltic Achondrite Best Initial ratio. The best estimate of the initial isotopic composition ($^{87}Sr/^{86}Sr$) of the basaltic achondrite meteorites, assumed to be similar to the primordial composition of the Earth's strontium.

CHUR
: Chondritic Uniform Reservoir. A notional composition which serves as a frame of reference in Sm/Nd isotopic studies. See De Paolo and Wasserburg (1979a).

CIPW norm
: The system of recalculating chemical analyses of rocks to 'normative' or 'hypothetical' minerals was devised by Cross, Iddings, Pirsson and Washington, hence the initials CIPW. The normative minerals and their abbreviations are listed in Table 64. Full details of the method of calculating the norm are given by Kelsey (1965).

CMAS
: A short-hand designation of the system $CaO-MgO-Al_2O_3-SiO_2$.

D
: A distribution coefficient (or partition coefficient), expressing the distribution of trace elements between crystals and magma. See Chapter 4.

D.I.
: 'Differentiation Index'. The normative sum Q + Or + Ab + Ne + Kp + Lc (Thornton and Tuttle 1960). Supposedly a measure of the degree of differentiation of a magma from a more parental composition.

EM1, EM2
: Components of the mantle source region of basalts ('EM = enriched mantle'). See Chapter 8.

Table 64. The standard minerals of the CIPW norm

Normative mineral		Idealized composition
Quartz	Q	SiO_2
Orthoclase	Or	$K_2O.Al_2O_3.6SiO_2$
Albite	Ab	$Na_2O.Al_2O_3.6SiO_2$
Anorthite	An	$CaO.Al_2O_3.2SiO_2$
Leucite	Lc	$K_2O.Al_2O_3.4SiO_2$
Nepheline	Ne	$Na_2O.Al_2O_3.2SiO_2$
Kaliophilite	Kp	$K_2O.Al_2O_3.2SiO_2$
Acmite	Ac	$Na_2O.Fe_2O_3.4SiO_2$
Sodium metasilicate	Ns	$Na_2O.SiO_2$
Potassium metasilicate	Ks	$K_2O.SiO_2$
Diopside	Di	$CaO.(Mg, Fe)O.2SiO_2$
Wollastonite	Wo	$CaO.SiO_2$
Hypersthene	Hy	$(Mg, Fe)O.SiO_2$
Olivine	Ol	$2(Mg, Fe)O.SiO_2$
Larnite (dicalcium silicate)	Cs	$2CaO.SiO_2$
Sphene (titanite)	Tn	$CaO.TiO_2.SiO_2$
Zircon	Z	$ZrO_2.SiO_2$
Corundum	C	Al_2O_3
Halite	Hl	$NaCl$
Thenardite	Th	$Na_2O.SO_3$
Sodium carbonate	Nc	$Na_2O.CO_2$
Magnetite	Mt	$FeO.Fe_2O_3$
Chromite	Cm	$FeO.Cr_2O_3$
Ilmenite	Il	$FeO.TiO_2$
Hematite	Hm	Fe_2O_3
Perovskite	Pf	$CaO.TiO_2$
Rutile	Ru	TiO_2
Apatite	Ap	$3CaO.P_2O_5.\frac{1}{3}CaF_2$
Fluorite	Fr	CaF_2
Pyrite	Pr	FeS_2
Calcite	Cc	$CaO.CO_2$

FMA (or AFM) A type of variation diagram in which the relative proportions of F ($FeO + Fe_2O_3$), M (MgO), and A ($Na_2O + K_2O$) are plotted on a triangle. In some variants F may be total Fe expressed as FeO, or $FeO + 0.9Fe_2O_3$.

f_{O_2} The fugacity of oxygen. See Chapter 5.

H_2O+ The water content of a rock released when it is heated from 110 °C up to nearly its melting temperature (i.e. water derived from the breakdown of minerals containing hydroxyl ions, e.g. amphibole or mica).

H_2O- The water content of a rock released when it is heated up to 105 °C or 110 °C (i.e. adsorbed water, or water present in zeolites, smectites, gypsum etc.).

HFSE High field-strength element.

HIMU A component of the mantle source material of basalts characterized by having a high U/Pb ratio (μ). See Chapter 8.

I-type	Granites whose magma source region contained mainly igneous material. See Chapter 9.
IAB	Island arc basalt.
LIL(E)	Large-ion lithophile element, e.g. K, Rb, Cs, Sr, Ba.
MORB	Mid-Ocean Ridge Basalt. Various categories may also be distinguished, such as N-type (normal) or E-type (enriched). See Chapter 8.
O≡F	An item which occurs in chemical analyses of fluorine-bearing rocks. It has the value of 0.421 times the F content, and is subtracted from the analysis total to correct for the other constituents being quoted as oxide percentages. Similar corrections are required for S and Cl in analyses of sulphide- or chloride-bearing rocks.
OIB	Oceanic island basalt.
PGE	Platinum-group elements.
ppb	Parts per billion (10^9).
ppm	Parts per million.
REE	Rare earth elements.
S-type	Granites whose magma source region contained mainly sedimentary material. See Chapter 9.
SMOW	Standard Mean Ocean Water. The reference material against which all oxygen isotopic compositions ($\delta^{18}O$) are measured.
SS	Solid solution series (in the labelling of phase diagrams).

REFERENCES

Ackermann, P. B. & Walker, F. 1960. Vitrification of arkose by Karroo dolerite near Heilbron, Orange Free State. *Quart. J. Geol. Soc. London*, **116**, 239–54.

Ahern, J. L. & Turcotte, D. L. 1979. Magma migration beneath an ocean ridge. *Earth Planet. Sci. Lett.*, **45**, 115–22.

Akaad, M. K. 1956. The Ardara granitic diapir of county Donegal, Ireland. *Quart. J. Geol. Soc. London*, **112**, 263–90.

Alderton, D. H. M. & Harmon, R. S. 1991. Fluid inclusion and stable isotope evidence for the origin of mineralizing fluids in south-west England. *Min. Mag.*, **55**, 605–11.

Alexander, P. O. & Paul, D. K. 1977. Geochemistry and strontium isotopic composition of basalts from the Eastern Deccan volcanic province, India. *Min. Mag.*, **41**, 165–72.

Alibert, C., Michard, A. & Albarède, F. 1983. The transition from alkali basalts to kimberlites: isotope and trace element evidence from melilitites. *Contr. Min. Petr.*, **82**, 176–86.

Allègre, C. J., Brévart, O., Dupré, B. & Minster, J. F. 1980. Isotopic and chemical effects produced in a continuously differentiating convecting Earth mantle. *Phil. Trans. Roy. Soc. Lond. A*, **297**, 447–77.

Allègre, C. J., Dupré, B., Richard, P. & Rousseau, D. 1982. Subcontinental versus suboceanic mantle, II. Nd–Sr–Pb isotopic comparison of continental tholeiites with mid-ocean ridge tholeiites, and the structure of the continental lithosphere. *Earth Planet. Sci. Lett.*, **57**, 25–34.

Allègre, C. J., Treuil, M., Minster, J. F., Minster, B. & Albarède, F. 1977. Systematic use of trace elements in igneous processes. Part I: Fractional crystallization processes in volcanic suites. *Contr. Min. Petr.*, **60**, 57–75.

Allen, J. C. & Boettcher, A. L. 1978. Amphibole in andesite and basalt: II. Stability as a function of P–T–fH$_2$O–fO$_2$. *Amer. Min.*, **63**, 1074–87.

Allred, A. L. 1961. Electronegativity values from thermochemical data. *J. Inorg. Nucl. Chem.*, **17**, 215–21.

Amit, O. & Eyal, Y. 1976. The genesis of Wadi Magrish migmatites (N-E Sinai). *Contr. Min. Petr.*, **59**, 95–110.

Andersen, S., Bohse, H. & Steenfelt, A. 1981. A geological section through the southern part of the Ilimaussaq intrusion. *Rapp. Grønlands Geol. Unders.*, **103**, 39–42.

Andersen, T. 1984. Secondary processes in carbonatites: petrology of 'rodberg' (hematite–calcite–dolomite carbonatite) in the Fen central complex, Telemark (South Norway). *Lithos*, **17**, 227–45.

Anderson, A. T. 1966. Mineralogy of the Labrieville anorthosite, Quebec. *Amer. Min.*, **51**, 1671–711.

Anderson, A. T. 1975. Some basaltic and andesitic gases. *Rev. Geophys. Space Phys.*, **13**, 37–55.

Anderson, A. T. 1976. Magma mixing: petrological process and volcanological tool. *J. Volc. Geotherm. Res.*, **1**, 3–33.

Anderson, A. T., Clayton, R. N. & Mayeda, T. K. 1971. Oxygen isotope thermometry of mafic igneous rocks. *J. Geol.*, **79**, 715–29.

Anderson, A. T. & Morin, M. 1969. Two types of massif anorthosites and their implications regarding the thermal history of the crust. *Mem. N. Y. State Mus. Sci. Serv.*, **18**, 57–69.

Anderson, A. T., Swihart, G. H., Artioli, G. & Geiger, C. A. 1984. Segregation vesicles, gas filter-pressing, and igneous differentiation. *J. Geol.*, **92**, 55–72.

Anderson, D. L. 1979. Chemical stratification of the mantle. *J. Geophys. Res.*, **84**, 6297–8.

Anderson, D. L. 1981a. Hotspots, basalts, and the evolution of the mantle. *Science*, **213**, 82–9.

Anderson, D. L. 1981b. Rise of deep diapirs. *Geology*, **9**, 7–9.

Anderson, D. L. 1982. Isotopic evolution of the mantle: the role of magma mixing. *Earth Planet. Sci. Lett.*, **57**, 1–12.

Anderson, D. L. 1984. Kimberlite and the evolution of the mantle, pp. 395–403 in Vol. 1 of Kornprobst (1984).

Anderson, D. L. 1989. *Theory of the Earth*. Blackwell, Oxford, 366 pp.

Anderson, R. N. and 10 co-authors. 1982. DSDP hole 504B, the first reference section over 1 km through layer 2 of the oceanic crust. *Nature*, **300**, 589–94.

Appleton, J. D. 1972. Petrogenesis of potassium-rich lavas from the Roccamonfina volcano, Roman region, Italy. *J. Petr.*, **13**, 425–56.

Aramaki, S. & Ui, T. 1982. Japan, pp. 259–92 in Thorpe (1982).

Arculus, R. J. & Delano, J. W. 1981. Siderophile element abundances in the upper mantle: evidence for a sulphide signature and equilibrium with the core. *Geochim. Cosmochim. Acta*, **45**, 1331–43.

Arculus, R. J. & Johnson, R. W. 1978. Criticism of generalised models for the magmatic evolution of arc–trench systems. *Earth Planet. Sci. Lett.*, **39**. 118–26.

Arculus, R. J. & Wills, K. J. A. 1980. The petrology of plutonic blocks and inclusions from the Lesser Antilles Island Arc. *J. Petr.*, **21**, 743–99.

Armstrong, R. L., Taubeneck, W. H. & Hales, P. O. 1977. Rb–Sr and K–Ar geochronometry of Mesozoic granitic rocks and their Sr isotopic composition, Oregon, Washington, and Idaho. *Bull. Geol. Soc. Amer.*, **88**, 397–411.

Arndt, N. T. & Nisbet, E. G. 1982. What is komatiite? Pp. 19–27 in *Komatiites* (ed. N. T. Arndt & E. G. Nisbet), George Allen & Unwin, London.

Arth, J. G. & Hanson, G. N. 1975. Geochemistry and origin of the early Precambrian crust of northeastern Minnesota. *Geochim. Cosmochim. Acta*, **39**, 325–62.

Ashwal, L. D. 1982. Mineralogy of mafic and Fe–Ti oxide-rich differentiates of the Marcy anorthosite massif, Adirondacks, New York. *Amer. Min.*, **67**, 14–27.

Ashwal, L. D. 1993. *Anorthosites*. Springer-Verlag, Berlin, 422 pp.

Ashwal, L. D., Morrison, D. A., Phinney, W. C. & Wood, J. 1983. Origin of Archaean anothosites: evidence from the Bad Vermilion Lake anorthosite complex, Ontario. *Contr. Min. Petr.*, **82**, 259–73.

Avé Lallemant, H. G. 1967. Structural and petrofabric analysis of an 'Alpine-type' peridotite: the lherzolite of the French Pyrenees. *Leidse Geol. Meded.*, **42**, 1–57.

Bachinski, S. W. & Scott, R. B. 1979. Rare-earth and other trace element contents and the origin of minettes (mica-lamprophyres). *Geochim. Cosmochim. Acta*, **43**, 93–100.

Badham, J. P. N. & Morton, R. D. 1976. Magnetite–apatite intrusions and calc-alkaline magmatism, Camsell River, N.W.T. *Can. J. Earth Sci.*, **13**, 348–54.

Bailey, D. K. & Macdonald, R. 1987. Dry peralkaline felsic liquids and carbon dioxide flux through the Kenya rift zone, pp. 91–105 in *Magmatic processes: physicochemical principles* (ed. B. O. Mysen), Geochemical Society Spec. Publ. No. 1.

Bailey, E. B. & Anderson, E. M. 1925. The geology of Staffa, Iona, and western Mull. *Mem. Geol. Surv. Scotland*, 107 pp. Edinburgh, HMSO.

Bailey, E. B., Clough, C. T., Wright, W. B., Richey, J. E. & Wilson, G. V. 1924. Tertiary and post-Tertiary geology of Mull, Loch Aline, and Oban. *Mem. Geol. Surv. Scotland*, 445 pp. (with corresponding Geological Survey map on the scale of 1:63360). Edinburgh, HMSO.

Bailey, E. B. & Maufe, H. B. 1960. The geology of Ben Nevis and Glen Coe and the surrounding country (explanation of Sheet 53). 2nd (revised) edition. *Mem. Geol. Surv. Scotland*, 307 pp.

Bailey, E. H., Blake, M. C. & Jones, D. L. 1970. On-land Mesozoic oceanic crust in California Coast Ranges. *Prof. Paper U.S.G.S.*, **700–C**, 70–81.

Bailey, R. A., Dalrymple, G. B. & Lanphere, M. A. 1976. Volcanism, structure, and geochronology of Long Valley caldera, Mono County, California. *J. Geophys. Res.*, **81**, 725–44.

Baker, B. H. & McBirney, A. R. 1985. Liquid fractionation. Part III: Geochemistry of zoned magmas and the compositional effects of liquid fractionation. *J. Volc. Geotherm. Res.*, **24**, 55–81.

Baker, I. 1968. Intermediate oceanic volcanic rocks and the 'Daly Gap'. *Earth Planet. Sci. Lett.*, **4**, 103–6.

Baker, P. E., Gass, I. G., Harris, P. G. & Lemaitre, R. W. 1964. The volcanological report of the Royal Society expedition to Tristan da Cunha, 1962. *Phil. Trans. Roy. Soc. Lond.*, **256A**, 439–575.

Baldridge, W. S., McGetchin, T. R., Frey, F. A. & Jarosewich, E. 1973. Magmatic evolution of Hekla, Iceland. *Contr. Min. Petr.*, **42**, 245–58.

Ballard, R. D. & Moore, J. G. 1977. *Photographic atlas of the mid-Atlantic ridge rift valley*. Springer-Verlag, New York, 114 pp.

Ballhaus, C., Berry, R. F. & Green, D. H. 1991. High pressure experimental calibration of the olivine–orthopyroxene–spinel oxygen geobarometer: implications for the oxidation state of the upper mantle. *Contr. Min. Petr.*, **107**, 27–40.

Bamford, D. & Prodehl, C. 1977. Explosion seismology and the continental crust–mantle boundary. *J. Geol. Soc. London*, **134**, 139–51.

Barazangi, M. & Isacks, B. L. 1976. Spatial distribution of earthquakes and subduction of the Nazca plate beneath South America. *Geology*, **4**, 686–92.

Barbieri, F., Ferrara, G., Santacroce, R., Treuil, M. & Varet, J. 1975. A transitional basalt–pantellerite sequence of fractional crystallization, the Boina Centre (Afar Rift, Ethiopia). *J. Petr.*, **16**, 22–56.

Barker, D. S. 1977. Northern Trans-Pecos magmatic province: introduction and comparison with the Kenya rift. *Bull. Geol. Soc. Amer.*, **88**, 1421–7.

Barker, D. S., Long, L. E., Hoops, G. K. & Hodges, F. N. 1977. Petrology and Rb–Sr isotope geochemistry of intrusions in the Diablo Plateau, northern Trans-Pecos magmatic province, Texas and New Mexico. *Bull. Geol. Soc. Amer.*, **88**, 1437–46.

Barker, F., Wones, D. R., Sharp, W. N. & Desborough, G. A. 1975. The Pikes Peak batholith, Colorado Front Range, and a model for the origin of the gabbro–

anorthosite–syenite–potassic granite suite. *Precambrian Research*, **2**, 97–160.

Barreiro, B. 1983. Lead isotopic compositions of South Sandwich Island volcanic rocks and their bearing on magma genesis in intra-oceanic island arcs. *Geochim. Cosmochim. Acta*, **47**, 817–22.

Barrière, M. 1976. Flowage differentiation: limitation of the 'Bagnold' effect to the narrow intrusions. *Contr. Min. Petr.*, **55**, 139–45.

Barrière, M. 1981. On curved laminae, graded layers, convection currents and dynamic crystal sorting in the Ploumanac'h (Brittany) subalkaline granite. *Contr. Min. Petr.*, **77**, 214–24.

Bartels, K. S., Kinzler, R. J. & Grove, T. L. 1991. High pressure phase relations of primitive high-alumina basalts from Medicine Lake volcano, northern California. *Contr. Min. Petr.*, **108**, 253–70.

Barth, T. F. W. & Ramberg, I. B. 1966. The Fen circular complex, pp. 225–57 in Tuttle & Gittins (1966).

Bartlett, R. W. 1969. Magma convection, temperature distribution, and differentiation. *Amer. J. Sci.*, **267**, 1067–82.

Barton, H. M. & Doig, R. 1977. Sr-isotopic studies of the origin of the Morin anorthosite complex, Quebec, Canada. *Contr. Min. Petr.*, **61**, 219–30.

Barton, M. & Hamilton, D. L. 1982. Water-undersaturated melting experiments bearing upon the origin of potassium-rich magmas. *Min. Mag.*, **45**, 267–78.

Barton, M. & Sidle, W. C. 1994. Petrological and geochemical evidence for granitoid formation: the Waldoboro pluton complex, Maine. *J. Petr.*, **35**, 1241–74.

Basu, A. R. & Tatsumoto, M. 1979. Samarium–neodymium systematics in kimberlites and in the minerals of garnet–lherzolite inclusions. *Science*, **205**, 398–401.

Basu, A. R. & Tatsumoto, M. 1980. Nd-isotopes in selected mantle-derived rocks and minerals and their implications for mantle evolution. *Contr. Min. Petr.*, **75**, 43–54.

Bateman, R. 1985. Aureole deformation by flattening around a diapir during in situ ballooning: the Cannibal Creek granite. *J. Geol.*, **93**, 293–310.

Batiza, R. 1982. Abundances, distribution and sizes of volcanoes in the Pacific Ocean and implications for the origin of non-hotspot volcanoes. *Earth Planet. Sci. Lett.*, **60**, 195–206.

Baxter, A. N. 1978. Ultramafic and mafic nodule suites in shield-forming lavas from Mauritius. *J. Geol. Soc. London*, **135**, 565–81.

Bea, F. 1991. Geochemical modeling of low melt-fraction anatexis in a peraluminous system: the Peña Negra complex (central Spain). *Geochim. Cosmochim. Acta*, **55**, 1859–74.

Bea, F., Pereira, M. D., Corretgé, L. G. & Fershtater, G. B. 1994. Differentiation of strongly peraluminous, perphosphorus granites: the Pedrobernardo pluton, central Spain. *Geochim. Cosmochim. Acta*, **58**, 2609–27.

Beard, J. S. & Lofgren, G. E. 1991. Dehydration melting and water-saturated melting of basaltic and andesitic greenstones and amphibolites at 1, 3, and 6.9 kb. *J. Petr.*, **32**, 365–401.

Becker, H. 1993. Garnet peridotite and eclogite Sm–Nd mineral ages from the Lepontine dome (Swiss Alps): new evidence for Eocene high-pressure metamorphism in the central Alps. *Geology*, **21**, 599–602.

Beckinsale, R. D., Sanchez-Fernandez, A. W., Brook, M., Cobbing, E. J., Taylor, W. P. & Moore, N. D. 1985. Rb–Sr whole rock isochron and K-Ar age determinations for the Coastal Batholith of Peru, pp. 177–202 in Pitcher *et al.* (1985).

Berckhemer, H. 1969. Direct evidence for the composition of the lower crust and the Moho. *Tectonophysics*, **8**, 97–105.

Berg, J. H. 1977. Regional geobarometry in the contact aureoles of the anorthositic Nain complex, Labrador. *J. Petr.*, **18**, 399–430.

Berg, R. B. 1969. Petrology of anorthosites of the Bitterroot Range, Montana. *Mem. N.Y. State Mus. Sci. Serv.*, **18**, 387–98.

Bergstøl, S. 1972. The jacupirangite at Kodal, Vestfold, Norway. *Mineralium Deposita*, **7**, 233–46.

Bernstein, S. 1994. High-pressure fractionation in rift-related basaltic magmatism: Faeroe plateau basalts. *Geology*, **22**, 815–18.

Best, M. G. 1975. Migration of hydrous fluids in the upper mantle and potassium variation in calc-alkalic rocks. *Geology*, **3**, 429–32.

Best, M. G., Armstrong, R. L., Graustein, W. C., Embree, G. F. & Ahlborn, R. C. 1974. Mica granites of the Kern Mountains pluton, eastern White Pine County, Nevada: remobilized basement of the Cordilleran miogeosyncline? *Bull. Geol. Soc. Amer.*, **85**, 1277–86.

Bhattacharji, S. 1967. Mechanics of flow differentiation in ultramafic and mafic sills. *J. Geol.*, **75**, 101–12.

Bickford, M. E, Chase, R. B., Nelson, B. L., Shuster, R. D. & Arruda, E. C. 1981. U–Pb studies of zircon cores and overgrowths, and monazite: implications for age and petrogenesis of the northeastern Idaho batholith. *J. Geol.*, **89**, 433–57.

Billings, M. P. 1925. On the mechanics of dike intrusion. *J. Geol.*, **33**, 140–50.

Billings, M. P. 1945. Mechanics of igneous intrusion in New Hampshire. *Amer. J. Sci.*, **243A**, 40–68.

Blake, D. H. 1966. The net-veined complex of the Austurhorn intrusion, southeastern Iceland. *J. Geol.*, **74**, 891–907.

Blattner, P., Dietrich, V. & Gansser, A. 1983. Contrasting ^{18}O enrichment and origins of High Himalayan and Transhimalayan intrusives. *Earth Planet. Sci. Lett.*, **65**, 276–86.

Blattner, P. & Reid, F. 1982. The origin of lavas and ignimbrites of the Taupo volcanic zone, New Zealand, in the light of oxygen isotope data. *Geochim. Cosmochim. Acta*, **46**, 1417–29.

Blaxland, A. B. & Parsons, I. 1975. Age and origin of the Klokken gabbro-syenite intrusion, South Greenland: Rb–Sr study. *Bull. Geol. Soc. Denmark*, **24**, 27–32.

Blaxland, A. B. & Upton, B. G. 1978. Rare-earth distribution in the Tugtutoq younger giant dyke complex: evidence bearing on alkaline magma genesis in south Greenland. *Lithos*, **11**, 291–99.

Blaxland, A. B., Van Breemen, O. & Steenfelt, A. 1976. Age and origin of agpaitic magmatism at Ilimaussaq, south Greenland: Rb–Sr study. *Lithos*, **9**, 31–8.

Blaxland, A. B., Van Breemen, O., Emeleus, C. H. & Anderson, J. G. 1978. Age and origin of the major syenite centres in the Gardar province of south Greenland: Rb–Sr studies. *Bull. Geol. Soc. Amer.*, **89**, 231–44.

Bloch, S. & Bischoff, J. L. 1979. The effect of low-temperature alteration of basalt on the oceanic budget of potassium. *Geology*, **7**, 193–6.

Bloomfield, A. L. & Arculus, R. J. 1989. Magma mixing in the San Francisco volcanic field, AZ. Petrogenesis of the O'Leary Peak and Strawberry Crater volcanics. *Contr. Min. Petr.*, **102**, 429–53.

Blot, C. 1981. Deep root of andesitic volcanoes: new evidence of magma generation at depth in the Benioff zone. *J. Volc. Geothermal Res.*, **10**, 339–64.

Boctor, N. Z. & Boyd, F. R. 1981. Oxide minerals in a layered kimberlite–carbonatite sill from Benfontein, South Africa. *Contr. Min. Petr.*, **76**, 253–9.

Bodinier, J. L., Dupuy, C. & Dostal, J. 1988. Geochemistry and petrogenesis of Eastern Pyrenean peridotites. *Geochim. Cosmochim. Acta*, **52**, 2893–907.

Boettcher, A. L., Burnham, C. W., Windom, K. E. & Bohlen, S. R. 1982. Liquids, glasses, and the melting of silicates to high pressures. *J. Geol.*, **90**, 127–38.

Boettcher, A. L. & O'Neil, J. R. 1980. Stable isotope, chemical, and petrographic studies of high-pressure amphiboles and micas: evidence for metasomatism in the mantle source regions of alkali basalts and kimberlites. *Amer. J. Sci.*, **280A**, 594–621.

Bohlen, S. R., Boettcher, A. L. & Wall, V. J. 1982. The system albite–H_2O–CO_2: a model for melting and activities of water at high pressures. *Amer. Min.*, **67**, 451–62.

Bonatti, E. (ed.). 1988. Zargabad Island and the Red Sea Rift. *Tectonophysics*, **150**, 1–251.

Boone, G. M. 1962. Potassic feldspar enrichment in magma: origin of syenite in Deboullie district, northern Maine. *Bull. Geol. Soc. Amer.*, **73**, 1451–76.

Booth, B. 1979. Assessing volcanic risk. *J. Geol. Soc. Lond.*, **136**, 331–40.

Bottinga, Y. & Allègre, C. J. 1978. Partial melting under spreading ridges. *Phil. Trans. Roy. Soc. Lond. A*, **288**, 501–25.

Bottinga, Y. & Weill, D. F. 1970. Densities of liquid silicate systems calculated from partial molar volumes of oxide components. *Amer. J. Sci.*, **269**, 169–82.

Bottinga, Y. & Weill, D. F. 1972. The viscosity of magmatic silicate liquids: a model for calculation. *Amer. J. Sci.*, **272**, 438–75.

Bottinga, Y., Weill, D. F. & Richet, P. 1984. Density calculations for silicate liquids: reply to a critical comment by Ghiorso and Carmichael. *Geochim. Cosmochim. Acta*, **48**, 409–14.

Boudier, F. & Nicolas, A. (eds). 1988. The ophiolites of Oman. *Tectonophysics*, **151**, 1–401.

Bowden, P., Black, R., Martin, R. F., Ike, E. C., Kinnaird, J. A. & Batchelor, R. A. 1987. Niger–Nigerian alkaline ring complexes: a classic example of African Phanerozoic anorogenic mid-plate magmatism, pp. 357–79 in *Alkaline igneous rocks* (eds J. C. Fitton & B. G. J. Upton), Geological Society Spec. Publ. No. 30, Blackwell, Oxford.

Bowen, N. L. 1913. The melting phenomena of the plagioclase feldspars. *Amer. J. Sci.*, 4th Series, **35**, 577–99.

Bowen, N. L. 1921. Diffusion in silicate melts. *J. Geol.*, **29**, 295–317.

Bowen, N. L. 1928. *The evolution of the igneous rocks.* Princeton University Press (reprinted by Dover Publications, New York, 1956), 334 pp.

Bowen, N. L. & Andersen, O. 1914. The binary system MgO–SiO_2. *Amer. J. Sci.*, 4th series, **37**, 487–500.

Bowen, N. L. & Schairer, J. F. 1935. The system MgO–FeO–SiO_2. *Amer. J. Sci.*, 5th series, **29**, 151–217.

Bowen, N. L. & Schairer, J. F. 1938. Crystallization equilibrium in nepheline–albite–silica mixtures with fayalite. *J. Geol.*, **46**, 397–411.

Bowes, D. R & Wright, A. E. 1967. The explosion pipes near Kentallen, Scotland, and their geological setting. *Trans. Roy. Soc. Edinburgh*, **67**, 109–43.

Bowman, H. R., Asaro, F. & Perlman, I. 1973. On the uniformity of composition in obsidians and evidence for magmatic mixing. *J. Geol.*, **81**, 312–27.

Boyd, F. R. 1961. Welded tuffs and flows in the rhyolite plateau of Yellowstone Park, Wyoming. *Bull. Geol. Soc. Amer.*, **72**, 387–426.

Boyd, F. R. 1973. A pyroxene geotherm. *Geochim. Cosmochim. Acta*, **37**, 2533–46.

Boyd, F. R. 1987. Origin and structure of continental cratons. *Carnegie Inst. Wash. Yearbook*, **86**, 97–100.

Boyd, F. R. & Meyer, H. O. A. (eds). 1979a. Kimberlites, diatremes and diamonds: their geology, petrology and geochemistry. *Proc. 2nd Int. Kimberlite Conf.*, vol. 1. American Geophysical Union, Washington, DC, 400 pp.

Boyd, F. R. & Meyer, H. O. A. (eds). 1979b. The mantle sample: inclusions in kimberlites and other volcanics. *Proc. 2nd Int. Kimberlite Conf.*, vol. 2. American Geophysical Union, Washington, DC, 424 pp.

Boyd, F. R. & Nixon, P. H. 1973. Structure of the upper mantle beneath Lesotho. *Carnegie Inst. Wash. Yearbook*, **72**, 431–45.

Boyd, F. R. & Nixon, P. H. 1975. Origins of the ultramafic nodules from some kimberlites of northern Lesotho and the Monastery mine, South Africa. *Phys. Chem. Earth*, **9**, 431–54.

Boyd, F. R. & Nixon, P. H. 1979. Garnet lherzolite xenoliths from the kimberlites of East Griqualand, South Africa. *Carnegie Inst. Wash. Yearbook*, **78**, 488–92.

Bradley, J. 1965. Intrusion of major dolerite sills. *Trans. Roy. Soc. New Zealand (Geology)*, **3**, 27–55.

Brammall, A. & Harwood, H. F. 1932. The Dartmoor

granites: their genetic relationships. *Quart. J. Geol. Soc. London*, **88**, 171–237.

Branch, C. D. 1966. Volcanic cauldrons, ring complexes, and associated granites of the Georgetown inlier, Queensland. *Bull. Bur. Min. Resources, Geol. Geophysics., Australia*, **76**, 159 pp.

Brey, G. 1978. Origin of olivine melilitites – chemical and experimental constraints. *J. Volc. Geotherm. Res.*, **3**, 61–88.

Brey, G., Brice, W. R., Ellis. D. J., Green, D. H., Harris, K. L. & Ryabchikov, I. D. 1983. Pyroxene–carbonate reactions in the upper mantle. *Earth Planet. Sci. Lett.*, **62**, 63–74.

Brey, G. & Green, D. H. 1975. The role of CO_2 in the genesis of olivine melilitite. *Contr. Min. Petr.*, **49**, 93–103.

Brey, G. P. & Köhler, T. 1990. Geothermobarometry in four-phase lherzolites II. New thermobarometers, and practical assessment of existing thermobarometers. *J. Petr.*, **31**, 1353–78.

Bridgwater, D. & Harry, W. T. 1968. Anorthosite xenoliths and plagioclase megacrysts in Precambrian intrusions of South Greenland. *Meddelelser om Grønland*, **185**(2), 1–243.

Brooks, C. & Hart, S. R. 1978. Rb–Sr mantle isochrons and variations in the chemistry of Gondwanaland's lithosphere. *Nature*, **271**, 220–3.

Brooks, C., Hart, S. R., Hofmann, A. & James, D. E. 1976. Rb–Sr mantle isochrons from oceanic regions. *Earth Planet. Sci. Lett.*, **32**, 51–61.

Brooks, C. K. & Gill, R. C. O. 1982. Compositional variation in the pyroxenes and amphiboles of the Kangerdlugssuaq intrusion, east Greenland: further evidence for the crustal contamination of syenite magma. *Min. Mag.*, **45**, 1–9.

Brooks, C. K., Larsen, L. M. & Nielsen, T. F. D. 1991. Importance of iron-rich tholeiitic magmas at divergent plate margins: a reappraisal. *Geology*, **19**, 269–72.

Brown, G. M., Holland, J. G., Sigurdsson, H., Tomblin, J. F. & Arculus, R. J. 1977. Geochemistry of the Lesser Antilles volcanic island arc. *Geochim. Cosmochim. Acta*, **41**, 785–801.

Brown, P. E. 1971. The origin of the granitic sheets and veins in the Loch Coire migmatites, Scotland. *Min. Mag.*, **38**, 446–50.

Brown, W. L., Moreau, C. & Demaiffe, D. 1989. An anorthosite suite in a ring-complex: crystallization and emplacement of an anorogenic type from Abontorok, Aïr, Niger. *J. Petr.*, **30**, 1501–40.

Buddington, A. F. 1959. Granite emplacement with special reference to North America. *Bull. Geol. Soc. Amer.*, **70**, 671–747.

Buddington, A. F. 1972. Differentiation trends and parental magmas for anorthositic and quartz mangerite series, Adirondacks, New York. *Mem. Geol. Soc. Amer.*, **132**, 477–88.

Buddington, A. F. & Lindsley, D. H. 1964. Iron–titanium oxide minerals and synthetic equivalents. *J. Petr.*, **5**, 310–57.

Burnham, C. W. 1959. Contact metamorphism of magnesian limestones at Crestmore, California. *Bull. Geol. Soc. Amer.*, **70**, 879–920.

Burnham, C. W. 1967. Hydrothermal fluids at the magmatic stage. In *Geochemistry of hydrothermal ore deposits* (ed. H. L. Barnes), Holt, Rinehart & Winston, New York, 670 pp.

Burnham, C. W. 1979. Magmas and hydrothermal fluids, pp. 71–136 in *Geochemistry of hydrothermal ore deposits* (2nd edn) (ed. H. L. Barnes). Wiley-Interscience, New York.

Burnham, C. W. & Davis, N. F. 1971. The role of H_2O in silicate melts. I. P–V–T relations in the system $NaAlSi_3O_8$–H_2O to 10 kilobars and 1000 °C. *Amer. J. Sci.*, **270**, 54–79.

Burnham, C. W., Holloway, J. R. & Davis, N. F. 1969. The specific volume of water in the range 1000 to 8900 bars, 20° to 900 °C. *Amer. J. Sci.*, **267A**, 70–95.

Burnham, C. W. & Jahns, R. H. 1962. A method for determining the solubility of water in silicate melts. *Amer. J. Sci.*, **260**, 721–45.

Busrewil, M. T., Pankhurst, R. J. & Wadsworth, W. J. 1973. The igneous rocks of the Boganclogh area, N.E. Scotland. *Scot. J. Geol.*, **9**, 165–76.

Bussell, M. A. 1985. The centred complex of the Rio Huaura: a study of magma mixing and differentiation in high-level magma chambers, pp. 128–55 in Pitcher *et al.* (1985).

Butler, B. C. M. 1961. Metamorphism and metasomatism of rocks of the Moine Series by a dolerite plug in Glenmore, Ardnamurchan. *Min. Mag.*, **32**. 866–97.

Button, A. 1976. Stratigraphy and the relations of the Bushveld floor in the Eastern Transvaal. *Trans. Geol. Soc. S. Africa*, **79**, 3–12.

Byerly, G. 1980. The nature of differentiation trends in some volcanic rocks from the Galapagos spreading center. *J. Geophys. Res.*, **85**, 3797–810.

Cameron, E. N., Jahns, R. H., McNair, A. H. & Page, L. R. 1949. Internal structure of granitic pegmatites. *Econ. Geol. Monograph*, **2**, 115 pp. Economic Geology Publishing Co., Urbana, Illinois.

Cameron, M., Sueno, S., Prewitt, C. T. & Papike, J. J. 1973. High-temperature crystal chemistry of acmite, diopside, hedenbergite, jadeite, spodumene, and ureyite. *Amer. Min.*, **58**, 594–618.

Campbell, I. H. & Griffiths, R. W. 1992. The changing nature of mantle hotspots through time: implications for the chemical evolution of the mantle. *J. Geol.*, **92**, 497–523.

Campbell, I. H., Griffiths, R. W. & Hill, R. I. 1989. Melting in an Archaean mantle plume: heads it's basalts, tails it's komatiites. *Nature*, **339**, 697–9.

Campbell, I. H., Roeder, P. L. & Dixon, J. M. 1978. Plagioclase buoyancy in basaltic liquids as determined with a centrifuge furnace. *Contr. Min. Petr.*, **67**, 369–77.

Campbell, I. H. & Turner, J. S. 1989. Fountains in magma chambers. *J. Petr.*, **30**, 885–923.

Carden, J. R. & Laughlin, A. W. 1974. Petrochemical variations within the McCartys basalt flow, Valencia county, New Mexico. *Bull. Geol. Soc. Amer.*, **85**, 1479–84.

Carlson, R. W. & Irving, A. J. 1994. Depletion and enrichment history of subcontinental lithospheric mantle: an Os, Sr, Nd and Pb isotopic study of ultramafic xenoliths from the northwestern Wyoming Craton. *Earth Planet. Sci. Lett.*, **126**, 457–72.

Carlson, R. W., Lugmair, G. W. & Macdougall, J. D. 1981. Columbia River volcanism: the question of mantle heterogeneity or crustal contamination. *Geochim. Cosmochim. Acta*, **45**, 2483–99.

Carlson, W. D. & Lindsley, D. H. 1988. Thermochemistry of pyroxenes on the join $Mg_2Si_2O_6$–$CaMgSi_2O_6$. *Amer. Min.*, **73**, 242–52.

Carmichael, I. S. E. 1964. The petrology of Thingmuli, a Tertiary volcano in eastern Iceland. *J. Petr.*, **5**, 435–60.

Carmichael, I. S. E. 1991. The redox state of basic and silicic magmas: a reflection of their source regions? *Contr. Min. Petr.*, **106**, 129–41.

Carmichael, I. S. E., Nicholls, J., Spera, F. J., Wood, B. J. & Nelson, S. A. 1977. High temperature properties of silicate liquids: applications to the equilibration and ascent of basic magma. *Phil. Trans. Roy. Soc. A*, **286**, 373–431.

Carmichael, I. S. E., Turner, F. J. & Verhoogen, J. 1974. *Igneous petrology*. McGraw Hill, New York, 739 pp.

Carr, M. J. 1984. Symmetrical and segmented variation of physical and geochemical characteristics of the Central American volcanic front. *J. Volc. Geothermal. Res.*, **20**, 231–52.

Carr, M. J., Rose, W. I. & Mayfield, D. G. 1979. Potassium content of lavas and depth to the seismic zone in Central America. *J. Volc. Geothermal Res.*, **5**, 387–401.

Carrigan, C. R. & Cygan, R. T. 1986. Implications of magma chamber dynamics for Soret-related fractionation. *J. Geophys. Res.*, **91**, 11451–61.

Carswell, D. A. & Gibb, F. G. F. 1980. The equilibration conditions and petrogenesis of European crustal garnet lherzolites. *Lithos*, **13**, 19–29.

Carswell, D. A. & Gibb, F. G. F. 1987a. Evaluation of mineral thermometers and barometers applicable to garnet lherzolite assemblages. *Contr. Min. Petr.*, **95**, 499–511.

Carswell, D. A. & Gibb, F. G. F. 1987b. Garnet lherzolite xenoliths in the kimberlites of northern Lesotho: revised P–T equilibration conditions and upper mantle geotherm. *Contr. Min. Petr.*, **97**, 473–87.

Castillo, P. 1988. The Dupal anomaly as a trace of the upwelling lower mantle. *Nature*, **336**, 667–70.

Cawthorn, R. G. 1975. The amphibole peridotite–metagabbro complex, Finero, Northern Italy. *J. Geol.*, **83**, 437–54.

Chacko, T., Creaser, R. A. & Poon, D. 1994. Spinel + quartz granites and associated metasedimentary enclaves from the Taltson magmatic zone, Alberta, Canada: a view into the root zone of a high-temperature S-type granitic batholith. *Min. Mag.*, **58A**, 161–2.

Chapman, C. A. 1968. A comparison of the Maine coastal plutons and the magmatic central complexes of New Hampshire, pp. 385–96 in *Studies of Appalachian geology: northern and maritime* (eds E. A. Zen, W. S. White, J. B. Hadley & J. B. Thompson Jr.), Interscience Publishers (John Wiley), New York, 475 pp.

Chappell, B. W. & White, A. J. R. 1974. Two contrasting granite types. *Pacific Geol.*, **8**, 173–4.

Chappell, B. W. & White, A. J. R. 1992. I- and S-type granites in the Lachlan fold belt. *Trans. Roy. Soc. Edinburgh, Earth Sciences*, **83**, 1–26.

Chappell, B. W., White, A. J. R. & Wyborn, D. 1987. The importance of residual source material (restite) in granite petrogenesis. *J. Petr.*, **28**, 1111–38.

Chase, C. G. 1981. Oceanic island Pb: two-stage histories and mantle evolution. *Earth Planet. Sci. Lett.*, **52**, 277–84.

Chatterjee, N. D., Leistner, H., Terhart, L., Abraham, K. & Klaska, R. 1982. Thermodynamic mixing properties of corundum-eskolaite, α-$(Al,Cr^{+3})_2O_3$, crystalline solutions at high temperatures and pressures. *Amer. Min.*, **67**, 725–35.

Chayes, F. 1963. Relative abundance of intermediate members of the oceanic basalt–trachyte association. *J. Geophys. Res.*, **68**, 1519–34.

Chayes, F. 1977. The oceanic basalt–trachyte relation in general and in the Canary Islands. *Amer. Min.*, **62**, 666–71.

Chidester, A. H. 1968. Evolution of the ultramafic complexes of northwestern New England, pp. 343–54 in *Studies of Appalachian Geology, Northern and Maritime* (eds E.A. Zen *et al.*), Interscience Publishers (John Wiley), New York, 475 pp.

Chidester, A. H. & Cady, W. M. 1972. Origin and emplacement of Alpine-type ultramafic rocks. *Nature Phys. Sci.*, **240**, 27–31.

Christensen, N. I. & Salisbury, M. H. 1975. Structure and constitution of the lower oceanic crust. *Rev. Geophys. Space Phys.*, **13**, 57–86.

Christiansen, R. L. & Lipman, P. W. 1972. Cenozoic volcanism and plate-tectonic evolution of the western United States. II. Late Cenozoic. *Phil. Trans. Roy. Soc. Lond.*, **271A**, 249–84.

Church, W. R. & Riccio, L. 1977. Fractionation trends in the Bay of Islands ophiolite of Newfoundland: polycyclic cumulate sequences in ophiolites and their classification. *Can. J. Earth Sci.*, **14**, 1156–65.

Clague, D. A. 1987. Hawaiian alkaline volcanism, pp. 227–52 in *Alkaline igneous rocks* (eds J. G. Fitton & B. G. J. Upton), Geol. Soc. Spec. Publ. No. 30, Blackwell Scientific Publications, Oxford.

Clague, D. A. & Frey, F. A. 1982. Petrology and trace element geochemistry of the Honolulu volcanics, Oahu: implications for the oceanic mantle below Hawaii. *J. Petr.*, **23**, 447–504.

Clague, D. A. & Straley, P. F. 1977. Petrologic nature of the oceanic Moho. *Geology*, **5**, 133–6.

Class, C., Goldstein, S. L., Galer, J. G. & Weis, D. 1993.

Young formation age of a mantle plume source. *Nature*, **362**, 715–21.

Clifford, T. N., Stumpfl, E. F. & McIver, J. R. 1975. A sapphirine–cordierite–bronzite–phlogopite paragenesis from Namaqualand, South Africa. *Min. Mag.*, **40**, 347–56.

Clough, C. T., Maufe, H. B. & Bailey, E. B. 1909. The cauldron-subsidence of Glencoe, and the associated igneous phenomena. *Quart. J. Geol. Soc. London*, **65**, 611–78.

Cobbing, E. J. & Pitcher, W. S. 1972. The coastal batholith of central Peru. *J. Geol Soc. London*, **128**, 421–60.

Cobbing, E. J., Pitcher, W. S., Wilson, J. J., Baldock, J. W., Taylor, W. P., McCourt, W. & Snelling, N. J. 1981. The geology of the Western Cordillera of northern Peru. *Overseas Memoir Inst. Geol. Sci.*, (London), **5**, 143 pp.

Coffin, M. F. & Eldholm, O. 1993. Scratching the surface: estimating dimensions of large igneous provinces. *Geology*, **21**, 515–18.

Cole, J. W. 1970. Structure and eruptive history of the Tarawera volcanic complex. *New Zealand J. Geol. Geophys.*, **13**. 879–902.

Cole, J. W. 1981. Genesis of lavas of the Taupo volcanic zone, North Island, New Zealand. *J. Volc. Geothermal Res.*, **10**, 317–37.

Cole, J. W., Graham, I. J. & Gibson, I. L. 1990. Magmatic evolution of late Cenozoic volcanic rocks of the Lau Ridge, Fiji. *Contr. Min. Petr.*, **104**, 540–54.

Coleman, R. G. 1980. Tectonic inclusions in serpentinites. *Archives des Sciences* (Geneva), **33**, 89–102.

Compton, R. R. 1955. Trondhjemite batholith near Bidwell Bar, California. *Bull. Geol. Soc. Amer.*, **66**, 9–44.

Condomines, M., Grönvold, K., Hooker, P. J., Muehlenbachs, K., O'Nions, R. K., Oskarsson, N. & Oxburgh, E. R. 1983. Helium, oxygen, strontium and neodymium isotopic relationships in Icelandic volcanics. *Earth Planet. Sci. Lett.*, **66**, 125–36.

Conrad, W. K. & Kay, R. W. 1984. Ultramafic and mafic inclusions from Adak Island: crystallization history, and implications for the nature of primary magmas and crustal evolution in the Aleutian arc. *J. Petr.*, **25**, 88–125.

Conrad, W. K., Nicholls, I. A. & Wall, V. J. 1988. Water-saturated and -undersaturated melting of metaluminous and peraluminous crustal compositions at 10 kb: evidence for the origin of silicic magmas in the Taupo Volcanic Zone, New Zealand, and other occurrences. *J. Petr.*, **29**, 765–803.

Conticelli, S. & Peccerillo, A. 1992. Petrology and geochemistry of potassic and ultrapotassic volcanism in central Italy: petrogenesis and inferences on the evolution of the mantle sources. *Lithos*, **28**, 221–40.

Coombs, D. S., Landis, C. A., Norris, R. J., Sinton, J. M., Borns, D. J. & Craw, D. 1976. The Dun Mountain ophiolite belt, New Zealand, its tectonic setting, constitution, and origin, with special reference to the southern portion. *Amer. J. Sci.*, **276**, 561–603.

Cooper, A. F. & Reid, D. L. 1991. Textural evidence for calcite carbonatite magmas, Dicker Willem, southwest Namibia. *Geology*, **19**, 1193–6.

Corry, C. E. 1988. Laccoliths; mechanics of emplacement and growth. *Geol. Soc. Amer. Spec. Paper*, **220**, 110 pp.

Cortini, M. & Hermes, O. D. 1981. Sr isotopic evidence for a multi-source origin of the potassic magmas in the Neapolitan area (S. Italy). *Contr. Min. Petr.*, **77**, 47–55.

Coryell, C. D., Chase, J. W. & Winchester, J. W. 1963. A procedure for the geochemical interpretation of terrestrial rare-earth abundance patterns. *J. Geophys. Res.*, **68**, 559–66.

Costesèque, P. & Schott, J. 1976. Fractionnement isotopique et thermodiffusion en milieu poreux pour quelques éléments important en géochimie. *C. R. Acad. Sci., Paris, Série D*, **282**, 1913–6.

Cox, K. G. 1980. A model for flood basalt vulcanism. *J. Petr.*, **21**, 629–50.

Cox, K. G., Bell, J. D. & Pankhurst, R. J. 1979. *The interpretation of igneous rocks*. George Allen & Unwin, London, 450 pp.

Cox, K. G., Johnson, R. L., Monkman, L. J., Stillman, C. J., Vail, J. R. & Wood, D. N. 1965. The geology of the Nuanetsi Igneous Province. *Phil. Trans. Roy. Soc. Lond. A*, **257**, 71–218.

Curtis, G. H. 1968. The stratigraphy of the ejectamenta of the 1912 eruption of Mt. Katmai and Novarupta, Alaska. *Mem. Geol. Soc. Amer.*, **116**, 153–210.

Czamanske, G. K. & Moore, J. G. 1977. Composition and phase chemistry of sulphide globules in basalt from the Mid-Atlantic Ridge rift valley near 37°N lat. *Bull. Geol. Soc. Amer.*, **88**, 587–99.

Dalton, J. A. & Wood, B. J. 1993. The compositions of primary carbonate melts and their evolution through wall-rock reaction in the mantle. *Earth Planet. Sci. Lett.*, **119**, 511–25.

Daly, R. A., Manger, G. E. & Clark, S. P. 1966. Density of rocks. *Mem. Geol. Soc. Amer.*, **97**, 19–26.

Danchin, R. V. & Boyd, F. R. 1976. Ultramafic nodules from the Premier kimberlite pipe, South Africa. *Carnegie Inst. Wash. Yearbook*, **75**, 531–8.

Dandurand, J. L., Fortuné, J. P., Pérami, R., Schott, J. & Tollon, F. 1972. On the importance of mechanical action and thermal gradient in the formation of metal-bearing deposits. *Mineralium Deposita*, **7**, 339–50.

Daniels, L. R. M. & Gurney, J. J. 1991. Oxygen fugacity constraints on the Southern African lithosphere. *Contr. Min Petr.*, **108**, 154–61.

Dasch, E. J. 1969. Strontium isotopes in weathering profiles, deep-sea sediments, and sedimentary rocks. *Geochim. Cosmochim. Acta*, **33**, 1521–52.

Davidson, C. F. 1964. On diamantiferous diatremes. *Econ. Geol.*, **59**, 1368–80.

Davidson, C. F. 1967. The kimberlites of the U.S.S.R., pp. 251–61 in Wyllie (1967).

Davies, G. F. & Richards, M. A. 1992. Mantle convection. *J. Geol.*, **100**, 151–206.

Davies, G. R., Nixon, P. H., Pearson, D. G. & Obata, M. 1993. Tectonic implications of graphitized diamonds from the Ronda peridotite massif, southern Spain. *Geology*, **21**, 471–4.

Davis, B. T. C. & Schairer, J. F. 1965. Melting relations in the join diopside–forsterite–pyrope at 40 kilobars and at one atmosphere. *Carnegie Inst. Wash. Yearbook*, **64**, 123–6.

Dawson, J. B. 1966. Oldoinyo Lengai – an active volcano with sodium carbonatite lava flows, pp. 155–68 in Tuttle & Gittins (1966).

Dawson, J. B. 1980. *Kimberlites and their xenoliths*. Springer-Verlag, Berlin, Heidelberg and New York. 252 pp.

Dawson, J. B. 1987. The kimberlite clan: relationship with olivine and leucite lamproites, and inferences for upper-mantle metasomatism, pp. 95–101 in *Alkaline igneous rocks* (eds J. G. Fitton & B. G. J. Upton), Geol. Soc. Spec. Publ. **30**, Blackwell Scientific Publications, Oxford.

Dawson, J. B. & Hawthorne, J. B. 1973. Magmatic sedimentation and carbonatitic differentiation in kimberlite sills at Benfontein, South Africa. *J. Geol. Soc. Lond.*, **129**, 61–84.

Dawson, J. B. & Smith, J. V. 1988. Metasomatised and veined upper-mantle xenoliths from Pello Hill, Tanzania: evidence for anomalously light mantle beneath the Tanzanian sector of the East African Rift Valley. *Contr. Min. Petr.*, **100**, 510–27.

Dawson, J. B., Pinkerton, H., Norton, G. E. & Pyle, D. M. 1990. Physicochemical properties of alkali carbonatite lavas: data from the 1988 eruption of Oldoinyo Lengai, Tanzania. *Geology*, **18**, 260–3.

Dawson, J. B., Pinkerton, H., Pyle, D. M. & Nyamweru, C. 1994. June 1993 eruption of Oldoinyo Lengai, Tanzania: exceptionally viscous and large cabonatite lava flows and evidence for coexisting silicate and carbonate magmas. *Geology*, **22**, 799–802.

De, A. 1974. Silicate liquid immiscibility in the Deccan traps and its petrogenetic significance. *Bull. Geol. Soc. Amer.*, **85**, 471–4.

De Albuquerque, C. A. R. 1979. Origin of the plutonic mafic rocks of southern Nova Scotia. *Bull. Geol. Soc. Amer.*, **90**, 719–31.

Deans, T. & Roberts, B. 1984. Carbonatite tuffs and lava clasts of the Tinderet foothills, western Kenya: a study of calcified natrocarbonatites. *J. Geol. Soc. London*, **141**, 563–80.

Defant, M. J. & Drummond, M. S. 1993. Mount St Helens: potential example of the partial melting of the subducted lithosphere in a volcanic arc. *Geology*, **21**, 547–50.

Delany, J. M. & Helgeson, H. C. 1978. Calculation of the thermodynamic consequences of dehydration in subducting oceanic crust to 100 kb and >800 °C. *Amer. J. Sci.*, **278**, 638–86.

Demaiffe, D. & Hertogen, J. 1981. Rare earth element geochemistry and strontium isotopic composition of a massif-type anorthositic charnockitic body: the Hidra massif (Rogaland, SW Norway). *Geochim. Cosmochim. Acta*, **45**, 1545–61.

Demant, A., Lestrade, P., Lubala, R. T., Kampunzu, A. B. & Durieux, J. 1994. Volcanological and petrological evolution of Nyiragongo volcano, Virunga volcanic field, Zaire. *Bull. Volc.*, **56**, 47–61.

De Paolo, D. 1981a. A neodymium and strontium isotopic study of the Mesozoic calc-alkaline granitic batholiths of the Sierra Nevada and Peninsular Ranges, California. *J. Geophys. Res.*, **86**, 10470–88.

De Paolo, D. J. 1981b. Trace element and isotopic effects of combined wallrock assimilation and fractional crystallization. *Earth Planet. Sci. Lett.*, **53**, 189–202.

De Paolo, D. 1983. Comment on 'Columbia River volcanism: the question of mantle heterogeneity or crustal contamination' by R. W. Carlson, G. W. Lugmair and J. D. Macdougall. *Geochim. Cosmochim. Acta*, **47**, 841–4.

De Paolo, D. J. 1985. Isotopic studies of processes in mafic magma chambers: I. The Kiglapait intrusion, Labrador. *J. Petr.*, **26**, 925–51.

De Paolo, D. J. & Johnson, R. W. 1979. Magma genesis in the New Britain island arc: constraints from Nd and Sr isotopes and trace element patterns. *Contr. Min. Petr.*, **70**, 367–79.

De Paolo, D. J. & Wasserburg, G. J. 1979a. Petrogenetic mixing models and Nd-Sr isotopic patterns. *Geochim. Cosmochim. Acta*, **43**, 615–27.

De Paolo, D. J. & Wasserburg, G. J. 1979b. Sm–Nd age of the Stillwater complex and the mantle evolution curve for neodymium. *Geochim. Cosmochim. Acta*, **43**, 999–1008.

De Paolo, D. J., Perry, F. V. & Baldridge, W. S. 1992. Crustal versus mantle sources of granitic magmas: a two-parameter model based on Nd isotopic studies. *Trans. Roy. Soc. Edinburgh, Earth Sciences*, **83**, 439–46.

De Vore, G. W. 1975. The role of partial melting of metasediments in the formation of the anorthosite–norite–syenite complex, Laramie Range, Wyoming. *J. Geol.*, **83**, 749–62.

De Waard, D. 1969. The anorthosite problem: the problem of the anorthosite–charnockite suite of rocks. *Mem. N. Y. State Mus. Sci. Serv.*, **18**, 71–90.

De Waard, D. 1970. The anorthosite–charnockite suite of rocks of Roaring Brook Valley in the Eastern Adirondacks (Marcy Massif). *Amer. Min.*, **55**, 2063–75.

De Waard, D. 1976. Anorthosite–adamellite–troctolite layering in the Barth Island structure of the Nain complex, Labrador. *Lithos*, **9**, 293–308.

De Waard, D. & Romey, W. D. 1969. Petrogenetic relationships in the anorthosite–charnockite series of Snowy Mountain Dome, South Central Adirondacks. *Mem. N. Y. State Mus. Sci. Serv.*, **18**, 307–15.

De Wit, M., Dutch, S., Kligfield, R., Allen, R. & Stern, C. 1977. Deformation, serpentinization and emplacement of a dunite complex, Gibbs Island, South Shetland Islands: possible fracture zone tectonics. *J. Geol.*, **85**, 745–62.

Dickenson, M. P. & Hess, P. C. 1986. The structural role and homogeneous redox equilibria of iron in peraluminous, metaluminous and peralkaline silicate melts. *Contr. Min. Petr.*, **92**, 207–17.

Dickin, A. P. 1981. Isotope geochemistry of Tertiary igneous rocks from the Isle of Skye, N.W. Scotland. *J. Petr.*, **22**, 155–89.

Dickinson, W. R. 1966. Table Mountain serpentinite extrusion in California Coast Ranges. *Bull. Geol. Soc. Amer.*, **77**, 451–72.

Dickinson, W. R. 1970. Relations of andesites, granites, and derivative sandstones to arc-trench tectonics. *Rev. Geophys. Space Sci.*, **8**, 813–60.

Dickinson, W. R. 1975. Potash–depth (K–h) relations in continental margin and intra-oceanic magmatic arcs. *Geology*, **3**, 53–6.

Didier, J. 1973. *Granites and their enclaves: the bearing of enclaves on the origin of granites.* Elsevier, Amsterdam, 393 pp.

Didier, J. & Barbarin, B. (eds). 1991. *Enclaves and granite petrology.* Elsevier, Amsterdam.

Dietz, R. S. 1964. Sudbury structure as an astrobleme. *J. Geol.*, **72**, 412–34.

Dimroth, E. 1970. Meimechites and carbonatites of the Castignon Lake complex, New Quebec. *Neues Jahrb. Min. Abh.*, **112**, 239–78.

Dingwell, D. B. 1987. Melt viscosities in the system $NaAlSi_3O_8$–H_2O–F_2O_{-1}, pp. 423–31 in *Magmatic processes: physicochemical principles* (ed. B. O. Mysen), Geochemical Society, Special Publication no. 1.

Dinham, C. H. & Haldane, D. 1932. The economic geology of the Stirling and Clackmannan Coalfield. *Mem. Geol. Surv. Scotland*, 242 pp.

Distler, V. V., Pertsev, N. N. & Boronikhin, V.A. 1983. Sulfide petrology of basalts from Deep Sea Drilling Project holes 504B and 505B. *Initial Rept. D.S.D.P.*, **69**, 607–17.

Dixon, S. & Rutherford, M. J. 1979. Plagiogranites as late-stage immiscible liquids in ophiolite and mid-ocean ridge suites: an experimental study. *Earth Planet. Sci. Lett.*, **45**, 45–60.

Dodge, F. C. W. & Calk, L. C. 1978. Fusion of granodiorite by basalt, central Sierra Nevada. *J. Res. U.S.G.S.*, **6**, 459–65.

Doe, B. R. & Delevaux, M. H. 1973. Variations in lead-isotope compositions in Mesozoic granitic rocks of California: a preliminary investigation. *Bull. Geol. Soc. Amer.*, **84**, 3513–26.

Doe, B. R., Leeman, W. P., Christiansen, R. L. & Hedge, C. E. 1982. Lead and strontium isotopes and related trace elements as genetic tracers in the Upper Cenozoic rhyolite–basalt association of the Yellowstone Plateau volcanic field. *J. Geophys. Res.*, **87**, 4785–806.

Doe, B. R. & Zartman, R. E. 1979. Plumbotectonics: the Phanerozoic, pp. 22–70 in *Geochemistry of hydrothermal ore deposits* (2nd edn) (ed. H. L. Barnes), Wiley-Interscience, New York.

Doell, R. R., Dalrymple, G. B., Smith, R. L. & Bailey, R. A. 1968. Paleomagnetism, potassium–argon ages, and geology of rhyolites and associated rocks of the Valles caldera, New Mexico. *Mem. Geol. Soc. Amer.*, **116**, 211–48.

Donaldson, C. H. 1977. Laboratory duplication of comb layering in the Rhum pluton. *Min. Mag.*, **41**, 323–36.

Drake, M. J. & Weill, D. F. 1975. Partition of Sr, Ba, Ca, Y, Eu^{2+}, Eu^{3+}, and other REE between plagioclase feldspar and magmatic liquid: an experimental study. *Geochim. Cosmochim. Acta*, **39**, 689–712.

Drewes, H., Fraser, G. D., Snyder, G. L. & Barnett, H. F. 1961. Geology of Unalaska Island and adjacent insular shelf, Aleutian Islands, Alaska. *Bull. U.S.G.S.*, **1028-S**, 583–676.

Duchesne, J. C. 1972. Iron–titanium oxide minerals in the Bjerkrem–Sogndal massif, south-western Norway. *J. Petr.*, **13**, 57–81.

Duchesne, J. C. & Demaiffe, D. 1978. Trace elements and anorthosite genesis. *Earth Planet. Sci. Lett.*, **38**, 249–72.

Duchesne, J. C., Denoiseux, B. & Hertogen, J. 1987. The norite–mangerite relationships in the Bjerkreim–Sokndal layered lopolith (Southwest Norway). *Lithos*, **20**, 1–17.

Duchesne, J. C., Wilmart, E., Demaiffe, D. & Hertogen, J. 1989. Monzonorites from Rogaland (Southwest Norway): a series of rocks coeval but not comagmatic with massif-type anorthosites. *PreCambrian Res.*, **45**, 111–28.

Duncan, R. A. 1981. Hotspots in the southern Oceans – an absolute frame of reference for the motion of the Gondwana continents. *Tectonophysics*, **74**, 29–42.

Dungan, M. A. & Rhodes, J. M. 1978. Residual glasses and melt inclusions in basalts from DSDP Legs 45 and 46: evidence for magma mixing. *Contr. Min. Petr.*, **67**, 417–31.

Dupré, B. & Allègre, C. J. 1980. Pb–Sr–Nd isotopic correlation and the chemistry of the North Atlantic mantle. *Nature*, **286**, 17–22.

Dupré, B. & Allègre, C. J. 1983. Pb–Sr isotope variation in Indian Ocean basalts and mixing phenomena. *Nature*, **303**, 142–6.

Du Toit, A. L. 1920. The Karroo dolerites of South Africa: a study in hypabyssal injection. *Trans. Geol. Soc. S. Africa*, **23**, 1–42.

Eales, H. V. & Booth, P. W. K. 1974. The Birds River gabbro complex, Dordrecht district. *Trans. Geol. Soc. S. Africa*, **77**, 1–15.

Ebadi, A. & Johannes, W. 1991. Beginning of melting and composition of first melts in the system Qz–Ab–Or–H_2O–CO_2. *Contr. Min. Petr.*, **106**, 286–95.

Edgar, A. D. & Charbonneau, H. E. 1993. Melting experiments on a SiO_2-poor, CaO-rich aphanitic kimberlite from 5–10 GPa and their bearing on sources of kimberlite magmas. *Amer. Min.*, **78**, 132–42.

Edwards, C. B. & Howkins, J. B. 1966. Kimberlites in Tanganyika with special reference to the Mwadui occurrence. *Econ. Geol.*, **61**, 537–54.

Edwards, G. R. 1992. Mantle decarbonation and Archaean high-Mg magmas. *Geology*, **20**, 899–902.

Eggler, D. H. 1974. Effect of CO_2 on the melting of peridotite. *Carnegie Inst. Wash. Yearbook*, **73**, 215–24.

Eggler, D. H. 1978. The effect of CO_2 upon partial melting of peridotite in the system $Na_2O-CaO-Al_2O_3-MgO-SiO_2-CO_2$ to 35 kb, with an analysis of melting in a peridotite–H_2O-CO_2 system. *Amer. J. Sci.*, **278**, 305–43.

Eggler, D. H. & Mysen, B. O. 1976. The role of CO_2 in the genesis of olivine melilitite: discussion. *Contr. Min. Petr.*, **55**, 231–6 (and reply by G. P. Brey & D. H. Green on pp. 237–9).

Eggler, D. H. & Wendlandt, R. F. 1979. Experimental studies on the relationship between kimberlite magmas and partial melting of peridotite, pp. 330–8 in Boyd & Meyer (1979a).

Egorov, L. S. 1970. Carbonatites and ultrabasic–alkaline rocks of the Maimecha–Kotui region, N. Siberia. *Lithos*, **3**, 341–59.

Ehlers, C. 1974. Layering in rapakivi granite, S.W. Finland. *Bull. Geol. Soc. Finland*, **46**, 145–9.

Eichelberger, J. C. 1975. Origin of andesite and dacite: evidence of mixing at Glass Mountain in California and at other circum-Pacific volcanoes. *Bull. Geol. Soc. Amer.*, **86**, 1381–91.

Ellam, R. M. & Hawkesworth, C. J. 1988. Elemental and isotopic variations in subduction related basalts: evidence for a three component model. *Contr. Min. Petr.*, **98**, 72–80.

Elthon, D. 1979. High magnesia liquids as the parental magma for ocean floor liquids. *Nature*, **278**, 514–18.

Emslie, R. F. 1965. The Michikamau anorthositic intrusion, Labrador. *Can. J. Earth Sci.*, **2**, 385–99.

Emslie, R. F. 1969. Crystallization and differentiation of the Michikamau intrusion. *Mem. N.Y. State Mus. Sci. Serv.*, **18**, 163–73.

Emslie, R. F. 1970. The geology of the Michikamau intrusion, Labrador (13L, 231). *Geol. Surv. Can. Paper*, **68-57**, 85 pp.

Emslie, R. F. 1971. Liquidus relations and subsolidus reactions in some plagioclase-bearing systems. *Carnegie Inst. Wash. Yearbook*, **69**, 148–55.

Emslie, R. F. 1978. Anorthosite massifs, rapakivi granites, and late Proterozoic rifting of North America. *Precambrian Research*, **7**, 61–98.

Emslie, R. F. 1980. Geology and petrology of the Harp Lake complex, central Labrador: an example of Elsonian magmatism. *Bull. Geol. Surv. Canada*, **293**, 136 pp.

Engel, A. E. J., Engel, C. G. & Havens, R. G. 1965. Chemical characteristics of oceanic basalts and the upper mantle. *Bull. Geol. Soc. Amer.*, **76**, 719–34.

England, R. W. 1992. The genesis, ascent, and emplacement of the Northern Arran granite, Scotland: implications for granitic diapirism. *Bull. Geol. Soc. Amer.*, **104**, 606–14.

Epstein, S. & Taylor, H. P. 1967. Variation of O^{18}/O^{16} in minerals and rocks, in *Researches in Geochemistry*, vol. 2 (ed. P. H. Abelson), pp. 29–62. John Wiley, New York, 663 pp.

Eriksson, S. C. 1989. Phalaborwa: a saga of magmatism, metasomatism and miscibility, in *Carbonatites: genesis and evolution* (ed. K. Bell), pp 221–54. Unwin-Hyman, London.

Ernst, W. G. 1976. *Petrologic phase equilibria*. W. H. Freeman & Co., San Francisco, 333 pp.

Ernst, W. G. 1978. Petrochemical study of lherzolitic rocks from the western Alps. *J. Petr.*, **19**, 341–92.

Ernst, W. G. 1981. Petrogenesis of eclogites and peridotites from the Western and Ligurian Alps. *Amer. Min.*, **66**, 443–72.

Evans, B. W. & Moore, J. G. 1968. Mineralogy as a function of depth in the prehistoric Makaopuhi tholeiitic lava lake, Hawaii. *Contr. Min. Petr.*, **17**, 85–115.

Ewart, A., Bryan, W. B. & Gill, J. B. 1973. Mineralogy and geochemistry of the younger volcanic islands of Tonga, S.W. Pacific. *J. Petr.*, **14**, 429–65.

Ewart, A. & Hawkesworth, C. J. 1987. The Pleistocene-Recent Tonga–Kermadec arc lavas: interpretation of new isotopic and rare earth data in terms of a depleted mantle source model. *J. Petr.*, **28**, 495–530.

Ewart, A. & Stipp, J. J. 1968. Petrogenesis of the volcanic rocks of the Central North Island, New Zealand, as indicated by a study of Sr^{87}/Sr^{86} ratios, and Sr, Rb, K, U and Th abundances. *Geochim. Cosmochim. Acta*, **32**, 699–736.

Exley, R. A., Sills, J. D. & Smith, J. V. 1982. Geochemistry of micas from the Finero spinel–lherzolite, Italian Alps. *Contr. Min. Petr.*, **81**, 59–63.

Falloon, T. J. & Green, D. H. 1989. The solidus of carbonated fertile peridotite. *Earth Planet. Sci. Lett.*, **94**, 364–70.

Falloon, T. J. & Green, D. H. 1990. Solidus of carbonated fertile peridotite under fluid-saturated conditions. *Geology*, **18**, 195–9.

Faure, G. 1986. *Principles of isotope geology* (2nd edn). John Wiley, New York.

Faure, G., Bowman, J. R., Elliott, D. H. & Jones, L. M. 1974. Strontium isotope composition and petrogenesis of the Kirkpatrick basalt, Queen Alexandra Range, Antarctica. *Contr. Min. Petr.*, **48**, 153–69.

Faure, G. & Powell, J. L. 1972. *Strontium isotope geology*. Springer-Verlag, Berlin, Heidelberg and New York, 188 pp.

Feigenson, M. D. & Spera, F. J. 1981. Dynamical model for temporal variation in magma type and eruption interval at Kohala volcano, Hawaii. *Geology*, **9**, 531–3.

Fenner, C. N. 1938. Contact relations between rhyolite and basalt on Gardiner River, Yellowstone Park. *Bull. Geol. Soc. Amer.*, **49**, 1441–84.

Ferguson, J. 1964. Geology of the Ilimaussaq alkaline intrusion, south Greenland. *Medd. Grønland*, **172**(4), 1–82.

Ferguson, J. 1970. The differentiation of agpaitic magmas: the Ilimaussaq intrusion, south Greenland. *Can. Min.*, **10**, 335–49.

Ferguson, J. & Currie, K. L. 1971. Evidence of liquid immiscibility in alkaline ultrabasic dikes at Callander Bay, Ontario. *J. Petr.*, **12**, 561–85.

Ferrara, G., Preite-Martinez, M., Taylor, H. P., Tonarini, S. & Turi, B. 1986. Evidence for crustal assimilation, mixing of magmas, and a ^{87}Sr-rich upper mantle. *Contr. Min. Petr.*, **92**, 269–80.

Finnerty, A. A. & Boyd, F.R. 1984. Evaluation of thermobarameters for garnet peridotites. *Geochim. Cosmochim. Acta*, **48**, 15–27.

Finnerty, A. A. & Boyd, F. R. 1992. Reply to comment by J. Ganguly on 'Evaluation of thermobarometers for garnet peridotites'. *Geochim. Cosmochim. Acta.* **56**, 843–4.

Fisher, R. V. & Schmincke, H. U. 1984. *Pyroclastic rocks*. Springer-Verlag, Berlin, 472 pp.

Fisher, R. V., Smith, A. L. & Roobol, M. J. 1980. Destruction of St. Pierre, Martinique, by ash-cloud surges, May 8 and 20, 1902. *Geology*, **8**, 472–6.

Fitton, J. G. 1987. The Cameroon line, West Africa: a comparison between oceanic and continental alkaline volcanism, pp. 273–91 in *Alkaline igneous rocks* (eds J. G. Fitton & B. G. J. Upton), Geol. Soc. Spec. Publ. **30**, Blackwell Scientific Publications, Oxford.

Flett, J. S. 1946. Geology of the Lizard and Meneage (explanation of sheet 359). *Mem. Geol. Surv. Gt. Britain*, 208 pp.

Flower, M. F. J. 1981. Thermal and kinematic control on ocean-ridge magma fractionation: contrasts between Atlantic and Pacific spreading axes. *J. Geol. Soc. Lond.*, **138**, 695–712.

Floyd, P. A. 1989. Geochemical features of intraplate oceanic plateau basalts, pp. 215–30 in *Magmatism in the ocean basins* (eds A. D. Saunders & M. J. Norry), Geol. Soc. Spec. Publ. **42**, Blackwell, Oxford.

Fodor, R. V., Corwin, C. & Sial, A. N. 1985. Crustal signatures in the Serra Geral flood-basalt province, southern Brazil: O- and Sr- isotope evidence. *Geology*, **13**, 763–5.

Foland, K. A. & Friedman, I. 1977. Application of Sr and O isotope relations to the petrogenesis of the alkaline rocks of the Red Hill complex, New Hampshire, U.S.A. *Contr. Min. Petr.*, **65**, 213–25.

Foland, K. A. & Henderson, C. M. B. 1976. Application of age and Sr isotope data to the petrogenesis of the Marangudzi ring complex, Rhodesia. *Earth Planet. Sci. Lett.*, **29**, 291–301.

Foland, K. A., Landoll, J. D., Henderson, C. M. B. & Jiangfeng, C. 1993. Formation of cogenetic quartz and nepheline syenites. *Geochim. Cosmochim. Acta.* **57**, 697–704.

Foley, S. 1992. Petrological characterization of the source components of potassic magmas: geochemical and experimental constraints. *Lithos*, **28**, 187–204.

Forester, R. W. & Taylor, H. P. 1977. ^{18}O/^{16}O, D/H, and ^{13}C/^{12}C studies of the Tertiary igneous complex of Skye, Scotland. *Amer. J. Sci.*, **277**, 136–77.

Fornari, D. J., Malahoff, A. & Heezen, B. C. 1978. Volcanic structure of the crest of the Puna Ridge, Hawaii: geophysical implications of submarine volcanic terrain. *Bull. Geol. Soc. Amer.*, **89**, 605–16.

Fowler, C. M. R. 1976. Crustal structure of the Mid-Atlantic Ridge crest at 37 °N. *Geophys. J. Roy. Astr. Soc.*, **47**, 459–91.

Francis, D. 1991. Some implications of xenolith glasses for the mantle sources of alkaline mafic magmas. *Contr. Min. Petr.*, **108**, 175–80.

Francis, D. & Ludden, J. 1990. The mantle source for olivine nephelinite, basanite, and alkaline olivine basalt at Fort Selkirk, Yukon, Canada. *J. Petr.*, **31**, 371–400.

Francis, E. H. 1982. Magma and sediment – I. Emplacement mechanism of late Carboniferous tholeiite sills in northern Britain. *J. Geol. Soc. London*, **139**, 1–20.

Freestone, I. C. & Hamilton, D. L. 1980. The role of liquid immiscibility in the genesis of carbonatites – an experimental study. *Contr. Min. Petr.*, **73**, 105–17.

French, W. J. 1966. Appinitic intrusions clustered around the Ardara pluton, county Donegal. *Proc. Roy. Irish Acad.*, **64B**, 303–22.

Freundt, A. & Schmincke, H. U. 1992. Mixing of rhyolite, trachyte and basalt magma erupted from a vertically and laterally zoned reservoir, composite flow P1, Gran Canaria. *Contr. Min. Petr.*, **112**, 1–19.

Frey, F. A., Green, D. H. & Roy, S. D. 1978. Integrated models of basalt petrogenesis: a study of quartz–tholeiites to olivine–melilitites from South-Eastern Australia utilizing geochemical and experimental petrological data. *J. Petr.*, **19**, 463–513.

Frisch, W. & Keusen, H. 1977. Gardiner intrusion, an ultramafic complex at Kangerdlugssuaq, East Greenland. *Bull. Grønlands Geol. Unders.*, **122**, 1–62.

Frith, R. A. & Currie, K. L. 1976. A model for the origin of the Lac St. Jean anorthosite massif. *Can. J. Earth Sci.*, **13**, 389–99.

Frost, R. B. 1985. On the stability of sulfides, oxides, and native metals in serpentinite. *J. Petr.*, **26**, 31–63.

Fujii, T., Kushiro, I., Nakamura, Y. & Koyaguchi, T. 1980. A note of silicate liquid immiscibility in Japanese volcanic rocks. *J. Geol. Soc. Japan*, **86**, 409–12.

Furman, T., Frey, F. A. & Meyer, P. S. 1992. Petrogenesis of evolved basalts and rhyolites at Austurhorn, southeastern Iceland: the role of fractional crystallization. *J. Petr.*, **33**, 1405–45.

Gamble, J. A. 1979. Some relationships between co-existing granitic and basaltic magmas and the genesis of hybrid rocks in the Tertiary central complex of Slieve Gullion, northeast Ireland. *J. Volc. Geotherm. Res.*, **5**, 297–316.

Garson, M. S. 1966. Carbonatites in Malawi, pp. 33–71 in Tuttle & Gittins (1966).

Gasparik, T. 1984. Two-pyroxene thermobarometry with new experimental data in the system CaO–MgO–Al$_2$O$_3$–SiO$_2$. *Contr. Min. Petr.*, **87**, 87–97.

Gass, I. G. 1980. The Troodos massif: its role in the unravelling of the ophiolite problem and its significance in the understanding of constructive plate margin processes, pp. 23–35 in Panayiotou (1980).

Gass, I. G. & Thorpe, R. S. 1976. Geological map of the Central Igneous Complex, Isle of Skye, Scotland. In *Igneous case study – The Tertiary Igneous rocks of Skye,*

NW Scotland (eds S. A. Drury *et al.*). Open University Press, Milton Keynes, 88 pp.

Gast, P. W. 1968. Trace element fractionation and the origin of tholeiitic and alkaline magma types. *Geochim. Cosmochim. Acta*, **32**, 1057–86.

Gerlach, D. C. & Grove, T. L. 1982. Petrology of Medicine Lake Highland volcanics: characterization of end members of magma mixing. *Contr. Min. Petr.*, **80**, 147–59.

Gerlach, D. C., Frey, F. A., Moreno-Roa, H. & Lopez-Escobar, L. 1988. Recent volcanism in the Puyehue-Cordon Caulle region, southern Andes, Chile (40.5 °S): petrogenesis of evolved lavas. *J. Petr.*, **29**, 333–82.

Gerlach, T. M. 1982. Interpretation of volcanic gas data from tholeiitic and alkaline mafic lavas. *Bull. Volc.*, **45**, 235–44.

Ghiorso, M. S. & Carmichael, I. S. E. 1985. Chemical mass transfer in magmatic processes. II. Applications in equilibrium crystallization, fractionation and assimilation. *Contr. Min. Petr.*, **90**, 121–41.

Ghiorso, M. S. & Sack, R. O. 1991. Fe–Ti oxide geothermometry: thermodynamic formulation and the estimation of intensive variables in silicic magmas. *Contr. Min. Petr.*, **108**, 485–510.

Gibson, I. L. & Walker, G. P. L. 1963. Some composite rhyolite/basalt lavas and related composite dykes in eastern Iceland. *Proc. Geol. Assoc.*, **74**, 301–18.

Giese, P. 1968. Die Struktur der Erdkruste im Bereich der Ivrea-Zone. *Schweiz. Min. Petr. Mitt.*, **48**, 261–84.

Gill, J. & Condomines, M. 1992. Short-lived radioactivity and magma genesis. *Science*, **257**, 1368–76.

Gill, J. B. 1981. *Orogenic andesites and plate tectonics*. Springer-Verlag, Berlin, 390 pp.

Gill, R. C. O. 1973. Mechanism for the salic magma bias of continental alkaline provinces. *Nature Phys. Sci.*, **242**, 41–2.

Gill, R. C. O., Pedersen, A. K. & Larsen, J. G. 1992. Tertiary picrites in West Greenland: melting at the periphery of a plume? Pp. 335–48 in *Magmatism and the causes of continental break-up* (eds B. C. Storey, J. Alabaster & R. J. Pankhurst), Geol. Soc. Spec. Publ. **68**. The Geological Society, London.

Gittins, J., Hewins, R. H. & Laurin, A. F. 1975. Kimberlitic–carbonatitic dikes of the Saguenay River valley, Quebec, Canada. *Phys. Chem. Earth*, **9**, 137–48.

Glazner, A. F. 1994. Foundering of mafic plutons and density stratification of continental crust. *Geology*, **22**, 435–8.

Gledhill, A. & Baker, P. E. 1973. Strontium isotope ratios in volcanic rocks from the South Sandwich Islands. *Earth Planet. Sci. Lett.*, **19**, 369–72.

Glennie, K. W., Boeuf, M. G. A., Hughes Clarke, M. W., Moody-Stuart, M., Pilaar, W. F. H. & Reinhardt, B. M. 1974. Geology of the Oman Mountains. *Verh. Kon. Nederlands geol. mijn. Genoots.*, **31**, 1–423.

Gold, D. P. 1967. Alkaline ultrabasic rocks in the Montreal area, Quebec, pp. 288–302 in Wyllie (1967).

Goldschmidt, V. M. 1937. The principles of distribution of chemical elements in minerals and rocks. *J. Chem. Soc. (Lond.)*, 655–73.

Govindaraju, K. 1989. 1989 compilation of working values and sample description for 272 geostandards. *Geostandards Newsletter*, **13** (Special Issue), 1–113.

Grant, J. A. 1968. Partial melting of common rocks as a possible source of cordierite–anthophyllite bearing assemblages. *Amer. J. Sci.*, **266**, 908–31.

Grant, N. K., Powell, J. L., Burkholder, F. R., Walther, J. V. & Coleman, M. L. 1976. The isotopic composition of strontium and oxygen in lavas from St. Helena, south Atlantic. *Earth Planet. Sci. Lett.*, **31**, 209–23.

Green, D. H. 1964. The petrogenesis of the high-temperature peridotite intrusion in the Lizard area, Cornwall. *J. Petr.*, **5**, 134–88.

Green, D. H. 1972. Archaean greenstone belts may include terrestrial equivalents of lunar maria? *Earth Planet. Sci. Lett.*, **15**, 263–70.

Green, D. H. 1973. Experimental melting studies on a model upper mantle composition at high pressure under water-saturated and water-undersaturated conditions. *Earth Planet. Sci. Lett.*, **19**, 37–53.

Green, D. H., Nicholls, I. A., Viljoen, M. & Viljoen, R. 1975. Experimental determination of the existence of peridotite liquids in earliest Archaean magmatism. *Geology*, **3**, 11–14.

Green, D. H. & Ringwood, A. E. 1967. The genesis of basaltic magmas. *Contr. Min. Petr.*, **15**, 103–90.

Green, T. H. 1969. Experimental fractional crystallization of quartz diorite and its application to the problem of anorthosite origin. *Mem. N.Y. State Mus. Sci. Serv.*, **18**, 23–9.

Green, T. H. 1976. Experimental generation of cordierite- or garnet-bearing granitic liquids from a pelitic composition. *Geology*, **4**, 85–8.

Green, T. H. 1982. Anatexis of mafic crust and high pressure crystallization of andesite, pp. 465–87 in Thorpe (1982).

Green, T. H. & Ringwood, A. E. 1968. Genesis of the calc-alkaline igneous rock suite. *Contr. Min. Petr.*, **18**, 105–62.

Greenland, L. P., Rose, W. I. & Stokes, J. B. 1985. An estimate of gas emissions and magmatic gas content from Kilauea volcano. *Geochim. Cosmochim. Acta*, **49**, 125–9.

Gregory, R. T. & Taylor, H. P. 1981. An oxygen isotope profile in a section of Cretaceous oceanic crust, Samail ophiolite, Oman: evidence for $\delta^{18}O$ buffering of the oceans by deep (>5 km) seawater-hydrothermal circulation at mid-ocean ridges. *J. Geophys. Res.*, **86**, 2737–55.

Greig, J. W. 1927. Immiscibility in silicate melts. *Amer. J. Sci.*, 5th Series, **13**, 1–44.

Greig, J. W. & Barth, T. F. W. 1938. The system, $Na_2O.Al_2O_3.2SiO_2$ (nephelite, carnegieite)–$Na_2O.Al_2O_3.6SiO_2$ (albite). *Amer. J. Sci.*, 5th Series, **35A**, 93–112.

Gribble, C. D. 1967. The basic intrusive rocks of Caledonian age of the Haddo House and Arnage districts, Aberdeenshire. *Scot. J. Geol.*, **3**, 125–36.

Gribble, C. D. 1968. The cordierite-bearing rocks of the

Haddo House and Arnage districts, Aberdeenshire. *Contr. Min. Petr.*, **17**, 315–30.

Gribble, C. D. & O'Hara, M. J. 1967. Interaction of basic magma with pelitic materials. *Nature*, **214**, 1198–201.

Griffin, B. J. & Varne, R. 1980. The Macquarie Island ophiolite complex: mid-Tertiary oceanic lithosphere from a major ocean basin. *Chem. Geol.*, **30**, 285–308.

Griffin, W. L., Heier, K. S., Taylor, P. N. & Weigand, P. W. 1974. General geology, age and chemistry of the Raftsund mangerite intrusion, Lofoten-Vesteralen. *Norges Geol. Unders*, **312**, 1–30.

Groome, D. R. & Hall, A. 1974. The geochemistry of the Devonian lavas of the northern Lorne Plateau, Scotland. *Min. Mag.*, **39**, 621–40.

Grout, F. F. 1937. Criteria of origin of inclusions in plutonic rocks. *Bull. Geol. Soc. Amer.*, **48**, 1521–71.

Grove, T. L. & Donnelly-Nolan, J. M. 1986. The evolution of young silicic lavas at Medicine Lake Volcano, California: implications for the origin of compositional gaps in calc-alkaline series lavas. *Contr. Min. Petr.*, **92**, 281–302.

Grove, T. L. & Kinzler, R. J. 1986. Petrogenesis of andesites. *Ann. Rev. Earth Planet. Sci.*, **14**, 417–54.

Grove, T. L., Kinzler, R. J., Baker, M. B., Donnelly-Nolan, J. M. & Lesher, C. E. 1988. Assimilation of granite by basaltic magma at Burnt Lava flow, Medicine Lake volcano, northern California: decoupling of heat and mass transfer. *Contr. Min. Petr.*, **99**, 320–43.

Gupta, A. K., Green, D. H. & Taylor, W. R. 1987. The liquidus surface of the system forsterite–nepheline–silica at 28 kb. *Amer. J. Sci.*, **287**, 560–5.

Gurney, J. J. & Ebrahim, S. 1973. Chemical composition of Lesotho kimberlites, pp. 280–4 in *Lesotho kimberlites* (ed. P. H. Nixon), Lesotho National Development Corporation, Maseru.

Gurney, J. J. & Harte, B. 1980. Chemical variations in upper mantle nodules from South African kimberlites. *Phil. Trans. Roy. Soc. Lond. A*, **297**, 273–93.

Gust, D. A. & Arculus, R. J. 1986. Petrogenesis of alkalic and calcalkalic volcanic rocks of Mormon Mountain volcanic field, Arizona. *Contr. Min. Petr.*, **94**, 416–26.

Haggerty, S. E. & Sautter, V. 1990. Ultradeep (greater than 300 kilometers) ultramafic upper mantle xenoliths. *Science*, **248**, 993–6.

Häkli, T. A. 1968. An attempt to apply the Makaopuhi nickel fractionation data to the temperature determination of a basic intrusive. *Geochim. Cosmochim. Acta*, **32**, 449–60.

Häkli, T. A. & Wright, T. L. 1967. The fractionation of nickel between olivine and augite as a geothermometer. *Geochim. Cosmochim. Acta*, **31**, 877–84.

Hall, A. 1965. On a granite–metadolerite contact at Curran Hill, County Donegal, Ireland. *Geol. Mag.*, **102**, 531–7.

Hall, A. 1966a. The Ardara pluton: a study of the chemistry and crystallization of a contaminated granite intrusion. *Proc. Roy. Irish Acad.*, **65B**, 203–35.

Hall, A. 1966b. A petrogenetic study of the Rosses granite complex, Donegal. *J. Petr.*, **7**, 202–20.

Hall, A. 1967. The chemistry of appinitic rocks associated with the Ardara pluton, Donegal, Ireland. *Contr. Min. Petr.*, **16**, 156–71.

Hall, A. 1971. The relationship between geothermal gradient and the composition of granitic magmas in orogenic belts. *Contr. Min. Petr.*, **32**, 186–92.

Hall, A. 1972a. Regional geochemical variation in the Caledonian and Variscan granites of western Europe. *Proc. 24th Internat. Geol. Congr., Montreal*, **2**, 171–80.

Hall, A. 1972b. New data on the composition of Caledonian granites. *Min. Mag.*, **38**, 847–62.

Hall, A. 1973a. Géochimie des granites Varisques du sud-ouest de l'Angleterre. *Bull. Soc. Geol. France* (7), **15**, 229–38.

Hall, A. 1973b. Geothermal control of granite compositions in the Variscan orogenic belt. *Nature Phys. Sci.*, **242**, 72–5.

Hall, A. 1982. The Pendennis peralkaline minette. *Min. Mag.*, **45**, 257–66.

Hall, A. 1987. The ammonium content of Caledonian granites. *J. Geol. Soc. London*, **144**, 671–4.

Hall, A. 1988. The distribution of ammonium in the granites of south-west England. *J. Geol. Soc. London*, **145**, 37–41.

Hall, A. 1990. Geochemistry of the Cornubian tin province. *Mineralium Deposita*, **25**, 1–6.

Hall, A., Bencini, A. & Poli, G. 1991. Magmatic and hydrothermal ammonium in the granites of the Tuscan magmatic province, Italy. *Geochim. Cosmochim. Acta*, **55**, 3657–64.

Hall, A., Jarvis, K. E. & Walsh, J. N. 1993. The variation of cesium and 37 other elements in the Sardinian granite batholith, and the significance of cesium for granite petrogenesis. *Contr. Min. Petr.*, **114**, 160–70.

Hall, A., Pereira, M. D. & Bea, F. 1995. The abundance of ammonium in the granites of central Spain, and the behaviour of the ammonium ion during anatexis and fractional crystallization. *Mineralogy and Petrology* (in press).

Hall, A., Stamatakis, M. G. & Walsh, J. N. 1994. Ammonium enrichment associated with diagenetic alteration in Tertiary pyroclastic rocks from Greece. *Chem. Geol.*, **118**, 173–83.

Hall, A. L. 1932. The Bushveld igneous complex of the central Transvaal. *Mem. Geol. Surv. S. Africa*, **28**, 560 pp.

Hamilton, D. L., Burnham, C. W. & Osborn, E. F. 1964. The solubility of water and effects of oxygen fugacity and water content on crystallization in mafic magmas. *J. Petr.*, **5**, 21–39.

Hamilton, D. L. & Mackenzie, W. S. 1965. Phase-equilibrium studies in the system $NaAlSiO_4$ (nepheline)–$KAlSiO_4$ (kalsilite)–SiO_2–H_2O. *Min. Mag.*, **34**, 214–31.

Hamilton, J. 1977. Sr isotope and trace element studies of the Great Dyke and Bushveld mafic phase and their relations to Early Proterozoic magma genesis in southern Africa. *J. Petr.*, **18**, 24–52.

Hamilton, W. 1965. Diabase sheets of the Taylor glacier region, Victoria Land, Antarctica. *Prof. Paper. U.S.G.S.*, *456B*, 71 pp.

Hamilton, W. & Myers, W. B. 1967. The nature of batholiths. *Prof. Paper U.S.G.S.*, *554-C*, 1–30.

Hanekom, H. J., Van Staden, C. M. V. H., Smit, P. J. & Pike, D. R. 1965. The geology of the Palabora igneous complex. *Mem. Geol. Surv. S. Africa*, **54**, 185 pp.

Hanuš, V. & Vaněk, J. 1978. Morphology of the Andean Wadati–Benioff zone, andesitic volcanism, and tectonic features of the Nazca plate. *Tectonophysics*, **44**, 65–77.

Hanuš, V. & Vaněk, J. 1979. Morphology and volcanism of the Wadati–Benioff zone in the Tonga–Kermadec system of recent subduction. *New Zealand J. Geol. Geophys.*, **22**, 659–71.

Hargraves, R. B. (ed.). 1980. *Physics of magmatic processes*. Princeton University Press, Princeton, 585 pp.

Harker, A. 1904. The Tertiary igneous rocks of Skye. *Mem. Geol. Surv. U.K.*, 481 pp. (with corresponding Geological Survey maps on the scale of 1:63360 and 1:50000).

Harley, S. L. 1984. The solubility of alumina in orthopyroxene coexisting with garnet in $FeO–MgO–Al_2O_3–SiO_2$ and $CaO–FeO–MgO–Al_2O_3–SiO_2$. *J. Petr.*, **25**, 665–96.

Harmon, R. S. 1983. Oxygen and strontium isotopic evidence regarding the role of continental crust in the origin and evolution of the British Caledonian granites, pp. 62–79 in *Migmatites, melting and metamorphism* (eds M. P. Atherton & C. D. Gribble), Shiva Publishing, Nantwich, 326 pp.

Harmon, R. S. & Gerbe, M. C. 1992. The 1982–1983 eruption at Galunggung volcano, Java (Indonesia): oxygen isotope geochemistry of a chemically zoned magma chamber. *J. Petr.*, **33**, 585–609.

Harmon, R. S. and 8 co-authors. 1984. Regional O-, Sr-, and Pb- isotope relationships in late Cenozoic calc-alkaline lavas of the Andean Cordillera. *J. Geol. Soc. London*, **141**, 803–22.

Harper, G. D. 1980. The Josephine ophiolite – remains of a late Jurassic marginal basin in northwestern California. *Geology*, **8**, 333–7.

Harris, C., Bell, J. D. & Atkins, F. B. 1982. Isotopic composition of lead and strontium in lavas and coarse-grained blocks from Ascension Island, South Atlantic. *Earth Planet. Sci. Lett.*, **60**, 79–85.

Harris, D. M. & Anderson, A. T. 1983. Concentrations, sources, and losses of H_2O, CO_2, and S in Kilauean basalt. *Geochim. Cosmochim. Acta*, **47**, 1139–50.

Harris, D. M. & Anderson, A. T. 1984. Volatiles H_2O, CO_2, and Cl in a subduction related basalt. *Contr. Min. Petr.*, **87**, 120–8.

Harris, P. G. 1957. Zone refining and the origin of potassic basalts. *Geochim. Cosmochim. Acta*, **12**, 195–208.

Harris, P. G. 1962. Increase of temperature in ascending basalt magma. *Amer. J. Sci.*, **260**, 783–6.

Harris, P. G. & Middlemost, E. A. K. 1970. The evolution of kimberlites. *Lithos*, **3**, 77–88.

Harrison, T. M., Aleinikoff, J. N. & Compston, W. 1987. Observations and controls on the occurrence of inherited zircon in Concord-type granitoids, New Hampshire. *Geochim. Cosmochim. Acta.*, **51**, 2549–58.

Hart, R. 1970. Chemical exchange between sea water and deep ocean basalts. *Earth Planet. Sci. Lett.*, **9**, 269–79.

Harte, B. 1978. Kimberlite nodules, upper mantle petrology, and geotherms. *Phil. Trans. Roy. Soc. Lond. A*, **288**, 487–500.

Haslam, H. W. 1971. Andalusite in the Mullach nan Coirean granite, Inverness-shire. *Geol. Mag.*, **108**, 97–102.

Haughton, D. R., Roeder, P. L. & Skinner, B. J. 1974. Solubility of sulfur in mafic magmas. *Econ. Geol.*, **69**, 451–67.

Hauri, E. H. & Hart, S. R. 1993. Re–Os isotope systematics of HIMU and EMII oceanic island basalts from the south Pacific Ocean. *Earth Planet. Sci. Lett.*, **114**, 353–71.

Hausen, D. M. 1954. Welded tuffs of Oregon and Idaho. *J. Miss. Acad. Sci.*, **5**, 209–20.

Hawkesworth, C. J. & Gallagher, K. 1993. Mantle hotspots, plumes and regional tectonics as causes of intraplate magmatism. *Terra Nova*, **5**, 552–9.

Hawkesworth, C. J. & Vollmer, R. 1979. Crustal contamination versus enriched mantle: $^{143}Nd/^{144}Nd$ and $^{87}Sr/^{86}Sr$ evidence from the Italian volcanics. *Contr. Min. Petr.*, **69**, 151–65.

Hawley, J. E. 1965. Upside-down zoning at Frood, Sudbury, Ontario. *Econ. Geol.*, **60**, 529–75.

Hawthorne, J. B. 1975. Model of a kimberlite pipe. *Phys. Chem. Earth*, **9**, 1–15.

Hay, R. L. 1983. Natrocarbonatite tephra of Kerimasi volcano, Tanzania. *Geology*, **11**, 599–602.

Hay, R. L. & O'Neil, J. R. 1983. Carbonatite tuffs in the Laetolil Beds of Tanzania and the Kaiserstuhl in Germany. *Contr. Min. Petr.*, **82**, 403–6.

Heath, S. A. & Fairbairn, H. W. 1969. Sr^{87}/Sr^{86} ratios in anorthosites and some associated rocks. *Mem. N.Y. State Mus. Sci. Serv.*, **18**, 99–110.

Heezen, B. C. & Fornari, D. J. 1978. Geological map of the Pacific Ocean 1/35,000,000. Sheet 20 of the Geological World Atlas, UNESCO, Paris.

Heier, K. S. & Thoresen, K. 1971. Geochemistry of high grade metamorphic rocks, Lofoten-Vesteralen, North Norway. *Geochim. Cosmochim. Acta*, **35**, 89–99.

Helz, R. T. 1976. Phase relations of basalts in their melting range at $P_{H_2O} = 5$ kb. Part II. Melt compositions. *J. Petr.*, **17**, 139–93.

Helz, R. T. 1980. Crystallization history of Kilauea Iki lava lake as seen in drill core recovered in 1967–1979. *Bull. Volc.*, **43**, 675–701.

Helz, R. T. 1987. Differentiation behaviour of Kilauea Iki lava lake, Kilauea volcano, Hawaii: an overview of past and current work, pp. 241–58 in *Magmatic processes: physicochemical principles* (ed. B. O. Mysen), Geochemical Society, Special Publication no. 1.

Helz, R. T., Kirschenbaum, H. & Marinenko, J. W. 1989. Diapiric transfer of melt in Kilauea Iki lava lake,

Hawaii: a quick, efficient process of igneous differentiation. *Bull. Geol. Soc. Amer.*, **101**, 578–94.

Henderson, P. 1982. *Inorganic geochemistry*. Pergamon Press, Oxford, 353 pp.

Hergt, J. M., Chappell, B. W., McCulloch, M. T., McDougall, I. & Chivas, A. R. 1989. Geochemical and isotopic constraints on the origin of the Jurassic dolerites of Tasmania. *J. Petr.*, **30**, 841–83.

Hermes, O. D. & Cornell, W. C. 1981. Quenched crystal mush and associated magma compositions as indicated by intercumulus glasses from Mt. Vesuvius, Italy. *J. Volc. Geotherm. Res.*, **9**, 133–49.

Herzberg, C. T. & O'Hara, M. J. 1985. Origin of mantle peridotite and komatiite by partial melting. *Geophys. Res. Letters*, **12**, 541–4.

Hess, H. H. & Poldervaart, A. (eds). 1967–68. *Basalts: the Poldervaart treatise on rocks of basaltic composition.* Wiley-Interscience, New York, 2 volumes (Vol. 1 – 1967, Vol. 2 – 1968).

Hess, P. C. 1992. Phase equilibria constraints on the origin of ocean floor basalts, pp. 67–102 in *Mantle flow and melt generation at mid-ocean ridges* (eds J. P. Morgan, D. K. Blackman & J. M. Sinton). American Geophysical Union, Geophysical Monograph 71.

Hildreth, W. 1981. Gradients in silicic magma chambers: implications for lithospheric magmatism. *J. Geophys. Res.*, **86**, 10153–92.

Hill, D. P., Bailey, R. A. & Ryall, A. S. 1985. Active tectonic and magmatic processes beneath Long Valley caldera, eastern California: an overview. *J. Geophys. Res.*, **90**, 11111–20.

Hilton, D. R., Hoogewerff, J. A., Van Bergen, M. J. & Hammerschmidt, K. 1992. Mapping magma sources in the east Sunda–Banda arcs, Indonesia: constraints from helium isotopes. *Geochim. Cosmochim. Acta*, **56**, 851–9.

Hodgson, F. D. I. 1973. Petrography and evolution of the Brandberg intrusion, South West Africa, *Spec. Paper Geol. Soc. S. Africa*, **3** (Symposium on granites, gneisses and related rocks, ed. L. A. Lister), 339–43.

Hoefs, J. Faure, G. & Elliot, D. H. 1980. Correlation of $\delta^{18}O$ and initial $^{87}Sr/^{86}Sr$ ratios in Kirkpatrick Basalt on Mt. Falla, Transantarctic Mountains. *Contr. Min. Petr.*, **75**, 199–203.

Hoernle, K. & Schmincke, H. U. 1993. The role of partial melting in the 15-Ma geochemical evolution of Gran Canaria: a blob model for the Canary hotspot. *J. Petr.*, **34**, 599–626.

Hogan, J. P. & Sinha, A. K. 1991. The effect of accessory minerals on the redistribution of lead isotopes during crustal anatexis: a model. *Geochim. Cosmochim. Acta.*, **55**, 335–48.

Holland, H. D. 1959. Some applications of thermochemical data to problems of ore deposits. I. Stability relations among the oxides, sulfides, sulfates and carbonates of ore and gangue metals. *Econ. Geol.*, **54**, 184–233.

Holloway, J. R. & Burnham, C. W. 1972. Melting relations of basalt with equilibrium water pressure less than total pressure. *J. Petr.*, **13**, 1–29.

Holloway, J. R. & Eggler, D. H. 1976. Fluid-absent

melting of peridotite containing phlogopite and dolomite. *Carnegie Inst. Wash. Yearbook*, **75**, 636–9.

Holtz, F. & Johannes, W. 1991. Genesis of peraluminous granites I. Experimental investigation of melt compositions at 3 and 5 kb and various H_2O activities. *J. Petr.*, **32**, 935–58.

Honjo, N. & Leeman, W. P. 1987. Origin of hybrid ferrolatite lavas from Magic Reservoir eruptive center, Snake River Plain, Idaho. *Contr. Min. Petr.*, **96**, 163–77.

Hooper, P. R. 1982. The Columbia River basalts. *Science*, **215**, 1463–8.

Hotz, P. E. 1952. Form of diabase sheets in south eastern Pennsylvania. *Amer. J. Sci.*, **250**, 375–88.

Huang, W. L. & Wyllie, P. J. 1981. Phase relationships of S-type granite with H_2O to 35 kbar: muscovite granite from Harney Peak, South Dakota. *J. Geophys. Res.*, **86**, 10515–29.

Huang, W. L., Wyllie, P. J. & Nehru, C. E. 1980. Subsolidus and liquidus phase relationships in the system $CaO–SiO_2–CO_2$ to 30 kbar with geological applications. *Amer. Min.*, **65**, 285–301.

Huheey, J. E. 1975. *Inorganic chemistry: principles of structure and reactivity*. Harper & Row, New York, 737 pp.

Hulme, G. 1974. The interpretation of lava flow morphology. *Geophys. J. Royal Astr. Soc.*, **39**, 361–83.

Hunter, R. H. & Sparks, R. S. J. 1987. The differentiation of the Skaergaard intrusion. *Contr. Min. Petr.*, **95**, 451–61.

Hunziker, J. C. & Zingg, A. 1982. Zur Genese der ultrabasichen Gesteine der Ivrea-Zone. *Schweiz. Min. Petr. Mitt.*, **62**, 483–6.

Huppert, H. E. & Sparks, R. S. J. 1980. The fluid dynamics of a basaltic magma chamber replenished by influx of hot, dense ultrabasic magma. *Contr. Min. Petr.*, **75**, 279–89.

Huppert, H. E. & Sparks, R. S. J. 1988. The generation of granitic magmas by intrusion of basalt into continental crust. *J. Petr.*, **29**, 599–624.

Hurlbut, C. S. 1935. Dark inclusions in a tonalite of southern California. *Amer. Min.*, **20**, 609–30.

Hutchison, W. W. 1970. Metamorphic framework and plutonic styles in the Prince Rupert region of the Central Coast Mountains, British Columbia. *Can. J. Earth Sci.*, **7**, 376–405.

Hyndman, D. W. 1981. Controls on source and depth of emplacement of granitic magma. *Geology*, **9**, 244–9.

Ikeda, Y. 1978. Intimate correlation in composition between granitic rocks and their country rocks in Japan. *J. Geol.*, **86**, 261–8.

Innocenti, F., Manetti, P., Mazzuoli, R., Pasquare, G. & Villari, L. 1982. Anatolia and north-western Iran, pp. 327–49 in Thorpe (1982).

Innocenti, F. Mazzuoli, R., Pasquare, G., Radicati di Brozolo, F. & Villari, L. 1976. Evolution of the volcanism in the area of interaction between the Arabian, Anatolian and Iranian plates (Lake Van, Eastern Turkey). *J. Volc. Geothermal Res.*, **1**, 103–12.

Irvine, T. N. 1974. Petrology of the Duke Island ultramafic complex, southeastern Alaska. *Mem. Geol. Soc. Amer.*, **138**, 240 pp.

Irvine, T. N. 1975. Crystallization sequences in the Muskox intrusion and other layered intrusions – II. Origin of chromitite layers and similar deposits of other magmatic ores. *Geochim. Cosmochim. Acta*, **39**, 991–1020.

Irvine, T. N. 1980. Magmatic infiltration metasomatism, double-diffusive fractional crystallization, and adcumulus growth in the Muskox and other layered intrusions, pp. 325–83 in *Physics of magmatic processes* (ed. R. B. Hargraves), Princeton University Press, Princeton, NJ.

Irvine, T. N. 1987. Layering and related structures in the Duke Island and Skaergaard intrusions: similarities, differences, and origins, pp. 185–245 in *Origins of igneous layering* (ed. I. Parsons), Reidel, Dordrecht.

Irvine, T. N. 1992. Emplacement of the Skaergaard intrusion. *Carnegie Inst. Wash. Yearbook*, **91**, 91–6.

Irvine, T. N. & Smith, C. H. 1967. The ultramafic rocks of the Muskox intrusion, North west Territories, Canada, pp. 38–49 in Wyllie (1967).

Irving, A. J. 1978. A review of experimental studies of crystal/liquid trace element partitioning. *Geochim. Cosmochim. Acta*, **42**, 743–70.

Irving, A. J. 1980. Petrology and geochemistry of composite ultramafic xenoliths in alkalic basalts and implications for magmatic processes within the mantle. *Amer. J. Sci.*, **280A**, 389–426.

Ishihara, S. 1978. Metallogenesis in the Japanese island arc system. *J. Geol. Soc. London*, **135**, 389–406.

Ishizaka, K. & Carlson, R. W. 1983. Nd–Sr systematics of the Setouchi volcanic rocks, southwest Japan: a clue to the origin of orogenic andesite. *Earth Planet. Sci. Lett.*, **64**, 327–40.

Ito, K. 1974. Petrological models of the oceanic lithosphere: geophysical and geochemical tests. *Earth Planet. Sci. Lett.*, **21**, 169–80.

Ito, K. & Kennedy, G. C. 1970. The fine structure of the basalt–eclogite transition. *Spec. Paper Min. Soc. Amer.*, **3**, 77–83.

Ito, K. & Kennedy, G. C. 1971. An experimental study of the basalt–garnet granulite–eclogite transition. *Geophysical Monograph Amer. Geophys. Union*, **14** (The structure and physical properties of the Earth's crust), 303–14.

Ito, K. & Kennedy, G. C. 1974. The composition of liquids formed by partial melting of eclogites at high temperatures and pressures. *J. Geol.*, **82**, 383–92.

Ivanov, V. M., Doilnitsyn, Ye. F., Lavrentyev, Yu. G. & Korolyuk, V. N. 1975. Experimental investigations of molecular-ionic-gravitational differentiation of basaltic melts. *Dokl. Akad. Sci. USSR, Earth Sci. Section* (English translation), **217**, 165–8.

Jacobsen, S. B. & Wasserburg, G. J. 1980. Sm–Nd isotopic evolution of chondrites. *Earth Planet. Sci. Lett.*, **50**, 139–55.

Jaffe, H. W. & Schumacher, J. C. 1985. Garnet and plagioclase exsolved from aluminium-rich orthopyroxene in the Marcy anorthosite, northeastern Adirondacks, New York. *Can. Min.*, **23**, 457–78.

Jahns, R. H. 1953. The genesis of pegmatites. I. Occurrence and origin of giant crystals. *Amer. Min.*, **38**, 563–98.

Jahns, R. H. 1955. The study of pegmatites, pp. 1025–1130 in *Economic geology*, 50th Anniversary volume (ed. A. M. Bateman).

Jahns, R. H. & Burnham, C. W. 1969. Experimental studies of pegmatite genesis: I. A model for the derivation and crystallization of granitic pegmatites. *Econ. Geol.*, **64**, 843–64.

Jahns, R. H. & Tuttle, O. F. 1963. Layered pegmatite-aplite intrusives. *Special Paper Min. Soc. Amer.*, **1**, 78–92.

Jakes, P. & White, A. J. R. 1972. Major and trace element abundances in volcanic rocks of orogenic areas. *Bull. Geol. Soc. Amer.*, **83**, 29–40.

Jakobsson, S. P. 1979. Outline of the petrology of Iceland. *Jökull*, **29**, 57–73.

James, R. S. & Hamilton, D. L. 1969. Phase relations in the system $NaAlSi_3O_8$–$KAlSi_3O_8$–$CaAl_2Si_2O_8$–SiO_2 at 1 kilobar water vapour pressure. *Contr. Min. Petr.*, **21**, 111–41.

Jamieson, R. A. 1980. Ophiolite emplacement as recorded in the dynamothermal aureole of the St. Anthony complex, north-western Newfoundland, pp. 620–7 in Panayiotou (1980).

Jaques, A. L. & Green, D. H. 1980. Anhydrous melting of peridotite at 0–15 kb pressure and the genesis of tholeiitic basalts. *Contr. Min. Petr.*, **73**, 287–310.

Jaques, A. L. and 6 co-authors. 1984. The diamond-bearing ultrapotassic (lamproitic) rocks of the West Kimberley region, Western Australia, pp. 225–54 in Vol. 1 of Kornprobst (1984).

Jaques, A. L., O'Neill, H. S., Smith, C. B., Moon, J. & Chappell, B. W. 1990. Diamondiferous peridotite xenoliths from the Argyle (AK1) lamproite pipe, western Australia. *Contr. Min. Petr.*, **104**, 255–76.

Jensen, B. B. 1973. Patterns of trace element partitioning. *Geochim. Cosmochim. Acta*, **37**, 2227–42.

Jin, C. 1981. The Mesozoic–Cenozoic volcanic rocks in Xizang and their bearing on geology, pp. 433–42 in *Geological and ecological studies of Qinghai-Xizang plateau*, Vol. 1. Science Press, Beijing, and Gordon & Breach, New York.

Johannes, W. 1978. Melting of plagioclase in the system Ab–An–H_2O and Qz–Ab–An–H_2O at P_{H_2O} = 5 kbars, an equilibrium problem. *Contr. Min. Petr.*, **66**, 295–303.

Johannes, W. 1984. Beginning of melting in the granite system Qz–Or–Ab–An–H_2O. *Contr. Min. Petr.*, **86**, 264–73.

Johannsen, A. 1938. *A descriptive petrography of the igneous rocks*, Vol. IV. University of Chicago Press, Chicago, 523 pp.

Johnson, C. M. 1989. Isotopic zonations in silicic magmas chambers. *Geology*, **17**, 1136–9.

Johnson, J. R. 1979. Transitional basalts and tholeiites from the East Pacific Rise, 9 °N. *J. Geophys. Res.*, **84**, 1635–51.

Johnson, R. B. 1961. Patterns and origin of radial dike swarms associated with West Spanish Peak and Dike Mountain, south central Colorado. *Bull. Geol. Soc. Amer.*, **72**, 579–90.

Johnston, A. D. 1986. Anhydrous P–T phase relations of near-primary high-alumina basalt from the South Sandwich Islands. *Contr. Min. Petr.*, **92**, 368–82.

Johnstone, G. S., Plant, J. & Watson, J. V. 1979. Caledonian granites in relation to regional geochemistry in northern Scotland, pp. 663–7 in *The Caledonides of the British Isles – reviewed* (ed. A. L. Harris *et al.*), Scottish Academic Press, Edinburgh. Spec. Publ. Geol. Soc. London, **8**.

Joplin, G. A. 1942. Petrological studies in the Ordovician of New South Wales. I. The Cooma complex. *Proc. Linnean Soc. N.S.W.*, **67**, 156–96.

Joplin, G. A. 1943. Petrological studies in the Ordovician of New South Wales. II. The northern extension of the Cooma complex. *Proc. Linnean Soc. N.S.W.*, **68**, 159–83.

Juteau, M., Michard, A., Zimmermann, J. L. & Albarède, F. 1984. Isotopic heterogeneities in the granitic intrusion of Monte Capanne (Elba Island, Italy) and dating concepts. *J. Petr.*, **25**, 532–45.

Kanasewich, E. R. 1968. The interpretation of lead isotopes and their geological significance, pp. 147–223 in *Radiometric dating for geologists* (eds E. I. Hamilton & R. M. Farquhar), Interscience, London and New York, 506 pp.

Karamata, S. 1980. Metamorphism beneath obducted ophiolite slabs, pp. 219–27 in Panayiotou (1980).

Karson, J. A., Elthon, D. L. & De Long, S. E. 1983. Ultramafic intrusions in the Lewis Hills massif, Bay of Islands ophiolite complex, Newfoundland: implications for igneous processes at oceanic fracture zones. *Bull. Geol. Soc. Amer.*, **94**, 15–29.

Katsui, Y., Oba, Y., Ando, S., Nishimura, S., Masuda, Y., Kurasawa, H. & Fujimaki, H. 1978. Petrochemistry of the Quaternary volcanic rocks of Hokkaido, north Japan. *J. Fac. Sci. Hokkaido Univ. Ser. IV*, **18** (3), 449–84.

Kay, R. W. 1979. Zone refining at the base of lithospheric plates: a model for a steady-state asthenosphere. *Tectonophysics*, **55**, 1–9.

Kay, R. W., Sun, S. S. & Lee-Hu, C. N. 1978. Pb and Sr isotopes in volcanic rocks from the Aleutian Islands and Pribilof Islands, Alaska. *Geochim. Cosmochim. Acta*, **42**, 263–73.

Kay, S. M., Ramos, V. A. & Marquez, M. 1993. Evidence in Cerro Pampa volcanic rocks for slab-melting prior to ridge–trench collision in southern South America. *J. Geol.*, **101**, 703–14.

Kearey, P. 1978. An interpretation of the gravity field of the Morin anorthosite complex, southwest Quebec. *Bull. Geol. Soc. Amer.*, **89**, 467–75.

Keith, D. W., Todd, S. G. & Irvine, T. N. 1982. Setting and compositions of the J–M platinum–palladium reef and other sulfide zones in the banded series of the Stillwater complex. *Carnegie Inst. Wash. Yearbook*, **81**, 281–6.

Keller, J. 1982. Mediterranean island arcs, pp. 307–25 in Thorpe (1982).

Keller, J. 1989. Extrusive carbonatites and their significance, pp. 70–88 in *Carbonatites: genesis and evolution* (ed. K. Bell), Unwin-Hyman, London.

Kelsey, C. H. 1965. Calculation of the C.I.P.W. norm. *Min. Mag.*, **34**, 276–82.

Kempe, D. R. C. & Deer, W. A. 1976. The petrogenesis of the Kangerdlugssuaq alkaline intrusion, east Greenland. *Lithos*, **9**, 111–23.

Kennedy, C. S. & Kennedy, G. C. 1976. The equilibrium boundary between graphite and diamond. *J. Geophys. Res.*, **81**, 2467–70.

Kennedy, G. C. 1955. Some aspects of the role of water in rock melts. *Spec. Paper Geol. Soc. Amer.*, **62**, 489–504.

Kersting, A. B., Arculus, R. J., Delano, J. W. & Loureiro, D. 1989. Electrochemical measurements bearing on the oxidation state of the Skaergaard layered intrusion. *Contr. Min. Petr.*, **102**, 376–88.

Kesler, S. E. 1968. Mechanisms of magmatic assimilation at a marble contact, northern Haiti. *Lithos*, **1**, 219–29.

Kesler, S. E. & Heath, S. A. 1968. The effect of dissolved volatiles on magmatic heat sources at intrusive contacts. *Amer. J. Sci.*, **266**, 824–39.

Kilpatrick, J. A. & Ellis, D. J. 1992. C-type magmas: igneous charnockites and their extrusive equivalents. *Trans. Roy. Soc. Edinburgh, Earth Sciences*, **83**, 155–64.

King, B. C. 1965. Petrogenesis of the alkaline igneous rock suites of the volcanic and intrusive centres of eastern Uganda. *J. Petr.*, **6**, 67–100.

Kingsley, L. 1931. Cauldron-subsidence of the Ossipee Mountains. *Amer. J. Sci.*, Series 5, **22**, 139–68.

Kita, I., Nitta, K., Nagao, K., Taguchi, S. & Koga, A. 1993. Difference in N_2/Ar ratio of magmatic gases from northeast and southwest Japan: new evidence for different states of plate subduction. *Geology*, **21**, 391–4.

Kjarsgaard, B. A. & Hamilton, D. L. 1989. The genesis of carbonatites by immiscibility, pp. 388–404 in *Carbonatites: genesis and evolution* (ed. K. Bell), Unwin-Hyman, London.

Knopf, A. 1957. The Boulder bathylith of Montana. *Amer. J. Sci.*, **255**, 81–103.

Knox, G. J. 1974. The structure and emplacement of the Rio Fortaleza centred acid complex, Ancash, Peru. *J. Geol. Soc. London*, **130**, 295–308.

Knudsen, C. 1991. *Petrology, geochemistry and economic geology of the Qaquarssuk carbonatite complex, southern west Greenland.* Gebruder Borntraeger, Berlin and Stuttgart, 110 pp.

Kobelski, B. J., Gold, D. P. & Deines, P. 1979. Variations in stable isotope compositions for carbon and oxygen in some South African and Lesothan kimberlites, pp. 252–71 in Boyd & Meyer (1979a).

Kogarko, L. N. 1987. Alkaline rocks of the eastern part of the Baltic Shield (Kola Peninsula), pp. 531–44 in *Alkaline* *igneous rocks* (eds J. G. Fitton & B. G. J. Upton), Spec. Publ. Geol. Soc., **30**, Blackwell Scientific Publications, Oxford.

Kolker, A. 1982. Mineralogy and geochemistry of Fe–Ti oxide and apatite (nelsonite) deposits and evaluation of the liquid immiscibility hypothesis. *Econ. Geol.*, **77**, 1146–58.

Korn, H. & Martin, H. 1954. The Messum igneous complex in South-West Africa. *Trans. Geol. Soc. S. Africa*, **57**, 83–124.

Kornprobst, J. (ed.). 1984. *Kimberlites.* (*Proc. 3rd Int. Kimberlite Conf.*). Elsevier, Amsterdam, 2 vols.

Kornprobst, J., Piboule, M., Roden, M. & Tabit, A. 1990. Corundum-bearing garnet clinopyroxenites at Beni Bousera (Morocco): original plagioclase-rich gabbros recrystallized at depth within the mantle. *J. Petr.*, **31**, 717–45.

Koster van Groos, A. F. & Wyllie, P. J. 1966. Liquid immiscibility in the system $Na_2O–Al_2O_3–SiO_2–CO_2$ at pressures to 1 kilobar. *Amer. J. Sci.*, **264**, 234–55.

Koster van Groos, A. F. & Wyllie, P. J. 1968. Liquid immiscibility in the join $NaAlSi_3O_8–Na_2CO_3–H_2O$ and its bearing on the genesis of carbonatites. *Amer. J. Sci.*, **266**, 932–67.

Koster van Groos, A. F. & Wyllie, P. J. 1973. Liquid immiscibility in the join $NaAlSi_3O_8–CaAl_2Si_2O_8–Na_2CO_3–H_2O$. *Amer. J. Sci.*, **273**, 465–87.

Koto, B. 1916. The great eruption of Sakura-jima in 1914. *J. College of Science, Imperial Univ. Tokyo*, **38**, article 3, 1–237.

Kouchi, A. & Sunagawa, I. 1985. A model for mixing basaltic and dacitic magmas as deduced from experimental data. *Contr. Min. Petr.*, **89**, 17–23.

Kramm, U., Kogarko, L. N., Kononova, V. A. & Vartiainen, H. 1993. The Kola alkaline province of the CIS and Finland: precise Rb–Sr ages define 380–360 Ma age range for all magmatism. *Lithos*, **30**, 33–44.

Kuroda, N., Shiraki, K. & Urano, H. 1978. Boninite as a possible calc-alkalic primary magma. *Bull. Volc.*, **41**, 563–75.

Kurz, M. D., Jenkins, W. J., Hart, S. R. & Clague, D. 1983. Helium isotopic variations in volcanic rocks from Lohi seamount and the island of Hawaii. *Earth Planet. Sci. Lett.*, **66**, 388–406.

Kurz, M. D., Jenkins, W. J., Schilling, J. G. & Hart, S. R. 1982. Helium isotopic variations in the mantle beneath the central North Atlantic Ocean. *Earth Planet. Sci. Lett.*, **58**, 1–14.

Kushiro, I. 1973. Origin of some magmas in oceanic and circum-oceanic regions. *Tectonophysics*, **17**, 211–22.

Kushiro, I. 1980a. Viscosity, density, and structure of silicate melts at high pressures, and their petrological applications, pp. 93–120 in Hargraves (1980).

Kushiro, I. 1980b. Changes with pressure of degree of partial melting and K_2O content of liquids in the system $Mg_2SiO_4–KAlSiO_4–SiO_2$. *Carnegie Inst. Wash. Yearbook*, **79**, 267–71.

Kushiro, I. 1987. A petrological model of the mantle wedge and lower crust in the Japanese island arcs, pp. 165–81 in *Magmatic processes: physicochemical principles* (ed. B. O. Mysen), Geochemical Society Spec. Publ. no 1.

Kushiro, I., Syono, Y. & Akimoto, S. 1967. Stability of phlogopite at high pressures and possible presence of phlogopite in the Earth's upper mantle. *Earth Planet. Sci. Lett.*, **3**, 197–203.

Kushiro, I. & Yoder, H. S. 1969. Melting of forsterite and enstatite at high pressures under hydrous conditions. *Carnegie Inst. Wash. Yearbook*, **67**, 153–8.

Kyle, P. R. 1980. Development of heterogeneities in the subcontinental mantle: evidence from the Ferrar Group, Antarctica. *Contr. Min. Petr.*, **73**, 89–104.

Lachenbruch, A. H. & Sass, J. H. 1978. Models of an extending lithosphere and heat flow in the Basin and Range province. *Mem. Geol. Soc. Amer.*, **152**, 209–50.

Lacroix, A. 1904. *La montagne Pelée et ses éruptions.* Masson, Paris, 662 pp.

Lagabrielle, Y. & Cannat, M. 1990. Alpine Jurassic ophiolites resemble the modern central Atlantic basement. *Geology*, **18**, 319–22.

Lange, R. A. & Carmichael, I. S. E. 1987. Densities of $Na_2O–K_2O–CaO–MgO–FeO–Fe_2O_3–Al_2O_3–TiO_2–SiO_2$ liquids: new measurements and derived partial molar properties. *Geochim. Cosmochim. Acta*, **51**, 2931–46.

Lange, R. L. & Carmichael, I. S. E. 1990. Thermodynamic properties of silicate liquids with emphasis on density, thermal expansion and compressibility. Chapter 2 (pp. 25–64) of *Modern methods of igneous petrology: understanding magmatic processes* (eds J. Nicholls & J. K. Russell). Reviews in Mineralogy, **24**. Mineralogical Society of America, Washington, DC.

Larrabee, D. M. 1966. Map showing distribution of ultramafic and intrusive mafic rocks from northern New Jersey to eastern Alabama. *Misc. Geol. Investig. U.S.G.S.*, Map 1–476.

Larsen, E. S. 1948. Batholith and associated rocks of Corona, Elsinore, and San Luis Rey quadrangles, southern California. *Mem. Geol. Soc. Amer.*, **29**, 182 pp.

Larsen, E. S. & Cross, W. 1956. Geology and petrology of the San Juan region, southwestern Colorado. *Prof. Paper U.S.G.S.*, **258**, 303 pp.

Larsen, L. A. & Smith, E. I. 1990. Mafic enclaves in the Wilson Ridge pluton, northwestern Arizona: implications for the generation of a calc-alkaline intermediate pluton in an extensional environment. *J. Geophys. Res.*, **95**, 17693–716.

Larsen, L. M. & Sørensen, H. 1987. The Ilimaussaq intrusion – progressive crystallization and formation of layering in an agpaitic magma, pp. 473–88 in *Alkaline igneous rocks* (eds J. G. Fitton & B. G. J. Upton), Spec. Publ. Geol. Soc. London, **30**. Blackwell Scientific Publications, Oxford.

Laurent, R. 1975. Occurrences and origin of the ophiolites of southern Quebec, northern Appalachians. *Can. J. Earth Sci.*, **12**, 443–55.

Leake, B. E. & Skirrow, G. 1960. The pelitic hornfelses

of the Cashel–Loch Wheelaun intrusion, County Galway, Eire. *J. Geol.*, **68**, 23–40.

Le Bas, M. J. 1977. *Carbonatite–nephelinite volcanism.* John Wiley & Sons, London and New York, 347 pp.

Le Bas, M. J. 1981. Carbonatite magmas. *Min. Mag.*, **44**, 133–40.

Le Bas, M. J. 1987. Nephelinites and carbonatites, pp. 53–83 in *Alkaline igneous rocks* (eds J. G. Fitton & B. G. J. Upton), Geol. Soc. Spec. Publ., **30**. Blackwell Scientific Publications, Oxford.

Le Bas, M. J. & Aspden, J. A. 1981. The comparability of carbonatitic fluid inclusions in ijolites with natrocarbonatite lava. *Bull. Volc.*, **44**, 429–38.

Leblanc, M., Dupuy, C., Cassard, D., Moutte, J., Nicolas, A., Prinzhoffer, A., Rabinovitch, M. & Routhier, P. 1980. Essai sur la genèse des corps podiformes de chromitite dans les péridotites ophiolitiques: étude des chromites de Nouvelle-Calédonie et comparaison avec celles de Méditerranée orientale, pp. 691–701 in Panayiotou (1980).

Leeman, W. P. 1982. Tectonic and magmatic significance of strontium isotopic variations in Cenozoic volcanic rocks from the western United States. *Bull. Geol. Soc. Amer.*, **93**, 487–503.

Leeman, W. P. & Dasch, E. J. 1978. Strontium, lead and oxygen isotopic investigation of the Skaergaard intrusion, east Greenland. *Earth Planet. Sci. Lett.*, **41**, 47–51.

Leeman, W. P. & Lindstrom, D. J. 1978. Partitioning of Ni^{2+} between basaltic and synthetic melts and olivines – an experimental study. *Geochim. Cosmochim. Acta*, **42**, 801–16.

Leeman, W. P., Vitaliano, C. J. & Prinz, M. 1976. Evolved lavas from the Snake River Plain: Craters of the Moon National Monument, Idaho. *Contr. Min. Petr.*, **56**, 35–60.

Le Fort, P. 1975. Himalayas: the collided range. Present knowledge of the continental arc. *Amer. J. Sci.*, **275-A**, 1–44.

Le Fort, P. 1981. Manaslu leucogranite: a collision signature of the Himalaya, a model for its genesis and emplacement. *J. Geophys. Res.*, **86**, 10545–68.

Le Maitre, R. W. 1976. The chemical variability of some common igneous rocks. *J. Petr.*, **17**, 589–637.

Lensch, G. 1968. Die Ultramafitite der Zone von Ivrea und ihre geologische Interpretation. *Schweiz. Min. Petr. Mitt.*, **48**, 91–102.

Le Roex, A. P., Cliff, R. A. & Adair, B. J. I. 1990. Tristan da Cunha, South Atlantic: geochemistry and petrogenesis of a basanite–phonolite lava series. *J. Petr.*, **31**, 779–812.

Lesher, C. E. & Walker, D. 1986. Solution properties of silicate liquids from thermal diffusion experiments. *Geochim. Cosmochim. Acta*, **50**, 1397–411.

Lewis, J. F. 1973. Petrology of the ejected plutonic blocks of the Soufriere Volcano, St. Vincent, West Indies. *J. Petr.*, **14**, 81–112.

Liebenberg, L. 1970. The sulphides in the layered sequence of the Bushveld Igneous Complex. *Spec. Publ. Geol. Soc. S. Africa*, **1**, 108–207.

Lippard, S. J., Shelton, A. W. & Gass, I. G. 1986. The ophiolite of northern Oman. *Geological Society of London, Memoir No. 11*, Blackwell Scientific Publications, Oxford.

Lirer, L., Pescatore, T., Booth, B. & Walker, G. P. L. 1973. Two Plinian pumice-fall deposits from Somma-Vesuvius, Italy. *Bull. Geol. Soc. Amer.*, **84**, 759–72.

Lockwood, J. P. 1972. Possible mechanisms for the emplacement of Alpine-type serpentinite. *Mem. Geol. Soc. Amer.*, **132**, 273–87.

Lodding, A. & Ott, A. 1966. Isotope thermotransport in liquid potassium, rubidium and gallium. *Z. Naturforschung*, **21A**, 1344–7.

Loney, R. A. 1968. Flow structure and composition of the Southern Coulee, Mono Craters, California – a pumiceous rhyolite flow. *Mem. Geol. Soc. Amer.*, **116**, 415–40.

Loomis, T. P. 1972. Diapiric emplacement of the Ronda high-temperature ultramafic intrusion, southern Spain. *Bull. Geol. Soc. Amer.*, **83**, 2475–96.

Lopez-Escobar, L., Frey, F. A. & Vergara, M. 1977. Andesites and high-alumina basalts from the central-south Chile High Andes: geochemical evidence bearing on their petrogenesis. *Contr. Min. Petr.*, **63**, 199–228.

Lundeen, M. T. 1978. Emplacement of the Ronda peridotite, Sierra Bermeja, Spain. *Bull. Geol. Soc. Amer.*, **89**, 172–80.

Lupton, J. E. 1983. Terrestrial inert gases: isotope tracer studies and clues to primordial components in the mantle. *Ann. Rev. Earth Planet. Sci.*, **11**, 371–414.

Lupton, J. E. & Craig, H. 1981. A major helium-3 source at 15 °S on the East Pacific Rise. *Science*, **214**, 13–18.

Luth, R. W., Virgo, D., Boyd, F. R. & Wood, B. J. 1990. Ferric iron in mantle-derived garnets: implications for thermobarometry and for the oxidation state of the mantle. *Contr. Min. Petr.*, **104**, 56–72.

Luth, W. C. 1968. The influence of pressure on the composition of eutectic liquids in the binary systems sanidine–silica and albite–silica. *Carnegie Inst. Wash. Yearbook*, **66**, 480–4.

Luth, W. C. 1969. The systems $NaAlSi_3O_8$–SiO_2 and $KAlSi_3O_8$–SiO_2 to 20 kb and the relationship between H_2O content, P_{H_2O}, and P_{total} in granitic magmas. *Amer. J. Sci.*, **267A**, 325–41.

Luth, W. C., Jahns, R. H. & Tuttle, O. F. 1964. The granite system at pressures of 4 to 10 kilobars. *J. Geophys. Res.*, **69**, 759–73.

Maaløe, S. 1973. Temperature and pressure relations of ascending primary magmas. *J. Geophys. Res.*, **78**, 6877–86.

Maaløe, S. & Aoki, K. 1977. The major element composition of the upper mantle estimated from the composition of lherzolites. *Contr. Min. Petr.*, **63**, 161–73.

Maaløe, S. & Jakobsson, S. P. 1980. The PT phase relations of a primary oceanite from the Reykjanes peninsula, Iceland. *Lithos*, **13**, 237–46.

Maaløe, S. & Petersen, T. S. 1981. Petrogenesis of oceanic andesites. *J. Geophys. Res.*, **86**, 10273–86.

Maaløe, S., Sørensen, I. & Hertogen, J. 1986. The trachybasaltic suite of Jan Mayen. *J. Petr.*, **27**, 439–66.

McBirney, A. R. 1985. Further considerations of double-diffusive stratification and layering in the Skaergaard intrusion. *J. Petr.*, **26**, 993–1001.

McBirney, A. R., Baker, B. H. & Nilson, R. H. 1985. Liquid fractionation. Part I: Basic principles and experimental simulations. *J. Volc. Geotherm. Res.*, **24**, 1–24.

McBirney, A. R. & Murase, T. 1984. Rheological properties of magmas. *Ann. Rev. Earth Planet. Sci.*, **12**, 337–57.

McBirney, A. R. & Nakamura, Y. 1974. Immiscibility in late-stage magmas of the Skaergaard intrusion. *Carnegie Inst. Wash. Yearbook*, **73**, 348–52.

McBirney, A. R. & Noyes, R. M. 1979. Crystallization and layering of the Skaergaard intrusion. *J. Petr.*, **20**, 487–554.

McBirney, A. R., Taylor, H. P. & Armstrong, R. L. 1987. Paricutin re-examined: a classic example of crustal assimilation in calc-alkaline magma. *Contr. Min. Petr.*, **95**, 4–20.

McBirney, A. R. & White, C. M. 1982. The Cascade province, pp. 115–35 in Thorpe (1982).

McBirney, A. R. and others. 1990. The differentiation of the Skaergaard intrusion. *Contr. Min. Petr.*, **104**, 235–54 (discussions of Hunter & Sparks 1987).

McCallum, I. S., Raedeke, L. D. & Mathez, E. A. 1980. Investigations of the Stillwater complex: part I. Stratigraphy and structure of the Banded Zone. *Amer. J. Sci.*, **280-A**, 59–87.

McCourt, W. J. 1981. The geochemistry and petrography of the Coastal Batholith of Peru, Lima segment. *J. Geol. Soc. Lond.*, **138**, 407–20.

McCulloch, M. T. & Bennett, V. C. 1994. Progressive growth of the Earth's continental crust and depleted mantle: geochemical constraints. *Geochim. Cosmochim. Acta*, **58**, 4717–38.

McCulloch, M. T., Gregory, R. T., Wasserburg, G. J. & Taylor, H. P. Jr. 1980. A neodymium, strontium, and oxygen isotopic study of the Cretaceous Samail ophiolite and implications for the petrogenesis and seawater-hydrothermal alteration of oceanic crust. *Earth Planet. Sci. Lett.*, **46**, 201–11.

McCulloch, M. T., Jaques, A. L., Nelson, D. R. & Lewis, J. D. 1983. Nd and Sr isotopes in kimberlites and lamproites from Western Australia: an enriched mantle origin. *Nature*, **302**, 400–3.

McCulloch, M. T. & Perfit, M. R. 1981. $^{143}Nd/^{144}Nd$, $^{87}Sr/^{86}Sr$ and trace element constraints on the petrogenesis of Aleutian island arc magmas. *Earth Planet. Sci. Lett.*, **56**, 167–79.

Macdonald, G. A. 1972. *Volcanoes*. Prentice-Hall, Englewood Cliffs, NJ, 510 pp.

Macdonald, G. A., Abbott, A. T. & Peterson, F. L. 1983. *Volcanoes in the sea. The geology of Hawaii* (2nd edn). University of Hawaii Press, Honolulu, 517 pp.

Macdonald, G. A. & Hubbard, D. H. 1975. *Volcanoes of the National Parks in Hawaii* (5th edn). Hawaii Natural History Association, Hawaii, 60 pp.

Macdonald, G. A. & Katsura, T. 1964. Chemical composition of Hawaiian lavas. *J. Petr.*, **5**, 82–133.

Macdonald, R., Kjarsgaard, B. A., Skilling, I. P., Davies, G. R., Hamilton, D. L. & Black, S. 1993. Liquid immiscibility between trachyte and carbonate in ash flow tuffs from Kenya. *Contr. Min. Petr.*, **114**, 276–87.

Macdonald, R., McGarvie, D. W., Pinkerton, H., Smith, R. L. & Palacz, Z. A. 1990. Petrogenetic evolution of the Torfajökull volcanic complex, Iceland. I. Relationship between the magma types. *J. Petr.*, **31**, 429–59.

McDowell, S. D. 1974. Emplacement of the Little Chief Stock, Panamint Range, California. *Bull. Geol. Soc. Amer.*, **85**, 1535–46.

McGetchin, T. R., Nikhanj, Y. S. & Chodos, A. A. 1973. Carbonatite–kimberlite relations in the Cane Valley diatreme, San Juan County, Utah. *J. Geophys. Res.*, **78**, 1854–69.

McGetchin, T. R., & Silver, L. T. 1970. Compositional relations in minerals from kimberlite and related rocks in the Moses Rock dike, San Juan county, Utah. *Amer. Min.*, **55**, 1738–71.

Macgregor, I. D. 1975. Petrologic and thermal structure of the upper mantle beneath South Africa in the Cretaceous. *Phys. Chem. Earth*, **9**, 455–66.

Macgregor, I. D. & Basu, A. R. 1976. Geological problems in estimating mantle geothermal gradients. *Amer. Min.*, **61**, 715–24.

Macgregor, I. D. & Basu, A. R. 1979. Petrogenesis of the Mount Albert ultramafic massif, Quebec: summary. *Bull. Geol. Soc. Amer.*, **90**, 898–900.

McKenzie, D. 1984. The generation and compaction of partially molten rock. *J. Petr.*, **25**, 713–65.

McKenzie, D. 1985. The extraction of magma from the crust and mantle. *Earth Planet. Sci. Lett.*, **74**, 81–91.

McKenzie, D. & Bickle, M. J. 1988. The volume and composition of melt generated by extension of the lithosphere. *J. Petr.*, **29**, 625–79.

Maclean, W. H. 1969. Liquidus phase relations in the $FeS–FeO–Fe_3O_4–SiO_2$ system, and their application in geology. *Econ. Geol.*, **64**, 865–84.

McMillan, N. J. & Dungan, M. A. 1988. Open system magmatic evolution of the Taos Plateau volcanic field, northern New Mexico: 3. Petrology and geochemistry of andesite and dacite. *J. Petr.*, **29**, 527–57.

Magaritz, M., Whitford, D. J. & James, D. E. 1978. Oxygen isotopes and the origin of high $^{87}Sr/^{86}Sr$ andesites. *Earth Planet. Sci. Lett.*, **40**, 220–30.

Mahood, G. & Hildreth, W. 1983. Large partition coefficients for trace elements in high-silica rhyolites. *Geochim. Cosmochim. Acta*, **47**, 11–30.

Mamyrin, B. A. & Tolstikhin, L. N. 1984. *Helium isotopes in nature*. Elsevier, Amsterdam, 273 pp.

Marsh, B. D. 1982a. On the mechanics of igneous diapirism, stoping, and zone melting. *Amer. J. Sci.*, **282**, 808–55.

Marsh, B. D. 1982b. The Aleutians, pp. 99–114 in Thorpe (1982).

Marsh, B. D. & Maxey, M. R. 1985. On the distribution

and separation of crystals in convecting magma. *J. Volc. Geotherm. Res.*, **24**, 95–150.

Martignole, J. 1974. L'évolution magmatique du complex de Morin et son apport au problème des anorthosites. *Contr. Min. Petr.*, **44**, 117–37.

Martignole, J. & Schrijver, K. 1970. Tectonic setting and evolution of the Morin anorthosite, Grenville province, Quebec. *Bull. Geol. Soc. Finland*, **42**, 165–209.

Martin, N. R. 1953. The structure of the granite massif of Flamanville, Manche, North-west France. *Quart. J. Geol. Soc. London*, **108**, 311–41.

Mason, E. A., Munn, R. J. & Smith, F. J. 1966. Thermal diffusion in gases, pp. 33–91 in *Advances in atomic and molecular physics* (eds D. R. Bates & I. Esterman), Vol. 2. Academic Press, New York.

Matsuhisa, Y., Goldsmith, J. R. & Clayton, R. N. 1979. Oxygen isotopic fractionation in the system quartz–albite–anorthite–water. *Geochim. Cosmochim. Acta*, **43**, 1131–40.

Matsui, Y., Onuma, N., Nagasawa, H., Higuchi, H. & Banno, S. 1977. Crystal structure control in trace element partition between crystal and magma. *Bull. Soc. Franc. Min. Crist.*, **100**, 315–24.

Mattioli, G. S., Baker, M. B., Rutter, M. J. & Stolper, E. M. 1989. Upper mantle oxygen fugacity and its relationship to metasomatism. *J. Geol.*, **97**, 521–36.

Medaris, L. G. & Carswell, D. A. 1990. Petrogenesis of Mg–Cr garnet peridotites in European metamorphic belts, pp. 260–90 in *Eclogite facies rocks* (ed. D. A. Carswell). Blackie, Glasgow.

Medenbach, O. & El Goresy, A. 1982. Ulvospinel in native iron-bearing assemblages and the origin of these assemblages in basalts from Ovifak, Greenland, and Buhl, Federal Republic of Germany. *Contr. Min. Petr.*, **80**, 358–66.

Meen, J. K. 1987. Formation of shoshonites from cal-calkaline basalt magmas: geochemical and experimental constraints from the type locality. *Contr. Min. Petr.*, **97**, 333–51.

Meen, J. K. 1990. Elevation of potassium content of basaltic magma by fractional crystallization: the effect of pressure. *Contr. Min. Petr.*, **104**, 309–31.

Mehnert, K. R. 1968. *Migmatites and the origin of granitic rocks.* Elsevier, Amsterdam, 393 pp.

Mehnert, K. R., Büsch, W. & Schneider, G. 1973. Initial melting at grain boundaries of quartz and feldspar in gneisses and migmatites. *Neues Jb. Min. Mh., Jahrgang 1973*, 165–83.

Meijer, A. 1976. Pb and Sr isotopic data bearing on the origin of volcanic rocks from the Mariana island-arc system. *Bull. Geol. Soc. Amer.*, **87**, 1358–69.

Melson, W. G. & Van Andel, T. H. 1966. Metamorphism in the mid-Atlantic Ridge, 22 °N latitude. *Marine Geology*, **4**, 165–86.

Menzies, M. A. 1984. Chemical and isotopic heterogeneities in orogenic and ophiolitic peridotites, pp. 231–40 in *Ophiolites and oceanic lithosphere* (eds I. G. Gass, S. J. Lippard & A. W. Shelton), Geological Society Special

Publication **13**. Blackwell Scientific Publications, Oxford.

Menzies, M. A. & Hawkesworth, C. J. (eds). 1987. *Mantle metasomatism.* Academic Press, London.

Menzies, M. A. & Kyle, P. R. 1990. Continental volcanism: a crust-mantle probe, pp. 157–77 in *Continental mantle* (ed. M. A. Menzies), Oxford Science Publications, Oxford.

Menzies, M., Blanchard, D., Brannon, J. & Korotev, R. 1977. Rare earth geochemistry of fused ophiolitic and Alpine lherzolites. *Contr. Min. Petr.*, **64**, 53–74.

Mercier, J. C. C. 1979. Peridotite xenoliths and the dynamics of kimberlite intrusion, pp. 197–212 in Boyd & Meyer (1979b).

Mercier, J.-C. & Carter, N. L. 1975. Pyroxene geotherms. *J. Geophys. Res.*, **80**, 3349–62.

Mertzman, S. A. 1979. Strontium isotope geochemistry of a low potassium olivine tholeiite and two basalt–pyroxene andesite magma series from the Medicine Lake Highland, California. *Contr. Min. Petr.*, **70**, 81–8.

Meschede, M. 1986. A method of discriminating between different types of mid-ocean ridge basalts and continental tholeiites with the Nb–Zr–Y diagram. *Chem. Geol.*, **56**, 207–18.

Métais, D. & Chayes, F. 1963. Varieties of lamprophyre. *Carnegie Inst. Wash. Yearbook*, **62**, 156–7.

Meyboom, A. F. & Wallace, R. C. 1978. Occurrence and origin of ring-shaped dolerite outcrops in the Eastern Cape Province and western Transkei. *Trans. Geol. Soc. S. Africa*, **81**, 95–9.

Meyer, H. O. A. 1985. Genesis of diamond: a mantle saga. *Amer. Min.*, **70**, 344–55.

Meyer, P. S. & Sigurdsson, H. 1978. Interstitial acid glass and chlorophaeite in Iceland. *Lithos*, **11**, 231–41.

Michot, J. & Michot, P. 1969. The problem of anortho-sites: the South-Rogaland igneous complex, southwestern Norway. *Mem. N.Y. State Mus. Sci. Serv.*, **18**, 399–410.

Miller, C. F. & Bradfish, L. J. 1980. An inner Cordilleran belt of muscovite-bearing plutons. *Geology*, **8**, 412–16.

Miller, C. F., Hanchar, J. M., Wooden, J. L., Bennett, V. C., Harrison, T. M., Wark, D. A. & Foster, D. A. 1992. Source region of a granite batholith: evidence from lower crustal xenoliths and inherited accessory minerals. *Trans. Roy. Soc. Edinburgh, Earth Sciences*, **83**, 49–62.

Minnigh, L. D., Van Calsteren, P. W. C. & Den Tex, E. 1980. Quenching: an additional model for emplacement of the lherzolite at Lers (French Pyrenees). *Geology*, **8**, 18–21.

Mitchell, R. H. 1979. The alleged kimberlite–carbonatite relationship: additional contrary mineralogical evidence. *Amer. J. Sci.*, **279**, 570–89.

Mitchell, R. H., Platt, R. G. & Downey, M. 1987. Petrology of lamproites from Smoky Butte, Montana. *J. Petr.*, **28**, 645–77.

Miyashiro, A., Shido, F. & Ewing, M. 1970. Crystallization and differentiation in abyssal tholeiites and gabbros from mid-ocean ridges. *Earth Planet. Sci. Lett.*, **7**, 361–5.

Moberly, R. & Campbell, J. F. 1984. Hawaiian hotspot volcanism mainly during geomagnetic normal intervals. *Geology*, **12**, 459–63.

Möckel, J. R. 1969. Structural petrology of the garnet–

peridotite of Alpe Arami (Ticino, Switzerland). *Leidse Geol. Med.*, **42**, 61–130.

Mohr, P. 1983. Ethiopian flood basalt province. *Nature*, **303**, 577–84.

Moorbath, S. & Bell, J. D. 1965. Strontium isotope abundance studies and rubidium-strontium age determinations on Tertiary igneous rocks from the Isle of Skye, North-West Scotland. *J. Petr.*, **6**, 37–66.

Moorbath, S., Thorpe, R. S. & Gibson, I. L. 1978. Strontium isotope evidence for petrogenesis of Mexican andesites. *Nature*, **271**, 437–9.

Moore, J. G. 1959. The quartz diorite boundary line in the western United States. *J. Geol.*, **67**, 198–210.

Moore, J. G. 1965. Petrology of deep-sea basalt near Hawaii. *Amer. J. Sci.*, **263**, 40–52.

Moore, J. G., Batchelder, J. N. & Cunningham, C. G. 1977. CO_2-filled vesicles in mid-ocean basalt. *J. Volc. Geotherm. Res.*, **2**, 309–27.

Moore, J. G. & Evans, B. W. 1967. The role of olivine in the crystallization of the prehistoric Makaopuhi tholeiitic lava lake, Hawaii. *Contr. Min. Petr.*, **15**, 202–23.

Moore, J. G. & Melson, W. G. 1970. Nuées ardentes of the 1969 eruption of Mayon volcano, Philippines. *Bull. Volc.*, **33** (for 1969), 600–20.

Moore, R. O., Gurney, J. J., Griffin, W. L. & Shimizu, N. 1991. Ultra-high pressure garnet inclusions in Monastery diamonds: trace element abundance patterns and conditions of origin. *Eur. J. Min.*, **3**, 213–30.

Moores, E. M. 1969. Petrology and structure of the Vourinos ophiolitic complex of northern Greece. *Spec. Paper Geol. Soc. Amer.*, **118**, 1–74.

Moores, E. M. & Vine, F. J. 1971. The Troodos Massif, Cyprus and other ophiolites as oceanic crust: evaluation and implications. *Phil. Trans. Roy. Soc., Lond., A*, **268**, 443–66.

Morgan, W. J. 1971. Convection plumes in the lower mantle. *Nature*, **230**, 42–3.

Mori, T. & Green, D. H. 1976. Subsolidus equilibria between pyroxenes in the $CaO–MgO–SiO_2$ system at high pressures and temperatures. *Amer. Min.*, **61**, 616–25.

Morogan, V. 1994. Ijolite versus carbonatite as sources of fenitization. *Terra Nova*, **6**, 166–76.

Morris, G. B., Raitt, R. W. & Shor, G. G. 1969. Velocity anisotropy and delay-time maps of the mantle near Hawaii. *J. Geophys. Res.*, **74**, 4300–16.

Morris, J. & Tera, F. 1989. [10]Be and [9]Be in mineral separates and whole rocks from volcanic arcs: implications for sediment subduction. *Geochim. Cosmochim. Acta*, **53**, 3197–206.

Morris, J. D. & Hart, S. R. 1983. Isotopic and incompatible element constraints on the genesis of island arc volcanics from Cold Bay and Amak Island, Aleutians, and implications for mantle structure. *Geochim. Cosmochim. Acta*, **47**, 2015–30.

Morse, S. A. 1969a. The Kiglapait layered intrusion, Labrador. *Mem. Geol. Soc. Amer.*, **112**, 204 pp.

Morse, S. A. 1969b. Syenites. *Carnegie Inst. Wash. Yearbook*, **67**, 112–20.

Morse, S. A. 1970. Alkali feldspars with water at 5 kb pressure. *J. Petr.*, **11**, 221–51.

Morse, S. A. 1979. Kiglapait geochemistry. I: Systematics, sampling and density. II: Petrography. *J. Petr.*, **20**, 555–624.

Morse, S. A. 1980. *Basalts and phase diagrams*. Springer-Verlag, New York, Heidelberg and Berlin, 493 pp.

Morse, S. A. 1981. Kiglapait geochemistry. IV: The major elements. *Geochim. Cosmochim. Acta*, **45**, 461–79.

Morse, S. A. 1982. Kiglapait geochemistry. V: Strontium. *Geochim. Cosmochim. Acta*, **46**, 223–34.

Muan, A. 1975. Phase relations in chromium oxide-containing systems at elevated temperatures. *Geochim. Cosmochim. Acta*, **39**, 791–802.

Muan, A. & Osborn, E. F. 1965. *Phase equilibria among oxides in steelmaking*. Addison-Wesley, Reading, MA, 236 pp.

Muir, I. D., Tilley, C. E. & Scoon, J. H. 1957. Contributions to the petrology of Hawaiian basalts. 1. The picrite-basalts of Kilauea. *Amer. J. Sci.*, **255**, 241–53.

Murase, T. 1962. Viscosity and related properties of volcanic rocks at 800° to 1400°C. *J. Fac. Sci. Hokkaido Univ. (Ser. VII)*, **1**, 487–584.

Murase, T., Kushiro, I. & Fujii, T. 1977. Electrical conductivity of partially molten tholeiite. *Carnegie Inst. Wash. Yearbook*, **76**, 416–9.

Murase, T. & McBirney, A. R. 1973. Properties of some common igneous rocks and their melts at high temperatures. *Bull. Geol. Soc. Amer.*, **84**, 3563–92.

Murase, T., McBirney, A. R. & Melson, W. G. 1985. Viscosity of the dome of Mount St. Helens. *J. Volc. Geotherm. Res.*, **24**, 193–204.

Murata, K. J. & Richter, D. H. 1966. The settling of olivine in Kilauean magma as shown by lavas of the 1959 eruption. *Amer. J. Sci.*, **264**, 194–203.

Murray, C. G. 1972. Zoned ultramafic complexes of the Alaskan type: feeder pipes of andesitic volcanoes. *Mem. Geol. Soc. Amer.*, **132**, 313–35.

Myers, J. & Eugster, H. P. 1983. The system Fe–Si–O: oxygen buffer calibrations to 1500 K. *Contr. Min. Petr.*, **82**, 75–90.

Myers, J. D. & Marsh, B. D. 1981. Geology and petrogenesis of the Edgecumbe volcanic field, SE Alaska: the interaction of basalt and sialic crust. *Contr. Min. Petr.*, **77**, 272–87.

Myers, J. S. 1975. Cauldron subsidence and fluidization: mechanisms of intrusion of the Coastal Batholith of Peru into its own volcanic ejecta. *Bull. Geol. Soc. Amer.*, **86**, 1209–20.

Mysen, B. O. 1977. The solubility of H_2O and CO_2 under predicted magma genesis conditions and some petrological and geophysical implications. *Rev. Geophys. Space Phys.*, **15**, 351–61.

Mysen, B. O. 1981. Rare earth element partitioning between minerals and ($CO_2 + H_2O$) vapor as a function of pressure, temperature, and vapor composition. *Carnegie Inst. Wash. Yearbook*, **80**, 347–9.

Mysen, B. O. (ed.). 1987. Magmatic processes: physicochemical principles. *Geochemical Society, Special Publication no. 1*, 500 pp.

Mysen, B. O., Arculus, R. J. & Eggler, D. H. 1975. Solubility of carbon dioxide in melts of andesite, tholeiite, and olivine nephelinite composition to 30 kbar pressure. *Contr. Min. Petr.*, **53**, 227–39.

Mysen, B. O. & Boettcher, A. L. 1975. Melting of a hydrous mantle: I. Phase relations of natural peridotite at high pressures and temperatures with controlled activities of water, carbon dioxide and hydrogen; II. Geochemistry of crystals and liquids formed by anatexis of mantle peridotite at high pressures and high temperatures as a function of controlled activities of water, hydrogen, and carbon dioxide. *J. Petr.*, **16**, 520–93.

Mysen, B. O. & Kushiro, I. 1976. Compositional variation of coexisting phases with degree of melting of peridotite under upper mantle conditions. *Carnegie Inst. Wash. Yearbook*, **75**, 546–55.

Mysen, B. O. & Kushiro, I. 1977. Compositional variations of coexisting phases with degree of melting of peridotite in the upper mantle. *Amer. Min.*, **62**, 843–65.

Mysen, B. O. & Kushiro, I. 1979. Pressure dependence of nickel partitioning between forsterite and aluminous silicate melts. *Earth Planet. Sci. Lett.*, **42**, 383–8.

Mysen, B. O. & Popp, R. K. 1978. Solubility of sulfur in silicate melts as a function of f_{S_2} and silicate bulk composition at high pressures. *Carnegie Inst. Wash. Yearbook*, **77**, 709–13.

Mysen, B. O. & Popp, R. K. 1980. Solubility of sulfur in $CaMgSi_2O_6$ and $NaAlSi_3O_8$ melts at high pressure and temperature with controlled f_{O_2} and f_{S_2}. *Amer. J. Sci.*, **280**, 78–92.

Nabelek, P. I., O'Neil, J. R. & Papike, J. J. 1983. Vapor phase exsolution as a controlling factor in hydrogen isotope variation in granitic rocks: the Notch Peak granitic stock, Utah. *Earth Planet. Sci. Lett.*, **66**, 137–50.

Nakada, S. 1983. Zoned magma chamber of the Osuzuyama acid rocks, southwest Japan. *J. Petr.*, **24**, 471–94.

Naldrett, A. J., Bray, J. G., Gasparrini, E. L., Podolsky, T. & Rucklidge, J. C. 1970. Cryptic variation and the petrology of the Sudbury nickel irruptive. *Econ. Geol.*, **65**, 122–55.

Naldrett, A. J. & Kullerud, G. 1967. A study of the Strathcona mine and its bearing on the origin of the nickel–copper ores of the Sudbury district, Ontario. *J. Petr.*, **8**, 453–531.

Nash, W. P. 1972. Mineralogy and petrology of the Iron Hill carbonatite complex, Colorado. *Bull. Geol. Soc. Amer.*, **83**, 1361–82.

Naslund, H. R. 1976. Liquid immiscibility in the system $KAlSi_3O_8$–$NaAlSi_3O_8$–FeO–Fe_2O_3–SiO_2 and its application to natural magmas. *Carnegie Inst. Wash. Yearbook*, **75**, 592–7.

Navrotsky, A., Zimmermann, H. D. & Hervig, R. L. 1983. Thermochemical study of glasses in the system $CaMgSi_2O_6$–$CaAl_2SiO_6$. *Geochim. Cosmochim. Acta*, **47**, 1535–8.

Nelson, D. R. 1992. Isotopic characteristics of potassic rocks: evidence for the involvement of subducted sediments in magma genesis. *Lithos*, **28**, 403–20.

Nelson, D. R., McCulloch, M. T. & Sun, S. S. 1986. The origin of ultrapotassic rocks as inferred from Sr, Nd and Pb isotopes. *Geochim. Cosmochim. Acta*, **50**, 231–45.

Nesbitt, H. W. 1980. Genesis of the New Quebec and Adirondack granulites: evidence for their production by partial melting. *Contr. Min. Petr.*, **72**, 303–10.

Nicholls, I. A. 1971. Petrology of Santorini volcano, Cyclades, Greece. *J. Petr.*, **12**, 67–119.

Nicholls, J. & Carmichael, I. S. E. 1969. A commentary on the absarokite–shoshonite–banakite series of Wyoming, U.S.A. *Schweiz. Min. Petr. Mitt.*, **49**, 47–64.

Nicholls, J. & Stout, M. Z. 1982. Heat effects of assimilation, crystallization, and vesiculation in magmas. *Contr. Min. Petr.*, **81**, 328–39.

Nickel, K. G. & Green, D. H. 1985. Empirical geothermobarometry for garnet peridotites and implications for the nature of the lithosphere, kimberlites and diamonds. *Earth Planet. Sci. Lett.*, **73**, 158–70.

Nicolas, A. & Jackson, E. D. 1972. Répartition en deux provinces des péridotites des chaînes alpines logeant la Méditerranée: implications géotectoniques. *Schweiz. Min. Petr. Mitt.*, **52**, 479–95.

Nielsen, T. F. D. 1981. The ultramafic cumulate series, Gardiner complex, East Greenland. Cumulates in a shallow level magma chamber of a nephelinitic volcano. *Contr. Min. Petr.*, **76**, 60–72.

Nielson, D. R. & Stoiber, R. E. 1973. Relationship of potassium content in andesitic lavas and depth to the seismic zone. *J. Geophys. Res.*, **78**, 6887–92.

Nielson, J. E., Budahn, J. R., Unruh, D. M. & Wilshire, H. G. 1993. Actualistic models of mantle metasomatism documented in a composite xenolith from Dish Hill, California. *Geochim. Cosmochim. Acta*, **57**, 105–21.

Ninkovich, D. & Hays, J. D. 1972. Mediterranean island arcs and origin of high potash volcanoes. *Earth Planet. Sci. Lett.*, **16**, 331–45.

Nisbet, E. G. & Walker, D. 1982. Komatiites and the structure of the Archaean mantle. *Earth Planet. Sci. Lett.*, **60**, 105–13.

Nixon, G. T. 1988. Petrology of the younger andesites and dacites of Iztaccihuatl volcano, Mexico: I. Disequilibrium phenocryst assemblages as indicators of magma chamber processes. *J. Petr.*, **29**, 213–64.

Noble, D. C. & Korringa, M. K. 1974. Strontium, rubidium, potassium, and calcium variations in Quaternary lavas, Crater Lake, Oregon, and their residual glasses. *Geology*, **2**, 187–90.

Nockolds, S. R. 1954. Average chemical compositions of some igneous rocks. *Bull. Geol. Soc. Amer.*, **65**, 1007–32.

Nohda, S. & Wasserburg, G. J. 1981. Nd and Sr isotopic study of volcanic rocks from Japan. *Earth Planet. Sci. Lett.*, **52**, 264–76.

Norton, D., Taylor, H. P. & Bird, D. K. 1984. The geometry and high-temperature brittle deformation of the Skaergaard intrusion. *J. Geophys. Res.*, **89**, 10178–92.

Nyström, J. O. & Henriquez, F. 1994. Magmatic features of iron ores of the Kiruna type in Chile and Sweden: ore textures and magnetite geochemistry. *Econ. Geol.*, **89**, 820–39.

Oakeshott, G. B. 1968. Diapiric structures in Diablo Range, California. In *Diapirism and diapirs* (eds J. Brannstein & G. D. O'Brien), *Mem. Amer. Ass. Petrol. Geol.*, **8**, 228–43.

Obata, M. 1976. The solubility of Al_2O_3 in orthopyroxenes in spinel and plagioclase peridotites and spinel pyroxenite. *Amer. Min.*, **61**, 804–16.

Obata, M. 1980. The Ronda peridotite: garnet–, spinel–, and plagioclase–lherzolite facies and the P–T trajectories of a high-temperature mantle intrusion. *J. Petr.*, **21**, 533–72 (also **23**, 296–8).

Obata, M. & Dickey, J. S. 1976. Phase relations of mafic layers in the Ronda peridotite. *Carnegie Inst. Wash. Yearbook*, **75**, 562–6.

Oen, I. S. 1960. The intrusion mechanism of the late-Hercynian, post-tectonic granite plutons of northern Portugal. *Geol. en Mijnbouw*, **39**, 257–96.

O'Hara, M. J. 1968. The bearing of phase equilibria studies in synthetic and natural systems on the origin and evolution of basic and ultrabasic rocks. *Earth Sci. Revs*, **4**, 69–133.

O'Hara, M. J. 1980. Nonlinear nature of the unavoidable long-lived isotopic, trace and major element contamination of a developing magma chamber. *Phil. Trans. Roy. Soc. Lond. A*, **297**, 215–27.

O'Hara, M. J. & Mathews, R. E. 1981. Geochemical evolution in an advancing, periodically replenished, periodically tapped, continuously fractionated magma chamber. *J. Geol. Soc. London*, **138**, 237–77.

O'Hara, M. J., Richardson, S. W. & Wilson, G. 1971. Garnet–peridotite stability and occurrence in crust and mantle. *Contr. Min. Petr.*, **32**, 48–68.

O'Hara, M. J. & Yoder, H. S. 1967. Formation and fractionation of basic magmas at high pressures. *Scot. J. Geol.*, **3**, 67–117.

Olafsson, M. & Eggler, D. H. 1983. Phase relations of amphibole, amphibole–carbonate, and phlogopite–carbonate peridotite: petrologic constraints on the asthenosphere. *Earth Planet. Sci. Lett.*, **64**, 305–15.

Oldenburg, C. M., Spera, F. J., Yuen, D. A. & Sewell, G. 1989. Dynamic mixing in magma bodies: theory, simulations, and implications. *J. Geophys. Res.*, **94**, 9215–36.

O'Neill, H. S. C. & Wall, V. J. 1987. The olivine–orthopyroxene–spinel oxygen geobarometer, the nickel precipitation curve, and the oxygen fugacity of the Earth's upper mantle. *J. Petr.* **28**, 1169–91.

O'Nions, R. K. & Grönvold, K. 1973. Petrogenetic relationship of acid and basic rocks in Iceland: Sr-isotopes and rare-earth elements in late and postglacial volcanics. *Earth Planet. Sci. Lett.*, **19**, 397–409.

O'Nions, R. K. & Pankhurst, R. J. 1974. Petrogenetic significance of isotope and trace element variations in volcanic rocks from the Mid-Atlantic. *J. Petr.*, **15**, 603–34.

Onuma, N., Higuchi, H., Wakita, H. & Nagasawa, H. 1968. Trace element partition between two pyroxenes and the host lava. *Earth Planet. Sci. Lett.*, **5**, 47–51.

Osborn, E. F. 1959. Role of oxygen pressure in the crystallization and differentiation of basaltic magma. *Amer. J. Sci.*, **257**, 609–47.

Osborn, E. F. 1969. The complementariness of orogenic andesite and alpine peridotite. *Geochim. Cosmochim. Acta*, **33**, 307–24.

Osborn, E. F. 1979. The reaction principle, pp. 133–69 in Yoder (1979).

Osborn, E. F. & Schairer, J. F. 1941. The ternary system pseudowollastonite–akermanite–gehlenite. *Amer. J. Sci.*, **239**, 715–63.

Osborn, E. F. & Tait, D. B. 1952. The system diopside–forsterite–anorthite. *Amer. J. Sci.*, Bowen volume, 413–33.

Oversby, V. M. & Ewart, A. 1972. Lead isotopic compositions of Tonga–Kermadec volcanics and their petrogenetic significance. *Contr. Min. Petr.*, **37**, 181–210.

Owens, B. E., Rockow, M. W. & Dymek, R. F. 1993. Jotunites from the Grenville Province, Quebec: petrological characteristics and implications for massif anorthosite petrogenesis. *Lithos*, **30**, 57–80.

Page, N. J. 1977. Stillwater complex, Montana: rock succession, metamorphism and structure of the complex and adjacent rocks. *Prof. Paper U.S.G.S.*, **999**, 79 pp.

Page, N. J. 1979. Stillwater complex, Montana – structure, mineralogy, and petrology of the Basal zone with emphasis on the occurrence of sulfides. *Prof. Paper U.S.G.S.*, **1038**, 69 pp.

Panayiotou, A. (ed.). 1980. *Ophiolites: Proceedings International Ophiolite Symposium, Cyprus 1979*. Geological Survey Dept, Republic of Cyprus, Nicosia, 781 pp.

Pankhurst, R. J. 1969. Strontium isotope studies related to petrogenesis in the Caledonian basic igneous province of N.E. Scotland. *J. Petr.*, **10**, 115–43.

Pankhurst, R. J., Beckinsale, R. D. & Brooks, C. K. 1976. Strontium and oxygen isotope evidence relating to the petrogenesis of the Kangerdlugssuaq alkaline intrusion, East Greenland. *Contr. Min. Petr.*, **54**, 17–42.

Pankhurst, R. J., Hole, M. J. & Brook, M. 1988. Isotope evidence for the origin of Andean granites. *Trans. Roy. Soc. Edinburgh, Earth Sciences*, **79**, 123–33.

Parmentier, E. M., Turcotte, D. L. & Torrance, K. E. 1975. Numerical experiments on the structure of mantle plumes. *J. Geophys. Res.*, **80**, 4417–24.

Parsons, I. 1979. The Klokken gabbro–syenite complex, south Greenland: cryptic variation and origin of inversely graded layering. *J. Petr.*, **20**, 653–94.

Paterson, B. A., Stephens, W. E., Rogers, G., Williams, I. S., Hinton, R. W. & Herd, D. A. 1992. The nature of zircon inheritance in two granite plutons. *Trans. Roy. Soc. Edinburgh, Earth Sciences*, **83**, 459–71.

Patiño Douce, A. E. 1993. Titanium substitution in biotite:

an empirical model with applications to thermometry, O_2 and H_2O barometries, and consequences for biotite stability. *Chem. Geol.*, **108**, 133–62.

Patiño Douce, A. E. & Johnston, A. D. 1991. Phase equilibria and melt productivity in the pelitic system: implications for the origin of peraluminous granitoids and aluminous granulites. *Contr. Min. Petr.*, **107**, 202–18.

Pauling, L. 1970. *General chemistry* (3rd edn), W. H. Freeman & Co., San Francisco.

Peacock, M. A. 1931. Classification of igneous rock series. *J. Geol.*, **39**, 54–67.

Pearce, J. A. 1976. Statistical analysis of major element patterns in basalts. *J. Petr.*, **17**, 15–43.

Pearce, J. A. & Cann, J. R. 1973. Tectonic setting of basic volcanic rocks determined using trace element analyses. *Earth Planet. Sci. Lett.*, **19**, 290–300.

Pearson, D. G., Davies, G. R. & Nixon, P. H. 1993. Geochemical constraints on the petrogenesis of diamond facies pyroxenites from the Beni Bousera peridotite massif, north Morocco. *J. Petr.*, **34**, 125–72.

Pedersen, A. K. 1979. Basaltic glass with high-temperature equilibrated immiscible sulphide bodies with native iron from Disko, Central west Greenland. *Contr. Min. Petr.*, **69**, 397–407.

Pedersen, A. K. & Rønsbo, J. G. 1987. Oxygen deficient Ti oxides (natural magneli phases) from mudstone xenoliths with native iron from Disko, central west Greenland. *Contr. Min. Petr.*, **96**, 35–46.

Pegram, W. J. & Allègre, C. J. 1992. Osmium isotopic compositions from oceanic basalts. Earth Planet. *Sci. Lett.*, **111**, 59–68.

Perfit, M. R., Brueckner, H., Lawrence, J. R. & Kay, R. W. 1980. Trace element and isotopic variations in a zoned pluton and associated volcanic rocks, Unalaska Island, Alaska: a model for fractionation in the Alaskan calcalkaline suite. *Contr. Min. Petr.*, **73**, 69–87.

Perfit, M. R. & Fornari, D. J. 1983. Geochemical studies of abyssal lavas recovered by DSRV Alvin from eastern Galapagos Rift, Inca transform, and Ecuador Rift 2. Phase chemistry and crystallization history. *J. Geophys. Res.*, **88**, 10530–50.

Persikov, E. S. 1977. Viscosity of water-rich granite melts at high pressures. *High Temperatures – High Pressures*, **9**, 700–1.

Peterman, Z. E. 1979. Strontium isotope geochemistry of late Archaean to late Cretaceous tonalites and trondhjemites, pp. 133–47 in *Trondhjemites, dacites, and related rocks* (ed. F. Barker), Elsevier, Amsterdam.

Peterman, Z. E., Carmichael, I. S. E. & Smith, A. L. 1970. Sr^{87}/Sr^{86} ratios of Quaternary lavas of the Cascade Range, Northern California. *Bull. Geol. Soc. Amer.*, **81**, 311–8.

Peterson, T. D. 1989. Peralkaline nephelinites. II. Low pressure fractionation and the hypersodic lavas of Old-oinyo L'engai. *Contr. Min. Petr.*, **102**, 336–46.

Petö, P. 1974. Plutonic evolution of the Canadian Cordillera. *Bull. Geol. Soc. Amer.*, **85**, 1269–76.

Petrini, R., Civetta, L., Piccirillo, E. M., Bellieni, G.,

Comin-Chiaramonti, P., Marques, L.S. & Melfi, A. J. 1987. Mantle heterogeneity and crustal contamination in the genesis of low-Ti continental flood basalts from the Parana Plateau (Brazil): Sr–Nd isotope and geochemical evidence. *J. Petr.*, **28**, 701–26.

Philpotts, A. R. 1966. Origin of the anorthosite–mangerite rocks in southern Quebec. *J. Petr.*, **7**, 1–64.

Philpotts, A. R. 1967. Origin of certain iron–titanium oxide and apatite rocks. *Econ. Geol.*, **62**, 303–15.

Philpotts, A. R. 1971. Immiscibility between feldspathic and gabbroic magmas. *Nature Phys. Sci.*, **229**, 107–9.

Philpotts, A. R. 1981. Liquid immiscibility in silicate melt inclusions in plagioclase phenocrysts. *Bull. Mineral.*, **104**, 317–24.

Philpotts, A. R. 1982. Composition of immiscible liquids in volcanic rocks. *Contr. Min. Petr.*, **80**, 201–18.

Philpotts, A. R. & Asher, P. M. 1993. Wallrock melting and reaction effects along the Higganum diabase dyke in Connecticut: contamination of a continental flood basalt feeder. *J. Petr.*, **34**, 1029–58.

Phinney, W. C., Morrison, D. A. & Maczuga, D. E. 1988. Anorthosites and related megacrystic units in the evolution of Archaean crust. *J. Petr.*, **29**, 1283–323.

Pichler, H. 1970. Volcanism in eastern Sicily and the Aeolian Islands, pp. 261–83 in *Geology and history of Sicily* (eds W. Alvarez & K. H. A. Gohrbandt). Petroleum Exploration Society of Libya, Tripoli, 291 pp.

Pichler, H., Günther, D. & Kussmaul, S. 1980. Thira Island. 1:50,000 Geological Map. I.G.M.E., Athens.

Pineau, F. & Javoy, M. 1983. Carbon isotopes and concentrations in mid-ocean ridge basalts. *Earth Planet. Sci. Lett.*, **62**, 239–57

Pinkerton, H. & Sparks, R. S. J. 1978. Field measurements of the rheology of lava. *Nature*, **276**, 383–5.

Pitcher, W. S. 1952. The migmatitic older granodiorite of Thorr district, Co. Donegal. *Quart. J. Geol. Soc. London*, **108**, 413–46.

Pitcher, W. S. 1953. The Rosses granitic ring-complex, County Donegal, Eire. *Proc. Geol. Assoc.* (London), **64**, 153–82.

Pitcher, W. S. 1978. The anatomy of a batholith. *J. Geol. Soc. London*, **135**, 157–82.

Pitcher, W. S. 1993. *The nature and origin of granite*. Blackie, Glasgow, 321 pp.

Pitcher, W. S., Atherton, M. P., Cobbing, E. J. & Beckinsale, R. D. (eds). 1985. *Magmatism at a plate edge: the Peruvian Andes*. Blackie, Glasgow, 328 pp.

Pitcher, W. S. & Berger, A. R. 1972. *The geology of Donegal: a study of granite emplacement and unroofing*. Wiley-Interscience, New York and London, 435 pp.

Podmore, F. & Wilson, A. H. 1987. A reappraisal of the structure, geology and emplacement of the Great Dyke, Zimbabwe, pp. 317–30 in *Mafic dyke swarms* (eds H. C. Halls & W. F. Fahrig), Geol. Soc. Canada Spec. Paper 34.

Poulson, S. R., Kubilius, W. P. & Ohmoto, H. 1991. Geochemical behavior of sulfur in granitoids during intrusion of the South Mountain batholith, Nova Scotia, Canada. *Geochim. Cosmochim. Acta*, **55**, 3809–30.

Presnall, D. C. 1966. The join forsterite–diopside–iron oxide and its bearing on the crystallization of basaltic and ultramafic magmas. *Amer. J. Sci.*, **264**, 753–809.

Presnall, D. C. 1969. The geometrical analysis of partial fusion. *Amer. J. Sci.*, **267**, 1178–94.

Presnall, D. C. & Bateman, P. C. 1973. Fusion relations in the system $NaAlSi_3O_8$–$CaAl_2Si_2O_8$–$KAlSi_3O_8$–SiO_2–H_2O and generation of granitic magmas in the Sierra Nevada batholith. *Bull. Geol. Soc. Amer.*, **84**, 3181–202.

Presnall, D. C., Dixon, J. R., O'Donnell, T. H. & Dixon, S. A. 1979. Generation of mid-ocean ridge tholeiites. *J. Petr.*, **20**, 3–35.

Presnall, D. C. & Hoover, J. D. 1984. Composition and depth of origin of primary mid-ocean ridge basalts. *Contr. Min. Petr.*, **87**, 170–8.

Price, R. C. & Compston, W. 1973. The geochemistry of the Dunedin volcano: strontium isotope chemistry. *Contr. Min. Petr.*, **42**, 55–61.

Propach, G. 1976. Models of filter differentiation. *Lithos*, **9**, 203–9.

Pushkar, P. 1968. Strontium isotope ratios in volcanic rocks of three island arc areas. *J. Geophys. Res.*, **73**, 2701–14.

Pushkar, P. & Stoeser, D. B. 1975. $^{87}Sr/^{86}Sr$ ratios in some volcanic rocks and some semifused inclusions of the San Francisco volcanic field. *Geology*, **3**, 669–71.

Quick, J. E. 1981. Petrology and petrogenesis of the Trinity peridotite, an upper mantle diapir in the eastern Klamath Mountains, Northern California. *J. Geophys. Res.*, **86**, 11837–63.

Rankin, A. H. & Le Bas, M. J. 1974. Liquid immiscibility between silicate and carbonate melts in naturally occurring ijolite magma. *Nature*, **250**, 206–9.

Read, H. H. 1931. The geology of central Sutherland. *Mem. Geol. Surv. Scotland*, 238 pp.

Read, H. H. 1935. The gabbros and associated xenolithic complexes of the Haddo House district, Aberdeenshire. *Quart. J. Geol. Soc. London*, **91**, 591–638.

Reid, A. M., Donaldson, C. H., Dawson, J. B., Brown, R. W. & Ridley, W. I. 1975. The Igwisi Hills extrusive 'kimberlites'. *Phys. Chem. Earth*, **9**, 199–218.

Reid, I., Orcutt, J. A. & Prothero, W. A. 1977. Seismic evidence for a narrow zone of partial melting underlying the East Pacific Rise at 21 °N. *Bull. Geol. Soc. Amer.*, **88**, 678–82.

Reid, J. B., Murray, D. P., Hermes, O. D. & Steig, E. J. 1993. Fractional crystallization in granites of the Sierra Nevada: how important is it? *Geology*, **21**, 587–90.

Reuther, H. & Hinz, W. 1980. Thermotransport in lithium silicate glasses. *Physica Status Solidi*, **59A**, K87–K89.

Reynolds, I. M. 1985. Contrasted mineralogy and textural relationships in the uppermost titaniferous magnetite layers of the Bushveld complex in the Bierkraal area north of Rustenburg. *Econ. Geol.*, **80**, 1027–48.

Rhodes, J. M., Dungan, M. A., Blanchard, D. P. & Long, P. E. 1979. Magma mixing at mid-ocean ridges:

evidence from basalts drilled near 22 °N on the mid-Atlantic Ridge. *Tectonophysics*, **55**, 35–61.

Richardson, S. H., Gurney, J. J., Erlank, A. J. & Harris, J. W. 1984. Origin of diamonds in old enriched mantle. *Nature*, **310**, 198–202.

Richet, P. 1984. Viscosity and configurational entropy of silicate melts. *Geochim. Cosmochim. Acta*, **48**, 471–83.

Richey, J. E. 1932. Tertiary ring structures in Britain. *Trans. Geol. Soc. Glasgow*, **19**, 42–140.

Richey, J. E. & Thomas, H. H. 1930. The geology of Ardnamurchan, north-west Mull and Coll. *Mem. Geol. Surv. Scotland*, 393 pp. (with corresponding Geological Survey map on the scale of 1:50,000).

Ringwood, A. E. 1974. The petrological evolution of island arc systems. *J. Geol. Soc. Lond.*, **130**, 183–204.

Ringwood, A. E. 1975. *Composition and petrology of the Earth's mantle*. McGraw-Hill, New York, 618 pp.

Ringwood, A. E. 1991. Phase transformations and their bearing on the constitution and dynamics of the mantle. *Geochim. Cosmochim. Acta*, **55**, 2083–110.

Ringwood, A. E., Kesson, S. E., Hibberson, W. & Ware, N. 1992. Origin of kimberlites and related magmas. *Earth Planet. Sci. Lett.*, **113**, 521–38.

Robertson, J. K. & Wyllie, P. J. 1971. Rock–water systems, with special reference to the water-deficient region. *Amer. J. Sci.*, **271**, 252–77.

Robie, R. A., Hemingway, B. S. & Fisher, J. R. 1978. Thermodynamic properties of minerals and related substances at 298.15 K and 1 bar (10^5 Pascals) pressure and at higher temperatures. *Bull. U.S.G.S.*, **1452**, 456 pp.

Rock, N. M. S. 1976. The comparative strontium isotopic composition of alkaline rocks: new data from southern Portugal and East Africa. *Contr. Min. Petr.*, **56**, 205–28.

Roden, M. K., Hart, S. R., Frey, F. A. & Melson, W. G. 1984. Sr, Nd and Pb isotopic and REE geochemistry of St. Paul's Rocks: the metamorphic and metasomatic development of an alkali basalt mantle source. *Contr. Min. Petr.*, **85**, 376–90.

Roedder, E. 1951. Low temperature liquid immiscibility in the system K_2O–FeO–Al_2O_3–SiO_2. *Amer. Min.*, **36**, 282–6.

Roedder, E. 1965. Liquid CO_2 inclusions in olivine-bearing nodules and phenocrysts from basalts. *Amer. Min.*, **50**, 1746–82.

Roedder, E. 1979. Silicate liquid immiscibility in magmas, pp. 14–57 in Yoder (1979).

Roedder, E. & Weiblen, P. W. 1970. Lunar petrology of silicate melt inclusions, Apollo 11 rocks. *Geochim. Cosmochim. Acta*, **34** (Supplement 1), Vol. 1, 801–37.

Rollinson, H. R. 1993. *Using geochemical data: evaluation, presentation, interpretation*. Longman Scientific & Technical, Harlow.

Rose, W. I., Anderson, A. T., Woodruff, L. G. & Bonis, S. B. 1978. The October 1974 basaltic tephra from Fuego volcano: description and history of the magma body. *J. Volc. Geotherm. Res.*, **4**, 3–53.

Rose, W. I., Stoiber, R. E. & Malinconico, L. L. 1982. Eruptive gas compositions and fluxes of explosive volcanoes: budget of S and Cl emitted from Fuego volcano, Guatemala, pp. 669–76 in Thorpe (1982).

Rosenhauer, M. 1976. Effect of pressure on the melting enthalpy of diopside under dry and H_2O-saturated conditions. *Carnegie Inst. Wash. Yearbook*, **75**, 648–51.

Rosholt, J. N., Zartman, R. E. & Nkomo, I. T. 1973. Lead isotope systematics and uranium depletion in the Granite Mountains, Wyoming. *Bull. Geol. Soc. Amer.*, **84**, 989–1002.

Ruckmick, J. C. & Noble, J. A. 1959. Origin of the ultramafic complex at Union Bay, southeastern Alaska. *Bull. Geol. Soc. Amer.*, **70**, 981–1018.

Ryabchikov, I. D., Babansky, A. D. & Dmitriev, Y. I. 1982. Genesis of calc-alkaline magmas: experiments with partial melting of mixed sediments and basalts from the Middle America trench, southern Mexico transect. *Initial Reports Deep Sea Drilling Project*, **66**, 699–702.

Ryan, J. G. & Langmuir, C. H. 1993. The systematics of boron abundances in young volcanic rocks. *Geochim. Cosmochim. Acta*, **57**, 1489–98.

Sabine, P. A. 1963. The Strontian granite complex, Argyllshire. *Bull. Geol. Surv. G.B.*, **20**, 6–42.

Sachs, P. M. & Strange, S. 1993. Fast assimilation of xenoliths in magmas. *J. Geophys. Res.*, **98**, 19741–54.

Sacks, I. S. 1984. Subduction geometry and magma genesis, pp. 34–46 in *Explosive volcanism; inception, evolution, and hazards*. National Academy Press, Washington, DC.

Saemundsson, K. 1979. Outline of the geology of Iceland. Jökull, 29, 7–28.

Saether, E. 1950. On the genesis of peralkaline rock provinces. *Rept. 18th Session Internat. Geol. Congr., London, Part II (Problems of geochemistry)*, 123–30.

Saether, E. 1957. The alkaline rock province of the Fen area in southern Norway. *Det Kgl. Norske Vid. Selsk. Skrifter 1957*, no. 1, 150 pp.

Sahama, Th. G. 1974. Potassium-rich alkaline rocks, pp. 96–109 in Sørensen (1974).

Sakuyama, M. & Kushiro, I. 1979. Vesiculation of hydrous andesitic melt and transport of alkalies by separated vapour phase. *Contr. Min. Petr.*, **71**, 61–6.

Salters, V. J. M. & Hart, S. R. 1989. The hafnium paradox and the role of garnet in the source of mid-ocean-ridge basalts. *Nature*, **342**, 420–2.

Sasaki, A. & Ishihara, S. 1979. Sulfur isotopic composition of the magnetite-series and ilmenite-series granitoids in Japan. *Contr. Min. Petr.*, **68**, 107–15.

Satake, H. & Matsuda, J. 1979. Strontium and hydrogen isotope geochemistry of fresh and metabasalt dredged from the Mid-Atlantic Ridge. *Contr. Min. Petr.*, **70**, 153–7.

Sato, M. & Wright, T. L. 1966. Oxygen fugacities directly measured in magmatic gases. *Science*, **153**, 1103–5.

Sawyer, E. W. 1991. Disequilibrium melting and the rate of melt-residuum separation during migmatization of mafic rocks from the Grenville Front, Quebec. *J. Petr.*, **32**, 701–38.

Saxena, S. K. & Eriksson, G. 1983. Theoretical computation of mineral assemblages in pyrolite and lherzolite. *J. Petr.*, **24**, 538–55.

Scarfe, C. M., Mysen, B. O. & Virgo, D. 1987. Pressure dependence of the viscosity of silicate melts, pp. 59–67 in Mysen (1987).

Scarpati, C., Cole, P. & Perrotta, A. 1993. The Neapolitan Yellow Tuff – a large volume multiphase eruption from Campi Flegrei, southern Italy. *Bull. Volc.*, **55**, 343–56.

Schairer, J. F. 1957. Melting relations of the common rock-forming oxides. *J. Amer. Ceram. Soc.*, **40**, 215–35.

Schairer, J. F. & Bowen, N. L. 1947. The system anorthite–leucite–silica. *Bull. Comm. Geol. Finlande*, **140**, 67–87.

Schairer, J. F. & Yoder, H. S. 1961. Crystallization in the system nepheline–forsterite–silica at one atmosphere pressure. *Carnegie Inst. Wash. Yearbook*, **60**, 141–4.

Schmid, R. 1968. Excursion guide for the Valle d'Ossola section of the Ivrea–Verbano zone (Prov. Novara, Northern Italy). *Schweiz. Min. Petr. Mitt.*, **48**, 305–14 + Plate I.

Schott, J. 1983. Thermal diffusion and magmatic differentiation: a new look at an old problem. *Bull. Mineral.*, **106**, 247–62.

Schreyer, R. W., Massonne, H. J. & Chopin, C. 1987. Continental crust subducted to depths near 100 km: implications for magma and fluid genesis in collision zones, pp. 155–63 in *Magmatic processes; physicochemical principles* (ed. B. O. Mysen). Geochemical Society Spec. Publ. no. 1.

Schubert, W. 1977. Reaktionen im alpinotypen Peridotitmassiv von Ronda (Spanien) und seinen partiellen Schmelzprodukten. *Contr. Min. Petr.*, **62**, 205–20.

Schulze, D. J. & Helmstaedt, H. 1988. Coesite–sanidine eclogites from kimberlite: products of mantle fractionation or subduction? *J. Geol.*, **96**, 435–43.

Scott, B. H. 1979. Petrogenesis of kimberlites and associated potassic lamprophyres from central West Greenland, pp. 190–205 in Boyd & Meyer (1979a).

Sederholm, J. J. 1907. On granite and gneiss: their origin, relations and occurrence in the Pre-Cambrian complex of Fennoscandia. *Bull. Comm. Geol. Finlande*, **23** (reprinted in *Selected Works, Granites and Migmatites*, J. J. Sederholm; Oliver & Boyd, Edinburgh and London, 1967).

Seifert, K. E., Voigt, A. F., Smith, M. F. & Stensland, W. A. 1977. Rare earths in the Marcy and Morin anorthosite complexes. *Can. J. Earth Sci.*, **14**, 1033–45.

Self, S. & Rampino, M. R. 1981. The 1883 eruption of Krakatau. *Nature*, **294**, 699–704.

Sen, C. & Dunn, T. 1994. Dehydration melting of a basaltic composition amphibolite at 1.5 and 2.0 GPa: implications for the origin of adakites. *Contr. Min. Petr.*, **117**, 394–409.

Shaw, H. F. & Wasserburg, G. J. 1984. Isotopic constraints on the origin of Appalachian mafic complexes. *Amer. J. Sci.*, **284**, 319–49.

Shaw, H. R. 1963. Obsidian-H_2O viscosities at 1000 and

2000 bars in the temperature range 700° to 900 °C. *J. Geophys. Res.*, **68**, 6337–43.

Shaw, H. R. 1965. Comments on viscosity, crystal settling, and convection in granitic magmas. *Amer. J. Sci.*, **263**, 120–52.

Shaw, H. R. 1969. Rheology of basalt in the melting range. *J. Petr.*, **10**, 510–35.

Shaw, H. R. 1972. Viscosities of magmatic silicate liquids: an empirical method of prediction. *Amer. J. Sci.*, **272**, 870–93.

Shaw, H. R. 1974. Diffusion of H_2O in granitic liquids: Part I. Experimental data; Part II. Mass transfer in magma chambers, pp. 139–70 in *Geochemical transport and kinetics.* (eds A. W. Hofmann, B. J. Giletti, H. S. Yoder & R. A. Yund), Carnegie Institute of Washington.

Shaw, H. R., Wright, T. L., Peck, D. L. & Okamura, R. 1968. The viscosity of basaltic magma: an analysis of field measurements in Makaopuhi lava lake, Hawaii. *Amer. J. Sci.*, **266**, 225–64.

Shervais, J. W. 1982. Ti–V plots and the petrogenesis of modern and ophiolitic lavas. *Earth Planet. Sci. Lett.*, **59**, 101–18.

Shervais, J. W. & Mukasa, S. B. 1991. The Balmuccia orogenic lherzolite massif, Italy, pp. 155–74 in *Orogenic lherzolites and mantle processes* (eds M. A. Menzies, C. Dupuy and A. Nicolas), *J. Petr.* Special Volume, Oxford University Press, Oxford.

Shibata, T. 1976. Phenocryst–bulk rock composition relations of abyssal tholeiites and their petrogenetic significance. *Geochim. Cosmochim. Acta*, **40**, 1407–17.

Shima, H. & Naldrett, A. J. 1975. Solubility of sulfur in an ultramafic melt and the relevance of the system Fe–S–O. *Econ. Geol.*, **70**, 960–7.

Shiraki, K. 1971. Metamorphic basement rocks of Yap Islands, western Pacific: possible oceanic crust beneath an island arc. *Earth Planet. Sci. Lett.*, **13**, 167–74.

Siena, F. & Coltorti, M. 1989. The petrogenesis of a hydrated mafic-ultramafic complex and the role of amphibole fractionation at Finero (Italian Western Alps). *Neues Jb. Min. Mh., Jahrgang 1989*, 255–74.

Sighinolfi, G. P. 1971. Investigations into deep crustal levels: fractionating effects and geochemical trends related to high grade metamorphism. *Geochim. Cosmochim. Acta*, **35**, 1005–21.

Sigmarsson, O., Hémond, C., Condomines, M., Fourcade, S. & Oskarsson, N. 1991. Origin of silicic magma in Iceland revealed by Th isotopes. *Geology*, **19**, 621–4.

Sigurdsson, H. & Sparks, R. S. J. 1981. Petrology of rhyolitic and mixed magma ejecta from the 1875 eruption of Askja, Iceland. *J. Petr.*, **22**, 41–84.

Simkin, T. & Fiske, R. S. 1983. *Krakatau 1883: the volcanic eruption and its effects.* Smithsonian Institution Press, Washington, DC, 464 pp.

Simkin, T., Siebert, L., McClelland, L., Bridge, D., Newhall, C. & Latter, J. H. 1981. *Volcanoes of the world.* Hutchinson Ross Publishing Co., Stroudsburg, Pennsylvania, 233 pp. Distributed by Academic Press.

Simmons, G. 1964. Gravity survey and geological interpretation, northern New York state. *Bull. Geol. Soc. Amer.*, **75**, 81–98.

Singer, B. S., Myers, J. D. & Frost, C. D. 1992. Mid-Pleistocene lavas from the Seguam volcanic center, central Aleutian arc: closed-system fractional crystallization of a basalt to rhyodacite eruptive suite. *Contr. Min. Petr.*, **110**, 87–112.

Sinton, J. M. 1980. Petrology and evolution of the Red Mountain ophiolite complex, New Zealand. *Amer. J. Sci.*, **280A**, 296–328.

Sisson, T. W. & Grove, T. L. 1993. Experimental investigations of the role of H_2O in calc-alkaline differentiation and subduction zone magmatism. *Contr. Min. Petr.*, **113**, 143–66.

Sisson, T. W. & Layne, G. D. 1993. H_2O in basalt and basaltic andesite glass inclusions from four subduction-related volcanoes. *Earth Planet. Sci. Lett.*, **117**, 619–35.

Size, W. B. 1972. Petrology of the Red Hill syenitic complex, New Hampshire. *Bull. Geol. Soc. Amer.*, **83**, 3747–60.

Size, W. B. 1979. Petrology, geochemistry and genesis of the type area trondhjemite in the Trondheim region, central Norwegian Caledonides. *Norges Geol. Unders.*, **351**, 51–76.

Skiba, W. 1952. The contact phenomena on the north-west of the Crossdoney complex, Co. Cavan. *Trans. Edinburgh Geol. Soc.*, **15**, 322–45.

Skinner, B. J. 1966. Thermal expansion. *Mem. Geol. Soc. Amer.*, **97**, 76–96.

Skinner, B. J. & Peck, D. L. 1969. An immiscible sulfide melt from Hawaii. *Econ. Geol.*, Monograph 4, 310–22.

Sleep, N. H. 1990. Hotspots and mantle plumes: some phenomenology. *J. Geophys. Res.*, **95**, 6715–36.

Smewing, J. D. 1980. An Upper Cretaceous ridge-transform intersection in the Oman ophiolite, pp. 407–13 in Panayiotou (1980).

Smewing, J. D. 1981. Mixing characteristics and compositional differences in mantle-derived melts beneath spreading axes: evidence from cyclically layered rocks in the ophiolite of North Oman. *J. Geophys. Res.*, **86**, 2645–59.

Smith, C. H. & Kapp, H. E. 1963. The Muskox intrusion, a recently discovered layered intrusion in the Coppermine River area, Northwest Territories, Canada. *Special Paper, Min. Soc. Amer.*, **1**, 30–5.

Smith, D. R. & Leeman, W. P. 1993. The origin of Mount St. Helens andesites. *J. Volcanology Geothermal Research*, **55**, 271–303.

Smith, H. S. & Erlank, A. J. 1982. Geochemistry and petrogenesis of komatiites from the Barberton greenstone belt, South Africa, pp. 347–97 in *Komatiites* (eds N. T. Arndt & E. G. Nisbet), George Allen & Unwin, London.

Smith, R. B. 1978. Seismicity, crustal structure, and intraplate tectonics of the interior of the western Cordillera. *Mem. Geol. Soc. Amer.*, **152**, 111–44.

Smith, R. L. & Bailey, R. A. 1966. The Bandelier tuff: a study of ash-flow eruption cycles from zoned magma chambers. *Bull. Volc.*, **29**, 83–103.

Smith, R. L. & Bailey, R. A. 1968. Resurgent cauldrons. *Mem. Geol. Soc. Amer.*, **116**, 613–62.

Smithson, S. B. & Ramberg, I. B. 1979. Gravity interpretation of the Egersund anorthosite complex, Norway: its petrological and geothermal significance. *Bull. Geol. Soc. Amer.*, **90**, 199–204.

Snyder, W. S., Dickinson, W. R. & Silberman, M. L. 1976. Tectonic implications of space–time patterns of Cenozoic magmatism in the western United States. *Earth Planet. Sci. Lett.*, **32**, 91–106.

Sørensen, H. (ed.). 1974. *The alkaline rocks.* John Wiley, London & New York, 622 pp.

Sparks, R. S. J., Meyer, P. & Sigurdsson, H. 1980. Density variation among mid-ocean ridge basalts: implications for magma mixing and the scarcity of primitive lavas. *Earth Planet. Sci. Lett.*, **46**, 419–30.

Speight, J. M., Skelhorn, R. R., Sloan, T. & Knapp, R. J. 1982. The dyke swarms of Scotland, pp. 449–59 in *Igneous rocks of the British Isles* (ed. D. S. Sutherland), John Wiley, Chichester.

Spera, F. J. 1980. Aspects of magma transport, pp. 265–323 in *Physics of magmatic processes* (ed. R. B. Hargraves), Princeton University Press, Princeton, NJ.

Spera, F. J. 1984. Carbon dioxide in petrogenesis III: role of volatiles in the ascent of alkaline magma with special reference to xenolith-bearing mafic lavas. *Contr. Min. Petr.*, **88**, 217–32.

Spera, F. J. & Bergman, S. C. 1980. Carbon dioxide in igneous petrogenesis: I. Aspects of the dissolution of CO_2 in silicate liquids. *Contr. Min. Petr.*, **74**, 55–66.

Spera, F. J., Yuen, D. A. & Kirschvink, S. J. 1982. Thermal boundary layer convection in silicic magma chambers: effect of temperature-dependent rheology and implications for thermogravitational chemical fractionation. *J. Geophys. Res.*, **87**, 8755–67.

Spooner, E. T. C., Beckinsale, R. D., Fyfe, W. S. & Smewing, J. D. 1974. O^{18} enriched ophiolitic metabasic rocks from E. Liguria (Italy), Pindos (Greece), and Troodos (Cyprus). *Contr. Min. Petr.*, **47**, 41–62.

Spry, A. 1958. Some observations on the Jurassic dolerite of the Eureka cone sheet near Zeehan, Tasmania, pp. 93–129 in *Dolerite: a symposium* (ed. S. W. Carey). University of Tasmania, Hobart.

Spulber, S. D. & Rutherford, M. J. 1983. The origin of rhyolite and plagiogranite in oceanic crust: an experimental study. *J. Petr.*, **24**, 1–25.

Stanley, R. S., Roy, D. L., Hatch, N. L. & Knapp, D. A. 1984. Evidence for tectonic emplacement of ultramafic and associated rocks in the pre-Silurian eugeosynclinal belt of western New England – vestiges of an ancient accretionary wedge. *Amer. J. Sci.*, **284**, 559–95.

Staudigel, H. & Hart, S.R. 1983. Alteration of basaltic glass: mechanisms and significance for the oceanic crust–seawater budget. *Geochim. Cosmochim. Acta*, **47**, 337–50.

Steck, A. & Tièche, J. C. 1976. Carte géologique de l'antiforme péridotitique de Finero avec des observations sur les phases de déformation et de recristallisation. *Schweiz. Min. Petr. Mitt.*, **56**, 501–12.

Steiner, J. C., Jahns, R. H. & Luth, W. C. 1975. Crystallization of alkali feldspar and quartz in the haplogranite system $NaAlSi_3O_8$–$KAlSi_3O_8$–SiO_2–H_2O at 4 kb. *Bull. Geol. Soc. Amer.*, **86**, 83–98.

Stephenson, R. & Thomas, M. D. 1979. Three-dimensional gravity analysis of the Kiglapait layered intrusion, Labrador. *Can. J. Earth Sci.*, **16**, 24–37.

Stern, C. R., Huang, W. L. & Wyllie, P. J. 1975. Basalt–andesite–rhyolite–H_2O: crystallization intervals with excess H_2O and H_2O-undersaturated liquidus surfaces to 35 kilobars, with implications for magma genesis. *Earth Planet. Sci. Lett.*, **28**, 189–96.

Stern, C. R. & Wyllie, P. J. 1981. Phase relations of I-type granite with H_2O to 35 kilobars: the Dinkey Lakes biotite–granite from the Sierra Nevada batholith. *J. Geophys. Res.*, **86**, 10412–22.

Stern, R. J. 1979. On the origin of andesite in the Northern Mariana Island arc: implications from Agrigan. *Contr. Min. Petr.*, **68**, 207–19.

Stern, R. J. & Ito, E. 1983. Trace element and isotopic constraints on the source of magmas in the active Volcano and Mariana island arcs, western Pacific. *J. Volc. Geotherm. Res.*, **18**, 461–82.

Stewart, B. W. & De Paolo, D. J. 1990. Isotopic studies of processes in mafic magma chambers: II. The Skaergaard intrusion, East Greenland. *Contr. Min. Petr.*, **104**, 125–41.

Stille, P., Unruh, D. M. & Tatsumoto, M. 1986. Pb, Sr, Nd, and Hf isotopic constraints on the origin of Hawaiian basalts and evidence for a unique mantle source. *Geochim. Cosmochim. Acta*, **50**, 2303–19.

Stolper, E. 1980. A phase diagram for mid-ocean ridge basalts: preliminary results and implications for petrogenesis. *Contr. Min. Petr.*, **74**, 13–27.

Stolper, E. & Walker, D. 1980. Melt density and the average composition of basalt. *Contr. Min. Petr.*, **74**, 7–12.

Stone, M. 1975. Structure and petrology of the Tregonning–Godolphin granite, Cornwall. *Proc. Geol. Assoc., London*, **86**, 155–70.

Streckeisen, A. 1976. To each plutonic rock its proper name. *Earth Science Reviews*, **12**, 1–33.

Sturt, B. A., Speedyman, D. L. & Griffin, W. L. 1980. The Nordre Bumandsfjord ultramafic pluton, Seiland, North Norway. Part I: Field relations. *Norges Geol. Unders.*, **358**, 1–30.

Styles, M. T. & Kirby, G. A. 1980. New investigations of the Lizard complex, Cornwall, England and a discussion of the ophiolite model, pp. 517–26 in Panayiotou (1980).

Suen, C. J. & Frey, F. A. 1987. Origins of the mafic and ultramafic rocks in the Ronda peridotite. *Earth Planet. Sci. Lett.*, **85**, 183–202.

Sugimura, A. 1968. Spatial relations of basaltic magmas in island arcs, pp. 537–71 in Hess & Poldervaart (1967–68).

Sugisaki, R. 1976. Chemical characteristics of volcanic rocks: relation to plate movements. *Lithos*, **9**, 17–30.

Sun, S. S. 1980. Lead isotopic study of young volcanic

rocks from mid-ocean ridges, ocean islands and island arcs. *Phil. Trans. Roy. Soc. Lond. A*, **297**, 409–45.

Sutherland, F. L. 1981. Migration in relation to possible tectonic and regional controls in Eastern Australian volcanism. *J. Volc. Geotherm. Res.*, **9**, 181–213.

Suyehiro, K. & Sacks, I. S. 1983. An anomalous low velocity region above the deep earthquakes in the Japan subduction zone. *J. Geophys. Res.*, **88**, 10429–38.

Tait, R. E. & Harley, S. L. 1988. Local processes involved in the generation of migmatites within mafic granulites. *Trans. Roy. Soc. Edinburgh, Earth Sciences*, **79**, 209–22.

Takahashi, E. & Kushiro, I. 1983. Melting of a dry peridotite at high pressures and basalt magma genesis. *Amer. Min.*, **68**, 859–79.

Tarney, J. & Windley, B. F. 1977. Chemistry, thermal gradients and evolution of the lower continental crust. *J. Geol. Soc. Lond.*, **134**, 153–72.

Tatsumi, Y. 1981. Melting experiments on a high-magnesian andesite. *Earth Planet. Sci. Lett.*, **54**, 357–65.

Tatsumi, Y. & Ishizaka, K. 1982. Magnesian andesite and basalt from Shoda-Shima Island, Southwest Japan, and their bearing on the genesis of calc-alkaline andesites. *Lithos*, **15**, 161–72.

Tatsumoto, M. 1978. Isotopic composition of lead in oceanic basalt and its implication to mantle evolution. *Earth Planet. Sci. Lett.*, **38**, 63–87.

Tatsumoto, M., Knight, R. J. & Allègre, C. J. 1973. Time differences in the formation of meteorites as determined from the ratio of Lead-207 to Lead-206. *Science*, **180**, 1279–83.

Taubeneck, W. H. 1967. Notes on the Glen Coe cauldron subsidence, Argyllshire, Scotland. *Bull. Geol. Soc. Amer.*, **78**, 1295–316.

Taylor, H. P. 1967. The zoned ultramafic complexes of southeastern Alaska, pp. 97–121 in *Ultramafic and related rocks* (ed. P. J. Wyllie), John Wiley & Sons, New York and London.

Taylor, H. P. 1968a. The oxygen isotope geochemistry of igneous rocks. *Contr. Min. Petr.*, **19**, 1–71.

Taylor, H. P. 1977. Water/rock interactions and the origin of H_2O in granitic batholiths. *J. Geol. Soc. Lond.*, **133**, 509–58.

Taylor, H. P. 1979. Oxygen and hydrogen isotope relationships in hydrothermal mineral deposits, pp. 236–77 in *Geochemistry of hydrothermal ore deposits* (2nd edn) (ed. H. L. Barnes). Wiley-Interscience, New York.

Taylor, H. P. 1980. The effects of assimilation of country rocks by magmas on $^{18}O/^{16}O$ and $^{87}Sr/^{86}Sr$ systematics in igneous rocks. *Earth Planet. Sci. Lett.*, **47**, 243–54.

Taylor, H. P. 1987. Comparison of hydrothermal systems in layered gabbros and granites, and the origin of low-^{18}O magmas, pp. 337–57 in *Magmatic processes: physicochemical principles* (ed. B. O. Mysen). Geochemical Society Special Publ. No. 1.

Taylor, H. P. & Forester, R. W. 1979. An oxygen and hydrogen isotope study of the Skaergaard intrusion and its country rocks: a description of a 55 m.y. old fossil hydrothermal system. *J. Petr.*, **20**, 355–419.

Taylor, S. R. 1964. Abundance of chemical elements in the continental crust: a new table. *Geochim. Cosmochim. Acta*, **28**, 1273–85.

Taylor, S. R. 1968. Geochemistry of andesites, pp. 559–83 in *Origin and distribution of the elements* (ed. L. H. Ahrens). Pergamon Press, Oxford.

Taylor, S. R. & McLennan, S. M. 1985. *The continental crust: its composition and evolution*. Blackwell Scientific Publications, Oxford.

Taylor, S. R., Arculus, R., Perfit, M. R., & Johnson, R. W. 1981. Island arc basalts, pp. 193–213 in *Basaltic volcanism on the terrestrial planets ('Basaltic volcanism study project')*. Pergamon Press, New York, 1286 pp.

Taylor, W. R., Tompkins, L. A. & Haggerty, S. E. 1994. Comparative geochemistry of West African kimberlites: evidence for a micaceous kimberlite endmember of sublithospheric origin. *Geochim. Cosmochim. Acta*, **58**, 4017–37.

Tazieff, H. 1977. An exceptional eruption: Mt. Niragongo, Jan. 10th, 1977. *Bull. Volc.*, **40**, 189–200.

Ten Brink, U. S. & Brocher, T. M. 1987. Multichannel seismic evidence for a subcrustal intrusive complex under Oahu and a model for Hawaiian volcanism. *J. Geophys. Res.*, **92**, 13687–707.

Tepper, J. H., Nelson, B. K., Bergantz, G. W. & Irving, A. J. 1993. Petrology of the Chilliwack batholith, north Cascades, Washington: generation of calc-alkaline granitoids by melting of mafic lower crust with variable water fugacity. *Contr. Min. Petr.*, **113**, 333–51.

Thomas, H. H. 1922. On certain xenolithic Tertiary minor intrusions in the Island of Mull (Argyllshire). *Quart. J. Geol. Soc. London*, **78**, 229–60.

Thompson, R. N. 1982. Magmatism of the British Tertiary volcanic province. *Scot. J. Geol.*, **18**, 49–107.

Thompson, R. N., Dickin, A. P., Gibson, I. L. & Morrison, M. A. 1982. Elemental fingerprints of isotopic contamination of Hebridean Palaeocene mantle-derived magmas by Archaean sial. *Contr. Min. Petr.*, **79**, 159–68.

Thorarinsson, S. 1970. The Lakagigar eruption of 1783. *Bull. Volc.*, **33**, 910–29.

Thorarinsson, S. & Sigvaldason, G. E. 1973. The Hekla eruption of 1970. *Bull. Volc.*, **36**, 269–88.

Thornton, C. P. & Tuttle, O. F. 1960. Chemistry of igneous rocks. I. Differentiation index. *Amer. J. Sci.*, **258**, 664–84.

Thorpe, R. S. (ed.). 1982. *Andesites: orogenic andesites and related rocks*. John Wiley & Sons, Chichester and New York, 724 pp.

Thorpe, R. S. & Tindle, A. G. 1992. Petrology and petrogenesis of a Tertiary bimodal dolerite–peralkaline/subalkaline trachyte/rhyolite dyke association from Lundy, Bristol Channel, U.K. *Geol. J.*, **27**, 101–17.

Thy, P., Beard, J. S. & Lofgren, G. E. 1990. Experimental constraints on the origin of Icelandic rhyolites. *J. Geol.*, **98**, 417–21.

Thy, P. & Lofgren, G. E. 1994. Experimental constraints

on the low-pressure evolution of transitional and mildly alkalic basalts: the effect of Fe–Ti oxide minerals and the origin of basaltic andesites. *Contr. Min. Petr.*, **116**, 340–51.

Tiezzi, L. J. & Scott, R. B. 1980. Crystal fractionation in a cumulate gabbro, mid-Atlantic Ridge, 26 °N. *J. Geophys. Res.*, **85**, 5438–54.

Tilley, C. E. 1947. The gabbro–limestone contact zone of Camas Mor, Muck, Inverness-shire. *Bull. Comm. Geol. Finlande*, **140**, 97–105.

Tilley, C. E. 1951. The zoned contact-skarns of the Broadford area, Skye: a study of boron–fluorine metasomatism in dolomites. *Min. Mag.*, **29**, 621–66.

Tilley, C. E. 1952. Some trends of basaltic magma in limestone syntexis. *Amer. J. Sci.*, Bowen volume, 529–45.

Todd, S. G., Keith, D. W., Le Roy, L. W., Schissel, D. J., Mann, E. L. & Irvine, T. N. 1982. The J–M platinum–palladium reef of the Stillwater complex, Montana: I. Stratigraphy and petrology. *Econ. Geol.*, **77**, 1454–80.

Treiman, A. H. & Schedl, A. 1983. Properties of carbonatite magma and processes in carbonatite magma chambers. *J. Geol.*, **91**, 437–47.

Turner, D. C. 1963. Ring structures in the Sara-Fier Younger Granite Complex, northern Nigeria. *Quart. J. Geol. Soc. London*, **119**, 345–66.

Turner, D. C. & Bowden, P. 1979. The Ningi-Burra complex, Nigeria: dissected calderas and migrating magmatic centres. *J. Geol. Soc. Lond.*, **136**, 105–19.

Tuttle, O. F. & Bowen, N. L. 1958. Origin of granite in the light of experimental studies in the system $NaAlSi_3O_8$–$KAlSi_3O_8$–SiO_2–H_2O. *Mem. Geol. Soc. Amer.*, **74**, 153 pp.

Tuttle, O. F. & Gittins, J. (eds). 1966. *Carbonatites.* Interscience Publishers, New York and London, 591 pp.

Tyrrell, G. W. 1928. The geology of Arran. *Mem. Geol. Surv. Scotland*, 292 pp.

Tyrrell, H. J. V. 1961. *Diffusion and heat flow in liquids.* Butterworths, London, 329 pp.

Ulff-Møller, F. 1990. Formation of native iron in sediment-contaminated magma: I. A case study of the Hanekammen Complex on Disko Island, West Greenland. *Geochim. Cosmochim. Acta*, **54**, 57–70.

Upton, B. G. J. 1960. The alkaline igneous complex of Kungnat Fjeld, South Greenland. *Meddelelser om Grønland*, **123** (4), 1–145.

Upton, B. G. J. & Thomas, J. E. 1973. Precambrian potassic ultramafic rocks: South Greenland. *J. Petr.*, **14**, 509–34.

Upton, B. G. J. & Thomas, J. E. 1980. The Tugtutoq Younger Giant Dyke Complex, south Greenland: fractional crystallization of transitional olivine basalt magma. *J. Petr.*, **21**, 167–98.

Valiquette, G. & Archambault, G. 1970. Les gabbros et les syénites du complexe de Brome. *Can. Min.*, **10**, 485–510.

Van Der Wal, D. & Vissers, R. L. M. 1993. Uplift and emplacement of upper mantle rock in the western Mediterranean. *Geology*, **21**, 1119–22.

Van Kooten, G. K. 1980. Mineralogy, petrology, and geochemistry of an ultrapotassic basaltic suite, central Sierra Nevada, California, U.S.A. *J. Petr.*, **21**, 651–84.

Varne, R. 1968. The petrology of Moroto Mountain, eastern Uganda, and the origin of nephelinites. *J. Petr.*, **9**, 169–90.

Vasil'yev, Yu. R. & Zolotukhin, V. V. 1970. A find of meymechite dikes with vitreous margins. *Dokl. Akad. Nauk. S.S.S.R.*, **190**, 172–5 (English translation).

Verhoogen, J. 1949. Thermodynamics of a magmatic gas phase. *Univ. Calif. Publications, Bull. Dept. Geol. Sci.*, **28**, 91–135.

Verwoerd, W. J. 1978. Liquid immiscibility and the carbonatite–ijolite relationship: preliminary data on the join $NaFe^{3+}Si_2O_6$–$CaCO_3$ and related compositions. *Carnegie Inst. Wash. Yearbook*, **77**, 767–74.

Vidal, P., Cocherie, A. & Le Fort, P. 1982. Geochemical investigations of the origin of the Manaslu leucogranite (Himalaya, Nepal). *Geochim. Cosmochim. Acta*, **46**, 2279–92.

Vielzeuf, D. & Holloway, J. R. 1988. Experimental determination of the fluid-absent melting relations in the pelitic system. Consequences for crustal differentiation. *Contr. Min. Petr.*, **98**, 257–76.

Visser, W. & Koster van Groos, A. F. 1979a. Phase relations in the system K_2O–FeO–Al_2O_3–SiO_2 at 1 atmosphere with special emphasis on low temperature liquid immiscibility. *Amer. J. Sci.*, **279**, 70–91.

Visser, W. & Koster van Groos, A. F. 1979b. Effects of P_2O_5 and TiO_2 on liquid–liquid equilibria in the system K_2O–FeO–Al_2O_3–SiO_2. *Amer. J. Sci.*, **279**, 970–88.

Visser, W. & Koster van Groos, A. F. 1979c. Effect of pressure on liquid immiscibility in the system K_2O–FeO–Al_2O_3–SiO_2–P_2O_5. *Amer. J. Sci.*, **279**, 1160–75.

Vlasov, K. A., Kuz'menko, M. Z. & Es'kova, E. M. 1966. *The Lovozero alkali massif.* Oliver & Boyd, Edinburgh and London, 627 pp.

Vogt, P. 1962. Geologisch–petrographische Untersuchungen im Peridotitstock von Finero. *Schweiz. Min. Petr. Mitt.*, **42**, 59–125.

Vogt, P. R. 1974. Volcano height and plate thickness. *Earth Planet. Sci. Lett.*, **23**, 337–48.

Von Eckermann, H. 1961. The petrogenesis of the Alnö alkaline rocks. *Bull. Geol. Inst. Univ. Uppsala*, **40**, 25–36.

Voshage, H., Hunziker, J. C., Hofmann, A. W. & Zingg, A. 1987. A Nd and Sr isotopic study of the Ivrea zone, Southern Alps, Italy. *Contr. Min. Petr.*, **97**, 31–42.

Voshage, H., Hofmann, A. W., Mazzucchelli, M., Rivalenti, G., Sinigoi, S., Raczek, I. & Demarchi, G. 1990. Isotopic evidence from the Ivrea Zone for a hybrid lower crust formed by magmatic underplating. *Nature*, **347**, 731–6.

Wager, L. R. 1963. The mechanism of adcumulus growth in the layered series of the Skaergaard intrusion. *Min. Soc. Amer., Spec. Paper*, **1**, 1–9.

Wager, L. R. 1965. The form and internal structure of the Kangerdlugssuaq intrusion, East Greenland. *Min. Mag.*, **34**, 487–97.

Wager, L. R. & Brown, G. M. 1968. *Layered igneous rocks.* Oliver & Boyd, Edinburgh and London, 588 pp.

Wager, L. R. & Deer, W. A. 1938. A dyke swarm and crustal flexure in East Greenland. *Geol. Mag.*, **75**, 39–46.

Wager, L. R. & Mitchell, R. L. 1951. The distribution of trace elements during strong fractionation of basic magma – a further study of the Skaergaard intrusion, East Greenland. *Geochim. Cosmochim. Acta*, **1**, 129–208.

Wager, L. R., Vincent, E. A. & Smales, A. A. 1957. Sulphides in the Skaergaard intrusion, east Greenland. *Econ. Geol.*, **52**, 855–903.

Wahl, W. 1946. Thermal diffusion-convection as a cause of magmatic differentiation. I. *Amer. J. Sci.*, **244**, 417–41.

Walawender, M. J., Hoppler, H., Smith, T. E. & Riddle, C. 1979. Trace element evidence for contamination in a gabbronorite–quartz diorite sequence in the Peninsular Ranges batholith. *J. Geol.*, **87**, 87–97.

Walker, D. & De Long, S. E. 1982. Soret separation of mid-ocean ridge basalt magma. *Contr. Min. Petr.*, **79**, 231–40.

Walker, D., Shibata, T. & De Long, S. E. 1979. Abyssal tholeiites from the Oceanographer Fracture Zone. II. Phase equilibria and mixing. *Contr. Min. Petr.*, **70**, 111–25.

Walker, F. & Davidson, C. F. 1935. Marginal and contact phenomena of the Dorback granite. *Geol. Mag.*, **72**, 49–63.

Walker, G. P. L. 1973. Explosive volcanic eruptions – a new classification scheme. *Geol. Rundschau*, **62**, 431–46.

Walker, G. P. L. & Leedal, G. P. 1954. The Barnesmore granite complex, County Donegal. *Sci. Proc. Royal Dublin Soc.*, **26**, 207–43.

Walker, G. P. L. & Skelhorn, R. R. 1966. Some associations of acid and basic igneous rocks. *Earth Science Reviews*, **2**, 93–109.

Walker, R. J., Echeverria, L. M., Shirley, S. B. & Horan, M. F. 1991. Re–Os isotopic constraints on the origin of volcanic rocks, Gorgona Island, Colombia: Os isotopic evidence for ancient heterogeneities in the mantle. *Contr. Min. Petr.*, **107**, 150–62.

Wall, V. J., Clemens, J. D. & Clarke, D. B. 1987. Models for granitoid evolution and source composition. *J. Geol.*, **95**, 731–49.

Wallace, M. E. & Green, D. H. 1988. An experimental determination of primary carbonatite magma composition. *Nature*, **335**, 343–6.

Walsh, J. N., Beckinsale, R. D., Skelhorn, R. R. & Thorpe, R. S. 1979. Geochemistry and petrogenesis of Tertiary granitic rocks from the Island of Mull, northwest Scotland. *Contr. Min. Petr.*, **71**, 99–116.

Walsh, J. N. & Clarke, E. 1982. The role of fractional crystallization in the formation of granitic and intermediate rocks of the Beinn Chaisgidle centre, Mull, Scotland. *Min. Mag.*, **45**, 247–55.

Watson, E. B. 1977. Partitioning of manganese between forsterite and silicate liquid. *Geochim. Cosmochim. Acta*, **41**, 1363–74.

Watson, E. B. 1982. Basalt contamination by continental crusts: some experiments and models. *Contr. Min. Petr.*, **80**, 73–87.

Watson, E. B. & Naslund, H. R. 1977. The effect of pressure on liquid immiscibility in the system K_2O–FeO–Al_2O_3–SiO_2–CO_2. *Carnegie Inst. Wash. Yearbook*, **76**, 410–4.

Watson, K. D. 1967. Kimberlites of eastern North America, pp. 312–23 in Wyllie (1967).

Weaver, B. L. 1991. The origin of oceanic island basalt end-member compositions: trace element and isotopic constraints. *Earth Planet. Sci. Lett.*, **104**, 381–97.

Wedepohl, K. H. (ed.). 1969–74. *Handbook of geochemistry.* Springer-Verlag, Berlin, Heidelberg and New York. 2 vols.

Wedepohl, K. H. 1994. The composition of the continental crust. *Min. Mag.*, **58A**, 959–60.

Weis, D., Frey, F. A., Leyrit, H. & Gautier, I. 1993. Kerguelen Archipelago revisited: geochemical and isotopic study of the Southeast Province lavas. *Earth Planet. Sci. Lett.*, **118**, 101–19.

Wells, A. K. & Wooldridge, S. W. 1931. The rock groups of Jersey, with special reference to intrusive phenomena. *Proc. Geol. Assoc. (London)*, **42**, 178–215.

Wendlandt, R. F. 1982. Sulfide saturation of basalt and andesite melts at high pressures and temperatures. *Amer. Min.*, **67**, 877–85.

Wendlandt, R. F. & Eggler, D. H. 1980. The origins of potassic magmas: 2. Stability of phlogopite in natural spinel lherzolite and in the system $KAlSiO_4$–MgO–SiO_2–H_2O–CO_2 at high pressures and high temperatures. *Amer. J. Sci.*, **280**, 421–58.

Wendlandt, R. F. & Mysen, B. O. 1980. Melting phase relations of natural peridotite + CO_2 as a function of degree of partial melting at 15 and 30 kbar. *Amer. Min.*, **65**, 37–44.

Wenner, D. B. 1981. Oxygen isotopic compositions of the late orogenic granites in the Southern Piedmont of the Appalachian Mountains, U.S.A., and their relationship to subcrustal structures and lithologies. *Earth Planet. Sci. Lett.*, **54**, 186–99.

Wenner, D. B. & Taylor, H. P. 1974. D/H and O^{18}/O^{16} studies of serpentinization of ultramafic rocks. *Geochim. Cosmochim. Acta*, **38**, 1255–86.

Wheeler, E. P. 1969. Minor intrusives associated with the Nain anorthosite. *Mem. N.Y. State Mus. Sci. Serv.*, **18**, 189–206.

White, A. J. R. & Chappell, B. W. 1977 Ultrametamorphism and granitoid genesis. *Tectonophysics*, **43**, 7–22.

White, A. J. R., Chappell, B. W. & Cleary, J. R. 1974. Geologic setting and emplacement of some Australian Palaeozoic batholiths and implications for intrusive mechanisms. *Pacific Geol.*, **8**, 159–71.

White, B. S. & Wyllie, P. J. 1992. Solidus reactions in

synthetic lherzolite–H_2O–CO_2 from 20–30 kbar, with applications to melting and metasomatism. *J. Volc. Geotherm. Res.*, **50**, 117–30.

White, R. & McKenzie, D. 1989. Magmatism at rift zones: the generation of volcanic continental margins and flood basalts. *J. Geophys. Res.*, **94**, 7685–729.

White, W. M. 1985. Sources of oceanic basalts: radiogenic isotope evidence. *Geology*, **13**, 115–18.

Whitford, D. J., White, W. M. & Jezek, P. A. 1981. Neodymium isotopic composition of Quaternary island arc lavas from Indonesia. *Geochim. Cosmochim. Acta*, **45**, 989–95.

Whitford-Stark, J. L. 1983. Cenozoic volcanic and petrochemical provinces of mainland Asia. *J. Volc. Geotherm. Res.*, **19**, 193–222.

Whitney, J. A. 1988. The origin of granite: the role and source of water in the evolution of granitic magmas. *Bull. Geol. Soc. Amer.*, **100**, 1886–97.

Whitney, J. A., Jones, L. M. & Walker, R. L. 1976. Age and origin of the Stone Mountain granite, Lithonia district, Georgia. *Bull. Geol. Soc. Amer.*, **87**, 1067–77.

Whittaker, E. J. W. & Muntus, R. 1970. Ionic radii for use in geochemistry. *Geochim. Cosmochim. Acta*, **34**, 945–56.

Whitten, E. H. T. 1962. A new method for determination of the average composition of a granite massif. *Geochim. Cosmochim. Acta*, **26**, 545–60.

Wickham, S. M. 1987. The segration and emplacement of granitic magmas. *J. Geol. Soc., London*, **144**, 281–97.

Widom, E., Schmincke, H. U. & Gill, J. B. 1992. Processes and timescales in the evolution of a chemically zoned trachyte: Fogo A, São Miguel, Azores. *Contr. Min. Petr.*, **111**, 311–28.

Wiebe, R. A. 1980. Anorthositic magmas and the origin of Proterozoic anorthosite massifs. *Nature*, **286**, 564–7.

Wiebe, R. A. 1984. Commingling of magmas in the Bjerkrem–Sogndal lopolith (southwest Norway): evidence for the compositions of residual liquids. *Lithos*, **17**, 171–88.

Wiebe, R. A. 1986. Lower crustal cumulate nodules in Proterozoic dikes of the Nain complex: evidence for the origin of Proterozoic anorthosites. *J. Petr.*, **27**, 1253–75.

Wiebe, R. A. 1988. Structural and magmatic evolution of a magma chamber: the Newark Island layered intrusion, Nain, Labrador. *J. Petr.*, **29**, 383–411.

Wiebe, R. A. 1990. Evidence for unusually feldspathic liquids in the Nain complex, Labrador. *Amer. Min.*, **75**, 1–12.

Wiebe, R. A. 1994. Silicic magma chambers as traps for basaltic magmas: the Cadillac Mountain intrusive complex, Mount Desert Island, Maine. *J. Geol.*, **102**, 423–37.

Wiebe, R. A. & Wild, T. 1983. Fractional crystallization and magma mixing in the Tigalik layered intrusion, the Nain anorthosite complex, Labrador. *Contr. Min. Petr.*, **84**, 327–44.

Wilcox, R. E. 1944. Rhyolite–basalt complex on Gardiner River, Yellowstone Park, Wyoming. *Bull. Geol. Soc. Amer.*, **55**, 1047–80.

Wilcox, R. E. 1979. The liquid line of descent and variation diagrams, pp. 205–32 in Yoder (1979).

Wilkinson, J. F. G. 1966. Residual glasses from some alkali basaltic lavas from New South Wales. *Min. Mag.*, **35**, 847–60.

Wilkinson, J. F. G. 1982. The genesis of mid-ocean ridge basalt. *Earth Science Reviews*, **18**, 1–57.

Wilkinson, J. F. G. 1985. Undepleted mantle composition beneath Hawaii. *Earth Planet. Sci. Lett.*, **75**, 129–38.

Wilkinson, J. F. G. 1991. Mauna Loan and Kilauean tholeiites with low 'ferromagnesian-fractionated' 100 Mg/ (Mg + Fe^{2+}) ratios: primary liquids from the upper mantle? *J. Petr.*, **32**, 863–907.

Willemse, J. & Viljoen, E. A. 1970. The fate of argillaceous material in the gabbroic magma of the Bushveld complex. *Spec. Publ. Geol. Soc. S. Africa*, **1**, 336–66.

Williams, H. 1942. The geology of Crater Lake National Park, Oregon, with a reconnaissance of the Cascade Range southward to Mount Shasta. *Carnegie Inst. Washington Publication* **540**, 162 pp.

Williams, H. & McBirney, A. R. 1979. *Volcanology*. Freeman, Cooper & Co., San Francisco, 397 pp.

Wilshire, H. G. 1967. The Prospect alkaline diabase–picrite intrusion, New South Wales, Australia. *J. Petr.*, **8**, 97–163.

Wilshire, H. G. & Jackson, E. D. 1975. Problems in determining mantle geotherms from pyroxene compositions of ultramafic rocks. *J. Geol.*, **83**, 313–29.

Wilson, A. H. 1982. The geology of the Great 'Dyke', Zimbabwe: the ultramafic rocks. *J. Petr.*, **23**, 240–92.

Wilson, C. J. N. 1985. The Taupo eruption, New Zealand. II. The Taupo ignimbrite. *Phil. Trans. Roy. Soc. London A*, **314**, 229–310.

Wilson, C. J. N. & Walker, G. P. L. 1985. The Taupo eruption, New Zealand. I. General aspects. *Phil. Trans. Roy. Soc. London A*, **314**, 199–228.

Wilson, H. D. B. 1956. Structure of lopoliths. *Bull. Geol. Soc. Amer.*, **67**, 289–300.

Wilson, J. T. 1963. A possible origin of the Hawaiian Islands. *Can. J. Phys.*, **41**, 863–70.

Wilson, J. T. 1973. Mantle plumes and plate motions. *Tectonophysics*, **19**, 149–64.

Windom, K. E. & Boettcher, A. L. 1981. Phase relations for the joins jadeite–enstatite and jadeite–forsterite at 28 kb and their bearing on basalt genesis. *Amer. J. Sci.*, **281**, 335–51.

Winkler, H. G. F. 1974. *Petrogenesis of metamorphic rocks* (3rd edn). Springer-Verlag, Berlin, Heidelberg and New York, 320 pp.

Winkler, H. G. F. & Von Platen, H. 1960. Experimentelle Gesteinsmetamorphose. III. Anatektische Ultrametamorphose kalkhaltiger Tone. *Geochim. Cosmochim. Acta*, **18**, 294–316.

Wones, D. R. 1989. Signifiance of the assemblage titanite + magnetite + quartz in granitic rocks. *Amer. Min.*, **74**, 744–9.

Wood, B. J. & Fraser, D. G. 1976. *Elementary thermody-*

namics for geologists. Oxford University Press, Oxford, 303 pp.

Wooden, J. L., Czamanske, G. K., Fedorenko, V. A., Arndt, N. T., Chauvel, C., Bouse, R. M., King, B. W., Knight, R. J. & Siems, D. F. 1993. Isotopic and trace-element constraints on mantle and crustal contributions to Siberian continental flood basalts, Noril'sk area, Siberia. *Geochim. Cosmochim. Acta,* **57,** 3677–704.

Woodhead, J. D., Harmon, R. S. & Fraser, D. G. 1987. O, S, Sr, and Pb isotope variations in volcanic rocks from the Northern Mariana Islands: implications for crustal recycling in intra-oceanic arcs. *Earth Planet. Sci. Lett.,* **83,** 39–52.

Woodhead, J. D., Greenwood, P., Harmon, R. S. & Stoffers, P. 1993. Oxygen isotope evidence for recycled crust in the source of EM-type oceanic island basalts. *Nature,* **362,** 809–13.

Worst, B. G. 1958. The differentiation and structure of the Great Dyke of Southern Rhodesia. *Trans. Geol. Soc. S. Africa,* **61,** 283–354.

Woussen, G., Dimroth, E., Corriveau, L. & Archer, P. 1981. Crystallization and emplacement of the Lac St-Jean anorthosite massif (Quebec, Canada). *Contr. Min. Petr.,* **76,** 343–50.

Wright, J. E. & Haxel, G. 1982. A garnet–two–mica granite, Coyote Mountains, southern Arizona: geologic setting, uranium–lead isotopic systematics of zircon, and the nature of the granite source region. *Bull. Geol. Soc. Amer.,* **93,** 1176–88.

Wright, T. L. & Peck, D. L. 1978. Crystallization and differentiation of the Alae magma, Alae lava lake, Hawaii. *Prof. Paper U.S.G.S.,* **935–C,** 20 pp.

Wyllie, P. J. 1963. Effects of the changes in slope occurring on liquidus and solidus paths in the system diopside–anorthite–albite. *Min. Soc. Amer. Special Paper,* **1,** 204–12.

Wyllie, P. J. (ed.). 1967. *Ultramafic and related rocks.* John Wiley & Sons, New York, London and Sydney. 464 pp.

Wyllie, P. J. 1977. Crustal anatexis: an experimental review. *Tectonophysics,* **43,** 41–71.

Wyllie, P. J. 1978. Mantle fluid compositions buffered in peridotite–CO_2–H_2O by carbonates, amphibole, and phlogopite. *J. Geol.,* **86,** 687–713.

Wyllie, P. J. & Huang, W. L. 1976. Petrogenetic grid for siliceous dolomites extended to mantle peridotite compositions and to conditions for magma generation. *Amer. Min.,* **61,** 691–8.

Wyllie, P. J. & Wolf, M. B. 1993. Amphibolite dehydration-melting: sorting out the solidus, pp. 405–16 in *Magmatic processes and plate tectonics* (eds H. M. Pritchard *et al.*). Geological Society Spec. Publ., **76.**

Yagi, K. & Takeshita, H. 1987. Impact of hornblende crystallization for the genesis of calc-alkaline andesites, pp. 183–90 in *Magmatic processes: physicochemical principles* (ed. B. O. Mysen). Geochemical Society Special Publication No.1.

Yoder, H. S. 1968. Albite–anorthite–quartz–water at 5 kb. *Carnegie Inst. Wash. Yearbook,* **66,** 477–8.

Yoder, H. S. 1969. Experimental studies bearing on the origin of anorthosite. *Mem. N.Y. State Mus. Sci. Serv.,* **18,** 13–22.

Yoder, H. S. 1973. Contemporaneous basaltic and rhyolitic magmas. *Amer. Min.,* **58,** 153–71.

Yoder, H. S. 1976. *Generation of basaltic magma.* National Academy of Sciences, Washington, DC, 205 pp.

Yoder, H. S. (ed.). 1979. *The evolution of the igneous rocks: 50th anniversary perspectives.* Princeton University Press, Princeton, NJ, 588 pp.

Yoder, H. S. 1988. The great basaltic 'floods'. *S. Afr. J. Geol.,* **91,** 139–56.

Yoder, H. S. & Tilley, C. E. 1962. Origin of basalt magmas: an experimental study of natural and synthetic rock systems. *J. Petr.,* **3,** 342–532.

Zeck, H. P. 1992. Restite–melt and mafic–felsic magma mixing and mingling in an S-type dacite, Cerro del Hoyazo, southeastern Spain. *Trans. Roy. Soc. Edinburgh, Earth Sciences,* **83,** 139–44.

Zhu, B.-Q., Mao, C-X., Lugmair, G. W. & Macdougall, J. D. 1983. Isotopic and geochemical evidence for the origin of Plio–Pleistocene volcanic rocks near the Indo-Eurasian collisional margin at Tengchong, China. *Earth Planet. Sci. Lett.,* **65,** 263–75.

Zielinski, R. A. 1975. Trace element evaluation of a suite of rocks from Réunion Island, Indian Ocean. *Geochim. Cosmochim. Acta,* **39,** 713–34.

Zimmerman, J. 1972. Emplacement of the Vourinos ophiolite complex, northern Greece. *Mem. Geol. Soc. Amer.,* **132,** 225–39.

Zindler, A. & Hart, S. 1986. Chemical geodynamics. *Ann. Rev. Earth Planet. Sci.,* **14,** 493–571.

Zindler, A., Hart, S. R. & Brooks, C. 1981. The Shabogamo intrusive suite, Labrador: Sr and Nd isotopic evidence for contaminated mafic magmas in the Proterozoic. *Earth Planet. Sci. Lett.,* **54,** 217–35.

Zingg, A., Handy, M. R., Hunziker, J. C. & Schmid, S. M. 1990. Tectonometamorphic history of the Ivrea Zone and its relationship to the crustal evolution of the Southern Alps. *Tectonophysics,* **182,** 169–92.

Zwart, H. J. 1967. The duality of orogenic belts. *Geologie en Mijnbouw,* **46,** 283–309.

INDEX